P9-CJR-812

International Review of

Cytology

A Survey of

Cell Biology

VOLUME 156

International Review of Cytology
A Survey of Cell Biology

Edited by

Kwang W. Jeon
Department of Zoology
The University of Tennessee
Knoxville, Tennessee

Jonathan Jarvik
Department of Biological Sciences
Carnegie Mellon University
Pittsburgh, Pennsylvania

VOLUME 156

ACADEMIC PRESS
San Diego New York Boston London Sydney Tokyo Toronto

This book is printed on acid-free paper. ♾

Academic Press, Inc.
A Division of Harcourt Brace & Company
525 B Street, Suite 1900, San Diego, California 92101-4495

United Kingdom Edition published by
Academic Press Limited
24-28 Oval Road, London NW1 7DX

International Standard Serial Number: 0074-7696

International Standard Book Number: 0-12-364559-X

PRINTED IN THE UNITED STATES OF AMERICA
94 95 96 97 98 99 EB 9 8 7 6 5 4 3 2 1

CONTENTS

Cellular and Molecular Effects of Thymic Epithelial Cells on Thymocytes during Differentiation and Maturation

Yoshihiro Kinoshita and Fumihiko Hato

Actin-Binding Proteins in Cell Motility

Sadashi Hatano

The Cell Biology of Nematocysts

Glen M. Watson and Patricia Mire-Thibodeaux

The Impact of Altered Gravity on Aspects of Cell Biology

Dale E. Claassen and Brian S. Spooner

CONTRIBUTORS

Numbers in parentheses indicate the pages on which the authors' contributions begin.

Michel Buiré (1), *Institut des Sciences Végétales, CNRS, F-91198 Gif-sur-Yvette, France*

Dale E. Claassen (301), *Division of Biology, Kansas State University, Manhattan, Kansas 66506*

Sadashi Hatano[1] (199), *Department of Molecular Biology, School of Science, Nagoya University, Chikusa-ku, Nagoya 464-01, Japan*

Fumihiko Hato (159), *Department of Physiology, Osaka City University Medical School, Osaka 545, Japan*

Yoshihiro Kinoshita (159), *Department of Physiology, Osaka City University Medical School, Osaka 545, Japan*

Ádám Kondorosi (1), *Institut des Sciences Végétales, CNRS, F-91198 Gif-sur-Yvette, France, and Institute of Genetics, Biological Research Center, Hungarian Academy of Sciences, H-6701 Szeged, Hungary*

Éva Kondorosi (1), *Institut des Sciences Végétales, CNRS, F-91198 Gif-sur-Yvette, France*

Christian Maare (77), *Department of Oncology, University of Copenhagen, Herlev Hospital, DK-2730 Herlev, Denmark*

Patricia Mire-Thibodeaux[2] (275), *Department of Biology, University of Southwestern Louisiana, Lafayette, Louisiana 70504*

[1] Present address: Chiyogaoka 1-107-1001 Chikusa-ku, Nagoya 464, Japan.
[2] Present address: Pennington Biomedical Research Center, Louisiana State University, Baton Rouge, Louisiana 70808.

Dorthe Nielsen (77), *Department of Oncology, University of Copenhagen, Herlev Hospital, DK-2730 Herlev, Denmark*

Pascal Ratet (1), *Institut des Sciences Végétales, CNRS, F-91198 Gif-sur-Yvette, France*

Michael Schultze (1), *Institut des Sciences Végétales, CNRS, F-91198 Gif-sur-Yvette, France*

Torben Skovsgaard (77), *Department of Oncology, University of Copenhagen, Herlev Hospital, DK-2730 Herlev, Denmark*

Brian S. Spooner (301), *Division of Biology, Kansas State University, Manhattan, Kansas 66506*

Karsten Wassermann (77), *Department of Biology and Toxicology, National Institute of Occupational Health, DK-2100 Copenhagen Ø, Denmark*

Glen M. Watson (275), *Department of Biology, University of Southwestern Louisiana, Lafayette, Louisiana 70504*

Cell and Molecular Biology of *Rhizobium*–Plant Interactions

Michael Schultze,* Éva Kondorosi,* Pascal Ratet,* Michel Buiré,*
and Ádám Kondorosi*,†
*Institut des Sciences Végétales, CNRS, F-91198 Gif-sur-Yvette Cedex, France
†Institute of Genetics, Biological Research Center, Hungarian Academy of Sciences, H-6701 Szeged, Hungary

I. Introduction

Soil bacteria, referred to as rhizobia belonging to the genera *Rhizobium, Bradyrhizobium,* and *Azorhizobium,* have the unique ability to induce nitrogen-fixing nodules on the roots or stems of leguminous plants. The symbiotic interaction is host specific; for example, *R. meliloti* nodulates *Medicago, Melilotus,* and *Trigonella,* whereas *Pisum, Vicia, Lens,* and *Lathyrus* are the host plants for *R. leguminosarum* bv. *viciae*. Based upon host infectivity, the bacterial species had previously been classified into cross-inoculation groups. Molecular techniques revealed, however, that this classification should be revised since the inoculation groups also included distantly related bacteria. This has led to the recent grouping of rhizobia into three genera (Elkan, 1992).

Nodule development consists of several stages determined by different sets of genes both in the host and symbiont. At least at the very early steps of symbiosis, the bacterial and plant genes are activated consecutively by signal exchanges between the symbiotic partners. First, flavonoid signal molecules exuded by the host plant root induce the expression of nodulation (*nod, nol*) genes in *Rhizobium* in conjunction with the bacterial activator NodD protein. Then, in the second step, lipooligosaccharide Nod factors with various host-specific structural modifications are produced by the bacterial Nod proteins. The Nod factors induce various plant reactions, such as root hair deformation, initiation of nodule meristems, and induction of early nodulin genes, leading to nodule formation. Other classes of bacterial genes are required for successful infection and for nitrogen fixation. This chapter covers only the early events of communication between rhizobia and their host plants, that is, the perception of flavonoid signals by the bacteria, the production of Nod signals by rhizobia, and the early plant responses to the bacteria.

II. Stages of Nodule Development

The morphological steps of nodule development and differentiation of bacteria to bacteroids have been described in several excellent reviews (Dart, 1977; Newcomb, 1981; Vance, 1983; Sprent, 1989). Here, only the early steps of the symbiosis involved in nodule meristem formation and nodule organogenesis will be discussed. The first interaction between rhizobia and the host plant occurs at the root hairs. The emerging root hair zone, behind the apical meristem, is the most susceptible for *Rhizobium* infection and induction of mitotic activity in the root cortex. Bacteria show chemotaxis toward many substances exuded by the plant root, including amino acids and carbohydrates, as well as flavonoids that serve not only as *nod* gene inducers but also as chemoattractants for the bacteria (Peters and Verma, 1990).

Though chemotaxis contributes to the competitiveness of the rhizobia, it is not essential for nodulation and is not a major factor in the determination of host specificity. The rhizobia attach to the root hairs in a two-step process (Kijne, 1992). Only attachment of bacteria to emerging root hairs results in root hair curling, formation of shepherd's crooks, and entrapment of rhizobia, which seems to be a prerequisite for ingestion and successful infection of root hairs with bacteria (Wood and Newcomb, 1989). Differentiated root hairs respond also to rhizobia with a variety of growth deformations (ballooning, elongation, branching, spiraling) but usually without being infected. All these root hair responses are induced rapidly, within 12–24 hr, and are dependent on the *nod* gene functions, that is, production of Nod signals.

Root hair curling is followed by the infection of root hairs by bacteria at the root hair tip through invagination of the cell wall whose continued growth leads to the formation of infection threads (Brewin, 1991). These tubes penetrate through the cell layers toward the nodule primordium by growing and branching and meanwhile transporting the multiplying bacteria. At the same time, the nodule primordium enlarges to include cells from the adjacent pericycle and endodermis; outgrowth of the emerging nodule begins. Then the bacteria surrounded by plant membranes are released from the infection threads into the cytoplasm. For a successful invasion, rhizobial *exo, lps,* and *ndv* genes are required that also contribute to host-specific nodulation (Noel, 1992; Gray *et al.,* 1992; Leigh and Coplin, 1992). However, infection is a very rare event; only a small portion of root hairs become infected and not all infections give rise to nodules (Dart, 1977). Many infection threads abort before reaching the base of the root hair; others abort at the base of the hair or in the cortex. Abortion of infection is accompanied by a hypersensitive reaction that might be part of the mechanism by which the plant controls infection and regulates nodulation (Vasse *et al.,* 1993).

At the same time that root hairs are undergoing curling, cell division starts in the root cortex (Calvert *et al.,* 1984; Dudley *et al.,* 1987). Cortical cell division is dependent on *nod* gene activity but does not require invasion of plant cells by rhizobia. Certain cultivars of *Medicago sativa,* however, under the limitation of combined nitrogen, have the capacity to form nodules spontaneously in the absence of rhizobia (Nar$^+$ phenotype; Truchet *et al.,* 1989).

The type of nodules as well as their shape and size are determined by the leguminous plants. Based on the presence or absence of a persistent meristem, the nodules are classified into two major groups—indeterminate and determinate nodules, respectively. The indeterminate nodules, formed on temperate legumes such as pea, alfalfa, or vetch, maintain the meristem, which is a group of dividing cells at the distal end of the nodule whose derivatives either differentiate to other cell types or remain as part of the meristem. The continuous production of new cells at the distal end results in elongate and club-shaped nodules. The indeterminate nodules comprise different zones that represent all stages of nodule development: (1) the meristem, (2) the invasion zone consisting of numerous infection threads and freshly infected cells, (3) the early symbiotic zone where the infected cells are tightly packed with bacteroids, the nitrogen-fixing form of rhizobia, (4) the late symbiotic zone with a decreasing level of nitrogen-fixation activity, and (5) the senescence zone. Recently a new nomenclature correlated with the multiple stages of bacteroid development has been proposed for the zones in *Medicago* nodules; it introduces an additional zone, the interzone II–III, between the invasion and the early symbiotic zones, where final differentiation of bacteroids takes place (Vasse *et al.,* 1990).

In the determinate nodules, mitotic activity ceases early during nodule development. The increase of nodule size is primarily due to cell enlargement, which results in a spherical nodule form. The central tissue consists of the nitrogen-fixing zone and in the center the senescent region. Determinate nodules appear, for example, on the soybean (*Glycine max*), common bean (*Phaseolus vulgaris*), or *Lotus* and in general on tropical legumes. A peculiarity is found in the tropical legume *Sesbania rostrata* that is nodulated not only on its roots but also on the stem. The root nodules

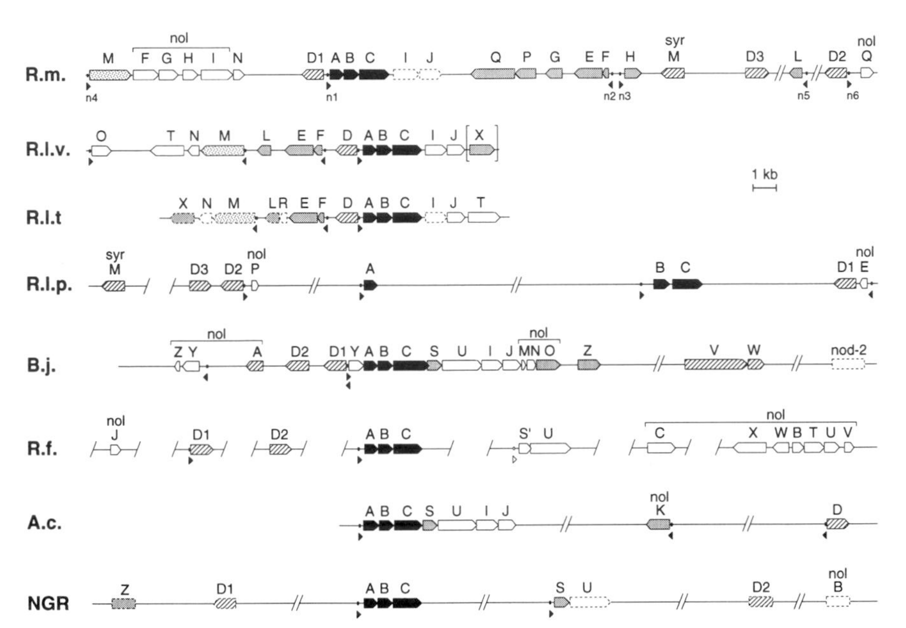

FIG. 1 Organization of nodulation genes in *R. meliloti* (R.m.) (Török *et al.*, 1984; Egelhoff *et al.*, 1985a,b; Jacobs *et al.*, 1985; Debellé and Sharma, 1986; Horvath *et al.*, 1986; Göttfert *et al.*, 1986; Rostas *et al.*, 1986; Aguilar *et al.*, 1987; Fisher *et al.*, 1987a; Cervantes *et al.*, 1989; Schwedock and Long, 1989, 1994; Barnett and Long, 1990; Honma *et al.*, 1990; Baev *et al.*, 1991; Baev and Kondorosi, 1992; Kondorosi *et al.*, 1991a,b; Rushing *et al.*, 1991; M. Cren and E. Kondorosi, unpublished), *R. leguminosarum* bv. *viciae* (R.l.v.) (Rossen *et al.*, 1984; Evans and Downie, 1986; Shearman *et al.*, 1986; Davis *et al.*, 1988; Surin and Downie, 1988; Canter-Cremers *et al.*, 1989; De Maagd *et al.*, 1989b; Economou *et al.*, 1990; Surin *et al.*, 1990), *R. leguminosarum* bv. *trifolii* (R.l.t.) (Innes *et al.*, 1985; Schofield and Watson, 1986; Surin *et al.*, 1990; Spaink *et al.*, 1989a; Lewis-Henderson and Djordjevic, 1991b; McIver *et al.*, 1993), *R. leguminosarum* bv. *phaseoli* type I (R.l.p.) (Davis and Johnston, 1990a; Vázquez *et al.*, 1991; Michiels *et al.*, 1993), *B. japonicum* (B.j.) (Nieuwkoop *et al.*, 1987; Göttfert *et al.*, 1990a,b, 1992; Sadowsky *et al.*, 1991; Wang and Stacey, 1991; Barbour *et al.*, 1992; Luka *et al.*, 1993; Dockendorff *et al.*, 1994; Stacey *et al.*, 1994), *R. fredii* (R.f.) (Appelbaum, 1988a; Krishnan and Pueppke, 1991a,b; Krishnan *et al.*, 1992; Meinhardt *et al.*, 1993; Boundy-Mills *et al.*, 1994), *A. caulinodans* (A.c.) (Goethals *et al.*, 1989, 1990, 1992b; Geelen *et al.*, 1993), and *Rhizobium* species NGR234 (NGR, near isogenic to MPIK3030, see Stanley and Cervantes, 1991) (Bachem *et al.*, 1985, 1986; Horvath *et al.*, 1987; Lewin *et al.*, 1990; Perret *et al.*, 1991; Relic *et al.*, 1994). The less detailed data available for *R. tropici* (previously named *R. leguminosarum* bv. *phaseoli* type II) (Martinez-Romero *et al.*, 1991; Vargas *et al.*, 1990; Van Rhijn *et al.*, 1993), *R. loti* (Collins-Emerson *et al.*, 1990; Young *et al.*, 1990; Scott *et al.*, 1994), *R. galegae* (Räsänen *et al.*, 1991) and *Bradyrhizobium* species isolated from *Parasponia* (Scott, 1986) are not represented. In *R. leguminosarum* bv. *viciae*, *nodX* was found only in strain TOM. The *nod* genes are shaded according to the known or proposed biochemical function of their products in precursor synthesis (stippled), synthesis of a basic Nod factor structure (black), host-specific modifications (gray), and *nod* gene regulation (hatched). Others and unknown functions are left white. Dashed bars represent *nod* genes that have not yet been fully characterized. The conserved *nod* box promoter elements are represented by small boxes; their direction is indicated underneath with arrowheads. In *R. fredii*, *nodS* contains small N-terminal deletions and is preceded by an inactive (white) Nod box. The linkage of the different *nod* gene clusters in *R. fredii* is not known.

seem to be intermediate between determinate and indeterminate nodules, the N_2-fixing nodules maintaining meristematic activity for a short time (Ndoye *et al.,* 1994).

III. Rhizobial Nodulation Genes

The nodulation genes are usually clustered either on indigenous plasmids or on the chromosome and are organized into several coordinately regulated operons (Fig. 1). Originally, they were classified into two groups. The common *nod* genes (*nodABC*) are present in all *Rhizobium* strains and are interchangeable among different species (Kondorosi *et al.,* 1984). The host-specific *nod* genes (called also *hsn*) are present only in some strains and, when present as allelic variants, cannot functionally complement each other (e.g., *nodEF*) (Horvath *et al.,* 1986). This classification is based on the phenotype of *nod* gene mutants and complementation studies. In addition, a number of other genes contribute to the efficiency of nodulation (*efn*) or are involved in the regulation of nodulation genes. With the role of many nodulation gene products known, they can be classified according to their biochemical function in Nod signal production (Table I). In the majority of rhizobial strains the common *nodABC* genes, which are responsible for the synthesis of the basic Nod signal structure,

TABLE I
Biochemical Functions of *nod* and *nol* Gene Products

Protein[a]	Function[b]	Selected references[c]
Synthesis of Nod factor precursors		
NodM	D-glucosamine synthase	Surin and Downie, 1988; Baev *et al.,* 1991, 1992; Marie *et al.,* 1992
Synthesis of a basic lipooligosaccharide structure		
NodA	Nod factor acylation	Kondorosi *et al.,* 1984; Rossen *et al.,* 1984; Török *et al.,* 1984; John *et al.,* 1993; Spaink *et al.,* 1993b; Röhrig *et al.,* 1994; Geremia *et al.,* 1994
NodB	Chitooligosaccharide deacetylase	
NodC	*N*-acetylglucosaminyltransferase, Synthesis of chitoligosaccharides	
Host-specific modifications of Nod factors		
NodE	β-ketoacylsynthase, synthesis of multiunsaturated fatty acid	Horvath *et al.,* 1986; Debellé and Sharma, 1986; Shearman *et al.,*

(*continued*)

TABLE I (*Continued*)

Protein[a]	Function[b]	Selected references[c]
NodF	Acyl carrier protein, synthesis of multiunsaturated fatty acid	1986; Schofield and Watson, 1986; Spaink *et al.*, 1991; Geiger *et al.*, 1991, 1994; Slabas *et al.*, 1992; Demont *et al.*, 1993
NodG	β-ketoacylreductase, synthesis of multiunsaturated fatty acid?	
NodH	Sulfotransferase, Nod factor sulfation	Horvath *et al.*, 1986; Debellé and Sharma, 1986; Roche *et al.*, 1991b
NodL	6-*O*-acetylation of Nod factors at the nonreducing end	Surin and Downie, 1988; Canter-Cremers *et al.*, 1989; Spaink *et al.*, 1991; Baev and Kondorosi, 1992; Bloemberg *et al.*, 1994
NodP,Q	Sulfate activation	Cervantes *et al.*, 1989; Schwedock and Long, 1989, 1990, 1992; Roche *et al.*, 1991b
NodS	Methyltransferase, *N*-methylation of Nod factors	Göttfert *et al.*, 1990b; Lewin *et al.*, 1990; Geelen *et al.*, 1993
NodX	6-*O*-acetylation of Nod factors at the reducing end	Götz *et al.*, 1985; Davis *et al.*, 1988; Firmin *et al.*, 1993
NodZ	2-*O*-methylfucosylation of Nod factors	Nieuwkoop *et al.*, 1987; Stacey *et al.*, 1994
NolK	Homology to sugar epimerase, D-arabinosylation of Nod factors?	Goethals *et al.*, 1992b; Holsters *et al.*, 1993
NolO	Efficiency of 2-*O*-methylfucosylation	Luka *et al.*, 1993
Auxiliary functions		
NodI,J	Nod factor secretion	Evans and Downie, 1986; McKay and Djordjevic, 1993; Vázquez *et al.*, 1993
NodN	Efficiency of Nod signal production	Surin and Downie, 1988; Baev *et al.*, 1991
NodO	Secreted protein, forms membrane pores	De Maagd *et al.*, 1989b; Economou *et al.*, 1990; Sutton *et al.*, 1993
NolF,G,H,I	Nod factor secretion?	Baev *et al.*, 1991; Göttfert, 1993; Saier *et al.*, 1994
Regulatory functions		
NodD	Activation of *nod* gene transcription, several alleles in some strains, determinant of host specificity	Mulligan and Long, 1985; Rossen *et al.*, 1985; Göttfert *et al.*, 1986; Horvath *et al.*, 1987; Spaink *et al.*, 1987a; Györgypal *et al.*, 1988; Honma *et al.*, 1990
SyrM	Flavone-independent *nod* gene activation	Mulligan and Long, 1989; Barnett and Long, 1990; Kondorosi *et al.*, 1991b

(*continued*)

TABLE I (*Continued*)

Protein[a]	Function[b]	Selected references[c]
NodV,W	Two-component regulatory system for *nod* gene activation, NodV: sensor, NodW: Transcriptional activator	Göttfert *et al.*, 1990a
NolA	*nod* gene regulation?	Sadowsky *et al.*, 1991
NolR	Repression of *nod* gene transcription	Kondorosi *et al.*, 1989, 1991a

[a] Only *nod* gene products with known or predicted function are listed.
[b] Functions deduced by genetic, biochemical, or sequence analysis.
[c] Related to identification and function of *nod* genes. Additional references in the text.

form a single operon which is often extended by other *nod* genes (Fig. 1). The Sym plasmid of *R. leguminosarum* bv. *phaseoli* type I strains, the American biovars of which were recently renamed to *R. etli* (Segovia *et al.*, 1993), represents an exception in that *nodBC* are separated from *nodA* by 20 kb (Girard *et al.*, 1991; Vázquez *et al.*, 1991). These *Rhizobium* strains may be especially prone to genomic rearrangements because of the presence of amplicons (Flores *et al.*, 1993).

IV. Transcriptional Regulation of Nodulation Genes

Induction of nodulation genes requires (1) flavonoids excreted by the host plant root, (2) the transcriptional activator NodD proteins, and (3) the NodD-binding cis regulatory element, the *nod* box. This highly conserved sequence (Rostas *et al.*, 1986) is present in the promoter region of the flavonoid-inducible *nod* transcriptional units (Fig. 1) and provides coordinated expression of *nod* operons forming the *nod* regulon. The NodD proteins contain a helix-turn-helix (HTH) DNA binding domain and bind to the *nod* box even in the absence of flavonoids (Hong *et al.*, 1987; Fisher *et al.*, 1988; Kondorosi *et al.*, 1989; Goethals *et al.*, 1992a). However, transcriptional activation occurs only in the presence of flavonoid inducers. The NodD proteins of various rhizobia are functionally different in their responsiveness to different sets of flavonoids and in their *nod* gene activation ability (Horvath *et al.*, 1987; Spaink *et al.*, 1987a; Györgypal *et al.*, 1991a,b). The structure of a given NodD protein determines which flavonoids act as *nod* gene inducers. Therefore, the flavonoid–NodD interactions represent the first major host-specific step in the establishment of symbiosis.

The *nod* gene functions are required for nodule induction and probably for the maintenance of nodule development in the case of indeterminate nodules. Expression of *nod* genes is detected in free-living rhizobia, in the presence of flavonoid signal molecules released by the symbiotic partner, and also in the invasion zone of the developing nodule but not in the symbiotic zone (Sharma and Signer, 1990; Schlaman *et al.*, 1991). The NodD protein in bacteroids was shown to be incapable of binding to the *nod* box (Schlaman *et al.*, 1992a).

The general pattern of *nod* gene regulation is similar in all rhizobial species, although significant differences exist, for example, in the copy number, regulation, and expression level of the *nodD* genes, and in the structure of the inducing plant signal molecules (Kondorosi, 1992; Schlaman *et al.*, 1992b; Göttfert, 1993). The behavior of the NodD proteins may be different not only between species but also within a single strain having multiple *nodD* copies. Moreover, in addition to NodD, other positive and negative transacting factors appear to contribute to *nod* gene regulation. In *R. meliloti*, the *nolR* repressor gene provides a negative control on *nod* gene expression by regulating the *nodD* as well as the common *nod* genes (Kondorosi *et al.*, 1989; 1991a). Expression of *nodD* can be controlled also by additional transacting elements that are again under the control of other regulatory elements. In addition to the *nodD*-mediated *nod* gene activation, a two-component regulatory system, encoded by the *nodVW* genes, is present in *B. japonicum*, (Göttfert *et al.*, 1990a) which is required for full expression of the common *nod* genes and *nodD1* (Sanjuan *et al.*, 1994).

Why are the *nod* gene regulatory circuits so complex? Although we are far from having a clear picture, it is likely that the cascade of the multiple positive and negative transacting factors might provide better adaptation to external stimuli, and that this fine adjustment of Nod factor production is required for optimal nodulation of the specific host plants.

A. Flavonoid Signals

Flavonoids are a broad class of plant secondary metabolites that exhibit versatile biological activities (McClure, 1975; Koes *et al.*, 1994). The natural *nod* gene inducers were found to be various flavonoids belonging to the subclasses of flavones, flavanones, isoflavones, and chalcones (reviewed in Rolfe, 1988; Györgypal *et al.*, 1991b). A recent list of *nod* gene inducers is given by Göttfert (1993).

Depending on the NodD proteins of the rhizobial strains, the same or similar compounds can act either as inducers or inhibitors, or can be ineffective in *nod* gene activation. The nature and the amount of compounds exuded depend on the plant and its developmental stage (Phillips, 1992) and are influenced by the presence of combined nitrogen (our unpublished results); for example, the range of flavonoids exuded from the seed coat or from the developing root varies significantly and it differs from the flavonoid pool inside the root. The exuded flavonoids of a given legume plant consist of a mixture of strong and weak *nod* gene inducers, as well as inhibitors and ineffective compounds (Innes *et al.*, 1985; Mulligan and Long, 1985; Rossen *et al.*, 1985; Firmin *et al.*, 1986; Peters *et al.*, 1986; Redmond *et al.*, 1986; Djordjevic *et al.*, 1987; Kosslak *et al.*, 1987; Spaink *et al.*, 1987a; Zaat *et al.*, 1987, 1988, 1989a; Banfalvi *et al.*, 1988; Bassam *et al.*, 1988; Göttfert *et al.*, 1988; Györgypal *et al.*, 1988, 1991a; Peters and Long, 1988; Sadowsky *et al.*, 1988; Hartwig *et al.*, 1989; 1990a,b; Maxwell *et al.*, 1989; Messens *et al.*, 1991; Hungria *et al.*, 1992; Kape *et al.*, 1992).

The initiation of symbiosis between the legume and rhizobia is controlled by the overall inducing effect of plant exudates in conjunction with the rhizobial NodD proteins. The effect of inducers can be additive or synergistic, but weak inducers in modest amounts might reduce the effect of strong inducers (Hartwig *et al.*, 1989). The synthesis and release of flavonoids are primarily controlled by the plant. However, bacteria are also able to influence flavonoid synthesis. It was demonstrated that *R. leguminosarum* bv. *viciae* (van Brussel *et al.*, 1990; Recourt *et al.*, 1991, 1992a,b) and bv. *phaseoli* (Dakora *et al.*, 1993b) as well as purified Nod factors (Spaink *et al.*, 1991) increased the production of inducing flavonoids in the host plants. Upon inoculation with *R. meliloti* alfalfa was shown to exude a *nod* gene-inducing isoflavonoid (Dakora *et al.*, 1993a), while a decrease of the same compound was observed within the roots (Tiller *et al.*, 1994).

In addition to the flavonoids, a few nonflavonoid *nod* gene inducers have been identified, including betaines (trigonellin and stachydrine) from *M. sativa* (Phillips *et al.*, 1992), nonphenolic compounds like hydroxycinnamic acids, which are intermediates in the isoflavonoid biosynthesis (Kape *et al.*, 1991), and monocyclic aromatic compounds, for example vanilline, which is present in wheat seedling extract (Le Strange *et al.*, 1990).

B. The *nod* Promoter

The 5′ regulatory region of inducible *nod* genes comprises (1) the consensus *nod* box sequence built up by three highly conserved stretches of nucleotides within a 47-bp-long sequence (Rostas *et al.*, 1986), (2) the AT(T)AG motif downstream of the *nod* box (Spaink *et al.*, 1987b), and

(3) the operator sequence identified until now only in *R. meliloti* (Kondorosi *et al.*, 1989) (Fig. 2).

The 47-bp-long *nod* box sequence is essential and sufficient for NodD binding without the need for additional downstream sequences such as the AT(T)AG motif (Kondorosi *et al.*, 1989). Gel retardation assays and analysis of NodD footprints suggest that the *nod* box is built up of two modules (Fisher and Long, 1989; Kondorosi *et al.*, 1989). Both modules were protected against DNase I by NodD and were essential for the binding; for example, removal of the first module drastically reduced NodD binding on the second module (Kondorosi *et al.*, 1989). The extent of the overall protected region is approximately 55 bp. However, in the central part, between the two modules, nucleotides are unprotected and display one to three hypersensitive cleavage sites, depending on the particular NodD–*nod* box interactions and on the presence of inducing flavonoids (Fisher and Long, 1989; Kondorosi *et al.*, 1989; Goethals *et al.*, 1992a), which suggests bending of the DNA linking the two modules. Changing the spaces between the conserved *nod* box elements affects NodD binding: The insertion of 4 nucleotides, which places the NodD-binding sequences on the opposite sides of the helix, inactivates the *nod* box whereas it remains active when 10 bp corresponding to a full turn of the double helix are inserted (Fisher and Long, 1993). This indicates that correct spacing and facing of binding sites on the DNA helix are essential for NodD binding. Moreover, it was demonstrated that bending of DNA is caused by the NodD protein, which probably acts as dimer and binds to the two separate binding sites within the *nod* box (Fisher and Long, 1993).

In *Azorhizobium caulinodans,* the *nod* box upstream from the common *nod* genes exhibits a low degree of homology to the consensus *nod* box. However, NodD protein from *Rhizobium* sp. NGR234 activated transcription from the *A. caulinodans nod* box and the *A. caulinodans* NodD

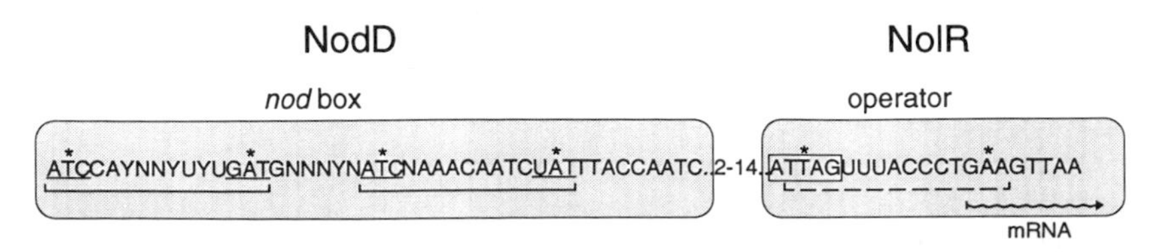

FIG. 2 Structure of the flavonoid-inducible promoter region. The consensus *nod* box sequence and the NolR-binding sequence identified in *R. meliloti* are shaded. The LysR type T-N_{11}-A motifs are indicated within the *nod* box (underlined) and in the NolR binding sequence (dashed line). The T and A nucleotides in the 3-bp inverted repeat (underlined) are labeled with an asterisk. The ATTAG sequence most likely serving as −10 element is boxed. U, purine; Y, pyrimidine.

protein was equally able to recognize its own and the consensus *nod* box sequence (Goethals *et al.*, 1992a). Both in the diverged and consensus *nod* box sequences, an ATC-N_9-GAT motif appears twice in the *nod* box, coinciding with the positions of the two *nod* box modules detected by the NodD footprints. Therefore, this sequence was suggested to be the binding target of the NodD proteins (Goethals *et al.*, 1992a). Moreover, the T N_{11}-A sequence, usually with inverted repeats around T and A, was found to be characteristic for the LysR-type-regulated promoters (Schell, 1993).

The AT(T)AG sequence is located 2–14 bp downstream of the *nod* boxes. Removal of this sequence strongly reduces the promoter activity, indicating that this element is an essential part of the promoter (Spaink *et al.*, 1987b). Transcription of *nod* genes, in the known cases, starts 9–11–(14) nucleotides downstream of the AT(T)AG sequence (Fisher *et al.*, 1987a,b; Wang and Stacey, 1991; Goethals *et al.*, 1992a), suggesting that this motif might function in the inducible *nod* promoters as the -10 promoter element recognized by the RNA polymerase.

In *R. meliloti*, downstream of the n1 and n6 *nod* boxes, a 21-bp-long sequence starting with the AT(T)AG motif was shown to bind the NolR repressor (Kondorosi *et al.*, 1989). This sequence, which comprises the transcriptional initiation site of the *nodABC* transcriptional unit (Fisher *et al.*, 1987a), is adjacent to the transcriptional initiation site of *nodD1* (Fisher *et al.*, 1987b). The 21-bp No1R binding sequence is only partially conserved in the other *nod* box regions of *R. meliloti* or other rhizobia. While the n2, n3, and n5 regions do not bind the repressor, the n4 region, being only slightly divergent from the n1 and n6 regions, does bind it (our unpublished results). The exact operator sequence has not been identified yet.

Synthetic oligonucleotides that comprise smaller overlapping parts of the 21-bp-long sequence do not bind the repressor, which indicates that probably a longer stretch of nucleotides is required for binding (our unpublished results). Interestingly, a T-N_{11}-A element is also present in the NolR-binding n1, n4, and n6 sequences, but not in the n2, n3, and n5 sequences, which do not bind the repressor. The T-N_{11}-A element was also found in the corresponding regions of many other rhizobia, primarily in the n1 common *nod* promoter regions. Further experiments are needed to elucidate whether the proposed T-N_{11}-A motif adjacent to the *nod* box indeed represents a target for NolR. The lack of interaction between NolR and the *nod* box indicates, however, that the T-N_{11}-A motif alone is not sufficient for NolR binding and that probably adjacent nucleotides confer recognition specificity. Recent unpublished data from our laboratory demonstrate that NolR binds also to its own promoter. Based on sequence comparison and mutational analysis of the putative binding motifs we propose the (A/T)TTAG-N_2-A(T/A) sequence to be the NolR target.

C. The Activator NodD Protein

1. Copy Number

The number of *nodD* gene copies varies from one to five in different rhizobia (Fig. 1). A single *nodD* gene is present in *R. leguminosarum* bv. *viciae* (Rossen *et al.*, 1985) and bv. *trifolii* (Innes *et al.*, 1985), in *B. parasponia* (Scott, 1986), in *A. caulinodans* (Goethals *et al.*, 1990), and probably in *R. galageae* (Räsänen *et al.*, 1991) and *R. loti* (Scott, 1994). Inactivation of *nodD* in these bacteria leads to a Nod^- phenotype.

Two *nodD* copies were found in *Rizobium* sp. NGR234/MPIK3030 (Bassam *et al.*, 1986; Rodriguez-Quiñones *et al.*, 1987; Horvath *et al.*, 1987; Bender *et al.*, 1988; Perret *et al.*, 1991), in *B. japonicum* (Appelbaum *et al.*, 1988a; Göttfert *et al.*, 1992), and in *R. fredii* (Appelbaum *et al.*, 1988b). Inactivation of *nodD1* abolished nodulation whereas that of *nodD2* did not have any obvious effect on nodulation. Therefore, it seems that in these strains only NodD1 acts as a transcriptional activator of the *nod* genes.

In *R. meliloti*, three functional *nodD* copies (Göttfert *et al.*, 1986; Honma and Ausubel, 1987; Györgypal *et al.*, 1988; Honma *et al.*, 1990) and a *nodD* homolog, *syrM* (Mulligan and Long, 1989; Barnett and Long, 1990; Kondorosi *et al.*, 1991b) play a role in *nod* gene activation. The organization of *nodD1* and *D2* genes is similar (Fig. 2). Both of them are divergently oriented from the *nod* boxes n1 and n6. No *nod* box sequence was found, however, in front of *nodD3* and *syrM* (Barnett and Long, 1990; Kondorosi *et al.*, 1991b; Rushing *et al.*, 1991). The *nodD1* and *nodD2* genes are more homologous to each other (Göttfert *et al.*, 1986) than to *nodD3* (Kondorosi *et al.*, 1991b; Rushing *et al.*, 1991). *syrM* shares homology with the *nodD* genes and genes belonging to the LysR family which is primarily centered around the DNA sequence encoding the putative HTH DNA-binding motif (Barnett and Long, 1990; Kondorosi *et al.*, 1991b). Mutation of *syrM* in strain 41 resulted in delayed nodulation of alfalfa (Kondorosi *et al.*, 1991b). The SyrM protein stimulates expression of *nodD3* (Rushing *et al.*, 1991; Maillet *et al.*, 1990), and, either directly or indirectly, the expression of the *nod* genes (Kondorosi *et al.*, 1991b; Dusha and Kondorosi, 1993), and also controls exopolysaccharide production (Mulligan and Long, 1989). Moreover, SyrM and NodD3 mediate the nitrogen control of *nod* gene expression via the general nitrogen regulatory system (NtrC) and NtrR that repress *nod* gene expression at high ammonia concentrations (Dusha and Kondorosi, 1993).

Three functional and significantly divergent copies of *nodD* are also present in *R. leguminosarum* bv. *phaseoli* (Davis and Johnston, 1990a). *nodD1* is not linked to the common *nod* genes and is preceded by *nolE* and a *nod* box sequence. The *nodD2* and *D3* genes are located approximately 40 kb from *nodD1* and are transcribed divergently. The promoter of *nodD2* overlaps with that of the *nolP,* which is under *nod* box control and transcribed divergently. Recently, in addition to the three *nodD* copies, the *syrM* gene was also found in *R. leguminosarum* bv. *phaseoli* (Michiels *et al.,* 1993), although its role in *nod* gene activation and nodulation has not been demonstrated yet.

R. tropicii strains contain 4–5 *nodD* copies (van Rhijn *et al.,* 1993), but only one allele seems to activate *nod* gene expression. At present, it is not clear what the contribution of the other copies is to nodulation.

2. Regulation of *nodD* Expression

The regulation and the expression level of *nodD* genes show great variation among species or even strains of the same species. In general, *nodD* is expressed constitutively in those rhizobia which harbor only a single *nodD* gene. *nodD* may be negatively autoregulated by its own product, as it is in *R. leguminosarum* bvs. *viciae* (Rossen *et al.,* 1985) and *trifolii* (Spaink *et al.,* 1989b). In other species carrying two or more *nodD* copies, such as NGR234 or *R. tropicii,* the single active *nodD* allele is expressed constitutively (Horvath *et al.,* 1987; van Rhijn *et al.,* 1993). In *B. japonicum,* the active *nodD1* gene, controlled by a single and divergent module of the *nod* box, is weakly expressed but is inducible by flavonoids in conjunction with its own product (Banfalvi *et al.,* 1988; Smit *et al.,* 1992). The expression is repressed by ammonia but, unlike *R. meliloti,* the ammonia control does not involve the general nitrogen regulatory system (Wang and Stacey, 1990).

As an extreme case, in *R. meliloti* and probably also in *R. leguminosarum* bv. *phaseoli,* each of the multiple *nodD* copies is expressed and regulated differently. In *R. leguminosarum* bv. *phaseoli, nodD1* is activated by itself and flavonoids whereas *nodD2* and *nodD3* are expressed constitutively and are probably controlled by a repressor (Davis and Johnston, 1990b).

In most *R. meliloti* strains, including strain 41, expression of all three *nodD* copies is regulated either directly or indirectly by the NolR repressor (Kondorosi *et al.,* 1989). This negative control, however, does not function in the widely used laboratory strain 1021 (see below). Both *nodD1* and *nodD2* are constitutively expressed. However, the level of expression of *nodD2* is lower than that of *nodD1* and is positively autoregulated (our unpublished results).

Expression of *nodD3* is under complex regulation but the data are somewhat contradictory. It is certain that *nodD3* is stimulated by SyrM (Mulligan and Long, 1989; Barnett and Long, 1990; Maillet *et al.,* 1990).

Expression of *nodD3* was reported to be stimulated by luteolin (Dusha *et al.*, 1989) and by NodD1 and NodD3 (Maillet *et al.*, 1990), whereas it was repressed by high levels of combined nitrogen (Dusha *et al.*, 1989; Dusha and Kondorosi, 1993). In contrast, Swanson *et al.* (1993) did not detect positive autoregulation by NodD3 and induction by luteolin. Expression of *syrM* is stimulated by SyrM, NodD2 (Kondorosi *et al.*, 1991b; Swanson *et al.*, 1993), and weakly by NodD3 (Kondorosi *et al.*, 1991b). Moreover, in *R. meliloti* strain 41, which expresses *syrM* at a level about sixfold higher than strain 1021, an additional putative transacting factor (CII) binding to the *syrM* promoter was found that might be a strain 41-specific SyrM activator (Kondorosi *et al.*, 1991b). The two *R. meliloti* strains may utilize different pathways in order to achieve an overall control of *nod* gene regulation. Low expression of *nodD* genes might be compensated for by higher expression of *syrM* and vice versa in strains 41 and 1021, respectively.

Although the available data are not directly comparable, the level of *nodD* expression shows differences of two to three orders of magnitude among the rhizobial species, for example, comparing *R. leguminosarum* bv. *viciae* (Rossen *et al.*, 1985) with *R. meliloti* (Mulligan and Long, 1985) or with *B. japonicum* (Banfalvi *et al.*, 1988).

3. Interaction of NodD with Flavonoids and the *nod* Box

NodD belongs to the LysR family of the bacterial regulatory proteins (Henikoff *et al.*, 1988). The 35-kDa protein contains the N-terminal highly conserved DNA binding domain. The C-terminal part is more divergent and implicated primarily in the interaction with flavonoids and host-specific control of *nod* gene activation. In this region, two modules were found to be homologous to the ligand-binding domains of vertebrate nuclear receptors (Györgypal and Kondorosi, 1991). Several compounds, including flavonoids, that bind to mammalian receptors also interact with NodD proteins, suggesting that the homologous modules may originate from a common ancestor protein and may be directly involved in ligand binding. Based on the extent and pattern of NodD footprints and size fractionation of the native protein, NodD might function as a dimer binding to the two *nod* box modules (Kondorosi *et al.*, 1989; Fisher and Long, 1989).

The interaction of NodD proteins with flavonoids differs not only between species or strains but also within a single strain among individual NodD alleles. In *R. meliloti,* the three NodD copies interact differently with flavonoids (Györgypal *et al.*, 1988); for example, luteolin is a *nod* gene inducer in combination with NodD1 (Peters *et al.*, 1986) but an inhibitor with NodD2. 4,4′-Dihydroxy-2′-methoxychalcone interacts with both NodD2 and NodD1 (Hartwig *et al.*, 1990b) but betaines only with NodD2 (Phillips *et al.*, 1992). In strain 41 NodD3 interacts with different flavonoids, including luteolin (Györgypal *et al.*, 1988, 1991a,b; Kondorosi *et al.*, 1991b). Only constitutive but not inducible *nod* gene activation was attributed to NodD3 of the strain 1021 by Mulligan and Long (1989) and Swanson *et al.* (1993), in contrast to the findings by Yu *et al.* (1993), who demonstrated both flavonoid-inducible and constitutive *nod* gene activation by NodD3, depending on the initiation site of the transcript.

NodD3 and SyrM overproduced from vector provide constitutive *nod* gene expression in the absence of inducer (Mulligan and Long, 1989; Györgypal *et al.*, 1988) and extend the capacity of the bacterium to nodulate a nonhost legume (Kondorosi *et al.*, 1991b). Inactivation of *syrM* on plasmid abolished the constitutive *nod* gene expression. SyrM is also member of the LysR activatory protein family and is conserved especially in the N-terminal region containing the DNA-binding domain, which suggests that it might be directly involved in *nod* gene regulation. Binding to the *nod* box of protein extracts from strain 41 carrying *syrM* and *nodD3* on plasmid was affected by luteolin; in the presence of luteolin the binding became stronger and differences in the DNaseI hypersensitive sites between the two *nod* box modules were detected in the presence and absence of luteolin (Kondorosi *et al.*, 1989) supporting, indeed, the interaction of NodD3 or eventually SyrM with luteolin. Recently a truncated *nodD* gene from *R. tropici*, encoding only the amino-terminal part, was shown to be able to induce *nod* genes and also conferred host range extension (Sousa *et al.*, 1993).

The fact that individual NodD proteins differ in their flavonoid specificity suggests a direct contact between NodD and the inducer. Although the biochemical proof is still missing, genetic data and NodD-*nod* box binding studies support the direct binding of specific flavonoids to the NodD proteins. The latter studies, based on retardation gel and DNaseI footprint experiments, indicate a conformational change of the NodD protein in the presence of inducing flavonoids and suggest activation of transcription of the inducible *nod* operons by the flavonoid-activated NodD (Kondorosi *et al.*, 1989; Goethals *et al.*, 1992a).

NodD proteins probably interact not only with inducing but also with inhibiting (iso)flavonoids or coumarins. These anti-inducers might also bind to NodD but do not convert it to the activated form. Thus, they may inhibit *nod* gene transcription by competing with the inducing flavonoids for the binding site (Firmin *et al.*, 1986; Djordjevic *et al.*, 1987; Peters and Long, 1988; Hartwig *et al.*, 1990b; Kosslak *et al.*, 1990; Cunningham *et al.*, 1991; Györgypal *et al.*, 1991a).

Single amino acid changes can severely perturb the ability of NodD proteins to induce *nod* genes, either positively or negatively, and alter both signal and host specificity. Construction of chimeric NodD proteins carrying the N-terminal and C-terminal regions from different species indicated that the more variable C-terminal part largely determines the interactions with flavonoids (Horvath *et al.*, 1987; Spaink *et al.*, 1987a). Point mutations and a series of hybrid NodD proteins revealed that many regions dispersed over the entire protein are involved in recognition of a particular flavonoid inducer and in activation of inducible *nod* promoters (Burn *et al.*, 1987, 1989; Spaink *et al.*, 1989c; McIver *et al.*, 1989).

The efficiency of *nod* gene activation depends also on the source of NodD. The level of *nod* gene induction was substantially lower in the presence of *R. meliloti nodD1* gene than in the presence of the *R. leguminosarum* bvs. *viciae* and *trifolii nodD* genes in otherwise isogenic bacteria (Spaink *et al.*, 1987a). DNaseI footprint experiments on the *R. meliloti* n1 *nod* box with protein extracts from *R. meliloti* overexpressing the *R. meliloti* NodD1 or the *R. leguminosarum* bv. *viciae* NodD proteins indicated significant differences in the NodD-*nod* box interaction, which may explain the altered activation of *nod* genes by these NodD proteins (our unpublished results).

Recently, it was shown that in *R. fredii* the gene *nolJ* which is not preceded by a *nod* box, is nevertheless activated by NodD1 in the presence of flavonoids (Boundy-Mills *et al.*, 1994). However, since interaction of NodD1 with sequences upstream of *nolJ* was not demonstrated, its upregulation might be indirect and involve other regulators.

The interaction between flavonoids and the NodD proteins might take place at the cytoplasmic membrane of the rhizobia. In *R. leguminosarum* bv. *viciae,* both the inducing flavonoid, naringenin, and the activator NodD protein were localized in the inner membrane (Recourt *et al.*, 1989; Schlaman *et al.*, 1989). In *R. meliloti,* where the proteins are produced in much lower quantities, the NodD proteins were, however, detected almost exclusively in the cytosol and a large proportion of NodD was found in a DNA-bound form (R. Fisher, personal communication and our unpublished results). It is not unlikely, however, that minor amounts of the *R. meliloti* NodD proteins are also incorporated into the cytoplasmic membrane. Further experiments are necessary to elucidate how flavonoids or the recently identified few nonflavonoid inducers are recognized by NodD and whether translocation of NodD from the cytosolic membrane is required for the *nod* box binding.

D. The Repressor NolR

In *R. meliloti, nod* gene regulation is under a dual control involving activators and the NolR repressor, which contribute to optimal nodulation

(Kondorosi *et al.*, 1989). An exception represents strain 1021 in which the *nolR* gene is inactivated by an insertional point mutation resulting in a change of the C-terminal amino acid sequence and abolishing DNA binding ability (Cren *et al.*, 1994). The *nolR* gene (Kondorosi *et al.*, 1991a) is present in single copy and located on the chromosome. The encoded protein has a size of 13 kDa, contains an HTH motif, and is homologous to the repressor XylR from *B. subtilis* (Kreuzer *et al.*, 1989) and to the NodD proteins of different *Rhizobium* and *Bradyrhizobium* species, as well as to SyrM from *R. meliloti* and NahR from *P. putida,* another member of the LysR family (Schell and Wender, 1986). NolR seems to bind in dimeric form to the conserved binding site downstream of the n1, n4, and n6 *nod* boxes (Kondorosi *et al.*, 1991a). DNAse I footprints suggest that NolR competes with the RNA polymerase for the binding site and controls transcription of both *nodD1* and the common *nod* genes (Kondorosi *et al.*, 1989).

Binding of NolR to the *nod* promoters correlates with the downregulation of the adjacent *nod* genes (unpublished results). Thus, NolR differentially regulates the *nod* genes affecting particularly those involved in the synthesis of the Nod factor core and precursor molecules and perhaps those participating in the excretion but not the ones required for the host-specific modifications of Nod factors. By controlling the key steps of Nod signal biosynthesis NolR would be expected to influence mainly the level but not the structure of the Nod signals.

It is likely that negative control of *nod* gene regulation also exists in other rhizobia. Gene expression studies suggest the presence of a repressor in *B. japonicum* (Banfalvi *et al.*, 1988), *R. leguminosarum* bvs. *phaseoli* (Davis and Johnston, 1990b), and *viciae* strain TOM (Firmin *et al.*, 1993). In fact, in strain TOM an nolR homolog was detected by Southern hybridization (our laboratory, unpublished).

E. NodD-Independent Control Elements in *B. japonicum*

1. NodV and NodW, a Two-Component Regulatory System

The *nodVW* genes of *B. japonicum* are required for nodulation of mung bean, cowpea, and siratro (Göttfert *et al.*, 1990a). Based on amino acid sequence homology, NodV is most likely a sensor kinase responding to a yet-unidentified external stimulus, while NodW is a response regulator. It was proposed that NodVW might activate expression of yet-unknown host range genes required for nodulation of these host plants. One candidate might be the *nodZ* gene, which is regulated independently from NodD and which is required for efficient nodulation on siratro (Stacey *et al.*,

1994). Recent results, however, indicate that NodW is essential for isoflavone-mediated induction of the *nodYABC* and *nodD1* operons (Sanjuan *et al.*, 1994). The preliminary data suggest a synergistic effect of NodD1 and NodW on *nod* gene activation. The mode of action of the NodVW regulatory system, however, is unknown, and it is not clear how NodW would interact with the *nod* promoter and the NodD-mediated transcriptional regulation. Recently, *nodVW* homologs (*nwsAB*) expressing at low level were identified (Grob *et al.*, 1993). Although not essential for nodulation, *nwsB* could suppress the nodulation defect of a *nodW* mutant.

2. NolA, a Putative DNA-Binding Protein

The *nolA* gene is required for genotype-specific nodulation of soybeans (Sadowsky *et al.*, 1991). Though the function of NolA is unknown, the N-terminal part of the protein shows similarity to the MerR transcriptional regulator of the mercury resistance genes. In both proteins, this region comprises a putative HTH DNA-binding domain. Therefore it was suggested that they may form a unique family of gene regulators. However, no experimental data are available yet which would support a regulatory function for NolA.

V. Bacterial Nod Signals

A. Structure of Nod Factors

The structure of the extracellular signal of *R. meliloti* that elicits root hair deformation on alfalfa, but not on vetch, was first reported by Lerouge *et al.* (1990) and was designated Nod factor, since its biosynthesis depends on the presence of functional *nod* genes. This molecule was shown to be an acylated and sulfated derivative of chitotetraose (Fig. 3). A backbone of four β-1,4-linked D-glucosamine residues is *N*-acylated with a C16:2$\Delta^{2,9}$ fatty acid chain at the nonreducing end moiety while the other glucosamine residues are *N*-acetylated. The sulfate group is located at the C-6 of the reducing terminus. This molecule, originally termed NodRm-1, was renamed NodRm-IV(C16:2,S) according to Roche *et al.* (1991b) and Spaink *et al.* (1991). Varying Nod factor preparations were found to be 6-*O*-acetylated at the nonreducing terminus (Roche *et al.*, 1991a). In some of the overproducing *Rhizobium* strains used for Nod signal purification, only a subset of the nodulation genes was amplified, not including the one conferring *O*-acetylation (Lerouge *et al.*, 1990; Schultze *et al.*, 1992), which explains why initially this modification was not detected. Strains containing extra copies of the positive regulatory genes, *syrM* and *nodD3*, which are supposed to overexpress all structural *nod* genes roughly at an equal level, indeed produced a majority of *O*-acetylated molecules (Roche *et al.*, 1991b). These factors showed a slightly higher biological activity than the nonacetylated forms (Truchet *et al.*, 1991; Roche *et al.*, 1991b).

Later it was shown that *R. meliloti* produces not only a single signal molecule, but excretes a whole family of structurally related lipooligosaccharides. The length of the oligosaccharide backbone varied between three and five residues (Roche *et al.*, 1991b; Schultze *et al.*, 1992). In addition, not only was a single type of acyl chain found, but fatty acids with different degrees of saturation were detected: C16:3$\Delta^{2,4,9}$, C16:1, as well as (ω-1)-hydroxylated C18 to C26 (Schultze *et al.*, 1992, Baev *et al.*, 1992; Demont *et al.*, 1993). Although most of the *R. meliloti* Nod factors are sulfated, a minor proportion (up to 10%) may be nonsulfated (Baev *et al.*, 1992). Thus, a remarkably large number of different lipochitooligosaccharides are excreted by a single *Rhizobium* strain. However, only few molecular species, the sulfated factors containing unsaturated acyl chains, displayed high biological activity toward host plants (Truchet *et al.*, 1991; Roche *et al.*, 1991b; Schultze *et al.*, 1992). Recently, the chemical synthesis of *R. meliloti* Nod factors has been achieved (Nicolaou *et al.*, 1992; Wang *et al.*, 1993, 1994).

Like *R. meliloti,* all other rhizobia investigated so far excrete a number of related Nod factors. Many of them have been at least partially purified, allowing the determination of their chemical structure (Fig. 3). All of them are lipochitooligosaccharides. The major Nod factors of different rhizobia have different structural features characteristic for each strain.

The major Nod metabolites of *R. leguminosarum* bv. *viciae* are nonsulfated penta- and tetrasaccharides, 6-*O*-acetylated at the nonreducing terminus and containing either a C18:4$\Delta^{2,4,6,11}$ or a C18:1Δ^{11} fatty acyl chain (Spaink *et al.*, 1991). The C18:4 acyl moiety had three trans double bonds conjugated with the carbonyl group and a nonconjugated cis double bond between carbons 11 and 12, whereas the C18:1 acyl chain had only the Δ^{11} double bond (Fig. 3) as in *cis*-vaccenic acid, a common component in membranes of Gram-negative bacteria. Only the Nod factors containing the highly unsaturated fatty acyl chain were active in eliciting a specific biological response on the host plant, *Vicia sativa*. In addition to the four purified factors, minor amounts of more hydrophobic molecules were detected, whose structure, however, has not been reported. Strain TOM of *R. leguminosarum* bv. *viciae* produces, in addition to the factors described above, lipopentasaccharides carrying *O*-acetyl groups at the C-6 of both the nonreducing and the reducing terminus (Firmin *et al.*, 1993). This

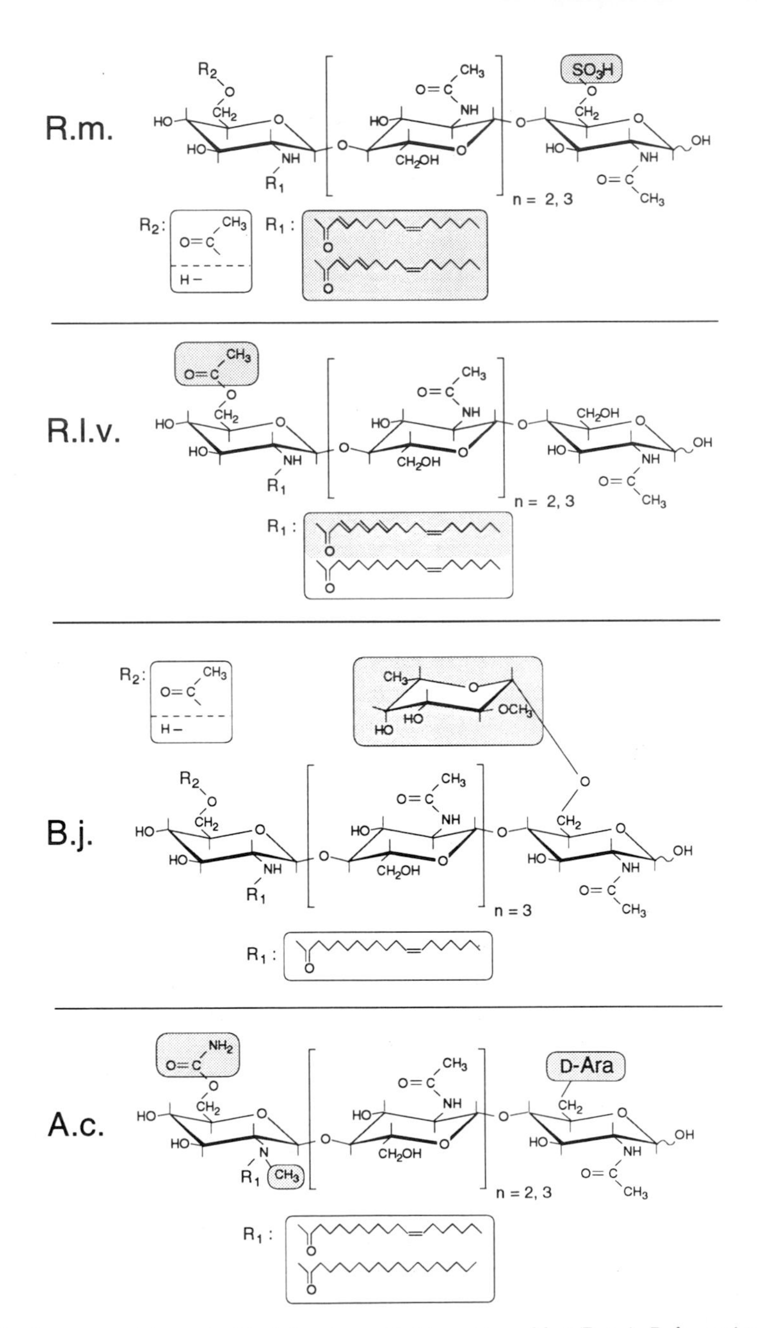

FIG. 3 Structure of the major Nod signals purified from *R. meliloti* (R.m.), *R. leguminosarum* bv. *viciae* (R.I.v.), *B. japonicum* (B.j.), *A. caulinodans* (A.c.), *R. tropici* (R.trop.) and *Rhizobium* strain NGR234. The structural modifications particular to each strain are shaded. These are the sulfate group and C16:2 and C16:3 acyl chains in NodRm factors, the 6-*O*-acetyl group and C18:4 acyl chain in NodRlv factors, the α-L-methylfucose in NodBj factors, the *N*-methyl group, 6-*O*-carbamoyl group and D-arabinose in NodAc factors, the *N*-methyl

property correlates with the ability of this strain to overcome nodulation resistance of Afghanistan peas (Winarno and Lie, 1979).

A large number of different Nod factors were identified in the three *Bradyrhizobium japonicum* isolates, USDA110, USDA135, and USDA61 (Sanjuan *et al.*, 1992; Carlson *et al.*, 1993; Stacey *et al.*, 1994). Two molecules seem to be common to all three strains: NodBj-V(Ac,C18:1,MeFuc) and NodBj-V(C18:1,MeFuc). They are C18:1Δ^{11}-acylated pentasaccharides carrying an α-linked 2-*O*-methyl-L-fucose modification at C-6 of the reducing end (Fig. 3). One of the molecules is *O*-acetylated at C-6 of the nonreducing end. While in USDA110 only these two Nod metabolites were detected, in USDA135 and USDA61, a number of additional lipooligosaccharides were observed in minor amounts (Carlson *et al.*, 1993). In USDA135, *O*-acetylated and non-*O*-acetylated factors having C16:0 or C16:1 acyl chains were detected. All of these compounds are 2-*O*-methylfucosylated. The largest diversity in Nod factors was observed in strain USDA61, (a type II strain recently

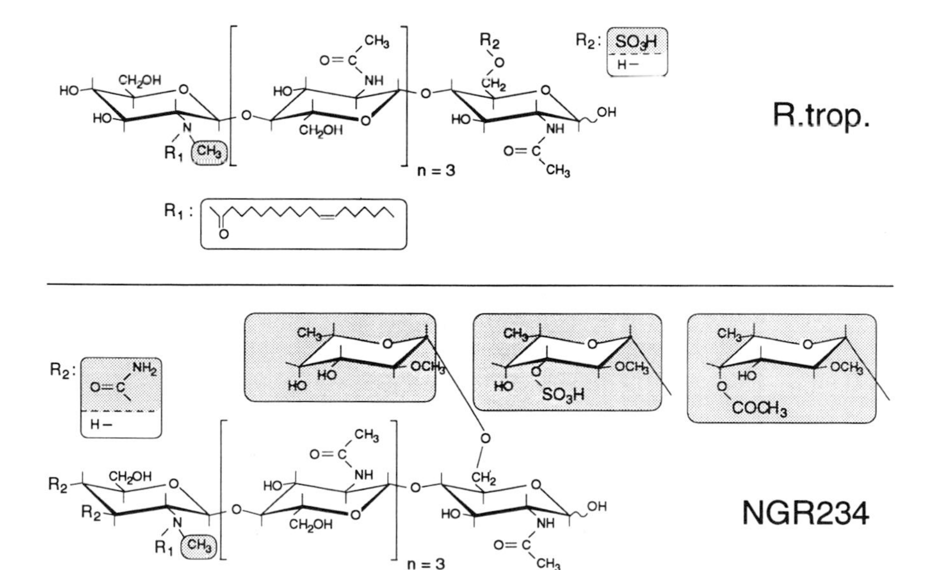

group and sulfate group in NodRt factors, and *N*-methyl group, *O*-carbamoyl groups and differently substituted α-L-methylfucose residues in NodNGR factors. The 6-*O*-acetylation at the reducing end of NodRlv factors in strain TOM is not shown. The conformation (pyranose versus furanose) of the D-arabinose group is not known yet (P. Mergaert and M. Holsters, personal communication).

renamed *B. elkanii; Kuykendall et al.*, 1992) which produced not only pentasaccharides but also tetrasaccharides. In contrast to USDA135, only a single type of acyl chain (C18:1) was detected. Moreover, the nonreducing end can be *N*-methylated and *O*-carbamoylated. Furthermore, some factors carry, in place of 2-*O*-methylfucose, a nonmodified α-L-fucosyl residue at the C-6, in conjunction with a glycerol group glycosidically linked to C-1 of the reducing end *N*-acetylglucosamine residue.

Nod factors of another microsymbiont of soybean, *R. fredii*, vary in the length of the oligosaccharide backbone between three and five residues. They have a C18:1 acyl chain and are modified at the reducing end by fucose or 2-*O*-methylfucose (Bec-Ferté *et al.*, 1993). Thus, these factors show a remarkable similarity to those produced by *B. japonicum*, NodRf-V(C18:1,MeFuc) being identical to one of the major Nod metabolites common to the three *Bradyrhizobium* isolates.

Ten different Nod factors were detected in *Azorhizobium caulinodans* (Mergaert *et al.*, 1993). The two major factors are pentasaccharides having either a C18:1 or a C18:0 fatty acyl chain. The nonreducing end is *N*-methylated and carries a carbamoyl group at C-6 (Fig. 3). In a subpopulation of the pentasaccharides, the carbamoyl group is absent while the reducing end can be substituted at C-6 by D-arabinose. In addition, tetrasaccharide factors were detected. The structural feature characteristic for *A. caulinodans* seems to be the arabinosyl modification. Since the Nod factors were analyzed in an overproducing strain carrying extra copies of only a subset of the nodulation genes, in wild-type strains of *A. caulinodans* possibly the majority of the Nod factors carry the arabinose group.

An intriguing situation is exemplified by the broad host range *Rhizobium* strain NGR234 (Price *et al.*, 1992). This strain combines several structural features observed in rhizobia of a narrow host range (Fig. 3). The *N*-acetylglucosamine backbone is a pentasaccharide. The majority of the Nod factors carry a C18:1Δ^{11} acyl chain. A minor fraction contains a C16:0 fatty acyl chain. The nonreducing end is *N*-methylated and can carry up to two carbamoyl groups at C-3 and C-4 positions. At the reducing terminus the C-6 is substituted with 2-*O*-methylfucose, which itself can be, in a mutually exclusive way, either 3-*O*-sulfated or 4-*O*-acetylated or nonmodified. Taken together, the Nod factors of this strain are the most highly derived ones found in rhizobia so far. The large structural diversity of Nod factors may partially contribute to the broad host range of NGR234.

Interestingly, another *Rhizobium* species of broad host range, *R. tropici*, also excretes sulfated as well as nonsulfated Nod factors (Poupot *et al.*, 1993). Similar to the NodNGR factors, they are *N*-methylated and C18:1Δ^{11}-acylated pentamers, but their sulfate group is attached to the terminal reducing glucosamine as in the NodRm factors (Fig. 3).

B. Biochemical Functions of *nod* Genes in the Production of Nod Factors

It has been known for a while that *nod* genes are involved in the synthesis of the rhizobial extracellular signals that induce various root responses such as deformation of root hairs (Van Brussel *et al.*, 1986; Zaat *et al.*, 1987; Faucher *et al.*, 1988, 1989; Banfalvi and Kondorosi, 1989; Baev *et al.*, 1991). The genetic data indicated that the factor produced by the common NodABC proteins is modified by the action of host-specific gene products (Faucher *et al.*, 1988, 1989; Banfalvi and Kondorosi, 1989). With the present knowledge of Nod factor structures for many *Rhizobium* strains, a detailed picture is emerging with regard to the function of structural nodulation genes in Nod metabolite production.

Important insights have come from the analysis of Nod factors produced by mutant rhizobia. In some cases a function has been determined also by *in vitro* assays using purified or partially purified *nod* gene products. In other cases, a function has not been directly demonstrated yet, but is inferred from sequence similarities with known proteins. The *nod* genes for which a role has been shown or proposed can be divided into four classes according to the following steps in Nod factor production (Table I): (1) synthesis of precursor molecules (*nodM*); (2) synthesis of a common lipooligosaccharide structure (*nodA*, *B* and *C*); (3) host-specific modifications of the basic structure (*nodE, F, G, H, L, P, Q, S, X, Z,* and *nolK, nolO*); (4) accessory functions like Nod signal excretion (*nodI, J* and *nolF, G, H,* and *I*).

1. Precursor Synthesis

NodM has been shown to be a glucosamine synthase. It functionally complements *glmS* mutants of *Escherichia coli* and *Rhizobium* (Baev *et al.*, 1991; Marie *et al.*, 1992). In fact, *nodM* mutant strains produced up to 20-fold lower amounts of Nod signals in *R. meliloti* (Baev *et al.*, 1992) and threefold lower amounts in *R. leguminosarum* (Spaink *et al.*, 1992). It seems surprising that, in addition to the housekeeping function, a symbiosis-specific form of glucosamine synthase is required for efficient Nod signal production. It has been proposed that Nod factor synthesis may take place in a particular compartment; for example, a multienzyme complex, which would explain the need for a specific enzyme (Baev *et al.*, 1992). If a compartmentalized biosynthetic pathway for Nod factors includes the synthesis of precursor molecules, then it is reasonable to expect it also for the further steps, that is, *N*-acetylation of glucosamine and activation to uridine diphosphate *N*-acetylglucosamine (UDP-

GlcNAc). However, no particular *nod* genes have been assigned to such functions yet. As for *nodM*, mutations in such genes would probably display weak phenotypes since they would be partially complemented by housekeeping enzymes (Gerhold *et al.*, 1989).

2. Synthesis of a Basic Lipooligosaccharide Structure

While many other oligosaccharides acting as elicitors or plant morphogens are produced during plant–microbe interactions by the degradation of polymers (see Darvill *et al.*, 1992), it is now emerging that rhizobial Nod signals are produced *de novo*. The common *nodABC* genes are involved in the synthesis of a basic lipooligosaccharide lacking host-specific modifications at the reducing and nonreducing end and carrying a $C18{:}1\Delta^{11}$ fatty acyl chain (Spaink *et al.*, 1991) that is a common component of the bacterial cell (MacKenzie *et al.*, 1979). The *nodABC* genes are also able to direct the synthesis of Nod factors when transferred to *E. coli* cells (Banfalvi and Kondorosi, 1989; Spaink *et al.*, 1992).

Chitin oligosaccharides appear to be synthesized from UDP-GlcNAc by NodC protein, which shows homology to yeast chitin synthase and cellulose synthase, both producing β-1,4-linked polymers (Bulawa and Wasco, 1991; Atkinson and Long, 1992; Bulawa, 1992; Debellé *et al.*, 1992). Recently, the *nodC*-dependent *in vitro* synthesis of chitin oligomers in cell-free extracts of *Rhizobium* and *E. coli* was demonstrated (Spaink *et al.*, 1993b; Geremia *et al.*, 1994). Purified NodB protein has been shown to deacetylate the nonreducing *N*-acetylglucosamine residue of chitooligosaccharides (John *et al.*, 1993). The deacetylated molecule then serves as an acceptor for a transacylation reaction catalyzed by NodA which might preferably incorporate the specific unsaturated fatty acids characteristic for some *Rhizobium* species (Röhrig *et al.*, 1994).

3. Host-Specific Modifications

a. Modifications at the Nonreducing End The role of several *nod* gene products involved in host-specific modifications of the basic lipooligosaccharide has been elucidated. *nodE* and *nodF* determine the specific fatty acyl modification of the Nod factors. Mutations in either of these genes do not prevent Nod factor synthesis but in *R. leguminosarum* bv. *viciae* and *R. meliloti* lead to an aberrant acylation with *cis*-vaccenic acid (C18:1) (Spaink *et al.*, 1991; Demont *et al.*, 1993). NodE is homologous to β-ketoacyl synthases (Bibb *et al.*, 1989; Tsay *et al.*, 1992) and NodF shows similarity to acyl carrier proteins (ACP) (Horvath *et al.*, 1986; Shearman *et al.*, 1986; Bibb *et al.*, 1989). Consistent with this is the finding that NodF carries a 4′-phosphopantetheine prosthetic group (Geiger *et al.*,

1991), as does the ACP of *E. coli*. Furthermore, in the absence of the *nodABC* genes, *nodEF* directs the production of phospholipids containing C18:4$\Delta^{2,4,6,11}$ fatty acids (Geiger *et al.*, 1994). Therefore, NodE and F seem to participate in the synthesis of specific unsaturated fatty acids, like C18:4 in *R. leguminosarum* bv. *viciae* and C16:2 in *R. meliloti*.

A model has been described by Spaink (1992) in which NodE and F replace housekeeping functions in fatty acid synthesis. During the synthesis of C18:4 or C18:3 fatty acids of *R. leguminosarum* bv. *viciae*, C16:3 and C16:2 fatty acyl chains, respectively, would be intermediates. Thus, synthesis of the fatty acids may be similar in *R. leguminosarum* bv. *viciae* and *R. meliloti*. In *R. meliloti*, however, the synthetic cycle would be interrupted one step earlier. This model is supported by the finding that transfer of the *nodEF* genes from *R. leguminosarum* to *R. meliloti* leads to the synthesis of Nod factors containing C18:4, C18:3 and C18:2 fatty acids (Demont *et al.*, 1993).

R. meliloti contains another gene, *nodG*, that is supposed to play a role in acyl chain specification based on its strong sequence similarity to β ketoacyl reductases (Slabas *et al.*, 1992; Rawlings and Cronan, 1992). In concert with *nodEF*, the *nodG* gene product might be necessary for the formation of the C16:2 and C16:3 unsaturated Nod factors of *R. meliloti*. However, no effect on Nod factor synthesis by mutations in *nodG* was found by Demont *et al.* (1993), although *nodG* mutants show a delayed nodulation phenotype (Horvath *et al.*, 1986). As in the case of several other nodulation genes, *nodG* function might be redundant and a mutation may be silent in particular genetic backgrounds.

The *nodL* gene has been identified in *R. leguminosarum* biovars (Surin and Downie, 1988; Canter-Cremers *et al.*, 1989) and *R. meliloti* (Baev and Kondorosi, 1992). It functions in *O*-acetylation at the nonreducing end of Nod factors (Spaink *et al.*, 1991; Bloemberg *et al.*, 1994). While in *R. leguminosarum* bv. *viciae*, essentially all Nod factors are *O*-acetylated, in *R. meliloti*, *O*-acetylation seems to be less efficient (Roche *et al.*, 1991a,b; Schultze *et al.*, 1992). This might be due to insufficient expression of *nodL* and/or a low gene dosage in the overexpressing strains used for Nod metabolite analysis. Besides its function in *O*-acetylation, *nodL* also strongly affects the efficiency of Nod signal production or excretion. The supernatant of *nodL* mutant strains contains five to tenfold fewer Nod factors than that of wild-type strains (Spaink *et al.*, 1992). It is not known whether the biosynthesis is reduced or whether the nonacetylated factors are less efficiently released from the bacterial cell.

N-Methylation in *A. caulinodans* has been attributed to the *nodS* gene product (Geelen *et al.*, 1993). *nodS* is present also in *Rhizobium* strain NGR234 (Lewin *et al.*, 1990), *B. japonicum* (Göttfert *et al.*, 1990b), and *R. fredii* (Krishnan *et al.*, 1992). In the latter two species, however, it is poorly expressed, which explains why the major Nod factors in these strains are not *N*-methylated (Carlson *et al.*, 1993; Bec-Ferté *et al.*, 1993). No *nod* gene function has yet been reported for the carbamoylation at the nonreducing end of Nod factors.

b. Modifications at the Reducing End The role for *nodP, Q,* and *H* in the sulfation of Nod factors has been clearly established. *nodP* and *Q,* homologous to the sulfate activation locus (*cysD, cysN,* and *cysC*) of *E. coli* (Leyh *et al.*, 1992), are involved in the production of 3′-phosphoadenosine-5′-phosphosulfate (PAPS) (Schwedock and Long, 1990; Atkinson *et al.*, 1992). NodH shows sulfotransferase activity that is able to sulfatate chitooligosaccharides in cell-free extracts (Atkinson *et al.*, 1992). Accordingly, *nodH* mutants of *R. meliloti* produce only nonsulfated Nod factors (Roche *et al.*, 1991b). *R. meliloti* contains multiple copies of the sulfate-activating genes, $nodP_1Q_1$, $nodP_2Q_2$, and a third locus, *saa,* which is required for the synthesis of sulfur-containing amino acids (Schwedock and Long, 1989, 1992). As a consequence of this redundancy, mutations in either of the *nodP* or *nodQ* genes are leaky, and diminish only the efficiency of Nod factor sulfatation (Faucher *et al.*, 1989; Roche *et al.*, 1991b; Schwedock and Long, 1992). A $nodQ_1nodQ_2$ double mutant, however, shows a phenotype similar to that of the *nodH* mutant (Schwedock and Long, 1992). Interestingly, in this mutant the total amount of excreted Nod factors is reduced 10-fold (Roche *et al.*, 1991b). A question arises that is similar to that described for *nodM* mutant phenotypes. Why are symbiosis-specific copies of housekeeping gene functions required? Compartmentalization might be the explanation in both cases.

O-Acetylation at the reducing end in strain TOM of *R. leguminosarum* bv. *viciae* is conferred by *nodX* (Firmin *et al.*, 1993). *nodX* is present also in *R. leguminosarum* bv. *trifolii* (Davis *et al.*, 1988), which explains the ability of *nodEF* mutants to nodulate not only European but also Afghanistan cultivars of pea (Djordjevic *et al.*, 1985).

nodZ is involved in 2-*O*-methylfucosylation of *B. japonicum* Nod factors (Stacey *et al.*, 1994). *nolO* identified in *B. japonicum* shows significant homology to *nodU* of *Bradyrhizobium* and other rhizobia in the amino acid sequence but not on the DNA level (Luka *et al.*, 1993). *nolO* mutant strains produce a mixture of Nod factors with and without the 2-*O*-methylfucosyl modification. Therefore, it seems to contribute to the efficiency of 2-*O*-methylfucosylation of NodBj factors.

The *nolK* gene product of *A. caulinodans* is likely to play a role in the D-arabinosyl modification (Holsters *et al.*, 1993) since it shows similarity to sugar epimerases (Goethals *et al.*, 1992b).

4. Auxiliary *nod* Gene Functions

a. Excretion of Nod Factors Several *nod* genes have been implicated in the transport of Nod factors. The *nodIJ* genes have been identified in most *Rhizobium* isolates (see Fig. 1), suggesting a common function. However, mutations in these genes do not affect nodulation in all cases. There is no reduction in nodulation with *nodIJ* mutants of *B. japonicum* on soybean (Göttfert *et al.*, 1989). Transposon insertions in the putative *nodIJ* genes of *R. meliloti* had no effect (Jacobs *et al.*, 1985) or led to a delay of nodulation on alfalfa (Debellé *et al.*, 1986). Nodulation on pea and vetch is delayed with *R. leguminosarum* bv. *viciae* (Downie *et al.*, 1985; Canter-Cremers *et al.*, 1988), and abolished on pea with transconjugants carrying the nodulation genes of strain TOM on plasmid (Davis *et al.*, 1988). With *nodIJ* mutants of *R. leguminosarum* bv. *trifolii*, nodulation is strongly inhibited in clovers and peas (Djordjevic *et al.*, 1985; Innes *et al.*, 1985; Canter-Cremers *et al.*, 1988; Huang *et al.*, 1988).

Recently, sequence similarity has been detected between NodI and J and a pair of proteins from a variety of Gram-negative bacteria that are involved in the energy-dependent excretion of growing polysaccharide chains through the inner membrane (Vázquez *et al.*, 1993). This finding is consistent with the previously reported similarity of NodI to ATP-binding proteins also involved in transport processes (Evans and Downie, 1986), and its association with the inner membrane (Schlaman *et al.*, 1990). Based on these similarities, NodI and J have been proposed to function in Nod factor export. Direct evidence for this was reported by McKay and Djordjevic (1993). Nod factors were undetectable in the supernatant of *nodI* and *nodJ* mutants of *R. leguminosarum* bv. *trifolii* carrying nodulation genes on a plasmid, while increased amounts were found in the cell pellet. By contrast, *nodIJ* mutants of *R. leguminosarum* bv. *viciae* were not detectably affected in Nod metabolite excretion (Spaink *et al.*, 1992). However, in the presence of only the essential *nodABC* and *nodEF* genes, *nodIJ* increased the amounts of Nod metabolites in the culture supernatant (Spaink *et al.*, 1993a). The C18:4 acylated molecules were specifically enhanced and, consistent with this, *nodIJ* did not affect the quantity of C18:1 Nod factors produced in the absence of *nodEF*. These results suggest that NodIJ may display specificity in the export or release of Nod factors from the bacterial cell. Furthermore, the data indicate that a contribution of NodI and J to Nod factor export is more pronounced in overproducing strains in which nodulation genes are present on plasmid (Evans and Downie, 1986; Davis *et al.*, 1988; McKay and Djordjevic, 1993; Spaink *et al.*, 1993a). The observation might be explained by the existence of a second transport system which may be saturated by excess amounts of Nod factors. In *R. leguminsosarum* bv. *trifolii*, a transport function does not seem to be redundant, which explains the stronger effects on nodulation observed with *nodIJ* mutants of this strain.

In *R. meliloti,* similarity of the predicted *nolFGHI* gene products (Baev *et al.,* 1991) to bacterial transport proteins has been reported and it was proposed that they may function in secretion of Nod factors (Göttfert, 1993; Saier *et al.,* 1994). Thus, the putative redundancy in transport functions may explain the lack of phenotype in *nodI* mutants of *R. meliloti.* Mutations in the *nolFGHI* genes lead to delayed nodulation on alfalfa and reduced Nod signal production (Baev *et al.,* 1991). Double mutants affected in both regions will have to be prepared in order to give a clear answer as to whether *nodIJ* and *nolFGHI* are functionally homologous.

b. Other* nod *Genes A number of *nod* genes have been described whose exact function, however, remains unclear. *nodN,* which is present in *R. leguminosarum* biovars and *R. meliloti* (Surin and Downie, 1988; Baev *et al.,* 1991), reduces Nod signal production (Baev *et al.,* 1991). The gene of *R. meliloti* can be functionally replaced by the one of *R. leguminosarum* bv. *viciae* and thus it has, like *nodABC* and *nodM,* a common function (Baev *et al.,* 1992).

nodT (Canter-Cremers *et al.,* 1989; Surin *et al.,* 1990) and *nodO* (De Maagd *et al.,* 1989b; Economou *et al.,* 1990) have been found only in *R. leguminosarum.* Mutations in *nodT* and *nodO* did not affect Nod factor production in *R. leguminosarum* bv. *viciae* (Spaink *et al.,* 1992). In *R. leguminosarum* bv. *trifolii, nodT* has been implicated in cultivar-specific nodulation on subterranean clover (*Trifolium subterraneum*) (Canter-Cremers *et al.,* 1989; Lewis-Henderson and Djordjevic, 1991a).

NodO is a secreted protein (De Maagd *et al.,* 1989a,b; Economou *et al.,* 1990; Scheu *et al.,* 1992) and is able to form ion channels in artificial lipid bilayers (Sutton *et al.,* 1993). The exact role it may play during nodulation, however, remains open. Mutations in *nodO* affect nodulation only in some chromosomal backgrounds and on some host plants; for example, *Vicia sativa* (De Maagd *et al.,* 1989a). Interestingly, *nodO* can partially complement *nodEF* mutations of *R. leguminosarum* bv. *viciae* for nodulation, and it appears to act synergistically also with *nodL* (Downie and Surin, 1990). It was proposed that *nodO* has a function complementary to Nod factors since nodulation of *nodE nodO* double mutants on pea is reduced to 5% of the normal levels while the single mutations have a much weaker effect (Sutton *et al.,* 1993). *nodE* mutants do not produce C18:4-acylated Nod factor, but only the C18:1 derivatives. These are inactive in meristem induction on vetch (Spaink *et al.,* 1991) and activate the expression of early nodulin genes only after a delay (Horvath *et al.,* 1993). Whether NodO acts in a direct way synergistically with C18:1 Nod factors—for example, by amplifying membrane responses or facilitating Nod signal uptake by the plant cell—or functions indirectly—for example, by supporting vitality of bacteria during the infection process, remains to be elucidated.

C. Subcellular Location of Nod Factor Synthesis

The subcellular location of a number of Nod proteins has been reported. NodA and NodB were localized in the cytosol (Schmidt *et al.*, 1986; 1988), NodA, however, is also associated with the inner membrane (Johnson *et al.*, 1989; Schlaman, 1992). While NodC was previously found in the outer membrane (John *et al.*, 1985, 1988), recent data appear to indicate a location predominantly in the inner membrane (Hubac *et al.*, 1992; Barny and Downie, 1993).

In combination with the known and putative biochemical functions of the Nod proteins, we propose the following model of Nod factor biosynthesis in association with the inner membrane. In analogy to membrane-localized synthesis of chitin oligomers in yeast (Cabib *et al.*, 1983), NodC polymerizes the chitooligosaccharide backbone from the reducing to the nonreducing end concomitant with its extrusion through the inner membrane. The acyl chain is synthesized by the integral membrane protein NodE (Spaink *et al.*, 1989a), the cytosolic NodF (Horvath *et al.*, 1986; Shearman *et al.*, 1986), and perhaps NodG (the location of which is yet unknown) in conjunction with other proteins, and is delivered to NodA. After deacetylation of the nonreducing GlcNAc residue by NodB, NodA catalyzes the acyl chain transfer. Host-specific modifications are introduced during synthesis at the nonreducing terminal by NodL located in the cytoplasm (Bloemberg *et al.*, 1994), and at the reducing terminus by NodH, which was reported to be membrane associated (Schmidt *et al.*, 1991). The subcellular location of NodP and NodQ is unknown and it remains to be shown where sulfate activation for Nod factor synthesis takes place. *R. leguminosarum* bv. *trifolii*, carrying the *nodG, E, F, H, P,* and *Q* genes of *R. meliloti,* released only sulfated Nod factors into the growth medium, but both sulfated and nonsulfated molecules were associated with the cell pellet, indicating that the sulfate transfer may occur during the export process (Djordjevic *et al.*, 1993). The release of the Nod factors from the inner membrane may be facilitated by NodI, which is associated with the inner membrane (Schlaman *et al.*, 1990), and the integral membrane protein NodJ (Evans and Downie, 1986; Vázquez *et al.*, 1993). Excretion through the outer membrane may occur via pores and may be enhanced by other transport systems involving also NolF, G,

H, and I in *R. meliloti*. Enzymes participating in precursor synthesis, like NodM, might also associate with the Nod factor synthetic complex, thereby locally elevating the concentration of activated GlcNAc molecules.

D. Structural Diversity of Nod Signals in Relation to Host Specificity

The differential activation of regulatory proteins by plant-derived signals constitutes a first major step in determining host specificity in the *Rhizobium*–plant interaction (see Section IV). In a second step, the signaling from the bacterium to the plant exploits the structural diversity of the lipooligosaccharides. Different *Rhizobium* strains excrete different sets of Nod factors and, in most cases, variations are also found between isolates of a single biovar. In many cases host specificity at the level of the genus, species, or cultivar can be explained by (1) variations in the Nod factor structure; (2) variations in the composition of the mixture of active and inactive or antagonistic Nod factors; (3) variations in the total concentration of Nod signals. Any direct or indirect changes in one of the three parameters may broaden, narrow, or alter the host range.

Transferring the host-specific *nod* genes from one strain to another can confer the host range of the donor strain to the recipient strain. Replacement of *nodE* in *R. leguminosarum* bv. *trifolii* by the one of *R. leguminosarum* bv. *viciae* changes the host range from *Trifolium* to *Vicia, Lathyrus,* and *Pisum* (Spaink *et al.*, 1989a). This is explained by the structure of the acyl chain of NodRlv factors, which is crucial for their biological activity (Spaink *et al.*, 1991).

Transfer of the *R. meliloti nodFEGHPQ* genes to *R. leguminosarum* bv. *trifolii* or bv. *viciae* confers to these strains the ability to nodulate alfalfa (Putnoky and Kondorosi, 1986) but prevents or strongly inhibits nodulation of the normal host plants, white clover and vetch, respectively (Debellé *et al.*, 1988; Faucher *et al.*, 1989). Mutations in *nodH* strongly inhibit the ability of *R. meliloti* to nodulate *Medicago sativa,* and lead to delayed nodulation on *Melilotus alba* but confer the ability to nodulate the nonhost plant, vetch (Horvath *et al.*, 1986; Faucher *et al.*, 1988). Therefore, sulfatation of the lipooligosaccharide signals is a major determinant of host range in *R. meliloti* (Roche *et al.*, 1991b).

Several examples are known in which the host range of a given bacterial strain was widened or narrowed (e.g., Djordjevic *et al.*, 1985). The following examples support the view that as the Nod factor increases in complexity, the host range is broadened.

Mutation of $nodQ_1$ extends the host range of *R. meliloti* to vetch (Cervantes *et al.*, 1989; Faucher *et al.*, 1989; Schwedock and Long, 1992). In fact, this mutant produces similar amounts of sulfated and nonsulfated factors (Roche *et al.*, 1991b).

Mutation of *nodL* strongly reduces the nodulation ability of *R. leguminosarum,* but this was found to depend on the plant species tested and the genomic background of the *Rhizobium* strain (Surin and Downie, 1988; Canter-Cremers *et al.*, 1989). In *R. meliloti,* it inhibited nodulation only weakly (Gerhold *et al.*, 1989). *nodX* mutants of *R. leguminosarum* lose the ability to nodulate Afghanistan peas (Davis *et al.*, 1988). *nodS* has been shown to be required for nodulation of NGR234 on the tropical tree *Leucaena leucocephala* (Lewin *et al.*, 1990) and can extend the host range of *R. fredii* to this plant species (Krishnan *et al.*, 1992). *nodZ* mutants of *B. japonicum* are not significantly impaired in nodulation of soybeans (*Glycine max*), but only inefficiently nodulate siratro (*Macroptilium atropurpureum*) (Nieuwkoop *et al.*, 1987; Stacey *et al.*, 1994). Taken together, these results indicate that *O*-acetylation as well as *N*-methylation at the nonreducing end, and *O*-acetylation and 2-*O*-methylfucosylation at the reducing terminus of Nod factors are able to widen the host range.

While it is clear that alterations in host specificity are determined to a large extent by variations in Nod factor structure, more subtle influences on host or cultivar specificity appear to be brought about by changes in the amount and/or ratio of active versus inactive Nod factors produced by a given strain. It has been proposed that some of the Nod factors excreted by rhizobia are not only inactive, but may act antagonistically to the active signals (Schultze *et al.*, 1992; Kondorosi *et al.*, 1993). Excess amounts of antagonists might inhibit nodulation on some of the host plants. In support of this hypothesis, the Nod factor profile of mutants of *B. japonicum* has been correlated with genotype-specific nodulation ability on a soybean cultivar (Stokkermans *et al.*, 1992). The non-nodulating mutants excreted higher amounts of those Nod factor species that were normally produced only in minor quantities. Another example of this type is represented by a *R. meliloti nodM* mutant in which the ratio of active versus inactive Nod factors is reduced (Baev *et al.*, 1992). Mutations in *nodM* showed delayed nodulation on alfalfa, but did not inhibit nodulation on another host plant, sweet clover (*Melilotus albus*) (Baev *et al.*, 1991). In *R. leguminosarum* bv. trifolii, nodM appears to act negatively on nodulation of subterranean clover (Lewis-Henderson and Djordjevic, 1991b).

Recently, Demont *et al.* (1994) demonstrated that the *nod* gene activators in *R. meliloti* induce the synthesis of different sets of Nod factors to a different extent. NodD3 or Syr*M,* not however NodD1 or NodD2, stimulate the production of Nod factors containing (ω-1)-hydroxylated fatty acids that might be required as symbiotic signals with certain plant genotypes.

Changes in the amount and profile of Nod factors have been observed in *R. leguminosarum* bv. *trifolii* in response to different culture conditions, such as pH and temperature (McKay and Djordjevic, 1993). Reduced Nod signal excretion has been discussed in relation to the temperature-sensitive nodulation phenotype on *Trifolium subterraneum* cv. 'Woogenellup' (Lewis-Henderson and Djordjevic, 1991a).

The examples described above demonstrate that host range selection at the level of Nod factor signaling can be influenced (1) directly by enzymes (*nod* gene products) modifying Nod factor structure; (2) indirectly by those nodulation genes participating in the efficiency of Nod signal production and excretion; (3) indirectly by other parameters such as the genomic background and metabolic state, and environmental factors such as nitrogen source, pH, and temperature.

In many cases, functional redundancy exists, that is, *nod* genes can be present in several copies. Alternatively, several different *nod* genes, each of which separately contributes only marginally to nodulation, may act synergistically. Sometimes different pathways seem to exist for establishing a successful symbiosis, for example, NodO alleviating the requirement for some of the host-specific modifications of NodRlv factors. It is conceivable that during evolution such a complex genetic system developed step by step through the adaptation of rhizobia to new plant genotypes.

VI. Plant Responses

A. Plant Genes

Morphogenesis of the symbiotic root nodule is triggered by the interaction of the host plant with its symbiotic bacteria. The nodule is a new organ of the plant with a defined structure (Dart, 1977; Newcomb, 1981; Vasse *et al.*, 1990; Hirsch, 1992; Franssen *et al.*, 1992). During its differentiation, a set of specific genes (called nodulin genes; Legocki and Verma, 1980) are activated in the root or in the developing nodule. For some nodulin genes, the expression is not restricted to the differentiating nodule but is also detected in other organs where they may play roles similar to those in the symbiotic organ. Two major types of nodulin genes have been defined by their pattern of expression and function (Verma *et al.*, 1986; Nap and Bisseling, 1990). Nodulins expressed before the beginning of N_2 fixation are called early nodulin (Enod). They are thought to participate in the early interaction between the bacteria and the plant (curling of root hairs, formation of infection threads), as well as being involved in nodule morphogenesis (cortical cell division, formation of the primordium, and establishment of a meristem). Genes expressed just before or during N_2 fixation are called late nodulins. These genes are involved mainly in the different metabolic activities necessary for the functioning (N_2 fixation) of the nodule. In this chapter we describe only the early nodulin genes.

1. Early Nodulin Genes

In roots or root hairs, a set of plant genes is specifically activated by the presence of symbiotic bacteria shortly after infection. Some of them can also be induced by Nod factor (like *Enod12*) whereas others are activated at later stages (e.g., *Enod2*). Among the *Enod* genes, one may expect those genes involved in triggering nodule development. Many of the early nodulins characterized so far, however, seem to be structural cell wall proteins. Putative nodulins participating in a signaling cascade are most likely of low abundance and therefore more difficult to identify.

Using two-dimensional gel electrophoresis, twelve *Rhizobium*-induced proteins were identified in root hairs of the cowpea (*Vigna unguiculata*) (Krause and Broughton, 1992). Three of the characterized proteins are transiently expressed and might participate in the very early steps of infection or recognition, or might be involved in root hair deformation. One of them seems to be host specific because it is induced only in the presence of *Rhizobium* sp. NGR234 and not by *R. fredii,* though both strains can induce N_2-fixing nodules on *V. unguiculata.*

In the pea (*Pisum sativum*), *in vitro* translation experiments of root hair RNA obtained after inoculation with *R. leguminosarum* bv. *viciae* revealed enhanced and induced expression of several genes (Gloudemans *et al.,* 1989). For example, nodulin RH44 is present in uninoculated root hairs but it is enhanced in the presence of the *Rhizobium.* In contrast, expression of RH42 is induced in root hairs following infection with *R. leguminosarum* bv. *viciae.* The early nodulins PsEnod12 and PsEnod5 (see later discussion) are also induced in pea root hairs after inoculation with *Rhizobium,* or after treatment with culture filtrate (Gloudemans *et al.,* 1989). Some of the root hair-induced nodulin genes are also expressed later in the indeterminate nodules of pea, cowpea, and alfalfa, the best-characterized example being *Enod12* (Scheres *et al.,* 1990a; Krause and Broughton, 1992; Pichon *et al.,* 1992; Allison *et al.,* 1993; Bauer *et al.,* 1994). In the following paragraphs, the *Enod* genes characterized so far are described.

Enod12 was isolated from *P. sativum* (*PsEnod12*) (Scheres *et al.,* 1990a), *M. truncatula* (*MtEnod12*) (Pichon *et al.,* 1992), and *M. sativa* (*MsEnod12*)

(Allison *et al.*, 1993). The Enod12 protein sequence is composed of a putative signal peptide followed by proline-rich pentapeptide repeats that presumably represent a cell wall protein. In pea and *M. sativa,* two gene copies (A and B) were identified while *M. truncatula* contains only a single copy. The *Enod12* transcripts were detected in root hairs and in cortical cells containing the infection thread as well as those in front of the infection thread, and at a later stage in infected and uninfected cells of the invasion zone of the nodule (Scheres *et al.*, 1990a,b; 1992). Based on these data, *Enod12* was suggested to be involved in the infection process.

Expression of the *MtEnod12-Gus* fusion in transgenic *M. Sativa* spp. *varia* was detectable as early as 3 hr after infection in the epidermal cells of the growing root hair zone (Pichon *et al.*, 1992). Otherwise it followed the same pattern as the *PsEnod12* gene. The expression was inducible by purified Nod factors (Horvath *et al.*, 1993). It was also expressed in *N*-(1-naphtyl)phthalamic acid-induced pseudonodules of pea cv. 'Afganistan' (Scheres *et al.*, 1992) and in meristematic foci in the roots of Nar$^+$ alfalfa plants (Pichon *et al.*, 1993). While the spatial and temporal expression of the two *Enod12* copies are identical in pea (Scheres *et al.*, 1990a), the alfalfa *MsEnod12A* and *B* genes are differentially regulated (Bauer *et al.*, 1994). Expression of *MsEnod12B* is inducible by Nod factors after 6 hr whereas *MsEnod12A* expression is detectable only later in the *Rhizobium*-induced nodules and thus is associated with the infection process. Recently, mutant plants lacking *Enod12* genes but capable of forming normal nitrogen-fixing nodules were found among hybrid progenies of *M. coerulea* and *M. quasifalcata* (Csanádi *et al.*, 1994). This indicates that *Enod12* is not essential for nodule organogenesis and functioning.

MsEnod10 is highly homologous to *Enod12A* and *B* in the putative signal peptide sequence but shows much less similarity in the structural part (Löbler and Hirsch, 1993). It is expressed in the invasion zone of the alfalfa nodule and can be detected on Northern blotting 4 to 6 days postinoculation.

PsEnod5 encodes a 14-kDa protein containing a putative N-terminal signal peptide and does not show homology to other proteins (Scheres *et al.*, 1990b). *PsEnod5* is expressed at a low level in the invasion zone directly adjacent to the apical meristem of nodules whereas the level of expression is high in the interzone II–III, where bacterial proliferation and enlargement of the infected cells occurs. However, it is not expressed in the symbiotic zone. Contrary to *Enod12, Enod5* expression is restricted to cells containing the infection thread and is not detectable in the noninfected primordium. *Enod5* is induced by Nod factors in pea root hairs but with different kinetics than *Enod12* (Horvath *et al.*, 1993).

Enod40 has been cloned from soybeans (Kouchi and Hata, 1993; Yang *et al.*, 1993), peas (Yang *et al.*, 1993), and alfalfa (Hirsch *et al.*, 1993;

Crespi *et al.*, submitted for publication). The members of the *Enod40* gene family do not show significant homology to any previously characterized sequence. A striking particularity of the majority of the cloned *Enod40* sequences is that the 0.7-kb polyadenylated transcript does not contain large open reading frames starting with an AUG. This would indicate that *Enod40* may function as an RNA molecule rather than coding for a peptide. Recent data from our laboratory indeed suggest that *Enod40* codes for a novel class of nontranslatable cytoplasmic RNA (Crespi *et al.*, submitted for publication).

During the early steps of infection, the *GmEnod40* gene is induced in the dividing cortical cells as well as in the pericycle cells opposite to the nodule primordia. At a later stage, it is highly expressed in the pericycle of the vascular tissue of soybean nodules, and at a low level in uninfected cells. The expression pattern of *Enod40* during pea, alfalfa, and soybean nodule development is similar. *Enod40* is expressed in nodule primordia induced by a *Bradyrhizobium* strain that is unable to invade plant cells or by purified Nod factors and therefore does not require infection (Yang *et al.*, 1993; Franssen *et al.*, 1993). In contrast, *Enod40* expression in the nodule vascular bundles depends on infection. Thus the induction of the *Enod40* can be triggered by two independent mechanisms, resulting in the expression of the gene at different times and locations during morphogenesis. Its expression in vascular bundles may require additional bacterial factors to trigger the late step of infection.

Enod2 was isolated from soybeans (Franssen *et al.*, 1987), peas (Van de Wiel *et al.*, 1990a), *Sesbania rostrata* (Strittmatter *et al.*, 1989; Dehio and De Bruijn, 1992), alfalfa (Dickstein *et al.*, 1988), and lupin (Szczyglowski and Legocki, 1990). This nodulin is also a (hydroxy)proline-rich protein consisting of repeats of the heptapeptide (PPHEKPP) that are separated by a less conserved sequence of eleven amino acids (P E/V Y Q/K PPHEKPP). Since the amino acid composition of Enod2 resembles that of proline-rich (PRP) proteins (see Schowalter, 1993), it is assumed that Enod2 is a cell wall component (Franssen *et al.*, 1987). Enod2 is a late early nodulin expressed much later than *Enod12* or *Enod40*. It is first expressed at the base of the nodule and in the parenchyma (inner cortex) cells (Van de Wiel *et al.*, 1990a,b; Allen *et al.*, 1991). It was proposed to contribute to the formation of an O_2 barrier (Tjepkema and Yocum, 1974; Witty *et al.*, 1987), providing a microaerobic condition inside the nodule for the function of the oxygen-sensitive nitrogenase. In roots but not in stems of *S. rostrata*, the *SrEnod2* gene is inducible by cytokinin (Dehio and De Bruijn, 1992), which may be involved in the signal transduction pathway required for *Enod2* induction. Using *GmEnod2-Gus* gene fusions, Lauridsen *et al.* (1993) showed that the cell-specific characteristics of the promoter were maintained in both determinate and indeterminate nodules.

Positive as well as tissue-specific elements were defined in the *GmEnod2* promoter. *Enod2* is expressed independently of infection and N_2 fixation, since it is detected in empty nodules (Dickstein *et al.*, 1988; Gloudemans *et al.*, 1987; Van de Wiel *et al.*, 1990b), in nodules induced by auxin transport inhibitors (Hirsch *et al.*, 1989; Van de Wiel *et al.*, 1990b), and in spontaneous nodules (Truchet *et al.*, 1989).

Another gene, *MsEnod8*, displaying the same time course of expression as *Enod2*, was also shown to be expressed in empty nodules (Dickstein *et al.*, 1993). It has the capacity to code for a 42-kDa protein with sequence similarity to anther-specific proteins of unknown function.

The *PsEnod3* and *PsEnod14* genes are members of the same gene family. They share extensive homology and are expressed in a similar way (Scheres *et al.*, 1990b). *PsEnod3* encodes an 8-kDa protein containing a putative N-terminal signal peptide. Four cysteine residues are arranged in two pairs in such a way that it can be a metal binding protein without the characteristics of DNA binding proteins. The *PsEnod3* transcript is detectable on Northern blots 10 days after *Rhizobium* inoculation, with a maximum at day 13 which is maintained thereafter. *PsEnod3* is expressed in the last cell layers of the interzone II–III and in the first cell layers of the symbiotic zone. *PsEnod3* transcript starts to accumulate where *PsEnod12* expression ceases and it disappears in those cell layers where the *Rhizobium nifH* RNA encoding the nitrogenase is first detectable. Thus, its expression is restricted to the symbiotic zone in cell layers where nitrogen fixation has not begun yet. Moreover, *PsEnod3* transcript is detectable only in infected cells.

GmEnod55 encodes a protein containing a domain rich in proline and serine flanked by hydrophobic, possibly membrane-spanning regions (De Blank *et al.*, 1993). It is expressed later than *GmEnod2* (Rossbach *et al.*, 1989) but before maximum expression of leghemoglobin. *GmEnod55* is expressed only in infected cells like *PsEnod5*, to which it shows weak homology in the N-terminal part.

Another early nodulin gene from soybean showing homology to *PsEnod5* is *GmEnod315* (GmN#315; Kouchi and Hata, 1993). The presence of a domain containing Pro-Ser alternating residues seems to indicate also a relationship to *GmEnod55*. It is also expressed in infected cells, but at a later stage of development than *PsEnod5*.

The *GmEnod93* gene (GmN#93; Kouchi and Hata, 1993) codes for a putative 11-kDa protein rich in Ala and Ser residues. It does not exhibit homology to any known sequences. *GmEnod93* expression is first detectable 3 days postinoculation by reverse transcription–polymerase chain reaction (RT–PCR) analysis, it increases until day 13 and stays constant afterward. *GmEnod93* transcripts are first detectable in the dividing cell cluster of the outer cortex adjacent to the infected site. In emerging nodules the expression was localized in the meristematic region, and in mature nodules it was detected in the infected cells. Since the expression of the gene continues to increase after the onset of nitrogen fixation, it is most likely limited to cells invaded by bacteria, even in the earliest stages of infection.

GmEnod70 (GmN#70; Kouchi and Hata, 1993) encodes a protein with a predicted molecular weight of 53 kDa. It contains at least ten hydrophobic domains and it might be a transmembrane protein. *GmEnod70* expression is first detectable by RT–PCR 5 days postinoculation and after 8 days by Northern blotting analysis. The appearance of the *GmEnod70* transcript coincides with the differentiation of the vascular connection, suggesting that its product might be involved in nutrition.

The early nodulins Nms30 (Dickstein *et al.*, 1988), N40′ of pea (Govers *et al.*, 1985, 1986) and Nvs-40 of *Vicia sativa* (Moerman *et al.*, 1987) are immunologically related and may represent the same genes in these different plant species. *Nms30* was detected in empty nodules of alfalfa (Dickstein *et al.*, 1988) and parallels the expression of *MsEnod2*.

The nodulin genes cloned so far have provided us with the molecular tools to study the different steps of nodule organogenesis, but no functional role can be unambiguously attributed to anyone. In addition, with the exception of *MsEnod12*, mutants for these nodulins which would give us indications for their exact roles in nodule development do not exist. More genes and mutant plants have to be isolated and analyzed in order to understand the molecular aspect of the nodule organogenesis.

2. Other Genes

a. Cell Cycle Machinery Nodule organogenesis requires the induction of cell division in the root cortex, suggesting that one possible role of the Nod factors may be the activation of the cell cycle machinery in the differentiated cells. A few cell cycle marker genes have been cloned already from leguminous plants, including *cdc2*, which, in association with different types of cyclins, plays a pivotal role in cell cycle progression. *cdc2* genes have been isolated from peas (Feiler and Jacobs, 1990) alfalfa (Hirt *et al.*, 1991), and soybeans (Miao *et al.*, 1993). In soybeans, two functional homologs of *cdc2* are differentially expressed in the different plant meristems: Expression of one of them is enhanced in roots after *Rhizobium* infection and during early nodule development (Verma, 1992), suggesting that it may play a role in nodule initiation and morphogenesis. Interestingly, the expression of this gene is also induced in roots by treatment with auxins (Miao *et al.*, 1993). Cyclin genes have been isolated from alfalfa (*cycMs2;* Hirt *et al.*, 1992) and soybeans (Hata *et al.*, 1991; Verma, 1992). In addition, the histone gene variant H3-1 was shown to be specifically expressed during the S-phase of the cell cycle in alfalfa cell suspensions (Kapros *et al.*, 1992). Recently, we have shown that Nod factors can activate cell cycle progression in *M. sativa* cell cultures (Savouré *et al.*, 1994). The biologically active Nod factors but not their inactive derivatives induced differential expression of the cell cycle-specific marker genes (*cdc2Ms, cycMs2,* and histone H3-1) as well as increased thymidine incorporation and $p34^{cdc2}$ kinase activity. These data indicate that the Nod factor can effectively activate the cell cycle progression.

b. Flavonoid Biosynthesis Genes Flavonoids control *Rhizobium nod* gene expression but they can also function as auxin transport inhibitors (Jacobs and Rubery, 1988), and might play a morphogenic role during nodule development. The expression of genes encoding phenylalanine ammonia-lyase (PAL) and chalcone synthase (CHS), two key enzymes of the flavonoid biosynthetic pathway, is induced during soybean nodule development (Estabrook and Sengupta-Gopalan, 1991). Moreover, *chs* transcripts are detected preferentially in root tips and nodule meristems of pea (Yang *et al.*, 1992) and in the root tip of bean plants (Rommeswinkel *et al.*, 1992). Also in alfalfa, *chs* as well as chalcone isomerase (*chi*) transcripts accumulated at high levels in young roots and root tips (Junghans *et al.*, 1993; McKhann and Hirsch, 1994). Therefore, flavonoid gene expression in the leguminous plant is correlated with root and nodule meristematic activity. It remains to be shown whether flavonoids are synthesized in these regions and whether they participate in the signaling pathway that leads to nodule organogenesis.

B. Plant Genetic Control of Nodulation

1. Nitrogen Control

The inhibition of nodule formation and N_2 fixation by combined nitrogen has been reviewed by Streeter (1988) and Carroll and Mathews (1990). Combined nitrogen exerts its effects on symbiosis at multiple stages. It controls the production of flavonoids by the plant; it influences bacterial adhesion to the plant cell, the rate of infection by rhizobia, the activity of nitrogenase, and the nodule mass per plant. In addition, rhizobial *nod* gene expression is negatively regulated by combined nitrogen. Thus, symbiotic nitrogen fixation is a facultative pathway which, probably due to its high energy cost, is not the preferred way of nitrogen assimilation.

Split root system experiments with various leguminous plants have shown that nitrate inhibition is a localized effect (Carroll and Matthews,

1990; Streeter, 1988 and references therein). Only the exposed root is sensitive to nitrate inhibition of early events of nodulation, provided that the amount of supplied nitrate is not too high. Thus nitrate inhibition of the early events of the symbiosis may not completely relate to a general physiological state of the plant (like the carbon to nitrogen ratio) but rather to the local presence of NO_3 in the vicinity of the infection site. Nitrate reductase mutant plants are still sensitive to nitrate inhibition of nodulation (Carroll and Mathews, 1990). The nodulation pattern of these mutants supports the hypothesis that nitrate itself has a major regulatory role, independently of its assimilation. In other words, the amount of nitrate in the medium, rather than its rate of assimilation, determines the degree of inhibition of nodulation in the plant.

There is some evidence that auxin and ethylene may be involved in the nitrogen control of nodulation. Indoleacetic acid (IAA) added with nitrate releases at least partially the effect of nitrate on the early events of infection (Dixon, 1969; Munns, 1977). Moreover, aminoethoxyvinylglycine was able to reduce the nitrate inhibition of nodulation, suggesting that ethylene may mediate this control (Ligero *et al.*, 1991). In contrast, no clear evidence for such a mechanism was obtained with peas (Lee and LaRue, 1992a).

The first effect of combined nitrogen on the establishment of the symbiosis can take place at the level of recognition between the host plant and *Rhizobium*. For example, root hair deformation, bacterial adhesion, and infection thread formation are inhibited by the presence of 18 m*M* KNO_3 in alfalfa (Truchet and Dazzo, 1982). These effects can act at different levels corresponding to the different steps of the infection: (1) the absence of *Rhizobium nod* gene induction by lack of inducer flavonoids and/or repression of *nod* gene expression, (2) the inability of the bacteria to bind to root hairs in order to initiate root hair deformation and infection thread formation, and (3) the lack of perception of the Nod signal due to the absence of a receptor or functional signal transduction pathway.

Combined nitrogen has a negative effect on the production of phenolic compounds in the root of leguminous plants, and the flavonoid supply can be a limiting factor for the establishment of symbiosis (Kapulnik *et al.*, 1987). Nitrogen control was demonstrated on flavolan production in *Lotus pedunculatus* roots (Pankhurst and Jones, 1979), on the production of isoflavonoids in roots of soybean (Cho and Harper, 1991a,b) and lupin (Wojtaszek *et al.*, 1993), and on *nod* gene-inducing flavonoids produced by alfalfa roots partially acting at the level of gene expression (Ratet *et al.*, 1994 and unpublished).

The amount of root lectins responsible for the binding of *Rhizobium* to the root hair is also affected by the concentration of combined nitrogen. In *Trifolium repens* L. seedlings, the proportion of active trifoliin A of plants grown with 1.5 m*M* KNO_3 was 40 to 50 times higher than in plants grown with 15 m*M* KNO_3 (Dazzo and Brill, 1978; Sherwood *et al.*, 1984). An increase in the number of aborted infection events observed at high concentrations of combined nitrogen (Munns, 1977) suggests that the early steps of infection are also under nitrogen control.

Plant mutants nodulating in the presence of nitrate have been identified in mutagenized populations of peas (Jacobsen, 1984; Jacobsen and Feenstra, 1984; Duc and Messager, 1989), soybeans (Carroll *et al.*, 1985a,b; Gremaud and Harper, 1989) and beans (Park and Buttery, 1988). These mutants nodulate more efficiently than their parents in the absence of nitrate and utilize nitrate normally, as assayed by nitrate reductase activity and growth on nitrate (Eskew *et al.*, 1989; Carroll and Mathews, 1990). In addition, it has been shown that these plants are still sensitive to the presence of high nitrate levels (Eskew *et al.*, 1989; Jacobsen and Feenstra, 1984). Thus, they are not affected in their response to nitrate but rather in autoregulation controlling the number of nodules formed by the plant (see Section VI,B,2).

2. Nodulation Mutants

Non-nodulating, supernodulating, and Nar^+ variants have been isolated in various species and revealed the different levels of control established by the plant. Genetic analysis of these mutants has shown that they are generally single recessive Mendelian characters affecting different stages of the symbiosis. We review what is known about the three types of mutants.

a. Non-nodulating Mutants An exhaustive list of symbiotic legume mutants was reported by Caetano-Anollés and Gresshoff (1991a). We describe here the best-characterized mutants.

Several of the mutants obtained from peas (Jacobsen, 1984; Kneen and LaRue, 1984; Engvild, 1987) have lost the ability to deform root hairs and to induce cell division in the root cortex. Some of these mutants might be affected in the perception or transduction of Nod signals, but so far there is no experimental evidence for this. The pea Nod^- mutant *sym5* has lost its ability to form nodules (Guinel and LaRue, 1991). The infection threads can progress to the inner cortex, but nodule primordia rarely appear. Since nodulation ability is restored in the presence of the ethylene inhibitor Ag^+, and the mutant line does not produce more ethylene than the wild-type parent, *P. sativum* cv. 'Sparkle,' it is probably affected in its sensitivity to ethylene, not in Nod signal perception (Fearn and LaRue, 1991). In the pleiotropic mutant E107 (*brz*), most infections are blocked when the infection thread arrives at the base of the epidermal cell (Guinel and LaRue, 1992), resulting in reduced nodulation. This mutant also shows stress symptoms, probably due to accumulation of toxic metal ions. Its nodulation is partly restored by treatment of the roots with the ethylene inhibitor Ag^+. The pea *sym8* and *sym9* mutants (Markwei and LaRue, 1992) are affected in the early steps of recognition. The *sym9* mutant responds to *R. leguminosarum* infection with severe root hair swelling. Neither root hair deformation nor infection thread formation nor cortical cell division can be detected in this mutant. The pea *sym2* mutant is also affected in early steps of the recognition. Based on genetic data, it was suggested that *sym2* encodes a receptor recognizing the *R. leguminosarum* bv. *viciae* Nod factors (Heidstra *et al.*, 1993). Interestingly, several of the Nod^- pea mutants isolated by Duc and Messager (1989) are also unable to establish symbiosis with vesicular-arbuscular mycorrhizal fungi, indicating that common components are used by the plant to build up symbiosis with the two partners (Duc *et al.*, 1989).

Several non-nodulating soybean mutants defining two complementation groups have been characterized (Caetano-Anollés and Gresshoff, 1991a, and references therein). The first complementation group includes mutants *nod49, nod772,* and *rj1*. These mutants are unable to induce root hair curling but can evoke few subepidermal cell divisions lacking infection threads (pseudoinfection). These results confirm the observations made with the pea mutants, indicating that root hair curling and primordium triggering can be uncoupled. The second complementation group is represented by the mutant *nod139*. It is unable to curl root hairs or to develop pseudoinfection and seems to be affected in an earlier step of recognition than the previous group of mutants, possibly in the perception or transduction of the Nod signals. All of these soybean mutants have the ability to form low numbers of normal, N_2-fixing nodules if they are inoculated at high bacterial titer, indicating that the developmental program is not affected in these plants (Mathews *et al.*, 1987). Grafting experiments demonstrated that the non-nodulation trait in the soybean resides in the plant root and not in the shoot (Delves *et al.*, 1986; Mathews *et al.*, 1992; Francisco and Akao, 1993).

A non-nodulating alfalfa mutant [MnNC-1008(NN)] (Peterson and Barnes, 1981; Barnes *et al.*, 1988) shows neither root hair curling nor cortical cell division upon infection by *R. meliloti* 1021 (Dudley and Long, 1989). Since this mutant line produces normal amounts of *nod* gene inducers (flavonoids), it was thought to be impaired in the early events of Nod factor perception. Interestingly, this mutant line also showed resistance to vesicular-arbuscular mycorrhizal colonization (Bradbury *et al.*, 1991), indicating that it is not the Nod signal reception but elements of a converging signal transduction chain that are affected. In support of this, we recently found that the mutant is still able to respond to Nod factors by plasma membrane depolarization (Felle *et al.*, in preparation).

Several mutants of white sweet clover (*Melilotus alba*), another host plant of *R. meliloti*, that are defective at different stages of nodule development have been characterized recently (Utrup *et al.*, 1993). Five different loci were identified which were affected either in root hair branching, infection thread development, or nodule formation.

b. Autoregulation and Supernodulating Mutants Plants can tightly control the extent of nodulation. Excision of nodules or root tips in red clover stimulates the formation of new nodules. Thus, established root meristematic foci can inhibit the development of new nodules (Nutman, 1952). Similar results were obtained with soybeans (Caetano-Anollés *et al.*, 1991b). Cell division was arrested at the level of cell clusters of about 20 to 30 cells. Thus, prior to the onset of nitrogen fixation, developing infection centers trigger a general plant response that inhibits further cell division activity in other parts of the plant. In soybeans, a first inoculation can suppress a second one applied 15 hr later (Pierce and Bauer, 1983). Similar control exists also in alfalfa, clover, and cowpeas (Caetano-Anollés and Gresshoff, 1991a). Autoregulation is also illustrated by the split root assay system in which inoculation of one side totally suppresses nodulation on the other side inoculated 1 week later. Suppression was already detectable after 24 hr (see Rolfe and Gresshoff, 1988). The total number of nodules remains constant, independent of the time and method of inoculation, suggesting a tight homeostatic control of nodule number throughout the root system (Caetano-Anollés and Gresshoff, 1991a).

Plant variants escaping this control (supernodulating mutants) were isolated in peas (Jacobsen, 1984; Jacobsen and Feenstra, 1984; Duc and Messager, 1989), soybeans (Carroll *et al.*, 1985a,b; Gremaud, and Harper, 1989), and beans (Park and Buttery, 1988). These mutants are in fact not affected in the mechanism controlling nodulation in the presence of NO_3 but rather in the autoregulation mechanism controlling the number of nodules formed per plant. In the soybean, 12 independent supernodulating (*nts*) mutants were isolated, which were shown to belong to a single complementation locus (Caetano-Anollés and Gresshoff, 1991a). The *nts* mutation is recessive and unlinked to the non-nodulation *nod49* and *nod139* loci. It was mapped in the soybean genome close to an RFLP marker (Landau-Ellis *et al.*, 1991), which may facilitate its cloning and molecular characterization. Grafting experiments showed that autoregulation is controlled by the shoot, since mutant shoots grafted on wild-type root stocks conferred the supernodulating phenotype (Delves *et al.*, 1986, 1987a,b; Francisco and Akao, 1993). Plants with a wild-type shoot and mutant root stocks showed wild-type levels of nodulation. Similar results were obtained in peas (Duc and Messager, 1989). The autoregulation signal may not be derived from the shoot apex but rather from the leaves (Delves *et al.*, 1992).

Split root experiments using the supernodulating mutants *nts382* show that nodulation on one side does not affect the nodulation on the other side (Olsson *et al.*, 1989). In the supernodulating plants, the number of infections is similar to that in the wild-type parent line, but the infection events proceed more frequently to nodule formation (Mathews *et al.*, 1989). This indicates that the *nts382* mutant fails to regulate nodulation due to a reduced capacity to arrest cell division in the cortex and the pericycle in a stage of development that occurs 72 hr after inoculation. In summary, in soybeans and peas, autoregulation is a systemic response blocking early steps of nodule organogenesis rather than the infection process.

Interestingly, it seems that in alfalfa the autoregulation mechanism acts earlier and occurs probably before infection. In addition, excision of preformed nodules allows further development of the arrested events of infection (Caetano-Anollés and Gresshoff, 1991b). These showed characteristics of a hypersensitive plant defense response (Vasse *et al.*, 1993). The relation between the systemic autoregulation effect and this localized hypersensitive response remains to be elucidated.

Empty nodules induced by *R. meliloti exo* mutants (Caetano-Anollés *et al.*, 1990) as well as spontaneous nodules formed in the absence of *Rhizobium* (Caetano-Anollés *et al.*, 1991a) are also able to trigger a systemic autoregulatory response, suggesting that the induction of the nodule primordium is sufficient to trigger the systemic response. Similar results were obtained with soybeans using approach-grafted wild-type and non-nodulating mutants: The *nod49* mutant, which is able to develop subepidermal cell division, also triggered an autoregulatory response in the wild type (Caetano-Anollés and Gresshoff, 1990).

In summary, the autoregulation response is induced during the early steps of the nodule developmental program. Initiation of the root nodule primordium seems to induce in the shoot the production of a signal molecule that systemically suppresses the development of root nodules.

3. Nodulation in the Absence of *Rhizobium* (Nar)

Nodule structures formed spontaneously in the absence of rhizobia possess similar histological features characteristic of indeterminate nodules (Truchet *et al.*, 1989; Joshi *et al.*, 1991). The Nar phenotype has been found in different *M. sativa* cultivars as well as in the diploid *M. coerulea* (Huguet *et al.*, 1993). The frequency and kinetics of the appearance of the Nar nodules is determined by the genotype of the *Medicago* lines

(Caetano-Anollés and Gresshoff, 1992; Huguet *et al.*, 1993; our laboratory). These data indicate that the nodule developmental pathway is controlled exclusively by the plant and is only triggered by the Nod signals. It is possible that endogenous lipooligosaccharide-type signal molecules exist in the plant, which is supported by detection of lipophilic compounds degradable by chitinases in *Lathyrus* plants (Spaink *et al.*, 1993c). Caetano-Anollés *et al.* (1993) suggested that the Nar trait is monogenic dominant and controlled by a gene dose effect. Huguet *et al.* (1993), however, proposed a more complex situation, so that the genetic determinant of this trait remains to be elucidated. It has been hypothesized that spontaneous nodules may represent modified carbon storage organs that have evolved into the nitrogen-fixing organ (Caetano-Anollés *et al.*, 1993).

C. The Role of Plant Hormones in Nodule Development

Plant hormones play a general role in plant differentiation and developmental processes, and are most likely involved in nodule organogenesis, which is a unique developmental program induced in legume roots by the host-specific rhizobial Nod signals. It is likely, however, that all elements of the nodule developmental pathway are encoded by the plant.

The most conclusive observations supporting this hypothesis were obtained on *M. sativa* grown under nitrogen starvation where (1) root nodules appeared spontaneously in the absence of *R. meliloti* with varying efficiency, depending on the *Medicago* cultivar (Truchet *et al.*, 1989; Caetano-Anollés and Gresshoff, 1992) or (2) nodule-like structures could be induced by polar auxin transport inhibitors such as *N*-(1-naphthyl)-phthalamic acid (NPA) and 2,3,5-triiodobenzoic acid (TIBA) (Hirsch *et al.*, 1989). These data indicate that the altered hormone levels, as a consequence of nitrogen limitation or perturbation of the hormone balance by auxin transport inhibitors, might provide, at least in some *Medicago* cultivars, signals for nodule organogenesis.

The most crucial step of nodule initiation, in which hormones are likely to be involved, is the activation of the cell division cycle in the differentiated cortical cells of the emerging root hair zone of the root. The role of hormones and gradients in the initiation of cortex proliferation was studied on excised root pieces from *Pisum sativum* (Libbenga *et al.*, 1973). Auxin alone induced only pericyclic cell division whereas addition of auxins and cytokinins induced three meristematic areas in the inner cortex opposite the three xylem radii. From this restricted division response, the existence of an endogenous division factor was proposed. Extracts from stele resulted in cell division throughout the cortex of root pieces. This was interpreted as evidence for the presence of a transverse gradient system of a cell division factor (stele factor) in the root cortex which may control induction of cell division and nodule primordium formation. This finding is strongly supported by the fact that the majority, over 85%, of the nodules examined and all lateral roots originated opposite the protoxylem poles (Phillips, 1971a).

It is possible that Nod factors, in one way or another, interact with plant hormones (especially with auxins and cytokinins) or perturb the endogenous hormone balance for high induction of nodule organogenesis. Treatment of *Medicago* roots with purified *R. meliloti* Nod factor led to the formation of meristematic foci and empty nodules (Truchet *et al.*, 1991). This indicated that no additional compounds produced by the bacterium, including hormones, are essential for nodule development.

The efficiency of nodule formation correlated with the auxin sensitivity of the *Medicago* cultivars: both spontaneous and *Rhizobium*-induced nodulation were more effective on the more auxin sensitive plants (Kondorosi *et al.*, 1993). Among the tested *Medicago* lines, those that exhibited the highest auxin sensitivity had the best embryogenic properties, also suggesting a correlation between nodule-organogenic and embryogenic potentials of the plants. The assumption that changes in hormone levels and sensitivity affect nodulation was tested by transformation of *Medicago* lines with the *rolABC* genes of *A. rhizogenes*. These genes were shown to increase auxin sensitivity (Shen *et al.*, 1989) and were suggested to alter the endogenous active hormone ratios (Estruch *et al.*, 1991a,b), although their exact role is still unclear (Nilsson *et al.*, 1993). The *rolABC* genes were introduced into two embryogenic *Medicago* lines via *A. tumefaciens*-mediated transformation in different combinations, and were expressed either from their own or from the cauliflower mosaic virus 35S promoter (Kondorosi *et al.*, 1993). The presence of *rol* genes affected the development of the plant and/or nodulation. In general, development of roots was positively influenced by the presence of the *rolB* genes, as expected from the work of Schmülling *et al.* (1988), whereas in most cases a negative effect of the *rolC* gene was observed. Both the kinetics of nodulation and the number of nodules were affected by the *rol* genes. Plants containing *rolABC* or particularly *rolB* nodulated faster and produced more nodules than the untransformed plants. The presence of *rolC* delayed nodulation, affected bacteroid development, and caused faster senescence.

In other *Rhizobium*–legume systems, purified Nod factors also induced division of cortical cells, initiating nodule primordia which, however, did not or only rarely developed into nodules (Spaink *et al.*, 1991; Mergaert *et al.*, 1993; Carlson *et al.*, 1993). Therefore, the possibility cannot be excluded that in these legumes additional factors, such as hormones supplied by the *Rhizobium* partner, contribute to nodule development.

Rhizobium species are known to produce indole-3-acetic acid (IAA), and a number of related compounds (e.g., Badenoch-Jones *et al.*, 1982) and flavonoids were shown to stimulate IAA production by rhizobia (Prinsen *et al.*, 1991). Reports on the presence of cytokinins in the *Rhizobium* culture media are contradictory (see Upadhyaya *et al.*, 1991). However, the Nod$^-$ phenotype of various mutants and deletion derivatives of *R. meliloti* was suppressed by introduction of the *tzs* (*trans*-zeatin secretion) gene of *A. tumefaciens* (Cooper and Long, 1994), suggesting the involvement of cytokinins in nodule organogenesis.

In the plants, the hormone ratios might be controlled by flavonoids present in the legume root. Flavonoids with particular structural parameters (e.g., hydroxyl groups on both the A and B rings, an unsaturated C2=C3 bond in the pyran ring, no large polar substituents), such as quercetin or apigenin, may function as natural auxin transport regulators and were shown to increase IAA uptake by hypocotyl segments in a manner similar to NPA (Jacobs and Rubery, 1988). Therefore, flavonoids produced in the root are involved not only in chemotaxis and activation of *nod* genes but might contribute also to regulation of the hormone balance in the root.

The inhibitory effect of the exogenously applied plant hormone, ethylene, on nodulation has been known for a long time. It was observed in beans (Grobbelaar *et al.*, 1971), peas, and white clover (Goodlass and Smith, 1979; Lee and LaRue, 1992b). In alfalfa and vetches, nodulation was stimulated by aminoethoxyvinylglycine (AVG), an inhibitor of ethylene biosynthesis (Peters and Christ-Estes, 1989; Zaat *et al.*, 1989b), which suggests an inhibitory effect of ethylene also in these species. However, in soybeans, the nodulation of only certain cultivars is sensitive to ethylene while others are not affected (Hunter, 1993; Z. P. Xie, personal communication).

Abscisic acid was reported to inhibit nodulation (Phillips, 1971b). Gibberellins (GA) and gibberellin-like substances are present in nodules formed on a range of leguminous plants (e.g., Radley, 1961; Dobert *et al.*, 1992); however, it is unclear whether GAs play a role in nodule initiation and development.

D. Plant Responses to Nod Factors

Many of the host-specific plant responses to rhizobia are triggered by purified Nod factors, suggesting that these are the predominant signals by which the bacteria influence root development. The Nod factors induce the deformation and branching of root hairs (Had) at pico- to nanomolar concentrations, the thick and short root phenotype (Tsr) on *Vicia*, and stimulation of root hair formation on *Vicia sativa, Vigna unguiculata,* and *Sesbania rostrata* (Lerouge *et al.,* 1990; Roche *et al.,* 1991b; Spaink *et al.,* 1991; Schultze *et al.,* 1992; Price *et al.,* 1992; Sanjuan *et al.,* 1992; Mergaert *et al.,* 1993; Carlson *et al.,* 1993).

At higher concentrations (>1 n*M*), Nod factors elicit cortical cell divisions and formation of meristems which can develop into nodule-like structures (Truchet *et al.,* 1991; Spaink *et al.,* 1991; Schultze *et al.,* 1992; Mergaert *et al.,* 1993; Carlson *et al.,* 1993; Relic *et al.,* 1993; Stacey *et al.,* 1994; Stokkermans and Peters, 1994). Recent evidence that purified Nod factors can complement *nodABC* mutants for the formation of N_2-fixing nodules (Relic *et al.,* 1993, 1994) seems to indicate that Nod signal-induced nodules are indeed functional when entered by rhizobia. Penetration of rhizobia might be aided by Nod factor-induced formation of cytoplasmic bridges through the central vacuole of outer cortical cells which was observed on *V. sativa* in response to *R. leguminosarum* Nod factors (Van Brussel *et al.,* 1992a). These structures, apparently preparing for the passage of true infection threads, were designated preinfection threads. Since a similar polarization occurs in cells of the inner cortex before the start of cell division and also in epidermal cells before the formation of root hairs (Wood and Newcomb, 1989), it was proposed that a common principle underlies all three processes—cell division, preinfection thread formation, and root hair formation (Rae *et al.,* 1992; van Brussel *et al.,* 1992a). The different morphological changes may be governed by gradients of at least two different stimuli, involving Nod factors as external signals as well as endogenous signals arising from the stele (stele factor) (Libbenga *et al.,* 1973).

Purified Nod factors have also been shown to induce several early nodulin genes (Franssen *et al.,* 1993; Pichon *et al.,* 1993; Horvath *et al.,* 1993; Bauer *et al.,* 1994) as well as cell cycle-related genes (Savouré *et al.,* 1994) (see Section VI,A). Moreover, increased production of flavonoids, including *nod* gene inducers (Ini), in response to Nod factors was reported (Spaink *et al.,* 1991; Schmidt *et al.,* 1994). This effect might be related to the induction of genes in the phenylpropanoid pathway at high concentrations of Nod factors (Savouré *et al.,* 1994, and in preparation).

The fastest host plant response to purified Nod signals was the depolarization of the plasma membrane potential in root hairs of *Medicago* (Ehrhardt *et al.,* 1992). This response is specific for the cognate NodRm-IV(C16:2,S) and is detectable at a concentration as low as 10^{-11} *M* (Felle *et al.,* in preparation). Whether it plays a role in a Nod signal transduction cascade or is just an accompanying effects is not known.

Some of the effects of Nod factors on plants seem to be incidental, that is, they do not appear to be required for nodule formation. Both Tsr and Ini on vetch were suppressed when roots were shielded from light (van

Brussel *et al.*, 1992b). In contrast, meristem induction took place only under protection from light.

Little is known about Nod signal perception and transduction. Lectins accumulate at the surface of root epidermal cells susceptible to infection by rhizobia (Díaz *et al.*, 1986). Therefore, it was proposed that, in addition to their role in cell attachment (see Kijne, 1992), lectins may also participate in Nod signal recognition (Lugtenberg *et al.*, 1991), possibly acting in concert with a cell surface receptor (see Hirsch, 1992). Since modifications at both the reducing and nonreducing terminus affect Nod signal activity, a receptor may have to bind to a large portion of the molecule. During the receptor–ligand interaction, the Nod factor might adopt a conformation in which both ends are brought into the vicinity (Whitfield and Tang, 1993).

The search for Nod factor binding proteins in membrane fractions of plant roots has so far been unsuccessful. However, specific binding sites for chitooligosaccharides have been identified on suspension-cultured rice and tomato cells (Shibuya *et al.*, 1993; Baureithel *et al.*, 1994). The tomato cells respond by a rapid alkalinization of the culture medium, not only to nanomolar concentrations of chitin oligomers (Felix *et al.*, 1993), but also to Nod factors isolated from different *Rhizobium* strains (Staehelin *et al.*, 1994b). The refractory behavior of cells pretreated with chitooligosaccharides to a second stimulation by Nod factors, and vice versa, suggested that the same reception system may be involved in the recognition of both types of molecules. Indeed, Nod factors were shown to compete with chitooligosaccharides in the binding to tomato cell membranes (Baureithel *et al.*, 1994).

Alfalfa cell suspension cultures appear to have a sensitive perception system for chitooligosaccharides since several defense-related genes are strongly induced by nanomolar concentrations of these molecules (Savouré *et al.*, in preparation). The same response is observed also in the presence of Nod factors, however only at much higher concentrations. On intact alfalfa roots, chitooligosaccharides do not show any kind of effect. Thus a response to chitooligosaccharides seems to be suppressed in intact plantlets. It is possible that legumes have a common binding protein for both, chitooligosaccharides as well as active and inactive Nod factors, but that it is the specific conformation of the receptor-ligand complex which determines the subsequent events of a divergent signal transduction pathway a branch of which might be specifically activated in cell cultures.

As an alternative to a specific Nod signal reception, differences in transport and stability might make it possible for Nod factors but not chitin oligomers to reach the inner root cortex in an intact form. It has in fact been shown that acylation and other modifications strongly increase the stability of the molecules in the presence of plant roots and protect against degradation by chitinases (Schultze *et al.*, 1993; Staehelin *et al.*, 1994a,b). Interestingly, sulfated Nod factors of *R. meliloti* are much less susceptible to hydrolysis by chitinases than nonsulfated factors, and this might explain part of the host-specific nature of Nod signal recognition in *Medicago* (Schultze *et al.*, 1993; Staehelin *et al.*, 1994a). *nodH* mutants of *R. meliloti* were able to induce nodules, although ineffective ones, on alfalfa when the plants were grown in vermiculite (Ogawa *et al.*, 1991; Schwedock and Long, 1992) and on genotypes of alfalfa selected for a high rate of spontaneous nodulation (Caetano-Anollés and Gresshoff, 1992). This suggests that *Medicago* may indeed possess receptors for both sulfated and nonsulfated Nod factors, and other parameters, such as Nod factor stability, may represent an additional level at which Nod signal recognition is specified.

Nod factor degrading activity is inducible by Nod signals (Staehelin *et al.*, unpublished), and elevated levels of chitinase were found in nodules (Staehelin *et al.*, 1992) and in necrotic cortical cells containing aborted infection threads (Vasse *et al.*, 1993). Therefore, hydrolysis of Nod factors by chitinases may also play a role in autoregulation of nodulation (see Section VI,B).

That there is more than specific receptors accounting for host-specific Nod signal activity is also suggested by the following observations. *R. meliloti* Nod factors are able to induce the expression of early nodulin genes in the nonhost pea plant, although at concentrations higher than those required for NodRlv factors (Horvath *et al.*, 1993). Moreover, even in nonlegumes such as carrot cells, Nod factors can evoke crucial developmental responses at low concentrations (De Jong *et al.*, 1993).

VII. Host-Specific Interactions Controlled by Multiple and Specific Signal Exchanges: A Model

The communication between rhizobia and their host plants appears to involve multiple signal exchanges that are under complex control (Fig. 4). A mixture of different flavonoids is produced by a given plant, and interaction of different *nod* gene regulatory proteins with both agonistic and antagonistic flavonoids seems to determine the strength and quality of *nod* gene induction. In addition, a number of nonflavonoid compounds are also able to induce *nod* genes. Similar to the *nod* gene inducers, the return signals produced by rhizobia consist of a complex mixture of Nod factors. A finely tuned balance in the concentration of active and antagonistic Nod signals seems to influence the recognition of a particular *Rhizobium*

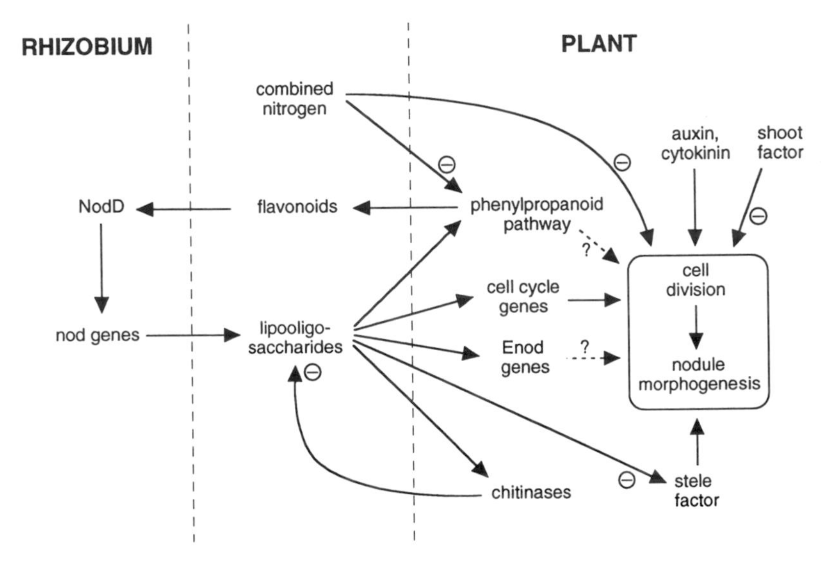

FIG. 4 Control elements in the early steps of *Rhizobium* plant communication: A model.

strain by the host plants. Furthermore, as for *nod* gene induction, alternative Nod signals may exist. For example, *N*-acetylglutamic acid, at micromolar concentrations, and diglycosyl diacylglycerides purified from *R. leguminosarum* bv. *trifolii* were reported to induce cortical cell divisions in white clover (*T. repens*) at subnanomolar concentrations (Philip-Hollingsworth *et al.*, 1991; Orgambide *et al.*, 1993). Similarly, division of plant protoplasts by *nodAB*-dependent but *nodC*-independent factors (Schmidt *et al.*, 1988) might be explained by the production of alternative glycolipids. Transgenic tobacco plants carrying the *R. meliloti nodA* and/or *nodB* genes showed developmental abnormalities, such as bifurcated leaves and compact growth (Schmidt *et al.*, 1993). These plants possibly also produce glycolipid morphogens which are not necessarily derived from chitooligosaccharides. The complexity of signal substances participating in the early steps of *Rhizobium*–plant communication may reflect the evolutionary adaptation of host and symbiont to each other. Multiple compounds and multiple mechanisms of recognition appear to act in an additive or synergistic way, establishing a host-specific symbiotic interaction.

The induction of cortical cell division and nodule morphogenesis is under the control of a plant developmental program. This is triggered by exogenous signals and seems to involve an interplay of auxins, cytokinins, and other elements, such as the stele factor. It is possible that specific genes; for example, cell cycle-related genes and early nodulin genes, are participating in nodule morphogenesis. In addition, flavonoids might contribute to the symbiotic interaction not only through *nod* gene induction, but possibly through a more direct effect on nodule formation.

Rhizobial *nod* gene activity as well as nodule induction is negatively regulated in the presence of combined nitrogen. Nodulation is also limited by autoregulatory mechanisms, which again may act at multiple levels, including shoot-derived factors, inactivation of the stele factor by Nod signals (Smit *et al.*, 1993), and Nod factor degradation by chitinases.

VIII. Conclusion

Unraveling the biochemical function of many of the *nod* gene products has contributed substantially to the understanding of host specificity in the *Rhizobium*–plant interactions. The existing literature on *nod* gene mutations and their effects on host range can now be interpreted in terms of the structural variability of lipooligosaccharide signals. A challenging task for future work will be the elucidation of Nod signal perception and transduction in plant cells.

References

Aguilar, O. M., Reiländer, H., Arnold, W., and Pühler, A. (1987). *Rhizobium meliloti nifN (fixF)* gene is part of an operon regulated by a *nifA*-dependent promoter and codes for a polypeptide homologous to the *nifK* gene product. *J. Bacteriol.* **169,** 5393–5400.

Allen, T., Raja, S., and Dunn, K. (1991). Cells expressing ENOD2 show differential spatial organization during the development of alfalfa root nodules. *Mol. Plant-Microbe Interact.* **4,** 139–146.

Allison, L. A., Kiss, G. B., Bauer, P., Poiret, M., Pierre, M., Savouré, A., Kondorosi, E. and Kondorosi, A. (1993). Identification of two alfalfa early nodulin genes with homology to members of the pea *Enodl2* gene family. *Plant Mol. Biol.* **21,** 375–380.

Appelbaum, E. R., Thompson, D. V., Idler, K., and Chartrain, N. (1988a). *Rhizobium japonicum* USDA 191 has two *nodD* genes that differ in primary structure and function. *J. Bacteriol.* **170,** 12–20.

Appelbaum, E. R., Barkei, J., Kosslak, R., Johansen, E., Thompson, D., and Maroney, M. (1988b). Regulation of *nodABC* expression in soybean rhizobia. *In* "Molecular Genetics of Plant-Microbe Interactions" (R. Palacios and D. P. S. Verma, Eds.), pp. 94–95. APS Press, St. Paul.

Atkinson, E. M., and Long, S. R. (1992). Homology of *Rhizobium meliloti* NodC to polysaccharide polymerizing enzymes. *Molec. Plant-Microbe Interact.* **5,** 439–442.

Atkinson, E. M., Ehrhardt, D. W., and Long, S. R. (1992). *In vitro* activity of the *nodH* gene product from *Rhizobium meliloti*. *In* "Sixth International Symposium on Molecular Plant–Microbe Interactions," Seattle, Poster abstract 45.

Bachem, C. W. B., Kondorosi, E., Banfalvi, Z., Horvath, B., Kondorosi, A., and Schell,

J. (1985). Identification and cloning of nodulation genes from the wide host range *Rhizobium* strain MPIK3030. *Mol. Gen. Genet.* **199,** 271–278.

Bachem, C. W. B., Banfalvi, Z., Kondorosi, E., Schell, J., and Kondorosi, A. (1986). Identification of host range determinants in the *Rhizobium* species MPIK3030. *Mol. Gen. Genet.* **203,** 42–48.

Badenoch-Jones, J., Summons, R. E., Entsch, B., Rolfe, B. G., Parker, C. W., and Letham, D. S. (1982). Mass spectrometric identification of indole compounds produced by *Rhizobium* strains. *Biomed. Mass Spect.* **9,** 429–437.

Baev, N., and Kondorosi, A. (1992). Nucleotide sequence of the *Rhizobium meliloti nodL* gene located in locus n5 of the nod regulon. *Plant. Mol. Biol.* **18,** 843–846.

Baev, N., Endre, G., Petrovics, G., Banfalvi, Z., and Kondorosi, A. (1991). Six nodulation genes of nod box locus 4 in *Rhizobium meliloti* are involved in nodulation signal production: *nodM* Codes for D-glucosamine synthetase. *Mol. Gen. Genet.* **228,** 113–124.

Baev, N., Schultze, M., Barlier, I., Cann, Ha, D., Virelizier, H., Kondorosi, É., and Kondorosi, Á. (1992). *Rhizobium nodM* and *nodN* genes are common nod genes: *nodM* encodes functions for efficiency of Nod signal production and bacteroid maturation. *J. Bacteriol.* **174,** 7555–7565.

Banfalvi, Z., and Kondorosi, A. (1989). Production of root hair deformation factors by *Rhizobium meliloti* nodulation genes in *Escherichia coli: hsnD (nodH)* is involved in the plant host-specific modification of the NodABC factor. *Plant Mol. Biol.* **13,** 1–12.

Banfalvi, Z., Nieuwkoop, A., Schell, M., Besl, L., and Stacey, G. (1988). Regulation of *nod* gene expression in *Bradyrhizobium japonicum. Mol. Gen. Genet.* **214,** 420–424.

Barbour, W. M., Wang, S. P., and Stacey, G. (1992). Molecular genetics of *Bradyrhizobium* symbioses. *In* "Biological Nitrogen Fixation" (G. Stacey, R. H. Burris, and H. J. Evans, Eds.), pp. 648–684. Chapman & Hall, New York.

Barnes, D. K., Vance, C. P., Heichel, G. H., Peterson, M. A., and Ellis, W. R. (1988). Registration of a non-nodulation and three ineffective nodulation alfalfa germplasms. *Crop Sci.* **28,** 721–722.

Barnett, M. J., and Long, S. R. (1990). DNA sequence and translational product of a new nodulation-regulatory locus: SyrM has sequence similarity to NodD proteins. *J. Bacteriol.* **172,** 3695–3700.

Barny, M. A., and Downie, J. A. (1993). Identification of the NodC protein in the inner but not the outer membrane of *Rhizobium leguminosarum. Mol. Plant-Microbe Interact.* **6,** 669–672.

Bassam, B. J., Rolfe, B. G., and Djordjevic, M. A. (1986). *Macroptilium atropurpureum* (siratro) host specificity genes are linked to a *nodD*-like gene in the broad host range *Rhizobium* strain NGR234. *Mol. Gen. Genet.* **203,** 49–57.

Bassam, B., Djordjevic, M. A., Redmond, J. W., Batley, M., and Rolfe, B. G. (1988). Identification of a *nodD*-dependent locus in the *Rhizobium* strain NGR234 activated by phenolic factors secreted by soybeans and other legumes. *Mol. Plant-Microbe Interact.* **1,** 161–168.

Bauer, P., Crespi, M., Szécsi, J., Allison, L. A., Schultze, M., Ratet, P., Kondorosi, E. and Kondorosi, A. (1994). Alfalfa *Enod12* genes are differentially regulated during nodule development by Nod factors and *Rhizobium* invasion. *Plant Physiol.* **705,** 585–592.

Baureithel, K., Felix, G., and Boller, T. (1994). Specific, high affinity binding of chitin fragments to tomato cells and membranes. *J. Biol. Chem.*, in press.

Bec-Ferté, M. P., Savagnac, A., Pueppke, S. G., and Promé, J. C. (1993). Nod factors from *Rhizobium fredii* USDA257. *In* "New Horizons in Nitrogen Fixation" (R. Palacios, J. Mora, and W. E. Newton, Eds.), pp. 157–158. Kluwer, Dordrecht, The Netherlands.

Bender, G. L., Nayudu, M., Le Strange, K. K., and Rolfe, B. G. (1988). The *nodD1* gene from *Rhizobium* strain NGR234 is a key determinant in the extension of host range to the nonlegume *Parasponia. Mol. Plant-Microbe Interact.* **1,** 259–266.

Bibb, M. J., Biró, S., Motamedi, H., Collins, J. F., and Hutchinson, C. R. (1989). Analysis of the nucleotide sequence of the *Streptomyces glaucescens tcml* genes provides key information about the enzymology of polyketide antibiotic biosynthesis. *EMBO J.* **8,** 2727–2736.

Bloemberg, G. V., Thomas-Oates, J. E., Lugtenberg, B. J. J., and Spaink, H. P. (1994). Nodulation protein NodL of *Rhizobium leguminosarum O*-acetylates lipooligosaccharides, chitin fragments and N-acetylglucosamine *in vitro*. *Mol. Microbiol.* **11,** 793–804.

Boundy-Mills, K. L., Kosslak, R. M., Tully, R. E., Pueppke, S. G., Lohrke, S., and Sadowsky, M. J. (1994). Induction of the *Rhizobium fredii nod* box-independent nodulation gene *nol*J requires a functional *nod*D1 gene. *Mol. Plant-Microbe Interact.* **7,** 305–308.

Bradbury, S. M., Peterson, R. L., and Bowley, S. R. (1991). Interactions between three alfalfa nodulation genotypes and two *Glomus* species. *New Phytol.* **119,** 115–120.

Brewin, N. J. (1991). Development of the legume root nodule. *Annu. Rev. Cell. Biol.* **7,** 191–226.

Broughton, W. J. (1993) (see EMBL Access No X73362).

Bulawa, C. E. (1992). *CSD2, CSD3,* and *CSD4,* genes required for chitin synthesis in *Saccharomyces cerevisiae:* The CSD2 gene product is related to chitin synthases and to developmentally regulated proteins in *Rhizobium* species and *Xenopus laevis. Mol. Cell. Biol.* **12,** 1764–1776.

Bulawa, C. E., and Wasco, W. (1991). Chitin and nodulation. *Nature* **353,** 710.

Burn, J., Rossen, L., and Johnston, A. W. B. (1987). Four classes of mutations in the *nodD* gene of *Rhizobium leguminosarum* biovar. *viciae* that affect its ability to autoregulate and/or activate other *nod* genes in the presence of flavonoid inducers. *Genes Dev.* **1,** 456–464.

Burn, J. E., Hamilton, W. D., Wootton, J. C., and Johnston, A. W. B. (1989). Single and multiple mutations affecting properties of the regulatory gene *nodD* of *Rhizobium*. *Mol. Microbiol.* **3,** 1567–1577.

Cabib, E., Bowers, B., and Roberts, R. L. (1983). Vectorial synthesis of a polysaccharide by isolated plasma membranes. *Proc. Natl. Acad. Sci. USA* **80,** 3318–3321.

Caetano-Anollés, G., and Gresshoff, P. M. (1990). Early induction of feedback regulatory responses governing nodulation in soybean. *Plant Sci.* **71,** 69–81.

Caetano-Anollés, G., and Gresshoff, P. M. (1991a). Plant genetic control of nodulation. *Annu. Rev. Microbiol.* **45,** 345–382.

Caetano-Anollés, G., and Gresshoff, P. M. (1991b). Alfalfa controls nodulation during the onset of *Rhizobium*-induced cortical cell division. *Plant Physiol.* **95,** 366–373.

Caetano-Anollés, G., and Gresshoff, P. M. (1992). Plant genetic suppression of the non-nodulation phenotype of *Rhizobium meliloti* host-range nodH mutants: gene-for-gene interaction in the alfalfa-*Rhizobium* symbiosis? *Theor. Appl. Genet.* **84,** 624–632.

Caetano-Anollés, G., Lagares, A., and Bauer, W. D. (1990). *Rhizobium meliloti* exopolysaccharide mutants elicit feedback regulation of nodule formation in alfalfa. *Plant Physiol.* **92,** 368–374.

Caetano-Anollés, G., Joshi, P. A., and Gresshoff, P. M. (1991a). Spontaneous nodules induce feedback suppression of nodulation in alfalfa. *Planta* **183,** 77–82.

Caetano-Anollés, G., Paparozzi, E. T., and Gresshoff, P. M. (1991b). Mature nodules and root tips control nodulation in soybean. *J. Plant Physiol.* **137,** 389–396.

Caetano-Anollés, G., Joshi, P. A., and Gresshoff, P. M. (1993). Nodule morphogenesis in the absence of *Rhizobium*. *In* "New Horizons in Nitrogen Fixation" (R. Palacios, J. Mora, and W. E. Newton, Eds.), pp. 297–302. Kluwer, Dordrecht, The Netherlands.

Calvert, H. E., Pence, M. K., Pierce, M., Malik, N. S. A., and Bauer, W. D. (1984). Anatomical analysis of the development and distribution of *Rhizobium* infections in soybean roots. *Can. J. Bot.* **62,** 2375–2384.

Canter-Cremers, H. C. J., Wijffelman, C. A., Pees, E., Rolfe, B. G., Djordjevic, M. A., and Lugtenberg, B. J. J. (1988). Host specific nodulation of plants of the pea cross-inoculation group is influenced by genes in fast growing *Rhizobium* downstream *nodC*. *J. Plant Physiol.* **132,** 398–404.

Canter-Cremers, H. C. J., Spaink, H. P., Wijfjes, A. H. M., Pees, E., Wijffelman, C. A., Okker, R. J. H., and Lugtenberg, B. J. J. (1989). Additional nodulation genes on the Sym plasmid of *Rhizobium leguminosarum* biovar *viciae*. *Plant Mol. Biol.* **13,** 163–174.

Carlson, R. W., Sanjuan, J., Bhat, U. R., Glushka, J., Spaink, H. P., Wijfjes, A. H. M., Van Brussel, A. A. N., Stokkermans, T. J. W., Peters, K., and Stacey, G. (1993). The structures and biological activities of the lipo-oligosaccharide nodulation signals produced by Type I and Type II strains of *Bradyrhizobium japonicum*. *J. Biol. Chem.* **268,** 18,372–18,381.

Carroll, B. J., and Mathews, A. (1990). Nitrate inhibition of nodulation in legumes. *In* "Molecular Biology of Symbiotic Nitrogen Fixation" (P. M. Gresshoff, Ed.), pp. 159–180. CRC Press, Boca Raton, FL.

Carroll, B. J., McNeil, D. L., and Gresshoff, P. M. (1985a). Isolation and properties of soybean [*Glycine max* (L.) Merr.] mutants that nodulate in the presence of high nitrate concentrations. *Proc. Natl. Acad. Sci. USA* **82,** 4162–4166.

Carroll, B. J., McNeil, D. L., and Gresshoff, P. M. (1985b). A supernodulation and nitrate-tolerant symbiotic (nts) soybean mutant. *Plant Physiol.* **78,** 34–40.

Cervantes, E., Sharma, S. B., Maillet, F., Vasse, J., Truchet, G., and Rosenberg, C. (1989). The *Rhizobium meliloti* host range *nodQ* gene encodes a protein which share homology with translation elongation and initiation factors. *Mol. Microbiol.* **3,** 745–755.

Cho, M. J., and Harper, J. E. (1991a). Effect of inoculation and nitrogen on isoflavonoid concentration in wild-type and nodulation-mutant soybean roots. *Plant Physiol.* **95,** 435–442.

Cho, M. J., and Harper, J. E. (1991b). Effect of localized nitrate application on isoflavonoid concentration and nodulation in split-root systems of wild-type and nodulation-mutant soybean plants. *Plant Physiol.* **95,** 1106–1112.

Collins-Emerson, J. M., Terzaghi, E. A., and Scott, D. B. (1990). Nucleotide sequence of *Rhizobium loti nodC*. *Nucleic Acids Res.* **18,** 6690.

Cooper, J. B., and Long, S. R. (1994). Morphogenetic rescue of *Rhizobium meliloti* nodulation mutants by *trans*-zeatin secretion. *Plant Cell* **6,** 215–225.

Cren, M., Kondorosi, A., and Kondorosi, E. (1994). An insertional point mutation inactivates NolR repressor in *Rhizobium meliloti* 1021. *J. Bacteriol.* **176,** 518–519.

Csanádi, G., Szécsi, J., Kaló, P., Kiss, P., Endre, G., Kondorosi, Á., Kondorosi, É., and Kiss, G. B. (1994). *Enod12,* an early nodulin gene, is not required for nodule formation and efficient nitrogen fixation in alfalfa. *Plant Cell,* **6,** 201–213.

Cunningham, S., Kollmeyer, W. D., and Stacey, G. (1991). Chemical control of interstrain competition for soybean nodulation by *Bradyrhizobium japonicum*. *Appl. Environ. Microbiol.* **57,** 1886–1892.

Dakora, F. D., Joseph, C. M., and Phillips, D. A. (1993a). Alfalfa (*Medicago sativa* L.) root exudates contain isoflavonoids in the presence of *Rhizobium meliloti*. *Plant Physiol.* **101,** 819–824.

Dakora, F. D., Joseph, C. M., and Phillips, D. A. (1993b). Common bean root exudates contain elevated levels of daidzein and coumestrol in response to *Rhizobium* inoculation. *Mol. Plant-Microbe Interact.* **6,** 665–668.

Dart, P. (1977). Infection and development of leguminous nodules. *In* "A Treatise on Dinitrogen Fixation" (R. W. F. Hardy and W. S. Silver, Eds.), Sect. III, pp. 367–472. Wiley, New York.

Darvill, A., Augur, C., Bergmann, C., Carlson, R. W., Cheong, J. J., Eberhard, S., Hahn, M. G., Ló, V. M., Marfà, V., Meyer, B., Mohnen, D., O'Neill, M. A., Spiro, M. D.,

Van Halbeek, H., York, W. S., and Albersheim, P. (1992). Oligosaccharins—oligosaccharides that regulate growth, development and defence responses in plants. *Glycobiology* **2,** 181–198.

Davis, E. O., and Johnston, A. W. B. (1990a). Analysis of three *nodD* genes in *Rhizobium leguminosarum* bivoar *phaseoli; nodD1* is preceded by *nolE,* a gene whose product is secreted from the cytoplasm. *Mol. Microbiol.* **4,** 921–932.

Davis, E. O., and Johnston, A. W. B. (1990b). Regulatory functions of three *nodD* genes of *Rhizobium leguminosarum biovar phaseoli. Mol. Microbiol.* **4,** 933–941.

Davis, E. O., Evans, I. J., and Johnston, A. W. B. (1988). Identification of *nodX,* a gene that allows *Rhizobium leguminosarum* biovar *viciae* strain TOM to nodulate Afghanistan peas. *Mol. Gen. Genet.* **212,** 531–535.

Dazzo, F. B., and Brill, W. J. (1978). Regulation by fixed nitrogen of host-symbiont recognition in the *Rhizobium*-clover symbiosis. *Plant Physiol.* **62,** 18–21.

De Blank, C., Mylona, P., Yang, W. C., Katinakis, P., Bisseling, T., and Franssen, H. (1993). Characterization of the soybean early nodulin cDNA clone GmENOD55. *Plant Mol. Biol.* **22,** 1167–1171.

De Jong, A. J., Heidstra, R., Spaink, H. P., Hartog, M. V., Meijer, E. A., Hendriks, T., Lo Schiavo, F., Terzi, M., Bisseling, T., Van Kammen, A., and De Vries, S. C. (1993). *Rhizobium* lipooligosaccharides rescue a carrot somatic embryo mutant. *Plant Cell* **5,** 615–620.

De Maagd, R. A., Spaink, H. P., Pees, E., Mulders, I. H. M., Wijfjes, A., Wijffelman, C. A., Okker, R. J. H., and Lugtenberg, B. J. J. (1989a). Localization and symbiotic function of a region on the *Rhizobium leguminosarum* Sym plasmid pRL1JI responsible for a secreted, flavonoid-inducible 50-kilodalton protein. *J. Bacteriol.* **171,** 1151–1157.

De Maagd, R. A., Wijfjes, A. H. M., Spaink, H. P., Ruiz-Sainz, J. E., Wijffelman, C. A., Okker, R. J. H., and Lugtenberg, B. J. J. (1989b). *nodO,* a new nod gene of the *Rhizobium leguminosarum* biovarviciae Sym plasmid pRL1JI, encodes a secreted protein. *J. Bacteriol.* **171,** 6764–6770.

Debellé, F., and Sharma, S. B. (1986). Nucleotide sequence of *Rhizobium meliloti* RCR2011 genes involved in host specificity of nodulation. *Nucleic Acids Res.* **14,** 7453–7472.

Debellé, F., Rosenberg, C., Vasse, J., Maillet, F., Martinez, E., Dénairié, J., and Truchet, G. (1986). Assignment of symbiotic developmental phenotypes to common and specific nodulation (*nod*) genetic loci of *Rhizobium meliloti. J. Bacteriol.* **168,** 1075–1086.

Debellé, F., Maillet, F., Vasse, J., Rosenberg, C., De Billy, F., Truchet, G., Dénarié, J., and Ausubel, F. M. (1988). Interference between *Rhizobium meliloti* and *Rhizobium trifolii* nodulation genes: Genetic basis of *R. meliloti* dominance. *J. Bacteriol.* **170,** 5718–5727.

Debellé, F., Rosenberg, C., and Dénarié, J. (1992). The *Rhizobium, Bradyrhizobium,* and *Azorhizobium* NodC proteins are homologous to yeast chitin synthases. *Molec. Plant-Microbe Interact.* **5,** 443–446.

Dehio, C., and De Bruijn, F. J. (1992). The early nodulin gene *SrEnod2* from *Sesbania rostrata* is inducible by cytokinin. *Plant J.* **2,** 117–128.

Delves, A. C., Mathews, A., Day, D. A., Carter, A. S., Carroll, B. J., and Gresshoff, P. M. (1986). Regulation of the soybean-*Rhizobium* nodule symbiosis by shoot and root factors. *Plant Physiol.* **82,** 588–590.

Delves, A. C., Higgins, A., and Gresshoff, P. M. (1987a). Supernodulation in interspecific grafts between *Glycine max* (soybean) and *Glycine soja. J. Plant Physiol.* **128,** 473–478.

Delves, A. C., Higgins, A. V., and Gresshoff, P. M. (1987b). Shoot control of supernodulation in a number of mutant soybeans, *Glycine max* (L.) Merr. *Aust. J. Plant. Physiol.* **14,** 689–694.

Delves, A. C., Higgins, A., and Gresshoff, P. (1992). Shoot apex removal does not alter autoregulation of nodulation in soybean. *Plant Cell Environ.* **15,** 249–254.

Demont, N., Debellé, F., Aurelle, H., Dénarié, J., and Promé, J. C. (1993). Role of the *Rhizobium meliloti nodF* and *nodE* genes in the biosynthesis of lipooligosacharidic nodulation factors. *J. Biol. Chem.* **268,** 20,134–20,142.

Demont, N., Ardourel, M., Maillet, F., Promé, D., Ferro, M., Promé, J. C., and Dénarié, J. (1994). The *Rhizobium meliloti* regulatory *nodD3* and *syrM* genes control the synthesis of a particular class of nodulation factors *N*-acylated by (ω-1)-hydroxylated fatty acids. *EMBO J.* **13,** 2139–2149.

Díaz, C. L., Van Spronsen, P. C., Bakhuizen, R., Logman, G. J. J., Lugtenberg, B. J. J. and Kijne, J. W. (1986). Correlation between infection by *Rhizobium leguminosarum* and lectin on the surface of *Pisum sativun* L. roots. *Planta* **168,** 350–359.

Dickstein, R., Bisseling, T., Reinhold, V. N., and Ausubel, F. M. (1988). Expression of nodule-specific genes in alfalfa root nodules blocked at an early stage of development. *Genes Dev.* **2,** 677–687.

Dickstein, R., Prusty, R., Peng, T., Ngo, W., and Smith, M. E. (1993). Enod8, a novel early nodule-specific gene, is expressed in empty alfalfa nodules. *Mol. Plant-Microbe Interact.* **6,** 715–721.

Dixon, R. O. D. (1969) Rhizobia. *Ann. Rev. Microbiol.* **23,** 137–158.

Djordjevic, M. A., Schofield, P. R., and Rolfe, B. G. (1985). Tn*5* mutagenesis of *Rhizobium trifolii* host-specific nodulation genes result in mutants with altered host-range ability. *Mol. Gen. Genet.* **200,** 463–471.

Djordjevic, M. A., Redmond, J. W., Batley, M., and Rolfe, B. G. (1987). Clovers secrete specific phenolic compounds which either stimulate or repress *nod* gene expression in *Rhizobium trifolii. EMBO J.* **6,** 1173–1179.

Djordjevic, M. A., McKay, I. A., De Boer, M., and Rolfe, B. G. (1993). Nod metabolites excreted by *Rhizobium leguminosarum* bv. *trifolii* also accumulate in the cell membrane. *In* "New Horizons in Nitrogen Fixation" (R. Palacios, J. Mora, and W. E. Newton, Eds.), p. 226. Kluwer, Dordrecht, The Netherlands.

Dobert, R. C., Rood, S. B., and Blevins, D. G. (1992). Gibberellins and the legume-*Rhizobium* symbiosis. I. Endogenous gibberellins of Lima bean (*Phaseolus lunatus* L.) stems and nodules. *Plant Physiol.* **98,** 221–224.

Dockendorff, T. C., Sharma, A. J., and Stacey, G. (1994). Identification and characterization of the *nolYZ* genes of *Bradyrhizobium japonicum. Mol. Plant-Microbe Interact.* **7,** 173–180.

Downie, J. A., and Surin, B. P. (1990). Either of two nod gene loci can complement the nodulation defect of a *nod* deletion mutant of *Rhizobium leguminosarum bv viciae. Mol. Gen. Genet.* **222,** 81–86.

Downie, J. A., Knight, C. D., Johnston, A. W. B., and Rossen, L. (1985). Identification of genes and gene products involved in the nodulation of peas by *Rhizobium leguminosarum. Mol. Gen. Genet.* **198,** 255–262.

Duc, G., and Messager, A. (1989). Mutagenesis of pea (*Pisum sativum* L.) and the isolation of mutants for nodulation and nitrogen fixation. *Plant Science* **60,** 207–213.

Duc, G., Trouvelot, A., Gianinazzi-Pearson, V., and Gianinazzi, S. (1989). First report of non-mycorrhizal plant mutants (Myc-) obtained in pea (*Pisum sativum* L.) and Fababean (*Vicia faba* L.). *Plant Science* **60,** 215–222.

Dudley, M. E., and Long, S. R. (1989). A non-nodulating alfalfa mutant displays neither root hair curling nor early cell division in response to *Rhizobium meliloti. Plant Cell* **1,** 65–72.

Dudley, M. E., Jacobs, T. W., and Long, S. R. (1987). Microscopic studies of cell divisions induced in alfalfa roots by *Rhizobium meliloti. Planta* **171,** 289–301.

Dusha, I., and Kondorosi, A. (1993). Genes at different regulatory levels are required for the ammonia control of nodulation in *Rhizobium meliloti. Mol. Gen. Genet.* **240,** 435–444.

Dusha, I., Bakos, A., Kondorosi, A., De Bruijn, F. J., and Schell, J. (1989). The *Rhizobium meliloti* early nodulation genes (*nodABC*) are nitrogen-regulated: Isolation of a mutant strain with efficient nodulation capacity on alfalfa in the presence of ammonium. *Mol. Gen. Genet.* **219,** 89–96.

Economou, A., Hamilton, W. D. O., Johnston, A. W. B., and Downie, J. A. (1990). The Rhizobium nodulation gene *nodO* encodes a Ca^{2+}-binding protein that is exported without N-terminal cleavage and is homologous to haemolysin and related proteins. *EMBO J.* **9,** 349–354.

Egelhoff, T. T., and Long, S. R. (1985a). *Rhizobium meliloti* nodulation genes: Identification of *nodDABC* gene products, purification of *nodA* protein, and expression of *nodA* in *Rhizobium meliloti*. *J. Bacteriol.* **164,** 591–599.

Egelhoff, T. T., Fisher, R. F., Jacobs, T. W., Mulligan, J. T., and Long, S. R. (1985b). Nucleotide sequence of *Rhizobiuim meliloti* 1021 nodulation genes: *nodD* is read divergently from *nodABC*. *DNA* **4,** 241–248.

Ehrhardt, D. W., Atkinson, E. M. and Long, S. R. (1992). Depolarization of alfalfa root hair membrane potential by *Rhizobium meliloti* Nod factors. *Science* **256,** 998–1000.

Elkan, G. H. (1992). Taxonomy of the rhizobia. *Can. J. Microbiol.* **38,** 446–450.

Engvild, K. C. (1987). Nodulation and nitrogen fixation mutants of pea (*Pisum sativum*). *Theor. Appl. Genet.* **74,** 711–713.

Eskew, D. L., Kapuya, J., and Danso, S. K. A. (1989). Nitrate inhibition of nodulation and nitrogen fixation by supernodulating nitrate-tolerant symbiosis mutants of soybean. *Crop Sci.* **29,** 1491–1496.

Estabrook, E. M., and Sengupta-Gopalan, C. (1991). Differential expression of phenylalanine ammonia-lyase and chalcone synthase during soybean nodule development. *Plant Cell* **3,** 299–308.

Estruch, J. J., Chriqui, D., Grossman, K., Schell, J., and Spena, A. (1991a). The plant oncogene *rolC* is responsible for the release of cytokinins from glucoside conjugates. *EMBO J.* **10,** 2889–2895.

Estruch, J. J., Schell, J., and Spena, A. (1991b). The protein encoded by the *rolB* plant oncogene hydrolyses indole glucosides. *EMBO J.* **10,** 3125–3128.

Evans, I. J., and Downie, J. A. (1986). The *nodI* gene product of *Rhizobium leguminosarum* is closely related to ATP-binding bacterial transport proteins; nucleotide sequence analysis of the *nodI* and *nodJ* genes. *Gene* **43,** 95–101.

Faucher, C., Maillet, F., Vasse, J., Rosenberg, C., Van Brussel, A. A. N., Truchet, G., and Dénarié, J. (1988). *Rhizobium meliloti* host range *nodH* gene determines production of an alfalfa-specific extracellular signal. *J. Bacteriol.* **170,** 5489–5499.

Faucher, C., Camut, S., Dénarié, J., and Truchet, G. (1989). The *nodH* and *nodQ* host range genes of *Rhizobium meliloti* behave as avirulence genes in *R. leguminosarum* bv. *viciae* and determine changes in the production of plant-specific extracellular signals. *Mol. Plant-Microbe Interac.* **2,** 291–300.

Fearn, J. C., and LaRue, T. A. (1991). Ethylene inhibitors restore nodulation to *sym* 5 mutants of *Pisum sativum* L. cv Sparkle. *Plant Physiol.* **96,** 239–244.

Feiler, H. S., and Jacobs, T. W. (1990). Cell division in higher plants: A *cdc2* gene, its 34-kDa product, and histone H1 kinase activity in pea. *Proc. Natl. Acad. Sci. USA* **87,** 5397–5401.

Felix, G., Regenass, M., and Boller, T. (1993). Specific perception of subnanomolar concentrations of chitin fragments by tomato cells: induction of extracellular alkalinization, changes in protein phosphorylation, and establishment of a refractory state. *Plant J.* **4,** 307–316.

Firmin, J. L., Wilson, K. E., Rossen, L., and Johnston, A. W. B. (1986). Flavonoid activation of nodulation genes in *Rhizobium* reversed by other compounds present in plants. *Nature* **324,** 90–92.

Firmin, J. L., Wilson, K. E., Carlson, R. W., Davies, A. E., and Downie, J. A. (1993). Resistance to nodulation of cv. Afghanistan peas is overcome by *nodX,* which mediates an *O*-acetylation of the *Rhizobium leguminosarum* lipooligosaccharide nodulation factor. *Mol. Microbiol.* **10,** 351–360.

Fisher, R. F., and Long, S. R. (1989). DNA footprint analysis of the transcriptional activator proteins NodD1 and NodD3 on inducible *nod* gene promoters. *J. Bacteriol.* **171,** 5492–5502.

Fisher, R. F., and Long, S. R. (1993). Interactions of NodD at the *nod* box: NodD binds to two distinct sites on the same face of the helix and induces a bend in the DNA. *J. Mol. Biol.* **233,** 336–348.

Fisher, R. F., Swanson, J. A., T, M. J. and Long, S. R. (1987a). Extended region of nodulation genes in *Rhizobium meliloti* 1021. II. Nucleotide sequence, transcription start sites and protein products. *Genetics* **117,** 191–201.

Fisher, R. F., Brierley, H. L., Mulligan, J. T., and Long, S. R. (1987b). Transcription of *Rhizobium meliloti* nodulation genes. Identification of a *nodD* transcription initiation site *in vitro* and *in vivo. J. Biol. Chem.* **262,** 6849–6855.

Fisher, R. F., Egelhoff, T. T., Mulligan, J. T., and Long, S. R. (1988). Specific binding of proteins from *Rhizobium meliloti* cell-free extracts containing NodD to DNA sequences upstream of inducible nodulation genes. *Genes Dev.* **2,** 282–293.

Flores, M., Brom, S., Stepkowski, T., Girard, M. L., Dávila, G., Romero, D., and Palacios, R. (1993). Gene amplification in *Rhizobium:* Identification and *in vivo* cloning of discrete amplifiable DNA regions (amplicons) from *Rhizobium leguminosarum* biovar *phaseoli. Proc. Natl. Acad. Sci. USA* **90,** 4932–4936.

Francisco, P. B., and Akao, S. (1993). Autoregulation and nitrate inhibition of nodule formation in soybean cv. Enrei and its nodulation mutants. *J. Exp. Bot.* **44,** 547–553.

Franssen, H., Nap, J. P., Gloudemans, T., Stiekema, W., Van Dam, H., Govers, F., Louwerse, J., Van Kammen, A., and Bisseling, T. (1987). Characterization of cDNA for nodulin-75 of soybean: A gene product involved in early stages of root nodule development. *Proc. Natl. Acad. Sci, USA* **84,** 4495–4499.

Franssen, H. J., Vijn, I., Yang, W. C., and Bisseling, T. (1992). Developmental aspects of the *Rhizobium*-legume symbiosis. *Plant Mol. Biol.* **19,** 89–107.

Franssen, H., Yang, W. C., Katinakis, P., and Bisseling, T. (1993). Characterization of *GmEnod40,* a gene expressed in soybean nodule primordia. *In* "New Horizons in Nitrogen Fixation" (R. Palacios, J. Mora, and W. E. Newton, Eds.), pp. 275–284. Kluwer, Dordrecht, The Netherlands.

Geelen, D., Mergaert, P., Geremia, R. A., Goormachtig, S., Van Montagu, M., and Holsters, M. (1993). Identification of *nodSUIJ* genes in Nod locus 1 of *Azorhizobium caulinodans:* evidence that nodS encodes a methyltransferase involved in Nod factor modification. *Mol. Microbiol.* **9,** 145–154.

Geiger, O., Spaink, H. P., and Kennedy, E. P. (1991). Isolation of the *Rhizobium leguminosarum* NodF nodulation protein: NodF carries a 4′-phosphopantetheine prosthetic group. *J. Bacteriol.* **173,** 2872–2878.

Geiger, O., Thomas-Oates, J. E., Glushka, J., Spaink, H. P., and Lugtenberg, B. J. J. (1994). Phospholipids of *Rhizobium* contain *nodE*-determined highly unsaturated fatty acid moieties. *J. Biol. Chem.* **269,** 11090–11097.

Geremia, R. A., Mergaert, P., Geelen, D., Van Montagu, M., and Holsters, M. (1994). The NodC protein of *Azorhizobium caulinodans* is an *N*-acetylglucosaminyltransferase. *Proc. Natl. Acad. Sci. USA* **91,** 2669–2673.

Gerhold, D., Stacey, G., and Kondorosi, A. (1989). Use of a promoter-specific probe to identify two loci from the *Rhizobium meliloti* nodulation regulon. *Plant Mol. Biol.* **12,** 181–188.

Girard, M. L., Flores, M., Brom, S., Romero, D., Palacios, R., and Dávila, G. (1991). Structural complexity of the symbiotic plasmid of *Rhizobium leguminosarum* bv. *phaseoli*. *J. Bacteriol.* **173,** 2411–2419.

Gloudemans, T, De Vries, S., Bussink, H. J., Malik, N. S. A., Franssen, H. J., Louwerse, J., and Bisseling, T. (1987). Nodulin gene expression during soybean (*Glycine max*) nodule development. *Plant Mol. Biol.* **8,** 395–403.

Gloudemans, T., Bhuvaneswari, T. V., Moerman, M., Van Brussel, T., Van Kammen, A., and Bisseling, T. (1989). Involvement of *Rhizobium leguminosarum* nodulation genes in gene expression in pea root hairs. *Plant Mol. Biol.* **12,** 157–167.

Goethals, K., Gao, M., Tomekpe, K., Van Montagu, M., and Holsters, M. (1989). Common *nodABC* genes in *nod* locus 1 of *Azorhizobium caulinodans:* Nucleotide sequence and plant-inducible expression. *Mol. Gen. Genet.* **219,** 289–298.

Goethals, K., Van den Eede, G., Van Montagu, M., and Holsters, M. (1990). Identification and characterization of a functional *nodD* gene in *Azorhizobium caulinodans* ORS571. *J. Bacteriol.* **172,** 2658–2666.

Goethals, K., Van Montagu, M., and Holsters, M. (1992a). Conserved motifs in a divergent *nod* box of *Azorhizobium caulinodans* ORS571 reveal a common structure in promoters regulated by LysR-type proteins. *Proc. Natl. Acad. Sci. USA* **89,** 1646–1650.

Goethals, K., Mergaert, P., Gao, M., Geelen, D., Van Montagu, M., and Holsters, M. (1992b). Identification of a new inducible nodulation gene in *Azorhizobium caulinodans*. *Mol. Plant-Microbe Interact.* **5,** 405–411.

Goodlass, G., and Smith, K. A. (1979). Effect of ethylene on root extension and nodulation of pea (*Pisum sativum* L.) and white clover (*Trifolium repens* L.). *Plant Soil* **51,** 387–395.

Göttfert, M. (1993). Regulation and function of rhizobial nodulation genes. *FEMS Microbiol. Rev.* **104,** 39–64.

Göttfert, M., Horvath, B., Kondorosi, E., Putnoky, P., Rodriguez-Quiñones, F., and Kondorosi, A. (1986). At least two *nodD* genes are necessary for efficient nodulation of alfalfa by *Rhizobium meliloti*. *J. Mol. Biol.* **191,** 411–420.

Göttfert, M., Weber, J., and Hennecke, H. (1988). Induction of a *nodA-lacZ* fusion in *Bradyrhizobium japonicum* by an isoflavone. *J. Plant Physiol.* **132,** 394–397.

Göttfert, M., Lamb, J. W., Gasser, R., Semenza, J., and Hennecke, H. (1989). Mutational analysis of the *Bradyrhizobium japonicum* common *nod* genes and further *nod* box-linked genomic DNA regions. *Mol. Gen. Genet.* **215,** 407–415.

Göttfert, M., Grob, P., and Hennecke, H. (1990a). Proposed regulatory pathway encoded by the *nodV* and *nodW* genes, determinants of host specificity in *Bradyrhizobium japonicum*. *Proc. Natl. Acad. Sci. USA* **87,** 2680–2684.

Göttfert, M., Hitz, S., and Hennecke, H. (1990b). Identification of nodS and nodU, two inducible genes inserted between the *Bradyrhizobium japonicum nodYABC* and *nodIJ* genes. *Mol. Plant-Microbe Interact.* **3,** 308–316.

Göttfert, M., Holzhäuser, D., Bäni, D., and Hennecke, H. (1992). Structural and functional analysis of two different *nodD* genes in *Bradyrhizobium japonicum* USDA110. *Mol. Plant-Microbe Interact.* **5,** 257–265.

Götz, R., Evans, I. J., Downie, J. A., and Johnston, A. W. B. (1985). Identification of the host-range DNA which allows *Rhizobium leguminosarum* strain TOM to nodulate cv. Afghanistan peas. *Mol. Gen. Genet.* **201,** 296–300.

Govers, F., Gloudemans, T., Moerman, M., Van Kammen, A., and Bisseling, T. (1985). Expression of plant genes during the development of pea root nodules. *EMBO J.* **4,** 861–867.

Govers, F., Moerman, M., Downie, J. A., Hooykaas, P., Franssen, H. J., Louwerse, J., Van Kammen, A., and Bisseling, T. (1986). Rhizobium *nod* genes are involved in inducing an early nodulin gene. *Nature* **323,** 564–566.

Gray, J. X., De Maagd, R. A., Rolfe, B. G., Johnston, A. W. B., and Lugtenberg,

B. J. J. (1992). The role of the *Rhizobium* cell surface during symbiosis. *In* "Molecular Signals in Plant-Microbe Communications" (D. P. S. Verma, Ed.), pp. 359–376. CRC Press, Boca Raton, FL.

Gremaud, M. F., and Harper, J. E. (1989). Selection and initial characterization of partially nitrate tolerant nodulation mutants of soybean. *Plant Physiol.* **89,** 169–173.

Grob, P., Michel, P., Hennecke, H., and Göttfert, M. (1993). A novel response-regulator is able to suppress the nodulation defect of a *Bradyrhizobium japonicum nodW* mutant. *Mol. Gen. Genet.* **241,** 531–541.

Grobbelaar, N., Clarke, B., and Hough, M. C. (1971). The nodulation and nitrogen fixation of isolated roots of *Phaseolus vulgaris* L. III. The effect of carbon dioxide and ethylene. *Plant Soil,* Special Vol., 215–223.

Guinel, F. C., and LaRue, T. A. (1991). Light microscopy study of nodule initiation in *Pisum sativum* L. cv Sparkle and its low-nodulating mutant E2 (Sym 5). *Plant Physiol.* **97,** 1206–1211.

Guinel, F. C., and LaRue, T. A. (1992). Ethylene inhibitors partly restore nodulation to pea mutant E107 (*brz*). *Plant Physiol.* **99,** 515–518.

Györgypal, Z., and Kondorosi, A. (1991). Homology of the ligand-binding regions of *Rhizobium* symbiotic regulatory protein NodD and vertebrate nuclear receptors. *Mol. Gen. Genet.* **226,** 337–340.

Györgypal, Z., Iyer, N., and Kondorosi, A. (1988). Three regulatory *nodD* alleles of diverged flavonoid-specificity are involved in host-dependent nodulation by *Rhizobium meliloti. Mol. Gen. Genet.* **212,** 85–92.

Györgypal, Z., Kondorosi, E., and Kondorosi, A. (1991a). Diverse signal sensitivity of NodD protein homologs from narrow and broad host range rhizobia. *Mol. Plant-Microbe Interact.* **4,** 356–364.

Györgypal, Z., Kiss, G. B., and Kondorosi, A. (1991b). Transduction of plant signal molecules by the *Rhizobium* NodD proteins. *Bioessays* **13,** 575–581.

Hartwig, U. A., Maxwell, C. A., Joseph, C. M., and Phillips, D. A. (1989). Interactions among flavonoid *nod* gene inducers released from alfalfa seeds and roots. *Plant Physiol.* **91,** 1138–1142.

Hartwig, U. A., Maxwell, C. A., Joseph, C. M., and Phillips, D. A. (1990a). Chrysoeriol and luteolin released from alfalfa seeds induce *nod* genes in *Rhizobium meliloti. Plant Physiol.* **92,** 116–122.

Hartwig, U. A., Maxwell, C. A., Joseph, C. M., and Phillips, D. A. (1990b). Effects of alfalfa *nod* gene-inducing flavonoids on *nodABC* transcription in *Rhizobium meliloti* strains containing different *nodD* genes. *J. Bacteriol.* **172,** 2769–2773.

Hata, S., Kouchi, H., Suzuka, I., and Ishii, T. (1991). Isolation and characterization of cDNA clones for plant cyclins. *EMBO J.* **10,** 2681–2688.

Heidstra, R., Kozik, A., Bisseling, T., and Lie, T. A. (1993). Pea Nod$^-$ mutants and Nod factor perception. *In* "New Horizons in Nitrogen Fixation" (R. Palacios, J. Mora, and W. E. Newton, Eds.), pp. 267–268. Kluwer, Dordrecht, The Netherlands.

Henikoff, S., Haughn, G. W., Calvo, J. M., and Wallace, J. C. (1988). A large family of bacterial activator proteins. *Biochemistry* **85,** 6602–6606.

Hirsch, A. M. (1992). Developmental biology of legume nodulation. *New Phytol.* **122,** 211–237.

Hirsch, A. M., Bhuvaneswari, T. V., Torrey, J. G., and Bisseling, T. (1989). Early nodulin genes are induced in alfalfa root outgrowths elicited by auxin transport inhibitors. *Proc. Natl. Acad. Sci. USA* **86,** 1244–1248.

Hirsch, A. M., Asad, S., Fang, Y., Wycoff, K., and Löbler, M. (1993). Molecular interactions during nodule development. *In* "New Horizons in Nitrogen Fixation" (R. Palacios, J. Mora, and W. E. Newton, Eds.), pp. 291–296. Kluwer, Dordrecht, The Netherlands.

Hirt, H., Páy, A., Györgyey, J., Bacó, L., Németh, K., Bögre, L., Schweyen, R. J., Heberle-Bors, E., and Dudits, D. (1991). Complementation of a yeast cell cycle mutant by an alfalfa cDNA encoding a protein kinase homologous to $p34^{cdc2}$. *Proc. Natl. Acad. Sci. USA* **88,** 1636–1640.
Hirt, H., Mink, M., Pfosser, M., Bögre, L., Györgyey, J., Jonak, C., Gartner, A., Dudits, D., and Heberle-Bors, E. (1992). Alfalfa cyclins: differential expression during the cell cycle and in plant organs. *Plant Cell* **4,** 1531–1538.
Holsters, M., Geelen, D., Goethals, K., Van Montagu, M., Geremia, R., Promé, J. C., and Mergaert, P. (1993). Nod factor production by *Azorhizobium caulinodans* strain ORS571. *In* "New Horizons in Nitrogen Fixation" (R. Palacios, J. Mora, and W. E. Newton, Eds.), pp. 191–196. Kluwer, Dordrecht, The Netherlands.
Hong, G. F., Burn, J. E., and Johnston, A. W. B. (1987). Evidence that DNA involved in the expression of nodulation (*nod*) genes in *Rhizobium* binds to the product of the regulatory gene *nodD. Nucleic Acids Res.* **15,** 9677–9690.
Honma, M. A., and Ausubel, F. M. (1987). *Rhizobium meliloti* has three functional copies of the *nodD* symbiotic regulatory gene. *Proc. Natl. Acad. Sci. USA* **84,** 8558–8562.
Honma, M. A., Asomaning, M., and Ausubel, F. M. (1990). *Rhizobium meliloti* nodD genes mediate host-specific activation of *nodABC. J. Bacteriol.* **172,** 901–911.
Horvath, B., Kondorosi, E., John, M., Schmidt, J., Török, I., Györgypal, Z., Barabas, I., Wieneke, U., Schell, J., and Kondorosi, A. (1986). Organization, structure and symbiotic function of *Rhizobium meliloti* nodulation genes determining host specificity for alfalfa. *Cell* **46,** 335–343.
Horvath, B., Bachem, C. W. B., Schell, J., and Kondorosi, A. (1987). Host-specific regulation of nodulation genes in *Rhizobium* is mediated by a plant-signal, interacting with the *nodD* gene product. *EMBO J.* **6,** 841–848.
Horvath, B., Heidstra, R., Lados, M., Moerman, M., Spaink, H. P., Promé, J. C., Van Kammen, A., and Bisseling, T. (1993). Lipo-oligosaccharides of *Rhizobium* induce infection-related early nodulin gene expression in pea root hairs. *Plant J.* **4,** 727–733.
Huang, S. Z., Djordjevic, M. A., and Rolfe, B. G. (1988). Characterization of aberrant infection events induced on *Trifolium subterraneum* by *Rhizobium trifolii* region II mutants. *J. Plant Physiol.* **133,** 16–24.
Hubac, C., Guerrier, D., Ferran, J., Trémolières, A., and Kondorosi, A. (1992). Lipid and protein composition of outer and inner membranes in wild-type strains and *nod* mutants of *Rhizobium meliloti. J. Gen. Microbiol.* **138,** 1973–1983.
Huguet, T., Pichon, M., Journet, E. P., de Billy, F., Barker, D., and Truchet, G. (1993). Alfalfa nodulation in the absence of *Rhizobium:* a physiologically and genetically complex process. *In* "New Horizons in Nitrogen Fixation" (R. Palacios, J. Mora, and W. E. Newton, Eds.), p. 347. Kluwer, Dordrecht, The Netherlands.
Hungria, M., Johnston, A. W. B., and Phillips, D. A. (1992). Effects of flavonoids released naturally from bean (*Phaseolus vulgaris*) on *nodD*-regulated gene transcription in *Rhizobium leguminosarum* bv. *viciae. Mol. Plant-Microbe Interact.* **5,** 199–203.
Hunter, W. J. (1993). Ethylene production by root nodules and effect of ethylene on nodulation in *Glycine max. Appl. Environ. Microbiol.* **59,** 1947–1950.
Innes, R. W., Kuempel, P. L., Plazinski, J., Canter-Cremers, H., Rolfe, B. G., and Djordjevic, M. A. (1985). Plant factors induce expression of nodulation and host-range genes in *Rhizobium trifolii. Mol. Gen. Genet.* **201,** 426–432.
Jacobs, M., and Rubery, P. H. (1988). Naturally occurring auxin transport regulators. *Science* **241,** 346–349.
Jacobs, T. W., Egelhoff, T. T., and Long, S. R. (1985). Physical and genetic map of a *Rhizobium meliloti* nodulation gene region and nucleotide sequence of *nodC. J. Bacteriol.* **162,** 469–476.

Jacobsen, E. (1984). Modification of symbiotic interaction of pea (*Pisum sativum* L.) and *Rhizobium leguminosarum* by induced mutations. *Plant Soil.* **82,** 427–438.

Jacobsen, E., and Feenstra, W. J. (1984). A new pea mutant with efficient nodulation in the presence of nitrate. *Plant Sci. Lett.* **33,** 337–344.

John, M., Schmidt, J., Wieneke, U., Kondorosi, E., Kondorosi, A., and Schell, J. (1985). Expression of the nodulation gene *nodC* of *Rhizobium meliloti* in *Escherichi coli:* role of the *nodC* gene product in nodulation. *EMBO J.* **4,** 2425–2430.

John, M., Schmidt, J., Wieneke, U., Krüssmann, H. D., and Schell, J. (1988). Transmembrane orientation and receptor-like structure of the *Rhizobium meliloti* common nodulation protein NodC. *EMBO J.* **7,** 583–588.

John, M., Röhrig, H., Schmidt, J., Wieneke, U., and Schell, J. (1993). *Rhizobium* NodB protein involved in nodulation signal synthesis is a chitooligosaccharide deacetylase. *Proc. Natl. Acad. Sci. USA* **90,** 625–629.

Johnson, D., Roth, L. E., and Stacey, G. (1989). Immunogold localization of the NodC and NodA proteins of *Rhizobium meliloti. J. Bacteriol.* **171,** 4583–4588.

Joshi, P. A., Caetano-Anollés, G., Graham, E. T., and Gresshoff, P. M. (1991). Ontogeny and ultrastructure of spontaneous nodules in alfalfa (*Medicago sativa*). *Protoplasma* **162,** 1–11.

Junghans, H., Dalkin, K., and Dixon, R. A. (1993). Stress responses in alfalfa (*Medicago sativa* L.). 15. Characterization and expression patterns of members of a subset of the chalcone synthase multigene family. *Plant Mol. Biol.* **22,** 239–253.

Kape, R., Parniske, M., and Werner, D. (1991). Chemotaxis and *nod* gene activity of *Bradyrhizobium japonicum* in response to hydroxycinnamic acids and isoflavonoids. *Appl. Environ. Microbiol.* **57,** 316–319.

Kape, R., Parniske, M., Brandt, S., and Werner, D. (1992). Isoliquiritigenin, a strong *nod* gene- and glyceollin resistance-inducing flavonoid from soybean root exudate. *Appl. Environ. Microbiol.* **58,** 1705–1710.

Kapros, T., Bögre, L., Nemeth, K., Bakó, L., Györgyey, J., Wu, S. C., and Dudits, D. (1992). Differential expression of histone H3 gene variants during cell cycle and somatic embryogenesis in alfalfa. *Plant Physiol.* **98,** 621–625.

Kapulnik, Y., Joseph, C. M., and Phillipps, D. A. (1987). Flavone limitations to root nodulation and symbiotic nitrogen fixation in alfalfa. *Plant Physiol.* **84,** 1193–1196.

Kijne, J. W. (1992). The *Rhizobium* infection process. *In* "Biological Nitrogen Fixation" (G. Stacey, R. H. Burris, and H. J. Evans, Eds.), pp. 394–398. Chapman & Hall, New York.

Kneen, B. E., and LaRue, T. A. (1984). Nodulation resistant mutants of *Pisum sativum* L. *J. Hered.* **75,** 238–240.

Koes, R. E., Quattrocchio, F., and Mol, J. N. M. (1994). The flavonoid biosynthetic pathway in plants: Function and evolution. *BioEssays* **16,** 123–132.

Kondorosi, A. (1992). Regulation of nodulation genes in rhizobia. *In* "Molecular Signals in Plant-Microbe Communications" (D. P. S. Verma, ed.), pp. 325–340. CRC Press, Boca Raton, FL.

Kondorosi, E., Banfalvi, Z., and Kondorosi, A. (1984). Physical and genetic analysis of a symbiotic region of *Rhizobum meliloti:* Identification of nodulation genes. *Mol. Gen. Genet.* **193,** 445–452.

Kondorosi, E., Gyuris, J., Schmidt, J., John, M., Duda, E., Hoffmann, B., Schell, J., and Kondorosi, A. (1989). Positive and negative control of *nod* gene expression in *Rhizobium meliloti* is required for optimal nodulation. *EMBO J.* **8,** 1331–1340.

Kondorosi, E., Pierre, M., Cren, M., Haumann, U., Buiré, M., Hoffmann, B., Schell, J., and Kondorosi, A. (1991a). Identification of NolR, a negative transacting factor controlling the nod regulon in *Rhizobium meliloti. J. Mol. Biol.* **222,** 885–896.

Kondorosi, E., Buiré, M., Cren, M., Iyer, N., Hoffmann, B., and Kondorosi, A. (1991b).

Involvement of the *syrM* and *nodD3* genes of *Rhizobium meliloti* in *nod* gene activation and in optimal nodulation of the plant host. *Mol. Microbiol.* **5,** 3035–3048.

Kondorosi, E., Schultze, M., Savouré, A., Hoffmann, B., Dudits, D., Pierre, M., Allison, L., Bauer, P., Kiss, G. B., and Kondorosi, A. (1993). Control of nodule induction and plant cell growth by Nod factors. *In* "Advances in Molecular Genetics of Plant-Microbe Interactions" (E. W. Nester and D. P. S. Verma, Eds.), Vol. 2, pp. 143–150. Kluwer, Dordrecht, The Netherlands.

Kosslak, R. M., Bookland, R., Barkei, J., Paaren, H. E., and Appelbaum, E. R. (1987). Induction of *Bradyrhizobium japonicum* common *nod* genes by isoflavones isolated from *Glycine max*. *Proc. Natl. Acad. Sci. USA* **84,** 7428–7432.

Kosslak, R. M., Joshi, R. S., Bowen, B. A., Paaren, H. E., and Appelbaum, E. R. (1990). Strain-specific inhibition of *nod* gene induction in *Bradyrhizobium japonicum* by flavonoid compounds. *Appl. Environ. Microbiol.* **56,** 1333–1341.

Kouchi, H., and Hata, S. (1993). Isolation and characterization of novel nodulin cDNAs representing genes expressed at early stages of soybean nodule development. *Mol. Gen. Genet.* **238,** 106–119.

Krause, A., and Broughton, W. J. (1992). Proteins associated with root-hair deformation and nodule initiation in *Vigna unguiculata*. *Mol. Plant-Microbe Interact.* **5,** 96–103.

Kreuzer, P., Gärtner, D., Allmansberger, R., and Hillen, W. (1989). Identification and sequence analysis of the *Bacillus subtilis* W23 *xylR* gene and *xyl* operator. *J. Bacteriol.* **171,** 3840–3845.

Krishnan, H. B., and Pueppke, S. G. (1991a). Sequence and analysis of the *nodABC* region of *Rhizobium fredii* USDA257, a nitrogen-fixing symbiont of soybean and other legumes. *Mol. Plant-Microbe Interact.* **4,** 512–520.

Krishnan, H. B., and Pueppke, S. G. (1991b). *nolC*, a *Rhizobium fredii* gene involved in cultivar-specific nodulation of soybean, shares homology with a heat-shock gene. *Mol. Microbiol.* **5,** 737–745.

Krishnan, H. B., Lewin, A., Fellay, R., Broughton, W. J., and Pueppke, S. G. (1992). Differential expression of *nodS* accounts for the varied abilities of *Rhizobium fredii* USDA257 and *Rhizobium* sp. strain NGR234 to nodulate *Leucaena* spp. *Mol. Microbiol.* **6,** 3321–3330.

Kuykendall, L. D., Saxena, B., Devine, T. E., and Udell, S. E. (1992). Genetic diversity in *Bradyrhizobium japonicum* Jordan 1982 and a proposal for *Bradyrhizobium elkanii* sp. nov. *Can. J. Microbiol.* **38,** 501–505.

Landau-Ellis, D., Angermüller, S., Shoemaker, R., and Gresshoff, P. M. (1991). The genetic locus controlling supernodulation in soybean (*Glycine max* L.) co-segregates tightly with a cloned molecular marker. *Mol. Gen. Genet.* **228,** 221–226.

Lauridsen, P., Franssen, H., Stougaard, J., Bisseling, T., and Marcker, K. A. (1993). Conserved regulation of the soybean early nodulin *ENOD2* gene promoter in determinate and indeterminate transgenic root nodules. *Plant J.* **3,** 483–492.

Le Strange, K. K., Bender, G. L., Djordjevic, M. A., Rolfe, B. G., and Redmond, J. W. (1990). The *Rhizobium* strain NGR234 *nodD1* gene product responds to activation by the simple phenolic compounds vanillin and isovanillin present in wheat seedling extracts. *Mol. Plant-Microbe Interact.* **3,** 214–220.

Lee, K. H., and LaRue, T. A. (1992a). Ethylene as a possible mediator of light- and nitrate-induced inhibition of nodulation of *Pisum sativum* L. cv Sparkle. *Plant Physiol.* **100,** 1334–1338.

Lee, K. H., and LaRue, T. A. (1992b). Exogenous ethylene inhibits nodulation of *Pisum sativum* L. cv Sparkle. *Plant Physiol.* **100,** 1759–1763.

Legocki, R. P., and Verma, D. P. S. (1980). Identification of "nodule-specific" host proteins (nodulins) involved in the development of Rhizobium-legume symbiosis. *Cell* **20,** 153–163.

Leigh, J. A., and Coplin, D. L. (1992). Exopolysaccharides in plant-bacterial interactions. *Annu. Rev. Microbiol.* **46,** 307–346.

Lerouge, P., Roche, P., Faucher, C., Maillet, F., Truchet, G., Promé, J. C., and Dénarié, J. (1990). Symbiotic host-specificity of *Rhizobium meliloti* is determined by a sulphated and acylated glucosamine oligosaccharide signal. *Nature (London)* **344,** 781–784.

Lewin, A., Cervantes, E., Chee-Hoong, W., and Broughton, W. J. (1990). *nodSU,* two new *nod* genes of the broad host range *Rhizobium* strain NGR234 encode host-specific nodulation of the tropical tree *Leucaena leucocephala. Mol. Plant-Microbe Interact.* **3,** 317–326.

Lewis-Henderson, W. R., and Djordjevic, M. A. (1991a). *nodT,* a positively-acting cultivar specificity determinant controlling nodulation of *Trifolium subterraneum* by *Rhizobium leguminosarum biovar trifolii. Plant Mol. Biol.* **16,** 515–526.

Lewis-Henderson, W. R., and Djordjevic, M. A. (1991b). A cultivar-specific interaction between *Rhizobium leguminosarum* bv. *trifolii* and subterranean clover is controlled by *nodM,* other bacterial cultivar specificity genes, and a single recessive host gene. *J. Bacteriol.* **173,** 2791–2799.

Leyh, T. S., Vogt, T. F., and Suo, Y. (1992). The DNA sequence of the sulfate activation locus from *Escherichia coli* K-12. *J. Biol. Chem.* **267,** 10,405–10,410.

Libbenga, K. R., Van Iren, F., Bogers, R. J., and Schraag-Lamers, M. F. (1973). The role of hormones and gradients in the initiation of cortex proliferation and nodule formation in *Pisum sativum* L. *Planta* **114,** 29–39.

Ligero, F., Caba, J. M., Lluch, C., and Olivares, J. (1991). Nitrate inhibition of nodulation can be overcome by the ethylene inhibitor aminoethoxyvinylglycine. *Plant Physiol.* **97,** 1221–1225.

Löbler, M., and Hirsch, A. M. (1993). A gene that encodes a prolin-rich nodulin with limited homology to PsENOD12 is expressed in the invasion zone of *Rhizobium meliloti*-induced alfalfa root nodules. *Plant Physiol.* **103,** 21–30.

Lugtenberg, B. J. J., Díaz, C., Smit, G., De Pater, S., and Kijne, J. W. (1991). Roles of lectin in the *Rhizobium*-legume symbioses. *In* "Advances in Molecular Genetics of Plant-Microbe Interactions" (H. Hennecke and D. P. S. Verma, Eds.), Vol. 1, pp. 174–181. Kluwer, Dordrecht, The Netherlands.

Luka, S., Sanjuan, J., Carlson, R. W., and Stacey, G. (1993). *nolMNO* genes of *Bradyrhizobium japonicum* are co-transcribed with *nodYABCSUIJ,* and *nolO* is involved in the synthesis of the lipooligosaccharide nodulation signals. *J. Biol. Chem.* **268,** 27,053–27,059.

MacKenzie, S. L., Lapp, M. S., and Child, J. J. (1979). Fatty acid composition of *Rhizobium* spp. *Can. J. Microbiol.* **25,** 68–74.

Maillet, F., Debellé, F., and Dénarié, J. (1990). Role of the *nodD* and *syrM* genes in the activation of the regulatory gene *nodD3,* and of the common and host-specific *nod* genes of *Rhizobium meliloti. Mol. Microbiol.* **4,** 1975–1984.

Marie, C., Barny, M. A., and Downie, J. A. (1992). *Rhizobium leguminosarum* has two glucosamine synthases, GlmS and NodM, required for nodulation and development of nitrogen-fixing nodules. *Mol. Microbiol.* **6,** 843–851.

Markwei, C. M., and LaRue, T. (1992). Phenotypic characterization of *sym8* and *sym9,* two genes conditioning non-nodulation in *Pisum sativum* 'Sparkle.' *Can. J. Microbiol.* **38,** 548–554.

Martínez-Romero, E., Segovia, L., Mercante, F. M., Franco, A. A., Graham, P., and Pardo, M. A. (1991). *Rhizobium tropici,* a novel species nodulating *Phaseolus vulgaris* L. beans and *Leucaena* sp. trees. *Int. J. Syst. Bacteriol.* **41,** 417–426.

Mathews, A., Carroll, B. J., and Gresshoff, P. M. (1987). Characterization of non-nodulation mutants of soybean [*Glycine max* (L.) Merr.]: *Bradyrhizobium* effects and absence of root hair curling. *J. Plant Physiol.* **131,** 349–361.

Mathews, A., Carroll, B. J., and Gresshoff, P. M. (1989). Development of *Bradyrhizobium* infections in supernodulating and non-nodulating mutants of soybean (*Glycine max* [L.] Merr.). *Protoplasma* **150,** 40–47.

Mathews, A., Carroll, B. J., and Gresshoff, P. M. (1992). Studies on the root control of non-nodulation and plant growth of non-nodulating mutants and a supernodulating mutant of soybean (*Glycine max* (L.) Merr.). *Plant Sci.* **83,** 35–43.

Maxwell, C. A., Hartwig, U. A., Joseph, C. M., and Phillips, D. A. (1989). A chalcone and 2 related flavonoids released from alfalfa roots induce *nod* genes of *Rhizobium meliloti*. *Plant Physiol.* **91,** 842–847.

McClure, J. W. (1975). Physiology and functions of flavonoids. *In* "The Flavonoids" (J. B. Harborne, T. J. Mabry, and H. Mabry, Eds.), pp. 970–1055. Academic Press, New York.

McIver, J., Djordjevic, M. A., Weinman, J. J., Bender, G. L., and Rolfe, B. G. (1989). Extension of host range of *Rhizobium leguminosarum* bv. *trifolii* caused by point mutations in *nodD* that result in alterations in regulatory function and recognition of inducer molecules. *Mol. Plant-Microbe Interact.* **2,** 97–106.

McIver, J., Djordjevic, M. A., Weinman, J. J., and Rolfe, B. G. (1993). Influence of *Rhizobium leguminosarum* biovar *trifolii* host specific nodulation genes on the ontogeny of clover nodulation. *Protoplasma* **172,** 166–179.

McKay, I. A., and Djordjevic, M. A. (1993). Production and excretion of Nod metabolites by *Rhizobium leguminosarum* bv. *trifolii* are disrupted by the same environmental factors that reduce nodulation in the field. *Appl. Env. Microbiol.* **59,** 3385–3392.

McKhann, H. I., and Hirsch, A. M. (1994). Isolation of chalcone synthase and chalcone isomerase cDNAs from alfalfa (*Medicago sativa* L.): Highest transcript levels occur in young roots and root tips. *Plant Mol. Biol.* **24,** 767–777.

Meinhardt, L. W., Krishnan, H. B., Balatti, P. A., and Pueppke, S. G. (1993). Molecular cloning and characterization of a sym plasmid locus that regulates cultivar-specific nodulation by soybean by *Rhizobium fredii* USDA257. *Mol. Microbiol.* **9,** 17–29.

Mergaert, P., Van Montagu, M., Promé, J. C., and Holsters, M. (1993). Three unusual modifications, a D-arabinosyl, an N-methyl, and a carbamoyl group, are present on the Nod factors of *Azorhizobium caulinodans* strain ORS571. *Proc. Natl. Acad. Sci. USA* **90,** 1551–1555.

Messens, E., Geelen, D., Van Montagu, M., and Holsters, M. (1991). 7,4′-Dihydroxyflavone is a major *Azorhizobium nod* gene-inducing factor present in *Sesbania rostrata* seedling exudate. *Mol. Plant-Microbe Interact.* **4,** 262–267.

Miao, G. H., Hong, Z., and Verma, D. P. S. (1993). Two functional soybean genes encoding $p34^{cdc2}$ protein kinases are regulated by different plant developmental pathways. *Proc. Natl. Acad. Sci. USA* **90,** 943–947.

Michiels, J., De Wilde, P., and Vanderleyden, J. (1993). Sequence of the *Rhizobium leguminosarum* biovar *phaseoli syrM* gene. *Nucleic Acids Res.* **21,** 3893.

Moerman, M., Nap, J. P., Govers, F., Schilperoort, R., Van Kammen, A., and Bisseling, T. (1987). *Rhizobium nod* genes are involved in the induction of two early nodulin genes in *Vicia sativa* root nodules. *Plant Mol. Biol.* **9,** 171–179.

Mulligan, J. T., and Long, S. R. (1985). Induction of *Rhizobium meliloti* nodC expression by plant exudate requires *nodD*. *Proc. Natl. Acad. Sci. USA* **82,** 6609–6613.

Mulligan, J. T., and Long, S. R. (1989). A family of activator genes regulates expression of *Rhizobium meliloti* nodulation genes. *Genetics* **122,** 7–18.

Munns, D. N. (1977). Mineral nutrition and the legume symbiosis. *In* "A Treatise on Dinitrogen Fixation" (R. W. F. Hardy, and A. H. Gibson, Eds.), Sect. IV, pp. 277–367, Wiley, New York.

Nap, J. P., and Bisseling, T. (1990). Nodulin function and nodulin gene regulation in root nodule development. *In* "Molecular Biology of Symbiotic Nitrogen Fixation" (P. M. Gresshoff, Ed.), pp. 181–229. CRC Press, Boca Raton, FL.

Newcomb, W. (1981). Nodule morphogenesis and differentiation. *In* "International Review of Cytology, Suppl. 13, Biology of the Rhizobiaceae" (K. L. Giles and A. G. Atherly, Eds.), pp. 247–298. Academic Press, New York.

Ndoye, I., De Billy, F., Vasse, J., Dreyfus, B., and Truchet, G. (1994). Root nodulation of *Sesbania rostrata*. *J. Bacteriol.* **176,** 1060–1068.

Nicolaou, K. C., Bockovich, N. J., Carcanague, D. R., Hummel, C. W., and Even, L. F. (1992). Total synthesis of the NodRm-IV factors, the *Rhizobium* nodulation signals. *J. Am. Chem. Soc.* **114,** 8701–8702.

Nieuwkoop, A. J., Banfalvi, Z., Deshmane, N., Gerhold, D., Schell, M. G., Sirotkin, K. M., and Stacey, G. (1987). A locus encoding host range is linked to the common nodulation genes of *Bradyrhizobium japonicum*. *J. Bacteriol.* **169,** 2631–2638.

Nilsson, O., Crozier, A., Schmülling, T., Sandberg, G., and Olsson, O. (1993). Indole-3-acetic acid homeostasis in transgenic tobacco plants expressing the *Agrobacterium rhizogenes rolB* gene. *Plant J.* **3,** 681–689.

Noel, K. D. (1992). Rhizobial polysaccharides required in symbiosis with legumes. *In* "Molecular Signals in Plant-Microbe Communications" (D. P. S. Verma, ed.), pp. 341–357. CRC Press, Boca Raton, FL.

Nutman, P. S. (1952). Studies on the physiology of nodule formation. III. Experiments on the excision of root-tip and nodules. *Ann. Bot.* **16,** 81–102.

Ogawa, J., Brierley, H. L., and Long, S. R. (1991). Analysis of *Rhizobium meliloti* nodulation mutant WL131: Novel insertion sequence IS*Rm3* in *nodG* and altered *nodH* protein product. *J. Bacteriol.* **173,** 3060–3065.

Olsson, J. E., Nakao, P., Bohlool, B., and Gresshoff, P. M. (1989). Lack of systemic suppression of nodulation in split root systems of supernodulating soybean (*Glycine max* [L.] Merr.) mutants. *Plant Physiol.* **90,** 1347–1352.

Orgambide, G., Hollingsworth, R., and Dazzo, F. (1993). Rhizobium trifolii produces a diglycosyl diacylglyceride signal molecule which can elicit host-specific responses on white clover roots at subnanomolar concentrations. *In* "New Horizons in Nitrogen Fixation" (R. Palacios, J. Mora, and W. E. Newton, Eds.), p. 248. Kluwer, Dordrecht, The Netherlands.

Pankhurst, C. E., and Jones, W. T. (1979). Effectiveness of *Lotus* root nodules. III. Effect of combined nitrogen on nodule effectiveness and flavolan synthesis in plant roots. *J. Exp. Bot.* **30,** 1109–1118.

Park, S. J., and Buttery, B. R. (1988). Nodulation mutants of white bean (*Phaseolus vulgaris* L.) induced by ethyl-methane sulphonate. *Can. J. Plant Sci.* **68,** 199–202.

Perret, X., Broughton, W. J., and Brenner, S. (1991). Canonical ordered cosmid library of the symbiotic plasmid of *Rhizobium species* NGR234. *Proc. Natl. Acad. Sci. USA* **88,** 1923–1927.

Peters, N. K., and Crist-Estes, D. K. (1989). Nodule formation is stimulated by the ethylene inhibitor aminoethoxyvinylglycine. *Plant Physiol.* **91,** 690–693.

Peters, N. K., and Long, S. R. (1988). Alfalfa root exudates and compounds which promote or inhibit induction of *Rhizobium meliloti* nodulation genes. *Plant Physiol.* **88,** 396–400.

Peters, N. K., and Verma, D. P. S. (1990). Phenolic compounds as regulators of gene expression in plant-microbe interactions. *Mol. Plant-Microbe Interact.* **3,** 4–8.

Peters, N. K., Frost, J. W. and Long, S. R. (1986). A plant flavone, luteolin, induces expression of *Rhizobium meliloti* nodulation genes. *Science* **233,** 977–980.

Peterson, M. A., and Barnes, D. K. (1981). Inheritance of ineffective nodulation and non-nodulation traits in alfalfa. *Crop Sci.* **21,** 611–616.

Philip-Hollingsworth, S., Hollingsworth, R. I., and Dazzo, F. B. (1991). *N*-acetylglutamic acid: An extracellular *nod* signal of *Rhizobium trifolii* ANU843 that induces root hair branching and nodule-like primordia in white clover roots. *J. Biol. Chem.* **266,** 16,854–16,858.

Phillips, D. A. (1971a). A cotyledonary inhibitor of root nodulation in *Pisum sativum*. *Physiol. Plant.* **25,** 482–487.

Phillips, D. A. (1971b). Abscisic acid inhibition of root nodule initation in *Pisum sativum*. *Planta* **100,** 181–190.

Phillips, D. A. (1992). Flavonoids: plant signals to soil microbes. *In* "Phenolic Metabolism in Plants" (Stafford, H. A. and Ibrahim R. K., eds.), pp. 201–231. Plenum, New York.

Phillips, D. A., Joseph, C. M., and Maxwell, C. A. (1992). Trigonelline and stachydrine released from alfalfa seeds activate NodD2 protein in *Rhizobium meliloti*. *Plant Physiol.* **99,** 1526–1531.

Pichon, M., Journet, E. P., Dedieu, A., De Billy, F., Truchet, G., and Barker, D. G. (1992). *Rhizobium meliloti* elicits transient expression of the early nodulin gene *ENOD*12 in the differentiating root epidermis of transgenic alfalfa. *Plant Cell* **4,** 1199–1211.

Pichon, M., Journet, E. P., Dedieu, A., De Billy, F., Huguet, T., Truchet, G., and Barker, D. G. (1993). Expression of the *Medicago truncatula Enod12* gene in response to *R. meliloti* Nod factors and during spontaneous nodulation in transgenic alfalfa. *In* "New Horizons in Nitrogen Fixation" (R. Palacios, J. Mora, and W. E. Newton, Eds.), pp. 285–290. Kluwer, Dordrecht, The Netherlands.

Pierce, M., and Bauer, W. D. (1983). A rapid regulatory response governing nodulation in soybean. *Plant Physiol.* **73,** 286–290.

Poupot, R., Martinez-Romero, E., and Promé, J. C. (1993). Nodulation factors from Rhizobium tropici are sulfated or nonsulfated chitopentasaccharides containing an *N*-methyl-*N*-acylglucosaminyl terminus. *Biochemistry* **32,** 10,430–10,435.

Price, N. P. J., Relic, B., Talmont, E., Lewin, A., Promé, D., Pueppke, S. G., Maillet, F., Dénarié, J., Promé, J. C., and Broughton, W. J. (1992). Broad-host-range *Rhizobium* species strain NGR234 secretes a family of carbamoylated, and fucosylated, nodulation signals that are *O*-acetylated or sulphated. *Mol. Microbiol.* **6,** 3575–3584.

Prinsen, E., Schmidt, J., John, M., Wieneke, U., De Greef, J., Schell, J., Chauvaux, N., and Van Onckelen, H. (1991). Stimulation of indole-3-acetic acid production in *Rhizobium* by flavonoids. *FEBS Lett.* **282,** 53–55.

Putnoky, P., and Kondorosi, A. (1986). Two gene clusters of *Rhizobium meliloti* code for early essential nodulation functions and a third influences nodulation efficiency. *J. Bacteriol.* **167,** 881–887.

Radley, M. (1961). Gibberellic-like substances in plants. *Nature* **191,** 684–685.

Rae, A. L., Bonfante-Fasolo, P., and Brewin, N. J. (1992). Structure and growth of infection threads in the legume symbiosis with *Rhizobium leguminosarum*. *Plant J.* **2,** 385–395.

Räsänen, L. A., Heikkilä-Kallio, U., Suominen, L., Lipsanen, P., and Lindström, K. (1991). Expression of Rhizobium galegae common *nod* clones in various backgrounds. *Mol. Plant-Microbe Interact.* **4,** 535–544.

Ratet, P., Esnault, R., Zuanazzi, J., Sallaud, C., Husson, P., Coronado, C., Dusha, I., Savouré, A., Dudits, D., Schultze, M., Bauer, P., Crespi, M., Jurkevitch, E., Kondorosi, E., and Kondorosi, A. (1994). Molecular control of the development of nitrogen-fixing symbiosis. *Nova Acta Leopoldina,* in press.

Rawlings, M., and Cronan, J. E. (1992). The gene encoding *Escherichia coli* acyl carrier protein lies within a cluster of fatty acid biosynthesis genes. *J. Biol. Chem.* **267,** 5751–5754.

Recourt, K., Van Brussel, A. A. N., Driessen, A. J. M. and Lugtenberg, B. J. J. (1989). Accumulation of a *nod* gene inducer, the flavonoid naringenin, in the cytoplasmic membrane of *Rhizobium leguminosarum* biovar *viciae* is caused by the pH-dependent hydrophobicity of naringenin. *J. Bacteriol.* **171,** 4370–4377.

Recourt, K., Schripsema, J., Kijne, J. W., Van Brussel, A. A. N., and Lugtenberg,

B. J. J. (1991). Inoculation of *Vicia sativa* subsp. *nigra* roots with *Rhizobium leguminosarum* biovar *viciae* results in release of *nod* gene activating flavanones and chalcones. *Plant Mol. Biol.* **16,** 841–852.

Recourt, K., Verkerke, M., Schripsema, J., Van Brussel, A. A. N., Lugtenberg, B. J. J., and Kijne, J. W. (1992a). Major flavonoids in uninoculated and inoculated roots of *Vicia sativa* subsp. *nigra* are four conjugates of the nodulation gene-inhibitor kaempferol. *Plant Mol. Biol.* **18,** 505–513.

Recourt, K., Van Tunen, A. J., Mur, L. A., Van Brussel, A. A. N., Lugtenberg, B. J. J., and Kijne, J. W. (1992b). Activation of flavonoid biosynthesis in roots of *Vicia sativa* subsp. *nigra* plants by inoculation with *Rhizobium leguminosarum* biovar *viciae*. *Plant Mol. Biol.* **19,** 411–420.

Redmond, J. W., Batley, M., Djordjevic, M. A., Innes, R. W., Kuempel, P. L., and Rolfe, B. G. (1986). Flavones induce expression of nodulation genes in *Rhizobium*. *Nature* **323,** 632–634.

Relic, B., Talmont, F., Kopcinska, J., Golinowski, W., Promé, J. C., and Broughton, W. J. (1993). Biological activity of *Rhizobium* sp. NGR234 Nod-factors on *Macroptilium atropurpureum*. *Mol. Plant-Microbe Interact.* **6,** 764–774.

Relic, B., Perret, X., Estrada-García, M. T., Kopcinska, J., Golinowski, W., Krishnan, H. B., Pueppke, S. G., and Broughton, W. J. (1994). Nod factors of *Rhizobium* are a key to the legume door. *Mol. Microbiol.* **13,** 171–178.

Roche, P., Lerouge, P., Ponthus, C., and Prome, J. C. (1991a). Structural determination of bacterial nodulation factors involved in the *Rhizobium meliloti*-alfalfa symbiosis. *J. Biol. Chem.* **266,** 10,933–10,940.

Roche, P., Debellé, F., Maillet, F., Lerouge, P., Faucher, C., Truchet, G., Denarie, J., and Prome, J. C. (1991b). Molecular basis of symbiotic host specificity in *Rhizobium meliloti: nodH* and *nodPQ* genes encode the sulfation of lipooligosaccharide signals. *Cell* **67,** 1131–1143.

Rodriguez-Quiñones, F., Banfalvi, Z., Murphy, P., and Kondorosi, A. (1987). Interspecies homology of nodulation genes in *Rhizobium*. *Plant Mol. Biol.* **8,** 61–75.

Röhrig, H., Schmidt, J., Wieneke, U., Kondorosi, E., Barlier, I., Schell, J., and John, M. (1994). Biosynthesis of lipooligosaccharide nodulation factors: *Rhizobium* NodA protein is involved in *N*-acylation of the chitooligosaccharide backbone. *Proc. Natl. Acad. Sci. USA* **91,** 3122–3126.

Rolfe, B. G. (1988). Flavones and isoflavones as inducing substances of legume nodulation. *BioFactors* **1,** 3–10.

Rolfe, G. B., and Gresshoff, P. M. (1988). Genetic analysis of legume nodule initiation. *Annu. Rev. Plant Physiol. Plant Mol. Biol.* **39,** 297–319.

Rommeswinkel, M., Karwatzki, B., Beerhues, L., and Wiermann, R. (1992). Immunofluorescence localization of chalcone synthase in roots of *Pisum sativum* L. and *Phaseolus vulgaris* L. and comparable immunochemical analysis of chalcone synthase from pea leaves. *Protoplasma* **166,** 115–121.

Rossbach, S., Gloudemans, T., Bisseling, T., Studer, D., Kaluza, B., Ebeling, S., and Hennecke, H. (1989). Genetic and physiologic characterization of a *Bradyrhizobium japonicum* mutant defective in early bacteroid development. *Mol. Plant-Microbe Interact.* **2,** 233–240.

Rossen, L., Johnston, A. W. B., and Downie, J. A. (1984). DNA sequence of the *Rhizobium leguminosarum* nodulation genes *nodAB* and *C* required for root hair curling. *Nucleic Acids Res.* **12,** 9497–9508.

Rossen, L., Shearman, C. A., Johnston, A. W. B., and Downie, J. A. (1985). The *nodD* gene of *Rhizobium leguminosarum* is autoregulatory and in the presence of plant exudate induces the *nodA,B,C* genes. *EMBO J.* **4,** 3369–3373.

Rostas, K., Kondorosi, E., Horvath, B., Simoncsits, A., and Kondorosi, A. (1986). Conservation of extended promoter regions of nodulation genes in *Rhizobium*. *Proc. Natl. Acad. Sci. USA* **83,** 1757–1761.

Rushing, B. G., Yelton, M. M., and Long, S. R. (1991). Genetic and physical analysis of the *nodD3* region of *Rhizobium meliloti*. *Nucleic Acids Res.* **19,** 921–927.

Sadowsky, M. J., Olson, E. R., Foster, V. E., Kosslak, R. M., and Verma, D. P. S. (1988). Two host-inducible genes of *Rhizobium fredii* and characterization of the inducing compound. *J. Bacteriol.* **170,** 171–178.

Sadowsky, M. J., Cregan, P. B., Göttfert, M., Sharma, A., Gerhold, D., Rodriguez-Quiñones, F., Keyser, H. H., Hennecke, H., and Stacey, G. (1991). The *Bradyrhizobium japonicum nolA* Gene and its involvement in the genotype-specific nodulation of soybeans. *Proc. Natl. Acad. Sci. USA* **88,** 637–641.

Saier, M. H., Tam, R., Reizer, A., and Reizer, J. (1994). Two novel families of bacterial membrane proteins concerned with nodulation, cell division and transport. *Mol. Microbiol.* **11,** 841–847.

Sanjuan, J., Carlson, R. W., Spaink, H. P., Bhat, U. R., Barbour, W. M., Glushka, J., and Stacey, G. (1992). A 2-*O*-methylfucose moiety is present in the lipooligosaccharide nodulation signal of *Bradyrhizobium japonicum*. *Proc. Natl. Acad. Sci. USA* **89,** 8789–8793.

Sanjuan, J., Grob, P., Göttfert, M., Hennecke, H., and Stacey, G. (1994). NodW is essential for full expression of the common nodulation genes in *Bradyrhizobium japonicum*. *Mol. Plant-Microbe Interact.* **7,** 364–369.

Savouré, A., Magyar, Z., Pierre, M., Brown, S., Schultze, M., Dudits, D., Kondorosi, A., and Kondorosi, E. (1994). Activation of the cell cycle machinery and the isoflavonoid biosynthesis pathway by active *Rhizobium meliloti* Nod signal molecules in *Medicago* microcallus suspensions. *EMBO J.* **13,** 1093–1102.

Schell, M. A. (1993). Molecular biology of the LysR family of transcriptional regulators. *Annu. Rev. Microbiol.* **47,** 597–626.

Schell, M. A., and Wender, P. (1986). Identification of the *nahR* gene product and nucleotide sequences required for its activation of the sal operon. *J. Bacteriol.* **166,** 9–14.

Scheres, B., Van de Wiel, C., Zalensky, A., Horvath, B., Spaink, H., Van Eck, H., Zwarrtkruis, F., Wolters, A. M., Gloudemans, T., Van Kammen, A., and Bisseling, T. (1990a). The ENOD12 gene product is involved in the infection process during the pea-*Rhizobium* interaction. *Cell* **60,** 281–294.

Scheres, B., Van Engelen, F., Van der Knaap, E., Van de Wiel, C., Van Kammen, A., and Bisseling, T. (1990b). Sequential induction of nodulin gene expression in the developing pea nodule. *Plant Cell* **2,** 687–700.

Scheres, B., McKhann, H. I., Zalensky, A., Löbler, M., Bisseling, T., and Hirsch, A. M. (1992). The PsENOD12 gene is expressed at two different sites in Afghanistan pea pseudonodules induced by auxin transport inhibitors. *Plant Physiol.* **100,** 1649–1655.

Scheu, A. K., Economou, A., Hong, G. F., Ghelani, S., Johnston, A. W. B., and Downie, J. A. (1992). Secretion of the *Rhizobium leguminosarum* nodulation protein NodO by haemolysin-type systems. *Mol. Microbiol.* **6,** 231–238.

Schlaman, H. R. M. (1992). Regulation of nodulation gene expression in *Rhizobium leguminosarum* biovar *viciae*. Ph.D. thesis, University of Leiden, Netherlands.

Schlaman, H. R. M., Spaink, H. P., Okker, R. J. H., and Lugtenberg, B. J. J. (1989). Subcellular localization of the *nodD* gene product in *Rhizobium leguminosarum*. *J. Bacteriol.* **171,** 4686–4693.

Schlaman, H. R. M., Okker, R. J. H., and Lugtenberg, B. J. J. (1990). Subcellular localization of the *Rhizobium leguminosarum nodI* gene product. *J. Bacteriol.* **172,** 5486–5489.

Schlaman, H. R. M., Horvath, B., Vijgenboom, E., Okker, R. J. H., and Lugtenberg, B. J. J. (1991). Suppression of nodulation gene expression in bacteroids of *Rhizobium leguminosarum* biovar *viciae*. *J. Bacteriol.* **173,** 4277–4287.

Schlaman, H. R. M., Lugtenberg, B. J. J., and Okker, R. J. H. (1992a). The NodD protein does not bind to the promoters of inducible nodulation genes in extracts of bacteroids of *Rhizobium leguminosarum* biovar *viciae*. *J. Bacteriol.* **174,** 6109–6116.

Schlaman, H. R. M., Okker, R. J. H., and Lugtenberg, B. J. J. (1992b). Regulation of nodulation gene expression by NodD in rhizobia. *J. Bacteriol.* **174,** 5177–5182.

Schmidt, J., John, M., Wieneke, U., Krüssmann, H. D., and Schell, J. (1986). Expression of the nodulation gene *nodA* in *Rhizobium meliloti* and localization of the gene product in the cytosol. *Proc. Natl. Acad. Sci. USA* **83,** 9581–9585.

Schmidt, J., Wingender, R., John, M., Wieneke, U., and Schell, J. (1988). *Rhizobium meliloti nodA* and *nodB* genes are involved in generating compounds that stimulate mitosis of plant cells. *Proc. Natl. Acad. Sci. USA* **85,** 8578–8582.

Schmidt, J., John, M., Wieneke, U., Stacey, G., Röhrig, H., and Schell, J. (1991). Studies on the function of *Rhizobium meliloti* nodulation genes. *In* "Advances in Molecular Genetics of Plant—Microbe Interactions" (H. Hennecke and D. P. S. Verma, Eds.), Vol. 1, pp. 150–155. Kluwer, Dordrecht, The Netherlands.

Schmidt, J., Röhrig, H., John, M., Wieneke, U., Stacey, G., Koncz, C., and Schell, J. (1993). Alteration of plant growth and development by *Rhizobium nodA* and *nodB* genes involved in the synthesis of oligosaccharide signal molecules. *Plant J.* **4,** 651–658.

Schmidt, P. E., Broughton, W. J., and Werner, D. (1994). Nod factors of *Bradyrhizobium japonicum* and *Rhizobium* sp. NGR234 induce flavonoid accumulation in soybean root exudate. *Mol. Plant-Microbe Interact.* **7,** 384–390.

Schmülling, T., Schell, J., and Spena, A. (1988). Single genes from *Agrobacterium rhizogenes* influence plant development. *EMBO J.* **7,** 2621–2629.

Schofield, P. R., and Watson, J. M. (1986). DNA sequence of *Rhizobium trifolii* nodulation genes reveals a reiterated and potentially regulatory sequence preceding *nodABC* and nodEF. Nucleic Acids Res. **14,** 2891–2903.

Schowalter, A. M. (1993). Structure and function of plant cell wall proteins. *Plant Cell* **5,** 9–23.

Schultze, M., Quiclet-Sire, B., Kondorosi, É., Virelizier, H., Glushka, J. N., Endre, G., Géro, S. D., and Kondorosi, Á. (1992). *Rhizobium meliloti* produces a family of sulfated lipooligosaccharides exhibiting different degrees of plant host specificity. *Proc. Natl. Acad. Sci. USA* **89,** 192–196.

Schultze, M., Kondorosi, E., Kondorosi, A., Staehelin, C., Mellor, R. B., and Boller, T. (1993). The sulfate group on the reducing end protects Nod signals of *Rhizobium meliloti* against hydrolysis by *Medicago* chitinases. *In* "New Horizons in Nitrogen Fixation" (R. Palacios, J. Mora, and W. E. Newton, Eds.), pp. 159–164. Kluwer, Dordrecht, The Netherlands.

Schwedock, J., and Long, S. R. (1989). Nucleotide sequence and protein products of two new nodulation genes of *Rhizobium meliloti, nodP* and *nodQ*. *Molec. Plant-Microbe Interact.* **2,** 181–194.

Schwedock, J., and Long, S. R. (1990). ATP sulphurylase activity of the *nodP* and *nodQ* gene products of *Rhizobium meliloti*. *Nature* (*London*) **348,** 644–647.

Schwedock, J. S., and Long, S. R. (1992). *Rhizobium meliloti* genes involved in sulfate activation: The two copies of *nodPQ* and a new locus, *saa*. *Genetics* **132,** 899–909.

Schwedock, J., and Long, S. R. (1994). An open reading frame downstream of *Rhizobium meliloti* nodQ1 shows nucleotide sequence similarity to an *Agrobacterium tumefaciens* insertion sequence. *Mol. Plant-Microbe Interact.* **7,** 151–153.

Scott, K. F. (1986). Conserved nodulation genes from the non-legume symbiont *Bradyrhizobium* sp. (*Parasponia*). *Nucleic Acids Res.* **14,** 2905–2919.

Scott, D. B. (1994). Mutational and structural analysis of *Rhizobium loti* nodulation genes. Submitted for publication (see EMBL Access No L06241).

Segovia, L., Young, J. P. W., and Martínez-Romero, E. (1993). Reclassification of American *Rhizobium leguminosarum* biovar *phaseoli* type I strains as *Rhizobium etli sp. nov. Int. J. Syst. Bact.* **43,** 374–377.

Sharma, S. B., and Signer, E. R. (1990). Temporal and spatial regulation of the symbiotic genes of *Rhizobium meliloti* in planta revealed by transposon Tn*5*-*gusA*. *Genes Dev.* **4,** 344–356.

Shearman, C. A., Rossen, L., Johnston, A. W. B., and Downie, J. A. (1986). The *Rhizobium leguminosarum* nodulation gene *nodF* encodes a polypeptide similar to acyl-carrier protein and is regulated by *nodD* plus a factor in pea root exudate. *EMBO J.* **5,** 647–652.

Shen, W. H., Davioud, E., David, C., Barbier-Brygoo, H., Tempé, J., and Guern, J. (1989). High sensitivity to auxin is a common feature of hairy root. *Plant Physiol.* **94,** 554–560.

Sherwood, J. E., Truchet, G. L., and Dazzo, F. B. (1984). Effect of nitrate supply on the in-vivo synthesis and distribution of trifoliin A, a *Rhizobium trifolii*-binding lectin, in *Trifolium repens* seedlings. *Planta* **162,** 540–547.

Shibuya, N., Kaku, H., Kutchitsu, K., and Maliarik, M. J. (1993). Identification of a novel high-affinity binding site for *N*-acetylchitooligosaccharide elicitor in the membrane fraction from suspension-cultured rice cells. *FEBS Lett.* **329,** 75–78.

Slabas, A. R., Chase, D., Nishida, I., Murata, N., Sidebottom, C., Safford, R., Sheldon, P. S., Kekwick, R. G. O., Hardie, D. G., and Mackintosh, R. W. (1992). Molecular cloning of higher-plant 3-oxoacyl-(acyl carrier protein) reductase—Sequence identities with the *nodG*-gene product of the nitrogen-fixing soil bacterium *Rhizobium meliloti*. *Biochem. J.* **283,** 321–326.

Smit, G., Puvanesarajah, V., Carlson, R. W., Barbour, W. M., and Stacey, G. (1992). *Bradyrhizobium japonicum nodD1* can be specifically induced by soybean flavonoids that do not induce the *nodYABCSUIJ* operon. *J. Biol. Chem.* **267,** 310–318.

Smit, G., Van Brussel, A. A. N., and Kijne, J. W. (1993). Inactivation of a root factor by infective *Rhizobium:* A molecular key to autoregulation of nodulation in *Pisum sativum*. *In* "New Horizons in Nitrogen Fixation" (R. Palacios, J. Mora, and W. E. Newton, Eds.), p. 371. Kluwer, Dordrecht, The Netherlands.

Sousa, C., Folch, J. L., Boloix, P., Megías, M., Nava, N., and Quinto, C. (1993). A *Rhizobium tropici* DNA region carrying the amino-terminal half of a *nodD* gene and a *nod*-box-like sequence confers host-range extension. *Mol. Microbiol.* **9,** 1157–1168.

Spaink, H. P. (1992). Rhizobial lipooligosaccharides: answers and questions. *Plant Mol. Biol.* **20,** 977–986.

Spaink, H. P., Wijffelman, C. A., Pees, E., Okker, R. J. H., and Lugtenberg, B. J. J. (1987a). *Rhizobium* nodulation gene *nodD* as a determinant of host specificity. *Nature* **328,** 337–340.

Spaink, H. P., Okker, R. J. H., Wijffelman, C. A., Pees, E., and Lugtenberg, B. J. J. (1987b). Promoters in the nodulation region of the *Rhizobium leguminosarum* Sym plasmid pRL1JI. *Plant Mol. Biol.* **9,** 27–39.

Spaink, H. P., Weinman, J., Djordjevic, M. A., Wijffelman, C. A., Okker, R. J. H., and Lugtenberg, B. J. J. (1989a). Genetic analysis and cellular localization of the *Rhizobium* host specificity-determining NodE protein. *EMBO J.* **8,** 2811–2818.

Spaink, H. P., Okker, R. J. H., Wijffelman, C. A., Tak, T., Goosen-De Roo, L., Pees, E., Van Brussel, A. A. N., and Lugtenberg, B. J. J. (1989b). Symbiotic properties of rhizobia containing a flavonoid-independent hybrid *nodD* product. *J. Bacteriol.* **171,** 4045–4053.

Spaink, H. P., Wijffelman, C. A., Okker, R. J. H., and Lugtenberg, B. J. J. (1989c). Localization of functional regions of the *Rhizobium nodD* product using hybrid *nodD* genes. *Plant Mol. Biol.* **12,** 59–73.

Spaink, H. P., Sheeley, D. M., Van Brussel, A. A. N., Glushka, J., York, W. S., Tak, T., Geiger, O., Kennedy, E. P., Reinhold, V. N., and Lugtenberg, B. J. J. (1991). A novel highly unsaturated fatty acid moiety of lipooligosaccharide signals determines host specificity of *Rhizobium*. *Nature (London)* **354,** 125–130.

Spaink, H. P., Aarts, A., Stacey, G., Bloemberg, G. V., Lugtenberg, B. J. J., and Kennedy, E. P. (1992). Detection and separation of *Rhizobium* and *Bradyrhizobium* Nod metabolites using thin-layer chromatography. *Mol. Plant-Microbe Interact.* **5,** 72–80.

Spaink, H. P., Aarts, A., Bloemberg, G. V., Folch, J., Geiger, O., Schlaman, H. R. M., Thomas-Oates, J. E., Van Brussel, A. A. N., Van de Sande, K., Van Spronsen, P., Wijfjes, A. H. M., and Lugtenberg, B. J. J. (1993a). Rhizobial lipooligosaccharide signals: Their biosynthesis and their role in the plant. *In* "Advances in Molecular Genetics of Plant-Microbe Interactions" (E. W. Nester and D. P. S. Verma, Eds.), Vol. 2, pp. 151–162. Kluwer, Dordrecht, The Netherlands.

Spaink, H. P., Wijfjes, A. H. M., Geiger, O., Bloemberg, G. V., Ritsema, T., and Lugtenberg, B. J. J. (1993b). The function of the rhizobial *nodABC* and *nodFEL* operons in the biosynthesis of lipooligosaccharides. *In* "New Horizons in Nitrogen Fixation" (R. Palacios, J. Mora, and W. E. Newton, Eds.), pp. 165–170. Kluwer, Dordrecht, The Netherlands.

Spaink, H. P., Wijfjes, A. H. M., Van Vliet, T. B., Kijne, J. W., and Lugtenberg, B. J. J. (1993c). Rhizobial lipooligosaccharide signals and their role in plant morphogenesis; are analogous lipophilic chitin derivatives produced by the plant? *Austr. J. Plant Physiol.* **20,** 381–392.

Sprent, J. I. (1989). Which steps are essential for the formation of functional legume nodules? *New Phytol.* **111,** 129–153.

Stacey, G., Luka, S., Sanjuan, J., Banfalvi, Z., Nieuwkoop, A. J., Chun, J. Y., Forsberg, L. S., and Carlson, R. W. (1994). *nodZ,* a unique host-specific nodulation gene, is involved in the fucosylation of the lipooligosaccharide nodulation signal of *Bradyrhizobium japonicum. J. Bacteriol.* **176,** 620–633.

Staehelin, C., Müller, J., Mellor, R. B., Wiemken, A., and Boller, T. (1992). Chitinase and peroxidase in effective (fix^+) and ineffective (fix^-) soybean nodules. *Planta* **187,** 295–300.

Staehelin, C., Schultze, M., Kondorosi, E., Mellor, R. B., Boller, T., and Kondorosi, A. (1994a). Structural modifications in *Rhizobium meliloti* Nod factors influence their stability against hydrolysis by root chitinases. *Plant J.,* **5,** 319–330.

Staehelin, C., Granado, J., Müller, J., Wiemken, A., Mellor, R. B., Felix, G., Regenass, M., Broughton, W. J., and Boller, T. (1994b). Perception of *Rhizobium* nodulation factors by tomato cells and inactivation by root chitinases. *Proc. Natl. Acad. Sci. USA* **91,** 2196–2200.

Stanley, J., and Cervantes, E. (1991). Biology and genetics of the broad host range *Rhizobium* sp. NGR234. *J. Appl. Bacteriol.* **70,** 9–19.

Stokkermans, T. J. W. and Peters, N. K. (1994). *Bradyrhizobium elkanii* lipooligosaccharide signals induce complete nodule structures on *Glycine soja* Siebold et Zucc. *Planta* **193,** 413–420.

Stokkermans, T. J. W., Sanjuan, J., Ruan, X., Stacey, G., and Peters, N. K. (1992). *Bradyrhizobium japonicum* rhizobitoxine mutants with altered host-range on *Rj4* soybeans. *Mol. Plant-Microbe Interact.* **5,** 504–512.

Streeter, J. (1988). Inhibition of legume nodule formation and N_2 fixation by nitrate. *CRC Crit. Rev. Plant Sci.* **7,** 1–23.

Strittmatter, G., Chia, T. F., Trinh, T. H., Katagiri, F., Kuhlemeier, C., and Chua, N. H. (1989). Characterization of nodule specific cDNA clones from *Sesbania rostrata* and expression of the corresponding genes during the initial stages of stem nodules and root nodules formation. *Mol. Plant-Microbe Interact.* **2,** 122–127.

Surin, B. P., and Downie, J. A. (1988). Characterization of the *Rhizobium leguminosarum* genes *nodLMN* involved in efficient host-specific nodulation. *Mol. Microbiol.* **2,** 173–183.

Surin, B. P., Watson, J. M., Hamilton, W. D. O., Economou, A., and Downie, J. A. (1990).

Molecular characterization of the nodulation gene, *nodT*, from two biovars of *Rhizobium leguminosarum*. *Mol. Microbiol.* **4,** 245–252.

Sutton, M. J., Lea, E. J. A., Crank, S., Rivilla, R., Economou, A., Ghelani, S., Johnston, A. W. B., and Downie, J. A. (1993). NodO: A nodulation protein that forms pores in membranes. *In* "Advances in Molecular Genetics of Plant-Microbe Interactions" (E. W. Nester and D. P. S. Verma, Eds.), Vol. 2, pp. 163–167. Kluwer, Dordrecht, The Netherlands.

Swanson, J. A., Mulligan, J. T., and Long, S. R. (1993). Regulation of *syrM* and *nodD3* in *Rhizobium meliloti*. *Genetics* **134,** 435–444.

Szczyglowski, K., and Legocki, A. B. (1990). Isolation and nucleotide sequence of cDNA clone encoding nodule-specific (hydroxy)proline-rich protein LENOD2 from yellow lupin. *Plant Mol. Biol.* **19,** 361–363.

Tiller, S. A., Parry, A. D., and Edwards, R. (1994). Changes in the accumulation of flavonoid and isoflavonoid conjugates associated with plant age and nodulation in alfalfa (*Medicago sativa*). *Physiol. Plant.* **91,** 27–36.

Tjepkema, J. D., and Yocum, C. S. (1974). Measurement of oxygen partial pressure with soybean nodules by oxygen microelectrodes. *Planta* **115,** 351–360.

Török, I., Kondorosi, E., Stepkowski, T., Pósfai, J., and Kondorosi, A. (1984). Nucleotide sequence of *Rhizobium meliloti* nodulation genes. *Nucleic Acids Res.* **12,** 9509–9524.

Truchet, G. L., and Dazzo, F. B. (1982). Morphogenesis of lucerne root nodules incited by *Rhizobium meliloti* in the presence of combined nitrogen. *Planta* **154,** 352–360.

Truchet, G., Barker, D. G., Camut, S., De Billy, F., Vasse, J., and Huguet, T. (1989). Alfalfa nodulation in the absence of *Rhizobium*. *Mol. Gen. Genet.* **219,** 65–68.

Truchet, G., Roche, P., Lerouge, P., Vasse, J., Camut, S., De Billy, F., Promé, J. C., and Dénarié, J. (1991). Sulphated lipooligosaccharide signals of *Rhizobium meliloti* elicit root nodule organogenesis in alfalfa. *Nature (London)* **351,** 670–673.

Tsay, J. T., Oh, W., Larson, T. J., Jackowski, S., and Rock, C. O. (1992). Isolation and characterization of the β-ketoacyl-acyl carrier protein synthase III gene (*fabH*) from *Escherichia coli* K-12. *J. Biol. Chem.* **267,** 6807–6814.

Upadhyaya, N. M., Letham, D. S., Parker, C. W., Hocart, C. H., and Dart, P. J. (1991). Do rhizobia produce cytokinin? *Biochem. Int.* **24,** 123–130.

Utrup, L. J., Cary, A. J., and Norris, J. H. (1993). Five nodulation mutants of white sweetclover (*Melilotus alba* Desr) exhibit distinct phenotypes blocked at root hair curling, infection thread development, and nodule organogenesis. *Plant Physiol.* **103,** 925–932.

Van Brussel, A. A. N., Zaat, S. A. J., Canter-Cremers, H. C. J., Wijffelman, C. A., Pees, E., Tak, T., and Lugtenberg, B. J. J. (1986). Role of plant root exudate and Sym plasmid-localized nodulation genes in the synthesis by *Rhizobium leguminosarum* of Tsr factor, which causes thick and short roots on common vetch. *J. Bacteriol.* **165,** 517–522.

Van Brussel, A. A. N., Recourt, K., Pees, E., Spaink, H. P., Tak, T., Wijffelman, C. A., Kijne, J. W., and Lugtenberg, J. J. B. (1990). A biovar-specific signal of *Rhizobium leguminosarum* bv. *viciae* induces increased nodulation gene-inducing activity in root exudate of *Vicia sativa* subsp. *nigra*. *J. Bacteriol.* **172,** 5394–5401.

Van Brussel, A. A. N., Bakhuizen, R., Van Spronsen, P. C., Spaink, H. P., Tak, T., Lugtenberg, B. J. J., and Kijne, J. W. (1992a). Induction of preinfection thread structures in the leguminous host plant by mitogenic lipooligosaccharides of *Rhizobium*. *Science* **257,** 70–72.

Van Brussel, A. A. N., Tak, T., Spaink, H. P., and Kijne, J. W. (1992b). Light and ethylene influence the expression of nodulation phenotypes induced by *Rhizobium* Nod factors on *Vicia sativa* ssp. *nigra*. *In* "Sixth International Symposium on Molecular Plant-Microbe Interactions," Seattle, Poster abstract 137.

Van de Wiel, C., Scheres, B., Franssen, H., Van Lierop, M. J., Van Lammeren, A., Van

Kammen, A., and Bisseling, T. (1990a). The early nodulin transcript ENOD2 is located in the nodule parenchyma (inner cortex) of pea and soybean root nodules. *EMBO J.* **9,** 1–7.

Van de Wiel, C., Norris, J. H., Bochenek, B., Dickstein, R., Bisseling, T., and Hirsch, A. M. (1990b). Nodulin gene expression and ENOD2 localization in effective, nitrogen-fixing and ineffective, bacteria-free nodules of alfalfa. *Plant Cell* **2,** 1009–1017.

Van Rhijn, P. J. S., Feys, B., Verreth, C., and Vanderleyden, J. (1993). Multiple copies of *nodD* in *Rhizobium tropici* CIAT899 and BR816. *J. Bacteriol.* **175,** 438–447.

Vance, C. P. (1983). *Rhizobium* infection and nodulation: A beneficial plant disease? *Annu. Rev. Microbiol.* **37,** 399–424.

Vargas, C., Martinez, L. J., Megias, M., and Quinto, C. (1990). Identification and cloning of nodulation genes and host specificity determinants of the broad host-range *Rhizobium leguminosarum* biovar *phaseoli* strain CIAT899. *Mol. Microbiol.* **4,** 1899–1910.

Vasse, J., De Billy, F., Camut, S., and Truchet, G. (1990). Correlation between ultrastructural differentiation of bacteroids and nitrogen fixation in alfalfa nodules. *J. Bacteriol.* **172,** 4295–4306.

Vasse, J., De Billy, F., and Truchet, G. (1993). Abortion of infection during the *Rhizobium meliloti*-alfalfa symbiotic interaction is accompanied by a hypersensitive reaction. *Plant J.* **4,** 555–566.

Vázquez, M., Dávalos, A., De las Peñas, A., Sánchez, F., and Quinto, C. (1991). Novel organization of the common nodulation genes in *Rhizobium leguminosarum* bv. *phaseoli* strains. *J. Bacteriol.* **173,** 1250–1258.

Vázquez, M., Santana, O., and Quinto, C. (1993). The NodI and NodJ proteins from *Rhizobium* and *Bradyrhizobium* strains are similar to capsular polysaccharide secretion proteins from Gram-negative bacteria. *Mol. Microbiol.* **8,** 369–377.

Verma, D. P. S. (1992). Signals in root nodule organogenesis and endocytosis of *Rhizobium*. *Plant Cell* **4,** 373–382.

Verma, D. P. S., Fortin, M. G., Stanley, J., Mauro, V. P., Purohit, S., and Morrison, N. (1986). Nodulins and nodulin genes of *Glycine max*. *Plant Mol. Biol.* **7,** 51–61.

Wang, S. P., and Stacey, G. (1990). Ammonia regulation of *nod* genes in *Bradyrhizobium japonicum*. *Mol. Gen. Genet.* **223,** 329–331.

Wang, S. P., and Stacey, G. (1991). Studies of the *Bradyrhizobium japonicum nodD1* promoter: a repeated structure for the nod box. *J. Bacteriol.* **173,** 3356–3365.

Wang, L. X., Li, C., Wang, Q. W., and Hui, Y. W. (1993). Total synthesis of the sulfated lipooligosaccharide signal involved in *Rhizobium meliloti*-alfalfa symbiosis. *Tetrahedron Lett.* **34,** 7763–7766.

Wang, L. X., Li, C., Wang, Q. W., and Hui, Y. Z. (1994). Chemical synthesis of NodRm-1: The nodulation factor involved in *Rhizobium meliloti*-legume symbiosis. *J. Chem. Soc. Perkin Trans. 1,* 621–628.

Whitfield, D. M., and Tang, T. H. (1993). Binding properties of carbohydrate *O*-sulfate esters based on *ab initio* 6-31 + G** calculations on methyl and ethyl sulfate anions. *J. Am. Chem. Soc.* **115,** 9648–9654.

Winarno, R., and Lie, T. A. (1979). Competition between *Rhizobium* strains in nodule formation: interaction between nodulating and non-nodulating strains. *Plant Soil* **51,** 135–142.

Witty, J. F., Skøt, L., and Revsbech, N. P. (1987). Direct evidence for changes in the resistance of legume root nodules in O_2 diffusion. *J. Exp. Bot.* **38,** 1129–1140.

Wood, S. M., and Newcomb, W. (1989). Nodule morphogenesis: the early infection of alfalfa (*Medicago sativa*) root hairs by *Rhizobium meliloti*. *Can. J. Bot.* **67,** 3108–3122.

Wojtaszek, P., Stobiecki, M., and Gulewicz, K. (1993). Role of nitrogen and plant growth regulators in the exudation and accumulation of isoflavonoids by roots of intact white lupin (*Lupinus albus* L.) plants. *J. Plant Physiol.* **142,** 689–694.

Yang, W. C., Canter-Cremers, H. C. J., Hogendijk, P., Katinakis, P., Wijffelman, C. A., Franssen, H., Van Kammen, A., and Bisseling, T. (1992). *In-situ* localization of chalcone synthase mRNA in pea root nodule development. *Plant J.* **2,** 143–151.

Yang, W. C., Katinakis, P., Hendriks, P., Smolders, A., De Vries, F., Spee, J., Van Kammen, A., Bisseling, T., and Franssen, H. (1993). Characterization of *GmENOD40,* a gene showing novel patterns of cell-specific expression during soybean nodule development. *Plant J.* **3,** 573–585.

Young, C., Collins-Emerson, J. M., Terzaghi, E. A., and Scott, D. B. (1990). Nucleotide sequence of *Rhizobium loti nodI*. *Nucleic Acids Res.* **18,** 6691.

Yu, G. Q., Zhu, J. B., Gu, J., Deng, X. B., and Shen, S. J. (1993). Evidence that the nodulation regulatory gene *nodD3* of *Rhizobium meliloti* is transcribed from two separate promoters. *Science in China* **36,** 225–236.

Zaat, S. A. J., Van Brussel, A. A. N., Tak, T., Pees, E., and Lugtenberg, B. J. J. (1987). Flavonoids induce *Rhizobium leguminosarum* to produce *nodDABC* gene-related factors that cause thick, short roots and root hair responses on common vetch. *J. Bacteriol.* **169,** 3388–3391.

Zaat, S. A. J., Wijffelman, C. A., Mulders, I. H. M., Van Brussel, A. A. N., and Lugtenberg, B. J. J. (1988). Root exudates of various host plants of *Rhizobium leguminosarum* contain different sets of inducers of *Rhizobium* nodulation genes. *Plant Physiol.* **86,** 1298–1303.

Zaat, S. A. J., Schripsema, J., Wijffelman, C. A., Van Brussel, A. A. N., and Lugtenberg, B. J. J. (1989a). Analysis of the major inducers of the *Rhizobium nodA* promoter from *Vicia sativa* root exudate and their activity with different *nodD* genes. *Plant Mol. Biol.* **13,** 175–188.

Zaat, S. A. J., Van Brussel, A. A. N., Tak, T., Lugtenberg, B. J. J., and Kijne, J. W. (1989b). The ethylene-inhibitor aminoethoxyvinylglycine restores normal nodulation by *Rhizobium leguminosarum* biovar. *viciae* on *Vicia sativa* subsp. *nigra* by suppressing the "thick and short roots" phenotype. *Planta* **177,** 141–150.

Cellular Resistance to Cancer Chemotherapy

Torben Skovsgaard,* Dorthe Nielsen,* Christian Maare,* and Karsten Wassermann†
*Department of Oncology, University of Copenhagen, Herlev Hospital, DK-2730 Herlev, Denmark
†Department of Biology and Toxicology, National Institute of Occupational Health, DK-2100 Copenhagen Ø, Denmark

I. Introduction

One of the most challenging aspects of cancer chemotherapy is the problem of resistance to clinical drugs. Although considerable insight has been gained, the underlying mechanisms of this phenomenon are still not fully understood. The reasons for clinical resistance may include pharmacokinetic or cell kinetic factors (Cazin *et al.*, 1992; Pallavicini, 1984). It is generally accepted, however, that cellular drug resistance is one of the major reasons that treatment fails.

The mechanisms of resistance depend on several factors and circumstances that have given rise to the following classifications: (1) natural (*de novo* or intrinsic) versus acquired resistance, (2) experimental versus clinical resistance, (3) resistance developed in rodent versus in human cell lines, (4) *in vitro* versus *in vivo* resistance and (5) low versus high degree of resistance. In addition, the dose schedule may affect resistance.

Even though the origins or the ways in which resistance develops differ, there is a common feature. Drug resistance is nearly always multifactorial, that is, it is a result of two or more mechanisms. The pattern of the different mechanisms of resistance is probably related to how resistance develops, but too little is known about this aspect. Basically, cellular resistance depends on the biological possibilities which are available for a mammalian cell to escape cellular injury from a cytotoxic drug. Consequently, the mechanisms of resistance found in one cellular system may also occur in another. Considering drug resistance from this point of view, all mechanisms described may be relevant. Consequently, this chapter starts with a description of different mechanisms of resistance, even though they may have been documented in only one of the classes of resistance. We have made no attempt to present a complete overview of all known biological mechanisms of drug resistance. Rather, we have concentrated on recent literature concerning major clinical drugs. More detailed information on general aspects of drug resistance can be found in the reviews cited in the individual sections of this chapter.

II. Drug Resistance

A. Genetic Mechanisms

A possible mechanism of drug resistance is gene amplification, which is defined as a mechanism by which cells can generate multiple copies of discrete regions of their genome. The amplified genes are usually located either within expanded chromosomal regions (homogeneously staining regions or abnormally banded regions) or in extrachromosomal elements (double minutes). There are now several examples of established cell lines in which amplification of a specific gene is responsible for resistance to a specific drug (e.g., methotrexate) (Stark, 1986; Schimke, 1984). Gene amplification has also been demonstrated in clinical samples (Trent *et al.*, 1984; Carman *et al.*, 1984). On the other hand, gene amplification is not always observed in resistant cell lines. Selection for gene amplification is only possible when the gene is active. Silent genes with low transcriptional activity or none require transcriptional activation before selection for amplification. The *mdr1* gene encoding for the multidrug resistance (MDR) protein, P-glycoprotein (Pgp), belongs to this group of genes (Noonan *et al.*, 1990). Thus, a number of MDR cell lines have shown increased expression of the *mdr1* gene with little, if any, gene amplification (Roninson, 1992). Among several hundred clinical samples with increased expression of Pgp, only a few cases of gene amplification have been reported (Michieli *et al.*, 1991; Lönn *et al.*, 1992; Lönn *et al.*, 1993).

Gene amplification always arises by mutation, whereas increased transcription may arise either by mutation (e.g., in genes coding for regulatory proteins) or by adaptation (Borst, 1991). Adaptation is induced by drugs and is dependent on the continuous presence of drugs. Mutations can occur at random in the presence or in the absence of drugs. Drug treatment subsequently selects the resistant cells, and their progeny will dominate in the recurring tumor. This mutation-selection theory has been discussed by Skipper *et al.* (1978) and by Goldie and Coldman (1979).

Several reports indicate that transcription of the *mdr1* gene can be induced in cell lines by adaptation. Thus, a variety of environmental stresses and circumstances which affect cellular differentiation have been shown to induce *mdr1* gene expression (Roninson, 1992). The anthracycline doxorubicin (DOX) has been shown to induce expression of Pgp after only 24 hr of contact with sensitive cells (Volm *et al.*, 1991; Chevillard *et al.*, 1992). The *mdr1* promoter has been shown to respond directly to treatment with anthracyclines or vinca alkaloids (Kohno *et al.*, 1989). In addition, exposure of cells to DNA-damaging agents such as cisplatin can lead to the induction of a variety of genes. Few of these DNA-damage-inducible genes have been identified in mammalian cells. The metallothionein and β-polymerase genes belong to this group (Papathanasiou and Fornace, 1991).

B. P-Glycoprotein-Mediated Drug Resistance

Classic multidrug resistance (MDR) is a well-characterized experimental phenomenon (extensively reviewed by Beck, 1987; Pastan and Gottesman, 1988; Bradley *et al.*, 1988; van der Bliek and Borst, 1989; Endicott and Ling, 1989). It consists of the simultaneous expression of cellular resistance to a series of structurally and functionally unrelated drugs. The phenotype is frequently observed in mammalian cell lines selected for resistance to a single agent (reviewed by Beck and Danks, 1991; Sugimoto and Tsuruo, 1991), and is characterized by (1) cross resistance between a series of chemically unrelated drugs of natural origin; (2) a decrease in drug accumulation (Biedler and Riehm, 1970; Kessel and Bosmann, 1970; Danø, 1971); (3) overproduction of a 170-kDa plasma membrane glycoprotein, Pgp (Juliano and Ling, 1976); (4) overexpression or amplification of the *mdr1* gene (Roninson *et al.*, 1992; Riordan *et al.*, 1985; van der Bliek *et al.*, 1986); (5) reversal of the phenotype by a variety of different compounds, e.g., verapamil (Tsuruo *et al.*, 1981; Skovsgaard *et al.*, 1984).

1. Drugs

The drugs most often involved in MDR are of fungal or plant origin and include the anthracycline group (primarily daunorubicin, DNR, and DOX), vinca alkaloids, and epipodophyllotoxins. Apart from drugs within these groups, a series of other unrelated compounds are able to induce MDR (actinomycin D, colchicine, amsacrine, paclitaxel, the anthracenedione derivatives, and ellipticine). All these drugs are hydrophobic and most are weak bases.

a. Anthracyclines The majority of intracellular drugs concentrate in the nucleus, where they intercalate into the DNA double helix and thereby inhibit DNA and RNA synthesis. Initially, the inhibitory effect on replication and transcription was assumed to explain the cytotoxicity of the drugs (Zunino *et al.*, 1975). Subsequent work has shown, however, that the cytotoxicity of the anthracyclines can be attributed to interaction with topoisomerase II (topo II) (Zwelling, 1985; Ross, 1985).

Although this interactioin with topo II is regarded today as by far the most important mechanism of action, several other mechanisms exist. All clinically active anthracyclines are anthraquinones which are able to undergo one or two electron reductions. One electron reduction of DNR or DOX leads to generation of a semiquinone free radical. This compound donates its electron to oxygen, forming a superperoxid. This superperoxid then reacts to generate hydrogen peroxide, which can undergo reductive cleavage to the very reactive and destructive hydroxyl radical OH·. The generation of hydroxyl radicals may play a role in the cardiac toxicity of these drugs, whereas its role in the death of tumor cells is controversial (Myers and Chabner, 1990). Free radical formation, however, has been detected in human breast cancer cell lines (Sinha *et al.*, 1987, 1989). The enzymes capable of causing the electron reduction are all flavin-centered dehydrogenases or reductases, and include cytochrome P-450. The enzymes superperoxide dismutase, glutathione peroxidase, and catalase act to reduce generation of free radicals.

Two-electron reduction leads to deoxyaglycone formation. The deoxyaglycone is far less cytotoxic than the parent compound. The formation of aglycone, however, may result in an intermediate quinone methide that may act as a monofunctional alkylating agent (Abdella and Fisher, 1985).

Anthracyclines, such as DOX, which containe a ketol side chain, can be oxidized by iron. This process leads to destruction of the side chain and the formation of Fe^{2+} hydrogen peroxide and the hydroxyl radical OH·. Finally, DOX has been postulated to be cytotoxic solely by interacting with the plasma membrane, without entering the cell (Tritton and Yee, 1982; Tokes *et al.*, 1982).

b. The Anthracenedione Mitoxantrone This drug acts by interaction with topo II. It has reduced potential to undergo one-electron reduction.

c. The Mitotic Inhibitors (Vinca Alkaloids, Colchicine, and Paclitaxel) These agents act by binding to tubulin. Vinca alkaloids and colchicine inhibit micotubule assembly, thereby blocking formation of the mitotic spindle and causing an accumulation of cells in mitosis. The binding sites of these two agents are distinct and both drugs can bind simultaneously. Paclitaxel enhances rather than inhibits tubulin polymerization.

d. Actinomycin D This drug binds to DNA and selectively inhibits RNA and protein synthesis.

e. The Podophyllotoxin Derivatives, Amsacrine, and Ellepticine These compounds act by interaction with topo II (see Section II, B.8).

2. Cross-Resistance, Collateral Sensitivity

In general, MDR cells may display cross-resistance to all the drugs mentioned here. In most cases, the cell lines display the highest degree of resistance to the selecting agent or a close analog. Cross-resistance to alkylating agents such as melphalan and nitrogen mustard or to antimetabolites such as methotrexate (MTX) is an exception and then usually occurs only at a low level.

Collateral sensitivity has been reported for nonionic detergents, steroids, and local anesthetics (Bech-Hansen *et al.*, 1976; Cano-Gauci and Riordan, 1991). It has been demonstrated for cytosine arabinoside (Danø, 1972) and carmustine (BCNU) (Jensen *et al.*, 1992).

3. The *mdr* Gene Families

Two and three genes have been identified in humans and rodents, respectively. All *mdr* genes share a close sequence homology and are linked to each other within several hundred kilobases. The human genes are designated *mdr1* and *mdr2;* the latter is sometimes refered to as *mdr3*. The human *mdr1* gene is responsible for resistance in MDR cells, whereas the function of *mdr2* is unknown. In mice, the two genes which correspond to the human *mdr1* gene are designated *mdr1a* (previously called *mdr3*) and *mdr1b,* as they both encode for functional Pgp (Hsu *et al.*, 1989). The three genes in hamsters are called *pgp1, 2,* and *3,* among which only *pgp1* encodes for functional Pgp (reviewed by Borst *et al.*, 1992). The human *mdr* genes are located adjacent to each other on chromosome 7q21.1 (Callen *et al.*, 1987). The mouse genes are mapped to chromosome 5 (Martinsson and Levan, 1987), and the hamster genes are mapped to chromosome 1q26 (Teeter *et al.*, 1986). The complete exon/intron structure of three mammalian Pgp genes is now known (Raymond and Gros, 1989; Chen *et al.*, 1990; Lincke *et al.*, 1991). All the genes consist of 28 exons and the positions of the introns are the same in each of the genes.

The promotor segments from human *mdr1* (Ueda *et al.*, 1987a), mouse *mdr1a* and *b,* and hamster *pgp1* have been isolated (Ikeguchi *et al.*, 1991; Hsu *et al.*, 1990; Raymond and Gros, 1990; Cohen *et al.*, 1991; Ueda *et al.*, 1987b). Two different human *mdr1* promotors have been identified (Rothenberg *et al.*, 1989; Ueda *et al.*, 1987a,b). While most cell lines that overexpress *mdr1* use the downstream promotor, both the upstream and the downstream promotor are used in a colchicine-resistant KB cell line (Ueda *et al.*, 1987a,b). The significance of the different transcription initiation sites is not known. Recent studies have focused on the downstream promotor. Studies have demonstrated a region 5 to 127 nucleotides downstream from the major transcription initiation site which is essential for initiation of proper transcription (Cornwell, 1990).

4. Structure of P-Glycoprotein, the Superfamily of ATP-Dependent Membrane Transporters

The primary sequence of Pgp has been determined from sequence data obtained from cDNA. The protein consists of 1276–1280 amino acids with a tandemly duplicated structure. Each half of the molecule contains a nucleotide-binding site and reveals six hydrophobic regions that fit all criteria for transmembrane domains. Several investigators have shown that Pgp possesses ATPase activity (Hamada and Tsuruo, 1988; Shimabuku *et al.*, 1991). Recent developments suggest that Pgp is a member of a superfamily of ATP-dependent membrane transporters called the ABC transporters (ATP-binding cassette) (Hyde *et al.*, 1990). These transporters have also been designated traffic ATPases (Ames *et al.*, 1990). More than 50 ABC transporters are now known (Higgins, 1992). The family includes bacterial transporters (Hyde *et al.*, 1990), the cystic fibrosis transmembrane conductance regulator (Riordan *et al.*, 1989), the *Plasmodium falciparum* drug resistance gene (Foote *et al.*, 1989), and genes apparently involved in peptide transport during antigene presentation (Spies *et al.*, 1992) and transport of an **a**-factor pheromone of yeast (STE6) (Raymond *et al.*, 1992).

5. The Function of P-Glycoprotein

The most striking feature of Pgp is its similarity to a varity of transport proteins. Several experiments have shown that a number of radiolabeled drugs and photoactivated drug analogs bind to Pgp (Cornwell *et al.*, 1986; Naito *et al.*, 1988). Since Danø in 1973 proposed the existence of an active drug extrusion in tumor cells resistant to DNR, increased activity of an energy-dependent drug efflux pump in MDR cells has been the working model for most investigators. A model for the function of Pgp has been proposed in which drugs bind directly to Pgp, which then acts as a carrier that actively extrudes drugs through a pore or channel in the membrane formed by transmembrane domains of the molecule using energy derived from ATP hydrolysis (Chen *et al.*, 1986).

The best evidence of the function of Pgp has been provided by transfection studies. The results of these studies have indicated that the Pgp density in the plasma membrane is sufficient to determine the level of MDR (Choi *et al.*, 1991; Lincke *et al.*, 1990). The only study concerning transport in transfected cells (Hammond *et al.*, 1989), however, failed to demonstrate a competitive inhibition by vincristine or anthracyclines on vinblastine efflux. Furthermore, Belli *et al.* (1990) have fused Pgp isolated from DOX-resistant cells with sensitive cells. Incorporation of Pgp in sensitive cells resulted in a significant increase in resistance (90-fold) and the fused cells effluxed DOX more rapidly within the first 5 min of observation. Steady-state accumulation was unchanged, however, and blocking of the efflux by inhibitors of energy production was not attempted. Thus, no study has yet provided data which incontestably indicate that Pgp functions as a drug transporter.

In comparing Pgp to bacterial transport proteins, the hypothesis has several weaknesses. Bacterial carrier molecules are characterized by being substrate specific, transporting the solutes in an osmotic-, energy-, and temperature-dependent, saturable manner; having characteristic binding constants; and being blocked by competitive and noncompetitive inhibitors. Pgp appears to recognize a wide range of different lipophilic compounds. The results concerning osmotic dependence in MDR membrane vesicles are conflicting. Thus, Horio *et al.* (1988) and Doige and Sharom (1992) have found osmotic dependence of vinblastine accumulation in MDR membrane vesicles, whereas Cornwell *et al.* (1986) and Sehested *et al.* (1989) were unable to demonstrate osmotic dependency of vinca alkaloid binding to vesicles of colchicine resistant-KB or DNR-resistant Ehrlich ascites cells, respectively.

Assuming that Pgp acts as a drug transporter mediating drug eflux, one would expect a close connection between Pgp expression, reduction in drug accumulation, and, if no other resistance mechanisms are operating, drug resistance. All MDR cells which overexpress Pgp ought to have increased drug efflux. We have previously reviewed all MDR cell lines which have been systematically investigated for both Pgp expression and transport properties (Nielsen and Skovsgaard, 1992). In this study, cell lines with increased expression of Pgp and decreased net accumulation were considered as supporting the hypothesis of Pgp as a multidrug transporter unless separate efflux experiments demonstrated the opposite. It is important to remember, however, that this assumption is an absolute minimum. If Pgp is to be considered as a genuine "drug-pump," a significant reduction in steady-state accumulation, and not merely a decreased influx, must be demonstrated (Fig. 1). Furthermore, the efflux must be blocked by the addition of inhibitors of energy production (both oxidative

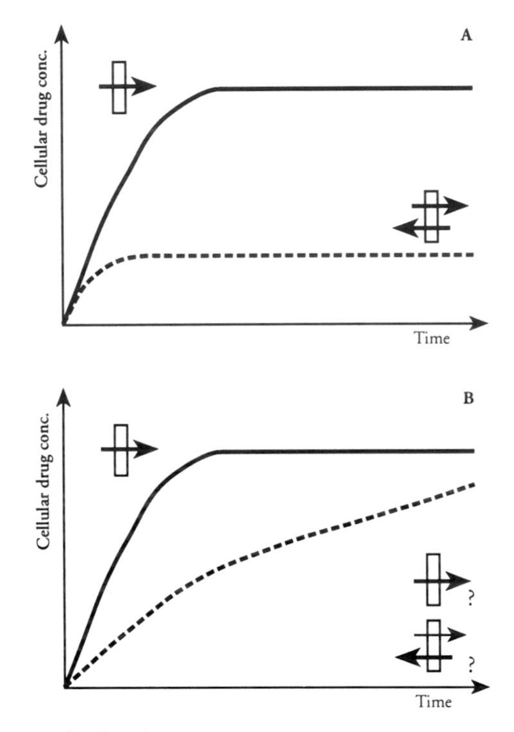

FIG. 1 Time course uptake showing (A) a significant decrease in steady-state accumulation of drug, and (B) reduced accumulation, which could be explained by a decreased influx. Solid line, wild-type tumor cells; dashed line, resistant tumor cells.

phosphorylation and glycolysis) and by analogs that act as competitive inhibitors.

Among cell lines developed for resistance to the classical MDR compounds (anthracyclines, vinca alkaloids, colchicine, and actinomycin D), 76 of 77 cell lines which overexpressed Pgp had decreased intracellular drug content at steady state or increased drug efflux. In a series of five SW948 DOX-resistant cell lines, Toffoli *et al.* (1991) were unable to detect any significant differences in either steady-state accumulation or efflux of DOX. The authors stressed that the results could be due to the detection limit of the experimental procedure being too high to detect variations in transport. In four instances, however, increased expression of Pgp and decreased steady-state accumulation were not followed by increased efflux (Ganapathi and Grabowski, 1988; Deffie *et al.*, 1988; Ramu *et al.*, 1989; Carlsen *et al.*, 1976). Among these, the findings by Carlsen *et al.* (1976) could be explained by tight binding of colchicine to tubulin. In addition, quenching could confound the results of Ramu *et al.* (1989). The results, however, are supported by the findings by Mickley *et al.* (1989) and Bates *et al.* (1989), who have investigated the effects of a variety of differentiating agents on human colon carcinoma and neuroblastoma cell lines, respectively. In an SW620 cell line, addition of butyrate resulted in increased expression of *mdr1, mdr1* mRNA, and Pgp, but the transport of vinblastine was not significantly changed and no increase in resistance to vinblastine or DOX was found.

Recently, the authors demonstrated increased efflux of colchicine in the SW620 cell line and found that butyrate modulated Pgp by phosphorylation, which could explain the changed specificity of Pgp (Bates *et al.*, 1992). The addition of retinoic acid to the neuroblastoma cell line SK-N-SH, however, resulted in increased expression of Pgp without any significant changes in drug accumulation. The levels of Pgp expression in these cell lines were comparable to levels associated with easily detectable decreases in drug accumulation in cell lines selected *in vitro* for resistance. In addition, Volm *et al.* (1991) have shown that the Pgp expression in a L1210 cell line increased within a few hours after treatment with DOX. In spite of a high Pgp expression, the authors detected only a slight increase in tumor cell resistance.

Comparing one resistant cell line with its sensitive counterpart can yield important information about changes accompanying development of resistance. By comparing sublines with different degrees of resistance, however, one can determine if any property is invariably associated with resistance (and other cellular changes) in a predictive and quantitative way. Only a limited number of series of MDR cell lines have been reported, many of which are not systematically described. In several of these, a relationship between Pgp, resistance, and decreased drug accumulation could be established. In other cases, however, no correlation between the Pgp expression and resistance or transport properties was observed (Kato *et al.*, 1990; Siegfried *et al.*, 1983; Twentyman *et al.*, 1990; Lothstein and Horwitz, 1986; Fojo *et al.*, 1985; Huet *et al.*, 1992).

Several explanations could account for the discrepancies. First, regarding Pgp and its activity, the only relevant relationship seems to be between the number of Pgp molecules on the cell surface and the efflux. Most studies, however, measure total Pgp or total *mdr* mRNA. Measuring total protein or mRNA is not completely satisfactory, as it has been shown that some of the protein can be inside the cell and not be functional (Broxterman *et al.*, 1989; Boiocchi and Toffoli, 1992). Second, the biological activity of Pgp could be modified. Finally, other cellular changes could influence transmembranous drug transportation.

The unidirectional influx of drugs has only been studied in nine cell lines with overexpression of Pgp. Among these, Yang *et al.* (1989a) in a preliminary report describe a DOX-resistant LoVo cell line with enhanced efflux but without significant changes in the influx. In eight of the cell lines, however, the influx has been found to be decreased (Fojo *et al.*, 1985; Carlsen *et al.*, 1976; Sirotnak *et al.*, 1986; Skovsgaard, 1978; Marsh *et al.*, 1986; Ramu *et al.*, 1989; Kessel and Wilberding, 1985; Soto *et al.*, 1992). Several studies have demonstrated that the influx can be enhanced by adding inhibitors of energy production (Skovsgaard, 1978; Kessel and Wilberding, 1985). Thus, in several MDR cells lines, both an increase in efflux and a decrease in influx have been demonstrated. In order to explaine this finding and the broad specificity of Pgp, a new model for the functioning of the molecule has been proposed (Gottesman, 1993).

According to this model, the two halves of Pgp form a single channel for a drug transport and the major function of Pgp is to remove the drugs directly from the outer lipid bilayer of the plasma membrane. This means that the drug must enter into the lipid bilayer and is only then recognized by Pgp, which extrudes it from the plasma membrane. Recent studies of Pgp in MDR cells have shown it to be bifunctional, associated both with drug transport and with cloride channel activity (Gill *et al.*, 1992; Valverde *et al.*, 1992). The transport and channel functions are found to be distinct and separable. Thus, the drug transport is found to be dependent on ATP hydrolysis while the channel function requires ATP binding, but not hydrolysis (Gill *et al.*, 1992). The channel function seems to be regulated by changes in cell volume.

To understand the mechanism of the action of Pgp, it is necessary to purify and reconstitute the protein. Ambudkar *et al.* (1992) have extracted Pgp by octyl glucoside, purified it by column chromatography, and reconstituted it into proteoliposomes. The authors report preparations of Pgp which are 80–90% pure (Gottesman, 1993) and exhibit a high level of substrate-stimultable ATPase activity that is similar to other ion-transporting pumps. The ultimate proof that Pgp is a transport system for multiple drugs, however, must await its purification to homogeneity and reconstitution into a vesicle system that is capable of drug transport.

6. Regulation of P-glycoprotein Function and Activity

Although increased expression of the *mdr* genes was originally associated with amplification of *mdr* DNA, recent studies have reported increased expression of *mdr1* mRNA to account for resistance in the absence of gene amplification (Shen *et al.*, 1986; Benard *et al.*, 1989). Studies have shown that the *mdr1* gene is regulated by both transcriptional and post-transcriptional control (Chin *et al.*, 1990; Gant *et al.*, 1992; Marino *et al.*, 1990).

Recently, Chin *et al.* (1992) have suggested that expression of specific oncogenes, including *ras* and mutant *p53*, may activate the expression of

Pgp. Furthermore, it has been suggested that expression of the *mdr1* gene can be regulated by a protein kinase C-mediated pathway (Chaudhary and Roninson, 1992).

It has been demonstrated that the affinity of Pgp for metabolites can be changed. Choi *et al.* (1988) have found that a single point mutation changing a glycine into a valine in the transmembrane segment of Pgp is responsible for a threefold increase in the relative resistance to colchicine.

Pgp undergoes extensive post-translational modification. It is synthesized as a 140-kDa precursor which is slowly converted over 2–4 hr to a 170-kDa glycoprotein (Richert *et al.*, 1988). In MDR KB cells, it has a half-life of 48–72 hr.

Very little is known about the mechanisms involved in the post-translational regulation of Pgp. It has been shown that the Pgp is phosphorylated and the function of the protein could be modulated by phosphorylation (Bates *et al.*, 1992; Ma *et al.*, 1991). Evidence has indicated a functional role for the calcium- and phospholipid-dependent protein kinase C (Hamada *et al.*, 1987; Chambers *et al.*, 1990). Recently, inhibitors of protein kinase C (e.g., staurosporine, calphostin C) have been shown to increase drug accumulation in MDR cells (Chambers *et al.*, 1992; Yu *et al.*, 1991; Sato *et al.*, 1990). Other findings, however, contradict the hypothesis. Overexpression of protein kinase C alone has been shown to produce an MDR phenotype without changes in Pgp expression (Fan *et al.*, 1992). Staurosporine has been shown to enhance drug accumulation by a direct interaction with Pgp (Miyamoto *et al.*, 1993). The activity of Pgp could also be changed by glycosylation of the molecule (Greenberger *et al.*, 1987; Kramer *et al.*, 1992). Thus, a variety of N-linked glycosylations of Pgp have been detected in different tissues and tumors (Ichikawa *et al.*, 1991). On the other hand, tunicamycin treatment, which blocks N-linked glycosylation, has been shown not to affect the drug resistance of MDR cells (Beck and Cirtain, 1982).

7. Expression of *mdr1* in Normal and Malignant Human Tissues

Studies on the expression of the *mdr1* gene in normal tissue have revealed a striking pattern of expression in transporting epithelia. Thus, the gene is expressed to a high level in liver (biliary ducts), kidney, small intestine, and colon (Fojo *et al.*, 1987; Thiebaut *et al.*, 1989; Hitchins *et al.*, 1988). This localization is consistent with a function that involves transport of substrates across an epithelial monolayer. The protein could have evolved to protect the organisms from toxic, hydrophobic chemicals. It has been suggested that bilirubin is a physiologic substrate for Pgp (Gosland *et al.*, 1991). Other localizations of Pgp in normal tissues are in capillary endothelial cells in the brain and testis (Thiebaut *et al.*, 1989; Cordon-Cardo *et al.*, 1989). A high level of Pgp has also been detected in the placenta (Sugawara *et al.*, 1988) and in adrenal cortical cells (Thiebaut *et al.*, 1987). The latter localization is especially significant since Pgp has been shown to transport cortisol, aldosterone, and dexamethasone (Ueda *et al.*, 1992; van Kalken *et al.*, 1993).

High expression of the *mdr1* gene is found in cancers derived from kidney, liver, colon, pancreas, and adrenal glands (Nooter and Herweijer, 1991; Fojo *et al.*, 1987; Goldstein *et al.*, 1989; Kanamaru *et al.*, 1989). In tumors derived from these tissues, especially in renal cell carcinoma, Pgp seems to act as a marker of differentiation (Kanamaru *et al.*, 1989). In colon cancer, Pgp has been found to be associated with enhanced local tumor aggressiveness (Weinstein *et al.*, 1991). Although the tissue of origin does not overexpress *mdr1* mRNA, some other untreated cancers often show high levels of *mdr1* mRNA. Such cancers include acute and chronic leukemias, non-Hodgkin's lymphoma, neuroblastoma, soft tissue sarcomas, and breast cancer (Goldstein *et al.*, 1989; Wishart *et al.*, 1990). In childhood neuroblastoma and sarcoma correlations have been made between expression of Pgp, failure to respond to chemotherapy, and poor prognosis (Chan *et al.*, 1990, 1991).

In addition, a correlation between expression of Pgp and response to chemotherapy has been suggested in acute nonlymphoblastic leukemia and breast cancer (Campos *et al.*, 1992; Verrelle *et al.*, 1991; Ro *et al.*, 1990). Increased expression of the *mdr1* gene is commonly seen in tumors which have relapsed during the course of chemotherapy or thereafter. Examples include breast cancer, lymphoma, leukemia, neuroblastoma, and multiple myeloma (Goldstein *et al.*, 1989; Dalton *et al.*, 1989; Herweijer *et al.*, 1990; Verrelle *et al.*, 1991; Ro *et al.*, 1990). Together these findings suggest that the occurrence of Pgp may be used as a marker of sensitivity for MDR drugs.

8. Reversal of Multidrug Resistance

There is a considerable body of experimental work showing that certain classes of membrane-active drugs are capable of circumventing MDR to varying degrees (Skovsgaard *et al.*, 1984; Ford and Hait, 1990). The compounds that modulate or reverse MDR represent a wide range of chemical structures, including detergents, calcium channel blockers, calmodulin inhibitors, steroids, immunosuppressives, antimalarials, and tranquilizers. Some of the most important of these agents are listed in Table I. Although many of the compounds in the table are hydrophobic, have two planar aromatic rings, and a tertiary nitrogen that is charged at physiological pH, there is no common chemical basis for the modulting action

TABLE I
Classes of Agents Shown to Reverse Multidrug Resistance

Type	Compound (example)	References
Calcium channel blockers	Verapamil, nifedipine, dihydropyridine analogs	Tsuruo *et al.*, 1981 Ramu *et al.*, 1984a Schuurhuis *et al.*, 1987
Antipsychotics	Calmodulin inhibitors, e.g., phenothiazines	Tsuruo *et al.*, 1982 Ganapathi *et al.*, 1984
Antiarrythmics	Quinidine, amiodarone	T. Skovsgaard and Friche, 1982 Tsuruo *et al.*, 1984 Chauffert *et al.*, 1986
Antihypertensives	Reserpine	Pearce *et al.*, 1989
Antimalarials	Chloroquine, quinine	Zamora and Beck, 1986 Zamora *et al.*, 1988
Antibiotics	Cephalosporins	Gosland *et al.*, 1989
Steroids	Progesterone	Yang *et al.*, 1989b
Modified steroids	Tamoxifen, torimefene	Ramu *et al.*, 1984b DeGregorio *et al.*, 1989
Immunosuppressants	Cyclosporins	Twentyman, 1992
Anthracycline analogs	*N*-Acetyl daunorubicin	Skovsgaard, 1980
Lysosomotropic agents	Monensin, nigericin	Sehested *et al.*, 1988
Detergents	Tween 80, Triton X-100	Riehm and Biedler, 1972

of the agents (Zamora *et al.*, 1988). Thus, a number of hydrophobic drugs have been shown to reverse MDR (Friche *et al.*, 1990; Hofsli and Nissen-Meyer, 1990). Using a computer program, Klopman *et al.* (1992) were able to identify a number of key structural features and thereby predict reversal activity of the different compounds.

The exact nature of the mechanisms of reversal is unclear. Calcium channel blockers and calmodulin inhibitors have been shown to inhibit the efflux of antitumor agents from resistant cells (Inaba *et al.*, 1979; Willingham *et al.*, 1986, Ganapathi *et al.*, 1984). A number of investigators have shown that photoactivated verapamil analogs, cyclosporin, and quinine bind to Pgp. Verapamil inhibits the binding of many cytostatic drugs as well as other chemosensitizers to Pgp. A correlation between the reversal of resistance and inhibition of photoaffinity labeling has been documented in membrane vesicles. These results suggest that chemosensitizers may act by competitive inhibition of Pgp (Cornwell *et al.*, 1987; Safa, 1988; Yusa and Tsuruo, 1989).

Regarding verapamil, several other mechanisms of action have been reported. The significance of these other actions on the ability of verapamil to reverse MDR, however, is unclear. The effect of the modulator could be explained by a nonspecific action on membranes (Cano-Gauci and Riordan, 1987). The compound has been shown to alter the hydrophobic or hydrophilic solubility of cytostatics in resistant cells and thereby to change the distribution of drugs (Schuurhuis *et al.*, 1989; Hindenburg *et al.*, 1987). In MDR cells, addition of verapamil causes a decrease in pH toward levels seen in sensitive cells (Keizer and Joenje, 1989). Both verapamil and cyclosporin have been shown to restore membrane-potential changes (depolarizations) in MDR cells (Vayuvegula *et al.*, 1988). In several resistant cell lines not overexpressing Pgp but with a defective accumulation, addition of verapamil has been shown to increase cytotoxicity (Cass *et al.*, 1989; Nygren *et al.*, 1991; Larsson and Nygren, 1990). These findings suggest that Pgp is not necessary for the action of verapamil.

The clinical experience with chemosensitizers for modification of intrinsic or required drug resistance is limited. The most important problem with studies like these relates to the rather high level of chemosensitizer (e.g., verapamil 1–10 μM) needed *in vitro* to reverse experimental MDR—levels that cannot be achieved clinically. A second problem is that clinical drug resistance is most likely due to a complex interplay of many factors (e.g., pharmacological factors, hypoxia, low growth fraction of the tumor cell population) in addition to a possible alteration in tumor cell resistance. Finally, since tumor cell resistance is probably multifactorial, other mechanisms apart from increased Pgp expression will often play a role. Generally, it remains to be determined whether distinctions can be made between tumor and normal tissue Pgp. If this is not the case, inhibition of Pgp in normal tissue may lead to increased toxicity and thereby make the use of Pgp inhibitors impossible.

The most extensive knowledge on reversing MDR in the clinic has been gathered from calcium channel blockers. Most studies have included verapamil. These studies have been complicated by the cardiac toxicity of the calcium channel blocker, which has been the dose-limiting toxicity (Ozols *et al.*, 1985; Presant *et al.*, 1986; Salmon *et al.*,1991; Miller *et al.*, 1991). In the initial studies there was no attempt to determine whether Pgp was overexpressed in patients' tumor cells. Recently, Salmon *et al.* (1991) have reported a 23% partial remission rate among 22 patients with refractory myeloma after treatment with a combination of verapamil, vincristine, DOX, and dexamethasone. Miller *et al.* (1991) have treated 18 patients with refractory malignant lymphoma with verapamil, cyclophosphamide, DOX, vincristine, and dexamethasone. The authors report a 72% remission rate, including 28% complete responders.

A phase II study of the effect of cyclosporin A and epirubicin in patients with colorectal cancer has shown disappointing results (only one partial response among 24 patients; Verweij *et al.*, 1991). Recently, however, very encouraging data have been reported for the use of cyclosporin A, including a very high proportion of complete responses in patients with refractory myeloma or acute leukemia (Sonneveld *et al.*, 1992; List *et al.*, 1992).Most clinical studies, however, have not been successful, suggesting that more specific chemosensitizers or new strategies are warranted. Several possibilities exist. New, less toxic or more active chemosensitizers should be considered. The R-verapamil stereoisomer is a promising compound since it is much less cardiotoxic than the L isomer, although it seems to be equally effective in reversing resistance (Plumb *et al.*, 1990). SDZ PSC 833 is a nonimmunosuppressive cyclosporine which *in vitro* is as potent as cyclosporin A (Keller *et al.*, 1992).

In vitro studies have shown synergistic interaction of verapamil and cyclosporin A and quinine, respectively, on diminishing drug resistance (Hu *et al.*, 1990; Lehnert *et al.*, 1991). Thus, a combination of several modulators could provide a more optimal therapeutic index.

Efforts to overcome MDR include introduction of non-cross-resistant treatment regimens with drugs not involved in MDR. If MDR cells show collateral sensitivity, certain agents may provide hope for eradication of resistant tumor populations. Due to collateral sensitivity, therapy that includes carmustine in regimens with MDR drugs could provide a benefit (Jensen *et al.*, 1992). Clinical application of most of the substances proposed, however, is not feasible.

Antisense oligonucleotides appear to be able to inhibit the synthesis of Pgp (Vasanthakumar and Ahmed, 1989) and have been shown to decrease resistance to DOX in multidrug-resistant, human adenocarcinoma cells (Corrias and Tonini, 1992). One can interfere with the function of Pgp by combining chemotherapy with antibodies or toxin-conjugated antibodies directed against Pgp. Thus, it has been demonstrated that *Pseudomonas* exotoxin conjugated to the monoclonal antibody MRK16 can kill MDR cells in tissue culture (FitzGerald *et al.*, 1987). In addition, MRK16 has been shown to reverse resistance in multidrug-resistant transgenic mice (Mickisch *et al.*, 1992). As for the "classic" chemonsensitizers, normal tissue expressing Pgp could be injured by the above-mentioned treatment. Thus, preclinical animal models are required to determine the efficacy and safety of the approaches.

A different, potentially useful therapeutic approach is based on decreasing the toxicity of chemotherapeutic drugs transported by Pgp by introducing a constitutively expressed *mdr1* gene into the normal tissues which are affected by these drugs, primarily the bone marrow. The development of retroviral vectors and gene transfer in mice have demonstrated the feasibility of this approach (Galski *et al.*, 1989).

Finally, the knowledge of the specificity of Pgp for drugs may lead to agents which circumvent the Pgp-mediated efflux. Examples are MX2, a morpholino anthracycline, which has been shown to be effective against DOX-resistant P388 cells (Watanabe *et al.*, 1991); ME2303, a fluorine-containing anthracycline derivative, effective against murine MDR cells *in vivo* (Tsuruo *et al.*, 1989); and *N*-(5,5-diacetoxypent-1-yl)-DOX, which has been shown to be approximately 100 times more potent than DOX in Pgp-expressing cells (Cherif and Farquhar, 1992). In addition, DOX-loaded liposomes or nanospheres (Vaage *et al.*, 1992; Cuvier *et al.*, 1992) could provide a benefit.

C. Non-P-Glycoprotein-Mediated Multidrug Resistance

Recently, several MDR cell lines have been derived which do not overexpress Pgp (Table II). These cell lines, however, are cross-resistant to all the classic MDR drugs, including vinca alkaloids (Mirski *et al.*, 1987; Zijlstra *et al.*, 1987; Bhalla *et al.*, 1985; Marsh *et al.*,1986; Slovak *et al.*, 1988; Reeve *et al.*, 1990; Garcia-Segura *et al.*, 1990). In all cell lines investigated except two (Mirski *et al.*, 1987; Marsh *et al.*, 1986), transport studies have demonstrated decreased accumulation, and increased energy-dependent efflux has been demonstrated in three of the sublines (Coley *et al.*, 1991; Versantvoort *et al.*, 1992a; Slapak *et al.*, 1992). In all instances, anthracyclines have been used to select for drug resistance.

A number of different proteins have been found to be overexpressed in the respective cell lines. In a small-cell lung cancer cell line, Cole *et al.* (1992) have discovered a gene encoding a protein called MRP (multidrug-resistance-associated protein). The deduced amino acid sequence of MRP (1531 amino acids) indicates that this protein belongs to the same transport system as Pgp (the ABC superfamily). A 190-kDa (p190) protein with some degree of homology with Pgp in a region adjacent to the second ATP binding site has been detected in three lung cancer cell lines (Reeve *et al.*, 1990; Versantvoort *et al.*, 1992a; Barrand *et al.*, 1993) and one leukemia cell line (Marquardt *et al.*, 1990). All these cell lines had increased expression of MRP, and a close correlation between the level of MRP expression and the level of expression of p190 has been demonstrated

(Krishnamachary and Center, 1993; Zaman *et al.*, 1993). Recently, Krishnamachary and Center (1993) have found that the MRP gene encodes for a 190-kDa membrane-bound glycoprotein.

A 110-kDa glycoprotein has been recognized in two lung cancer cell lines (Versantvoort *et al.*, 1992). Two series of cell lines have been reported in which changes in drug accumulation precede expression of Pgp (Slapak *et al.*, 1990; Baas *et al.*, 1990). Among these, the 110-kDa glycoprotein has been demonstrated in the lung cancer cell line described by Baas *et al.* (1990) (Versantvoort *et al.*, 1992b).

Several cell lines selected for resistance to the anthracene derivative mitoxantrone seem to be non-Pgp-mediated resistant. In a mitoxantrone-resistant human colon carcinoma cell line (WiDr) in which Pgp was not detected by immunoblotting, Dalton *et al.* (1988) found decreased accumulation and enhanced efflux. Taylor *et al.* (1991) described a subline of MCF-7 which was 1208-fold resistant to mitoxantrone and did not overexpress Pgp, but accumulated significantly less drug than the sensitive counterpart due to increased efflux. This mitoxantrone-resistant cell line seems to react with LRP-56, an antibody which recognizes the 110-kDa glycoprotein (Bellamy, 1993). Nakagawa *et al.* (1991) described a subline of MCF-7 which was 3000-fold resistant to mitoxantrone and had enhanced efflux, but did not overexpress Pgp. Furthermore, Dalton (1993) described a mitoxantrone-resistant subline of the human myeloma cell line 8226 which had reduced accumulation. In this cell line, a 160-kDa protein has been isolated.

1. Other Cellular Changes Affecting Drug Transport

The subcellular distribution of anthracyclines has been shown to shift from mainly nuclear in wild-type cells to mainly cytoplasmic with more free drug present in acidic vesicles or bound to membrane of vesicles in MDR cells (Gervasoni *et al.*, 1991; Keizer *et al.*, 1989a). Morphological studies (Egorin *et al.*, 1974; Chauffert *et al.*, 1984) have demonstrated a cytoplasmatic distribution of anthracyclines in MDR cells. Freeze-fracture studies have shown increased endosomal volume and membrane traffic in MDR cells (Sehested *et al.*, 1987). In addition, increased secretion of lysosomal enzymes has been demonstrated in vinblastine-resistant lymphoblastic leukemia cells (Warren *et al.*, 1991). Based on these findings, a "membrane trafficking" model for classical MDR has been suggested by Beck (1987). According to this hypothesis, the drugs are expelled via an exocytotic process which includes binding to Pgp. Demant *et al.* (1990), however, who used numerical computer simulations to analyze some possible mechanisms controlling DNR fluxes in a kinetic model, found the

TABLE II

Non P-Glycoprotein-Mediated Drug Resistance, Characterization of Cell Lines

Parent line	Primary resistance[a]	Cross-resistance[a]	Protein	Decreased accumulation	References
COR-L	DOX (23)	VCR (34), COL (10)	MRP, 190 kDa	+	Reeve *et al.*, 1990 Coley *et al.*, 1991 Barrand *et al.*, 1992
GLC 4	DOX (44)	VCF (6), COL (0.5) VP16 (38)	MRP, 190 kDa	+	Zijlstra *et al.*, 1987 de Jong *et al.*, 1989 Versantvoort *et al.*, 1992a Zaman *et al.*, 1993
H69	DOX (100)	DNR (85), VCR (112) COL (16), VP16 (12)	MRP, 190 kDa 36 kDa	–	Mirski *et al.*, 1987, Cole *et al.*, 1992
HL60	DOX (111)	DNR (50), VCR (8) ACTD (6)	–	+	Bhalla *et al.*, 1985 Cass *et al.*, 1989
HL60	DOX (80)	DNR (75), VCR (25) COL (10), ACTD (15)	MRP, 190 kDa	–	Marsh *et al.*, 1986 Marquardt *et al.*, 1990 Krishnamachary and Center, 1993
HT1080	DOX (222)	VCR (25), ACRD (17) VP 16, (837)	–	+	Slovak *et al.*, 1988, 1991

MOR	DOX (R)[b]	–	MRP, 190 kDa	?	Barrand *et al.*, 1993
MCF-7	MIT (1208)	Dox (8.3), VCR (22) VP 16 (26)	–	+	Taylor *et al.*, 1991
MCF-7	MIT (3000)	DOX (10, VBL (22) VP 16 (10)	–	+	Nakagawa *et al.*, 1991
WiDr	MIT (21)	DOX (8), VBL (2)	–	–	Dalton *et al.*, 1988
SW1573	DOX (5–10) (3 sublines)	DNR (4–5), VCR (2–16)	110 kDa	–	Baas *et al.*, 1990 Kuiper *et al.*, (1990)
	DOX (R)	DNR (4), VCR (17) VP 16 (45)	110 kDa	–	
MEL PC4	DOX (5–17) (3 sublines)	VCR (1–2), VP 16 (8–62)	–	–	Slapak *et al.*, 1990
	DOX (44)	VCR (16), VP 16 (74)	–	+	
MEL C7D	DOX (3,5) (2 sublines)	VCR (1,2), VP 16 (11)	–	–	Slapak *et al.*, 1990
	DOX (10)	VCR (3), VP 16 (15)	–	+	
8226	MIT (R)	–	160 kDa	+	Dalton, 1993

[a] Numbers in parenthesis refer to fold of resistance compared with the parent line.
[b] R = resistant.

exocytotic model to be physically impossible. Furthermore, Lankelma *et al.* (1990) found that exocytosis was not an important component of active drug efflux in multidrug-resistant ovarian carcinoma cells.

Recently, attention has focused on changes in the drug distribution in resistant cells not overexpressing Pgp. Hindenburg *et al.* (1989) have investigated HL60 and a DOX-resistant subline not overexpressing Pgp. In the sensitive cell line, DNR was located primarily in the nucleus, whereas a considerable proportion of the drug in the resistant subline was located in cytoplasmic vesicles. A similar shift from a nuclear localization in sensitive cells to a cytoplasmatic localization in resistant cells not expressing Pgp has been reported by Coley *et al.* (1990).

Alterations in composition of the plasma membrane could modulate drug accumulation and resistance. Thus, several authors have demonstrated alterations in the plasma membrane lipids in MDR cells selected for high degrees of resistance (Alon *et al.*, 1991). Although no consistent pattern has emerged, alterations in sphingomyelin, cardiolipin, phosphatidyl ethanolamine, and phosphatidyl serine have been detected (Ramu *et al.*, 1983; Wright *et al.*, 1985). The various changes in lipids could affect the structural order and the fluidity of plasma membranes, and thereby determine how the drug was distributed within the membrane and presented to Pgp.

Several authors have focused on the regulation of intracellular pH and membrane potential. The intracellular pH in resistant cells was higher than in sensitive cells (0.1–0.44 units), and in a series of cell lines with different degrees of resistance, pH increased with increasing resistance (Boscoboinik *et al.*, 1990; Keizer and Joenje 1989b; Thiebaut *et al.*, 1990). A lower membrane potential has been documented in resistant cells (Hasmann *et al.*, 1989). The anthracyclines and vinca alkaloids are lipophilic weak bases, and passive diffusion of the noncharged molecules across membranes in much faster than transport of the protonated forms (Skovsgaard, 1989). Therefore, changes in the pH gradients across the membrane may influence drug efflux and resistance (Skovsgaard, 1989; Alabaster *et al.*, 1989).

2. Other Resistance Mechanisms

Classic MDR is not the only resistance mechanism associated with the anticancer agents discussed here. Reflecting the many targets of the drugs, several other resistance mechanisms have been described. Thus, changes in tubulin have been associated with resistance to the vinca alkaloids and paclitaxel (Tsuruo *et al.*, 1986; Schibler and Cabral, 1986). Enhanced glutathionetransferase has been reported in several DOX-resistant cell lines (Batist *et al.*, 1986; Akman *et al.*, 1990). Buthionine sulfoximine, which inhibits glutathione synthesis, has been shown to partially reverse

DOX resistance (Doroshow *et al.*, 1980). Qualitative or quantitative changes in the activity of topo II have been detected in cell lines selected for resistance to anthracyclines, epipodophyllotoxins, and amsacrine (Friche *et al.*, 1991; Sullivan and Ross, 1991). Although all MDR cell lines investigated are cross-resistant to the epipodophyllotoxins, amsacrine, and ellipticine, cell lines selected for resistance to these compounds often show alterations in topo II without overexpression of Pgp. In fact, overexpression of Pgp has not been reported for cell lines selected for resistance to amsacrine (the number of cell lines selected for resistance to this compound is very limited).

3. Conclusion

There is now a substantial body of knowledge pertaining to the *mdr* genes and the structural nature of Pgp. Little is known, however, about the mechanisms involved in the regulation of the *mdr* genes or the post-translational regulation of Pgp. The ultimate proof that Pgp is a transport system for multiple drugs still awaits its purification to homogeneity and reconstitution into a vesicle system which is capable of drug transport.

A number of studies have been published recently on the detection of *mdr1* in human cancers. A correlation has been found between *mdr1* expression and refractoriness of chemotherapy in some hematological malignancies and solid tumors. Clinical studies of several MDR modulators have been initiated in advanced cancer patients, and few of these have shown promising results, but significant side effects. There is still a considerable lack of information, however, regarding the causative role of Pgp in the refractoriness of chemotherapy. A better knowledge of the functional activity of Pgp and the mechanism of reversal will, it is hoped, lead to better results in the clinic.

Proteins with a structure homologous to Pgp seem to participate in the non-Pgp-mediated MDR phenotype. The decreased accumulation and active efflux demonstrated in several cell lines suggest that these proteins could be capable of drug transport. Our understanding of the mechanisms involved, however, is very preliminary. Furthermore, clinical evidence of non-Pgp-mediated drug resistance is still missing.

D. DNA Topoisomerases and Drug Resistance

DNA topoisomerases are the targets of several different classes of antitumor drugs. DNA topoisomerase II (topo II) is targeted by epipodophyllotoxins (etoposide/VP-16, teniposide/VM-26), anthracyclines, aminoacridines (m-AMSA), anthracenediones (e.g., mitoxantrone), and ellipticines

(Ross, 1985; Ross *et al.*, 1978; Wozniak and Ross, 1983; Pommier *et al.*, 1984; Chen *et al.*, 1984; Tewey *et al.*, 1984; Nelson *et al.*, 1984; Pommier *et al.*, 1985). DNA topoisomerase I (topo I) is targeted by camptothecin and its derivatives (Hsiang *et al.*, 1985). Thus, it would seem reasonable to assume that changes in these enzymes contribute to drug resistance and, accordingly, several cell lines have been described in which alterations in topoisomerases can account for drug resistance.

1. Mechanism of Action of Topo II Inhibitors

Topo II is a nuclear enzyme that influences the supercoiling of DNA by inducing transient double-stranded breaks (Ross, 1985; Wang, 1985; Fernandes and Catapano, 1991; Osheroff *et al.*, 1991). Topo II is essential for the survival of eukaryotic cells and is important for DNA replication and transcription, mRNA processing, recombination, repair, and chromosomal condensation. It can relax, catenate, decatenate, knot, and unknot DNA.

Topo II alters DNA topology by passing an intact double strand of DNA through a transient double-stranded break. The catalytic cycle can be broken down into six steps (Osheroff, 1991): (1) enzyme–DNA binding; (2) prestrand passage DNA cleavage and religation; (3) DNA strand passage, which requires ATP binding; (4) poststrand passage DNA cleavage and religation; (5) ATP hydrolysis; and (6) enzyme turnover.

During this process an intermediate DNA–enzyme complex (the cleavable complex) is formed by a covalent bond between a tyrosyl residue on the enzyme and the phosphate moiety at the 5' terminal of DNA. By binding to this complex and stabilizing it, topo II-interacting drugs inhibit the rejoining action of the enzyme, resulting in DNA strand breaks and DNA–protein cross-links (Osheroff, 1989; Holm *et al.*, 1989; Robinson and Osheroff, 1990). A good correlation between DNA double-stranded breaks and cytotoxicity has been established (Glisson *et al.*, 1986; Covey *et al.*, 1988), although the DNA strand breaks may not be directly responsible for cytotoxicity; a subsequent event, such as interference with replication fork progression, may be the cause of cell death. Thus, DNA strand breaks are reversible upon the removal of drugs, and several observations suggest that continuing DNA and RNA synthesis is necessary for cytotoxicity to occur (Holm *et al.*, 1989; D'Arpa *et al.*, 1990; Fernandes and Catapano, 1991).

Two isoforms of topo II have been described. They have been purified and found to have molecular masses of 170 kDa (p170, or topo II alpha) and 180 kDa (p180, or topo II beta), respectively (Drake *et al.*, 1987). This isoforms have similar catalytic activities and ATP requirements, and qualitatively similar m-AMSA-stimulated covalent binding of DNA. The p180 topo II differs from the p170 isoform in being more heat labile, more resistant (three-fold) to VM-26 inhibition of catalytic activity (unknotting of P4 DNA), and more resistant to VM-26-induced DNA cleavage (3 to 25-fold).

By screening a Raji-HM_2 library with a *Drosophila* topo II cDNA probe, two classes of cDNA have been found (Chung *et al.*, 1989) and p170 and p180 have been recently mapped to human chromosomes 17 and 3, respectively (Tan *et al.*, 1992; Jenkins *et al.*, 1992). Close homology has been found between the N-terminal three-quaters of the two isoforms. The C-terminals differ, however, suggesting different cellular functions (Austin *et al.*, 1993).

The levels of topo II, and thus the sensitivity to topo II-inhibiting drugs, vary according to the cell cycle phase, the proliferative and developmental stage of the cells, and the degree of transformation. Maximum topo II levels are found in the G_2/M phase (Chow and Ross, 1987; Heck *et al.*, 1988) and maximum drug-induced DNA strand breaks in the G_2/M (Chow and Ross, 1987) or M and early G_1 phases (Estey *et al.*, 1987). Drug-induced cytotoxicity peaks in the S phase (Chow and Ross, 1987; Estey *et al.*, 1987; Markovits *et al.*, 1987). Cells in the plateau phase show a decrease in topo II content as well as decreased topo II activity and drug-induced DNA strand breaks, compared with cells in the log phase (Sullivan *et al.*, 1986, 1987; Zwelling *et al.*, 1987; Markovits *et al.*, 1987; Robbie *et al.*, 1988; Schneider *et al.*, 1988). In some experimental cell lines, equal amounts of topo II in log and plateau phases have been described (Sullivan *et al.*, 1986, 1987). Yet, the cytotoxic effect of VP-16 was greater in the log phase than in the plateau phase.

Increased topo II activity in response to transformation has been shown in regenerating rat liver (Duguet *et al.*, 1983), epidermal growth factor-stimulated human and mouse fibroblasts (Miskimins *et al.*, 1983), and concanavalin A-stimulated guineapig lymphoblasts (Taudou *et al.*, 1984).

The isoforms of topo II (p170 and p180) are influenced differently by the proliferative state of the cells. Generally, the ratio between p180 and p170 is higher in the plateau phase than the log phase, because of a decrease in p170, combined with an actual or relative increase in p180, indicating that p170 is mainly related to DNA synthesis and p180 to transcription (Drake *et al.*, 1987; Woessner *et al.*, 1990; Prosperi *et al.*, 1992). This may also explain why some cell lines have been found to have equal amounts of topo II in the log and plateau phases, as mentioned earlier.

a. Method for Determination of Topo II Activity Several assays have been employed for measuring topo II activity (Wang, 1985; Sullivan and Ross, 1991). Drug-induced DNA damage can be measured as DNA single-stranded breaks, DNA double-stranded breaks, or DNA–protein crosslinks by filter elution techniques. In a potassium–sodium dodecyl sulfate (K–SDS) assay, the covalent DNA–topo II complexes, which are increased in the presence of drugs, are measured by precipitating the complexes with potassium chloride and SDS.

The amount of topo II in cells or nuclear extracts can be quantitated immunologically. The catalytic activity of topo II from nuclear extracts can be quantified by gel electrophoresis techniques such as unknotting of phage P4 DNA or decatenation of kinetoplast DNA.

b. Topo II-Mediated Drug Resistance MDR caused solely by altered topo II is characterized by cross-resistance to epipodophyllotoxins and anthracyclines — but *not* vinca alkaloids — without decreased accumulation or increased expression of Pgp. It is often referred to as at-MDR, for "*at*ypical" MDR (compared with the "classic" P-glycoprotein (Pgp)-mediated MDR) or *a*ltered *t*opoisomerase II-MDR (Danks *et al.*, 1987; Beck, 1990).

Alterations in topo II-mediated drug activity can be divided into two groups: (1) quantitative changes (decreased cellular amount of enzyme) and (2) qualitative changes (decreased drug-induced DNA cleavage, differential expression of topo II isoforms, altered subcellular distribution of topo II, and changes in extrinsic factors modulating topo II activity).

Although many tumor cell lines with resistance due to alterations in topo II have been described, most cell lines resistant to topo II-inhibiting drugs have more than one mechanism of resistance. Many cell lines have reduced uptake of drug due to Pgp or other membrane-associated transport mechanisms, increased glutathione S-transferase, or increased repair of DNA damage (for a review of individual cell lines, see Sullivan and Ross, 1991).

c. Quantitative Changes in Topo II A decreased amount of topo II (measured as immunoreactive enzyme) in cell lines resistant to topo II-inhibiting drugs has been described in several cell lines. Four series of KB cell lines (Ferguson *et al.*, 1988; Liu *et al.*, 1989) selected for resistance to VP-16 all had decreased topo II content (20–66%) without qualitatively altered topo II (as shown by equal VP-16 inhibition of unknotting activity).

A Chinese hamster ovary (CHO) cell line resistant to 9-hydroxyellipticine showed a two- to threefold reduction in topo II content by Western blotting analysis and a 3.5-fold decrease in catalytic activity (Pommier *et al.*, 1987; Charcosset *et al.*, 1988).

Long *et al.* (1991) described one human colon and two human lung cancer cell lines selected for resistance to epipodophyllotoxins. These cell lines all had decreased amounts of topo II, topo II mRNA, and enzyme activity without altered drug sensitivity by the enzyme. Likewise, Webb *et al.* (1991) reported on a VP-16 resistant CHO cell line in which topo II activity was reduced five-fold, drug-induced DNA damage four-fold, enzyme amount 4.5-fold, and mRNA content five-fold. The cell line was three to seven-fold resistant to topo II inhibitors. Drug accumulation was unchanged. Thus, the decrease in topo II content seemed sufficient to explain the resistance in this cell line.

Two sublines of P388 leukemia resistant to DOX showed decreased amounts of immunoreactive topo II accompanied by decreased catalytic activity three- to fivefold and drug-induced DNA cleavage (Deffie *et al.*, 1989a). A decreased copy number of the topo II allele could account for the decrease in topo II content (Deffie *et al.*, 1989b). Two mRNA transcripts were identified. A 6.6-kb mRNA was present in both sensitive and resistant cell lines, but was reduced by seven- to eightfold in the resistant cell lines. A 5.6-kb mRNA was found only in the resistant cell lines, but the corresponding protein could not be detected by immunoblotting.

A P388 cell line selected for resistance with m-AMSA had a twofold reduction in both catalytic activity and content of topo II (Tan *et al.*, 1989). A Northern blot showed a 50% decrease in a 6-kb mRNA compared with sensitive cells, corresponding to the reduction in topo II content. Likewise, in resistant cells, a Southern blot showed a 50% reduction compared with sensitive cells in a fragment of DNA, and the topo II gene was found to be hypermethylated, implying that the decreased topo II content was caused by inactivation of one topo II allele by gene rearrangement and/or hypermethylation.

In a DOX-resistant CHO cell line and a mitoxantrone-resistant HeLa cell line, a decrease in topo II amount was found in spite of unchanged DNA and mRNA contents, raising the possibility that post-trancriptional events might account for the decrease in enzyme levels (Deffie *et al.*, 1992).

Thus, it seems clear that a decrease in cellular content of topo II in many instances contributes to MDR. The decrease in topo II amount can be caused by gene inactivation, leading to diminished transcription, or possibly by post-transcriptional events.

Also, in favor of the amount of topo II as determining for resistance, a CHO cell line mutagenized with ethyl methane sulfonate (EMS) had an increased amount of topo II and was found to be hypersensitive to anthracyclines, m-AMSA, mitoxantrone, VP-16, and ellipticine (with unchanged sensitivity to vincristine and cisplatin; Robson *et al.*, 1987a; Davies *et al.*, 1988). Likewise, Nitiss *et al.* (1992) recently reported that transfection of yeast with the topo II gene,leading to increased topo II expression, greatly increased sensitivity to topo II-active drugs.

d. Qualitative Changes in Topo II

i. Decreased Drug-Induced DNA Cleavage In a CHO cell line selected for resistance to VM-26, the cellular amounts and catalytic activity of topo II were equal in resistant and wild-type cells, but unlike wild-type topo II, the purified enzyme from the resistant cells was heat labile and showed diminished drug-stimulated DNA cleavage, whereas molecular masses and specific activities were equal (Glisson *et al.*, 1986; Sullivan *et al.*, 1989). Thus, it seemed that topo II from resistant cells cleaved DNA normally but was insensitive to drug inhibition of religation. Recently, Chan *et al.* (1993) reported a point mutation in the topo II gene. A base substitution at nucleotide 1478 changed the amino acid at position 493 i topo II from arginine to glutamine. As the normal allele was found to be lost, the cell line was homozygous for the mutant allele. This mutation might interfere with the drug-induced inhibition of DNA religation activity of the enzyme. Likewise, an HL60 cell line selected for resistance to m-AMSA had unchanged amounts of immunoreactive topo II compared with the wild type (Zwelling *et al.*, 1989). Purified topo II from resistant and wild-type cells had the same catalytic activity, whereas the enzyme from the resistant cell line was resistant to m-AMSA stimulation of the formation of complexes with DNA. Interestingly, while the resistant cells were 10- to 100-fold resistant to m-AMSA, they were only 1.6- and 2-fold resistant to DOX and VP-16, respectively; and while DNA strand breaks induced by m-AMSA were reduced in resistant cells, there was no difference in VP-16-induced strand breaks between resistant and sensitive cells, suggesting that the interaction of m-AMSA with topo II is different from that of VP-16 and DOX. The resistant and wild-type enzymes had the same molecular mass, but Southern blotting analyses gave different patterns after treatment with restriction enzymes. A point mutation in the topo II gene resulting in a change from arginine to lysine at amino acid position 486 was later identified (Hinds *et al.*, 1991).

Two sublines of CCRF–CEM cells selected for resistance to VM-26 also had unaltered amounts of topo II compared with the wild-type (Danks *et al.*, 1988; Wolverton *et al.*, 1989). Not only was topo II cleavage activity reduced (fewer drug-induced DNA strand breaks), the topo II catalytic activity was also reduced (decreased P4 unknotting activity). Reduction in topo II activity correlated well with the degree of resistance because, the more resistant cell line also had the greater reduction in VM-26 stimulated DNA cleavage. These cell lines were also found to have a greater requirement of ATP for topo II-dependent strand passage than the parent cell line (Danks *et al.*, 1989). Thus, alterations in ATP binding by topo II may also contribute to resistance to topo II inhibitors.

A point mutation, resulting in a change from arginine to glutamine at amino acid position 449, was found in one allele in both sublines (Bugg *et al.*, 1991).

ii. Topo II Isoforms In some cell lines, resistance may be caused by differential expression of the two different isoforms of topo II. In an m-AMSA-resistant P388 cell line with a reduced number of drug-induced DNA strand breaks and diminished topo II activity, the content of the p170 isoform was greatly reduced compared with the parent cell line, whereas there was no difference in content of the p180 isoform (Drake *et al.*, 1987). Harker *et al.* (1991) described a mitoxantrone-resistant HL60 cell line in which only the p170 isoform was present, whereas both isoforms were detected in the parent cell line. Instead, a 160-kDa immunoreactive species was found in the resistant cell line. Whether this was catalytically active was not determined. The p170 isoform was noted to be resistant to drug-induced DNA cleavage.

In a VP-16-resistant CCRF–CEM cell line, the amounts of the p170 and p180 isoforms were found to be roughly similar (within two fold) in resistant and parent cell lines (Patel and Fisher, 1993). In the resistant cell line, however, a chromosome 3p deletion was discovered, probably deleting one topo II p180 allele, and in the gene encoding the p170 isoform, a point mutation was found, changing lysine to asparagine at position 797. This is close to the catalytic site and may interfere with drug trapping by the cleavable complex. Thus, this cell line was shown to have changes in the genes for both topo II isoforms, both of which could contribute to drug resistance.

In contrast, a mechlorethamine-resistant Burkitt lymphoma cell line with unaltered content of p170 topo II, but greatly increased p180, was hypersensitive to topo II-active drugs (Tan *et al.*, 1987b; 1988). The topo II catalytic activity was increased three-fold. Thus, an altered ratio of expression of the two isoforms of topo II can also result in drug hypersensitivity (collateral sensitivity).

e. Altered Intracellular Distribution of Topo II It has been suggested that an altered intracellular distribution of topo II may contribute to drug resistance (Fernandes and Catapano, 1991). One of the VM-26-resistant CCRF–CEM cell lines mentioned earlier (Danks *et al.*, 1988) was found to contain threefold less immunoreactive topo II in the nuclear matrix, whereas the amount of total nuclear topo II (high-salt-soluble) was equal in resistant and sensitive cells. The catalytic activity of topo II in nuclear matrix preparations was reduced by six to seven-fold in the resistant cell line. Since DNA replication primarily takes place in the nuclear matrix, the reduction in matrix-associated topo II (without a concomitant reduction in total topo II) could explain the diminished topo II activity. A mutation in the enzyme may impair its incorporation into the nuclear matrix (Fernandes *et al.*, 1990). Likewise, Harker *et al.* (1991), in a mitoxantrone-resistant HL60 cell line, found that while whole-cell levels of topo II were decreased by 50%, compared with the parent cell line, catalytic activity in nuclear extracts was decreased four to five-fold. Topo II from both cell lines had equal drug-stimulated DNA cleavage. Since large amounts of decatenating activity were present in the hypotonic wash portions of nuclear preparations from the resistant cells, it seemed that resistance could, in part, be explained by an altered cellular distribution of topo II.

Recently, Sullivan *et al.* (1993) reported on a VP-16-resistant CHO cell line in which drug-induced DNA cleavage in whole cells was reduced, whereas it was unchanged in nuclear extracts. Differences in the cleavage pattern led the authors to conclude that resistance resulted from an alteration in a specific topo II population intimately associated with the nuclear matrix — a population that cleaves DNA into topological domains in the presence of topo II-active drugs.

f. Extrinsic Factors Modulation of topo II activity by extrinsic factors has also been proposed. In isolated nuclei from the murine mastocytoma cell line K21, the formation of an m-AMSA-induced topo II–DNA complex was reduced compared with whole cells (Darkin and Ralph, 1989). When protein from cytoplasmic extracts was added, complex formation was increased to the same level as in whole cells. A cytoplasmic protein was tentatively identified as a type of casein kinase (Darkin-Rattray and Ralph, 1991). Following growth arrest, the stimulatory effect of the cytoplasmic factor was lost, while the m-AMSA-induced topo II activity remained unchanged in isolated nuclei (Collet and Ralph, 1992). Thus, the resistance of quiescent cells to topo II-active drugs may be explained by decreased activation of topo II.

Sinha and Eliot (1991) found similar results in a DOX-resistant HL60 cell line. Nuclei from resistant cells were markedly more resistant to VP-16-induced DNA cleavage than wild-type nuclei. Cytosolic proteins from wild-type cells, but not from resistant cells, increased the drug-induced DNA cleavage in resistant nuclei. The nature of this cytosolic factor was not further characterized.

In a VP-16-resistant human melanoma cell line, Campain *et al.* (1993) found that drug-induced DNA cleavage was decreased in whole cells, but unchanged in isolated nuclei. Drug uptake was unchanged. Catalytic activity and drug-induced DNA strand breaks mediated by isolated topo II from resistant and sensitive cells were similar.

g. Phosphorylation of Topo II Though evidence has not yet been presented, changes in phosphorylation of topo II caused by mutations in protein kinases or phosphorylation sites on topo II could result in drug resistance, because phosphorylation has been shown to stimulate topo II activity *in vivo* (Ackerman *et al.*, 1989) and topo II has been shown to be phosphorylated *in vitro* by casein kinase II (Ackerman *et al.*,1988), protein kinase C (Rottmann *et al.*, 1987), and a cAMP-independent protein kinase (Sander *et al.*, 1984). Recently, it was shown that phosphorylation of topo II decreases its sensitivity to drug-induced DNA cleavage, while increasing the overall catalytic activity (DeVore *et al.*, 1992). The protein kinases could, in some instances, be identical to the extrinsic factors described earlier. (For a review of topo II phosphorylation, see Cardenas and Gasser, 1993.)

2. Mechanism of Action of Topo I Inhibitors

Like topo II, topo I influences DNA topology and is involved in DNA replication and transcription. It induces transient DNA single-stranded breaks and thereby likely provides the swivelling mechanism needed to advance the replication fork during DNA synthesis (Wang, 1985; Liu, 1989). During this process, topo I is bound to the 3′ end of the break. Topo I is not essential for cell viability and apparently its functions can be carried out by topo II (Trash *et al.*, 1984). Unlike topo II, its action is independent of ATP. The intracellular content does not vary with the proliferative state of the cell (Duguet *et al.*, 1983).

Owing, to the discovery that camptothecin acts through inhibition of topo I (Hsiang *et al.*, 1985), the drug has gained renewed interest (Sullivan and Ross, 1991; Slichenmyer *et al.*, 1993). Furthermore, several new water-soluble analogs, such as topotecan and CPT-11, without the unpredictable side effects of camptothecin, have been developed (Sullivan and Ross, 1991). The analogs CPT-11 and SK&F 104,864 have been found to be effective against cells exhibiting the MDR phenotype (Tsuruo *et al.*, 1988; Johnson *et al.*, 1989). The cytotoxicity is highly S-phase specific and only cycling cells are killed (Li *et al.*, 1972; Drewinko *et al.*, 1974).

Camptothecin acts by stabilizing a cleavable complex between topo I and DNA and, apparently through inhibition of the rejoining step of the enzyme and arrest of replication fork advancement, induces DNA single-stranded breaks and inhibits the catalytic activity (Hsiang *et al.*, 1985; 1989a; Covey *et al.*, 1989; Porter and Champoux, 1989). Though inhibition of topo I is well correlated with induction of DNA single-stranded breaks and cytotoxicity (Hsiang *et al.*, 1989b; Jaxel *et al.*, 1989), it seems that the interference of moving DNA replication forks with enzyme–DNA complexes is the crucial event because active DNA replication is necessary for cell death (Holm *et al.*, 1989; D'Arpa *et al.*, 1990; Yoshida *et al.*, 1993).

Camptothecin has no effect on topo II (Hsiang *et al.*, 1985), and yeast cells in which the gene encoding topo I has been deleted are resistant to camptothecin (Nitiss and Wang, 1988; Eng *et al.*, 1988). Transfection with a plasmid-borne human topo I gene was able to restore camptothecin sensitivity (Bjornsti *et al.*, 1989).

a. Topo I-Mediated Drug Resistance There are relatively few reports on camptothecin-resistant cell lines. As seen with topo II, it would appear that both quantitative and qualitative changes in topo I contribute to resistance to topo I inhibitors. The genetic basis for this has been elucidated (Tan *et al.*, 1989) in a P388 leukemia cell line resistant to camptothecin. The topo I mRNA was reduced in the resistant cell line, corresponding to the decrease in enzyme. Restriction digests of the topo I gene showed rearrangement in one allele and hypermethylation of the other, thus explaining the reduced transcription of the gene.

Qualitative changes in topo I have also been shown. Purified topo I from a cell line resistant to CPT-11, had the same catalytic activity as the wild-type enzyme, but a greater capacity for DNA cleavage in the absence of drug (Andoh *et al.*, 1987; Kjeldsen *et al.*, 1988). In the presence of drugs, there was no detectable inhibition of catalytic activity or induction of DNA strand breaks. It was recently reported that the mutant enzyme binds to DNA with approximately 10 times higher affinity than the wild type, and it may be this improved DNA binding ability that makes the enzyme resistant (Gromova *et al.*, 1993).

Purified topo I from a camptothecin-resistant CHO cell line also showed less inhibition of catalytic activity and induction of cleavable complexes by drug than the wild-type enzyme (Gupta *et al.*, 1988). Hybrids between resistant and sensitive cells were sensitive to camptothecin. Madelaine *et al.*, (1993) found that in the early stages of selection for resistance to camptothecin in a P388 cell line, a decrease in cellular topo I content was accompanied by reduced activity of the enzyme. At later stages, however, the enzyme had recovered the activity found in the wild-type cells. These workers also found rearrangement of one topo I allele accompanied by reduced topo I transcription.

In a P388 cell line made hyperresistant to camptothecin (1000-fold), no immunoreactive topo I was present and topo I activity was less than 4% of wild-type cells (Woessner *et al.*, 1993). A 130-kDa immunoreactive protein, possibly representing an altered, inactive form of topo I, was present. When grown in the absence of camptothecin, topo I content and activity increased. Topo I content was found to be inversely correlated to topo II content. These findings suggest that topo I may be influenced by camptothecin treatment.

b. DNA Topoisomerases in the Clinic Topo II has mostly been studied in samples from patients with treated or untreated hematologic malignancies (Brox *et al.*, 1986; Edwards *et al.*, 1987; Potmesil *et al.*, 1987; 1988; Gekeler *et al.*, 1992). Low levels of topo II activity and drug-induced DNA strand breaks were found in chronic leukemias (CML) and chronic lymphatic leukemia (CLL), variable levels were found in acute leukemia, and other hematologic cancers.

These findings do not seem to reflect the at-MDR phenotype, but are probably related to the proliferative status of the cells (Edwards *et al.*, 1987; Potmesil *et al.*, 1987; 1988). In line with this, Edwards *et al.* (1987) found that following stimulation of proliferation, patients' cells became as sensitive to VP-16 as cultured cells.

Thus, apart from indications that enzyme levels in chronic leukemias may be too low for topo II inhibitors to be effective, little is known about the occurrence of topo II-mediated resistance in human cancers, or how this may affect prognosis or treatment. In the few clinical studies available regarding topo I, the levels were similar or elevated compared with the corresponding normal tissues (Potmesil *et al.*, 1988; Giovanella *et al.*, 1989; van der Zee *et al.*, 1991; Hirabayashi *et al.*, 1992; Gekeler *et al.*, 1992). Whether any of the genetic changes found in the laboratory can be encountered in the clinic remains to be seen.

3. Future Prospects

In recent years, much progress has been made in our understanding of the role of topoisomerases as targets for anticancer agents and alterations in topoisomerases as a mechanism of drug resistance; however, much still remains to be clarified. The interaction between drug, topoisomerase, and DNA, and the events leading to cell death are not completely understood and need to be worked out in detail. Also, the relative importance of the two isoforms of topo II needs clarification. Structure–activity studies of the drug-enzyme–DNA interaction will help in the rational design of new topoisomerase-targeting drugs. Different topo II-inhibiting drugs have been found to bind to topo II at different sites on the enzyme, and to generate different patterns of DNA strand breakage (Zwelling *et al.*, 1992; Pommier, 1993; Capranico *et al.*, 1993). Concerning topo I, other drugs besides camptothecin derivatives have been found to inhibit the enzyme. Thus, actinomycin-D inhibits topo I (Trask and Muller, 1988). DOX, in addition to its effects on topo II, has also been found to inhibit topo I at concentrations achievable *in vivo* (Foglesong *et al.*, 1992). The quinone substitute, saintopin (Yamashita *et al.*, 1991) and certain indoquinolinediones (Riou *et al.*, 1991) have also shown simultaneous activity against topo I and topo II. Drugs with effects on both enzymes could be of interest because some of the actions of one enzyme can be carried out by the other, and topo I levels have been found to be normal or elevated in tumors with low topo II levels (Potmesil *et al.*, 1988; Gekeler *et al.*, 1992).

Though resistance against topo I- and topo II-active drugs has been described in many experimental cell lines, little is known about the importance of topoisomerase-related resistance mechanisms in the clinic. Thus, there is a growing need for clinical investigations concerning the frequency of alterations in topoisomerase activity and levels in different cancers before and after treatment with topo I- and II-active drugs, and their relation to prognosis and therapeutic response.

III. Antimetabolites

Antimetabolites are drugs that are chemically related to naturally occurring compounds and which interfere with the cellular metabolism, especially processes involved in synthesis of DNA. Traditionally this group is divided into folate antagonists, pyrimidine analogs, and purine analogs. Among the antifolates, methotrexate is the most important drug. Furthermore, since it was introduced in the clinic in 1948, it is also the most intensively studied and probably the best understood antineoplastic compound today. Consequently, in our description of mechanism of action and resistance, the greatest importance will be attached to methotrexate. Besides methotrexate, 5-fluorouracil (5-Fu), cytosine arabinoside, and the purine analogs will be described as representatives of other antimetabolites.

A. Methotrexate

1. Mechanisms of Action of Methotrexate

Methotrexate (MTX) is an analog of folic acid differing from the parent compound by substitution of an amino group for a hydroxyl group at position 4 of the pteridine ring and addition of a methyl group at the N^{10}-position, forming the molecule 2,4-diamino, N^{10}-methylpteroyl glutamic acid or amethopterin (Fig. 2). MTX is a bivalent anion with pK_a values estimated to be 3.76 and 4.83 for the α- and the γ-carboxyl groups, respectively (Hakala, 1985). Thus, both carboxyl groups are completely ionized at physiological pH.

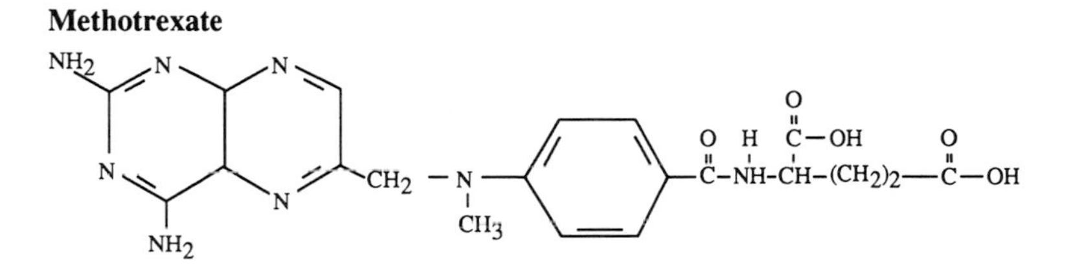

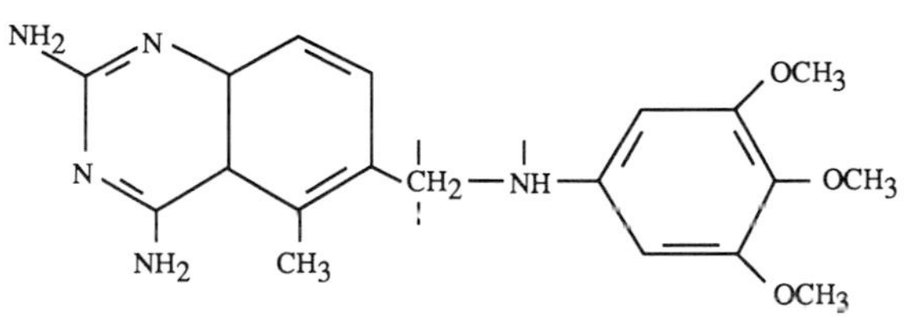

FIG. 2 Chemical structure of methotrexate and trimetrexate.

2. Membrane Transport

a. Influx MTX influx has been investigated in several experimental tumor cell lines. The studies indicate that MTX influx takes place as an active carrier-mediated transport (Goldman and Matherly, 1985; Dembo and Sirotnak, 1976; Henderson *et al.*, 1986a). MTX and reduced folate probably share a common carrier, whereas folic acid seems to be transported by a separate system, a low-affinity, high-capacity system (Sirotnak, 1987; Kamen and Bertino, 1980; Yang *et al.*, 1982). On the other hand, some findings indicate that folic acid is transported by the same route as reduced folate and MTX in L1210 cells under specific conditions (Yang *et al.*, 1983; Henderson, 1986b). The MTX transport system is also referred to as "the reduced folate system" and is characterized as a high-affinity, low-capacity system. This system has been shown to be independent of sodium, but is inhibited by organic and inorganic anions (Goldman and Matherly 1985).

That the entry takes place by a carrier is supported by the following findings: it shows saturation kinetics, demonstrates competitive inhibition by structural analogs, is highly temperature dependent, and is sensitive to sulfhydryl-inactivating agents such as ethylmaleimid and organic mercurials. The activity of the reduced folate transport system may depend on the substrate concentration in the culture medium. An upregulation of the carrier is observed when the concentration of reduced folate is low or during cell proliferation (Kane *et al.*, 1986). The carrier permits cells to accumulate 5-substituted reduced folate at concentrations several hundred times higher than the extracellular concentration (Anthony *et al.*, 1985). The receptor has been isolated in different mammalian cells and the molecular weight has been determined to be 40–50 kDa (Jansen *et al.*, 1989; Yang *et al.*, 1988).

The mechanism of the concentrative uptake against the electrochemical gradient does not depend on energy supply, but may be a counter transport by exchange of intracellular anions (inorganic phosphate, sulfate, or chloride anions as proposed by Goldman and Matherly, 1985; Henderson and Zevely, 1982, 1984). In various tumor cell lines, the value of K_m has been estimated to be 0.7 to 6 μM. In L 1210 cells, the K_m of the influx carrier has been determined as 4 μM, which should result in a maximum intracellular concentration of free drug of 6 μM (Goldman *et al.*, 1968). The carrier transport is competitively inhibited by N^5-methyltetrahydrofolate, the predominant naturally circulating folate, and by leucoverin (N^5-formyltetrahydrofolate).

b. Efflux The influx and the efflux transport systems are probably two independent systems. Thus, the two systems show different stereospecificity and a different inhibitory effect of probenecid (Dembo and Sirotnak, 1976; Goldman, 1969; Henderson and Zevely, 1985). Furthermore, changes in V_{max} for influx after development of drug resistance are not accompanied by changes in efflux (Sirotnak *et al.*, 1981b). The mechanism of efflux of MTX has been thoroughly reviewed by Goldman and Matherly (1985). The efflux shows inhibition by verapamil, bromosulfaphtalein, the epipodophyllotoxins, the vinca alcaloids, and probenecid. (Henderson and Tsuji, 1987). In contrast to influx, anions such as Cl^- do not influence efflux of MTX. Inhibition of oxidative phosphorylation by 2,4-dinitrophenol or sodium azide results in inhibition of MTX efflux, indicating that the efflux is energy dependent (Schlemmer and Sirotnak, 1992). According to Henderson *et al.*, (1986a), three routes of efflux exist: (1) one route is dependent on extracellular anions and inhibited by the *N*-hydroxysuccemid ester of MTX. This route may be common to the influx system. (2) Another route is inhibited by bromosulfaphtalein, which is probably the predominant route (Schlemmer and Sirotnak, 1992). (3) A route which is insensitive to the inhibitors described in 1 and 2, but is inhibited by probenecid.

3. Intracellular Actions

MTX exerts its cytotoxic effect by competitive inhibition of the cytosolic enzyme dihydrofolate reductase (DHFR) (Waltham *et al.*, 1988), which is the key enzyme in pyrimidine and purine biosynthesis (see Fig. 3). DHFR has been purified and consists of a single polypeptide with a MW of 21500 kDa in murine cells (Nunberg *et al.*, 1980). The human DHFR gene has been localized to chromosome 5 (Kanda, 1983). The affinity of MTX to DHFR is very high, with a K_d value of 10^{-11}–$10^{-12}M$. The binding is dependent on the concentrations of NADPH, NADH, and pH. The binding is very stable, but reversible, and consists essentially of one molecule of MTX to one molecule of enzyme. DHFR catalyzes the NADPH-dependent reduction of 7,8-dihydrofolate (FH_2) to the active form, 5,6,7,8-tetrahydrofolate (FH_4) and in a lesser degree, folate to tetrahydrofolate.

FH_4 is oxidized to FH_2 only when deoxyuridine monophosphate (dUMP) is converted to deoxythymidine monophosphate (dTMP) by thymidylate synthestase (TS). In the other one-carbon transfer reactions catalyzed by the formyltransferases (the GAR and AICAR reactions), FH_4 is regenerated. Inhibition of DHFR by MTX [or MTX (Glu)n] prevents to a substantial degree the regeneration of the active FH_4 from the inactive FH_2 and, subsequently, a depletion of the FH_4 pool will take place in addition to an elevation of the FH_2 pool. Since the normal level of DHFR is in great excess, however, > 95% of DHFR must be inhibited before the processes of converting dUMP to dTMP and *de novo* purine synthesis are impaired (Goldman and Matherly, 1985). Since the intracellular concentration of FH_2 accumulates behind the blocked DHFR, it will be able to compete for the binding sites on DHFR and thereby reverse the enzyme inhibition. To prevent this effect, a considerable free intracellular concentration of MTX must be maintained to ensure continued saturation of the enzyme. Increased activity of TS during DNA synthesis (in the S-phase of cell cyclus) increases the requirement for reduced folate, resulting in increased sensitivity to MTX.

On the contrary, low levels of TS activity as a result of exposure to 5-FU lead to relative resistance to MTX (Rodenhuis *et al.*, 1986; Curt *et al.*, 1985; Sengupta *et al.*, 1982). Leucoverin represents a form of reduced folate that circumvents the metabolic block of DHFR. It is also a competitive inhibitor of MTX influx because both compounds use the same carrier. As shown in Fig. 3, MTX (Glu)n also exerts a direct inhibitory effect on TS, the GAR-, and the AICAR-transformylases.

Like naturally occurring folates, MTX undergoes intracellular transformation to polyglutamated compounds by folylpolyglutamate synthetase (FPGS) (Fry *et al.*, 1982; Schilsky *et al.*, 1980). By this process, one to five additional γ-glutamyl residues of the native compound are formed. The actual intracellular concentration of polyglutamates is a result of the balance between FPGS and the corresponding hydrolase activity (Galivan *et al.*, 1987). As a result of the higher molecular weight, the polyglutamates

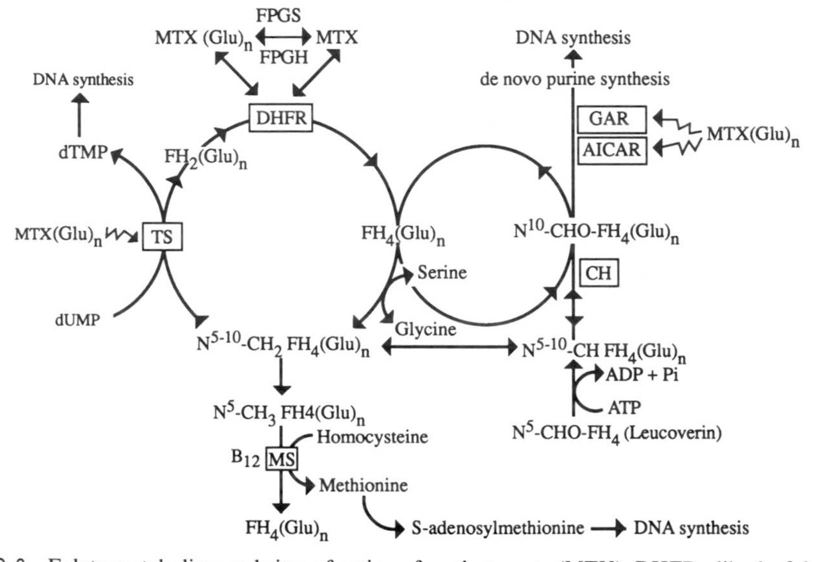

FIG. 3 Folate metabolism and sites of action of methotrexate (MTX); DHFR, dihydrofolate reductase; TS, thymidylate synthase; MS, methionine synthetase: CH, $N^{5\text{-}10}$-methyl-FH_4 cyclohydrolase; GAR, glycinamide ribotide transformylase; AICAR, aminoimidazole carboxamide ribotide transformylase: (Glu)n; polyglutamated metabolites of MTX or folate; FPGS, folylpolyglutamate synthetase; FPGH, folylpolyglutamate hydrolase; $N^{5\text{-}10}$-CH_2FH_4, 5,10-methylenetetrahydrofolic acid; N^{10}-CHO-FH_4, 10-formyltetrahydrofolate; N^5-CH_3-FH_4, 5-methyltetrahydrofolate; dUMP, deoxyuridine monophosphate; dTMP, deoxythymidine monophosphate.

are preferentially retained and they possess binding affinity at least equal to DHFR and to other folate-requiring cellular enzymes, such as TS and the transformylases (GAR and AICAR) (Allegra *et al.*, 1985b). Thus, polyglutamylation is probably an important determinant of MTX cytotoxicity. (McGuire *et al.*, 1985).

Trimetrexate (TMTX) is a 2,4-diaminoquinazoline folate analog. TMTX enters the cells by a different mechanism than the reduced folate system used by MTX. The influx is most probably by passive diffusion, but the finding of saturation, temperature sensitivity, and the sensitivity to metabolic inhibitors suggest the existence of a more complex transport route (Lin and Bertino, 1991). Trimetrexate is not polyglutamated and consequently it is rapidly effluxed from the cells. Like MTX, intracellular TMTX acts as a potent inhibitor of DHFR.

4. Mechanisms of Resistance on the Level of Drug Transport

Membrane transport processes determine the rate at which MTX penetrates the cell membrane. Since the target of MTX is the cytosolic DHFR, reduction in net uptake results in reduced cellular toxicity of MTX. Transport defects may be a result of decreased influx and/or increased efflux. Moreover, a reduced drug accumulation may be a result of reduced intracellular binding and/or reduced polyglutamylation.

5. Changes in Membrane Transport

a. Influx Resistance to MTX as a result of impaired drug transport was initially reported by Fischer (1962). Later, Kessel *et al.* (1965) could demonstrate that the sensitivity in different murine cell lines was narrowly correlated to the cellular uptake of MTX. Generally, a slower influx is observed as an early step in the development of resistance to MTX. The mechanism has been related to alterations in the reduced folate transport system; however, different mechanisms have been described. Thus, in some experimental models, the slower influx was due to increased value of K_m, leading to a reduced affinity to the carrier with almost no change in V_{max} (Sirotnak and Donsbach, 1976; Jackson *et al.*, 1975; Sirotnak *et al.*, 1968; Sirotnak, 1987). Selectively decreased levels of V_{max} have been described in another system (Niethammer and Jackson, 1975).

Finally, both increased K_m and decreased V_{max} have been described (Sirotnak, 1987). The last findings emphasize the fact that drug resistance in most cases is multifactorial. It seems surprising that the resistant cell with a defective transport system for MTX, which is common for the folate transport, can continue to grow in the presence of folate. Kinetic analysis of the folate transport in these cell lines has disclosed an alternative carrier system preferentially transporting folic acid (Henderson and Strauss, 1990; Jansen *et al.*, 1989). This transport system is not blocked by the inhibitors of the reduced folate carrier and it is characterized by being energy dependent and extremely pH dependent.

b. Efflux Changes in the efflux system leading to drug resistance have not been described thus far. According to Sirotnak *et al.*, (1981a), none of 14 MTX-resistant L1210 clonal cell lines exhibited altered efflux of [^{3}H]MTX.

6. Increased Dihydrofolate Reductase Activity

Gene amplification is a common finding in drug resistance (reviewed by Borst, 1991). Possibly this finding reflects the higher rate of gene amplification in cancer cells than normal cells (Wright *et al.*, 1990) due to genetic instability of cancer cells (Goldie and Goldman, 1979). That gene amplification plays a major role in MTX resistance was recognized by the observation of homogeneously staining regions (HSR) in MTX-resistant cells and subsequently by the demonstration of amplification of the gene coding for DHFR (reviewed by Borst, 1991; Dolnick and Bertino, 1982). As a result of the increased gene copy number, the target DHFR will be overproduced and thus allow cells to survive in spite of cytotoxic concentrations of MTX. Apart from HSR, the amplification can also be reflected in double-minute chromosomes, which are small chromatin particles that replicate and regregate in the absence of centromers and are highly unstable compared with HSR (Haber *et al.*, 1981). Possibly, gene amplification initially takes place as double-minute chromosomes, whereas in a highly resistant cell it occurs as HSR (Meltzer *et al.*, 1985). The exact mechanism of gene amplification is still not clarified. Mechanisms and examples of gene amplifications are reviewed in Borst (1991) and Hamlin (1992).

Data indicate that MTX resistance by gene amplification may be induced within a single cell cycle (Mariani and Schimke, 1984). Other data indicate that MTX resistance by gene amplification may also be induced by exposure to different unspecific injuries, such as UV light, irradiation, hypoxia (Kleinberger *et al.*, 1986), or nonantimetabolite cytostatics (Kleinberger *et al.*, 1986; Newman *et al.*, 1988; Rosowsky *et al.*, 1987; Mandelbaum-Shavit and Ramu, 1987). Generally DHFR isolated from resistant cells with gene amplification is identical to that of sensitive cells (Domin *et al.*, 1982a); however, increased DHFR activity may be accompanied by altered binding affinity, as described in the following paragraphs.

The increased activity of DHFR is not always a result of gene amplification. Thus, a fourfold increase in cellular DHFR has been observed and is probably a result of alterations on the level of mRNA translation (Cowan *et al.*, 1986; Domin *et al.*, 1982b). An MTX-resistant promyelocytic leukemia cell line had a 20-fold increase in DHFR activity without evidence of amplification (Dedhar and Goldie, 1985). When higher levels of DHFR activity are observed, however, it is usually due to gene amplification. Very high levels of DHFR have also been observed. A 300-fold increase in DHFR activity has been described in a murine lymphoblastoid cell line (Dolnick *et al.*, 1979) that was due to a several hundredfold elevation of the gene copies.

7. Decreased Polyglutamation

As described earlier, MTX is a substrate for the enzyme folylpolyglutamate synthestase (FPGS) by which polyglutamates are formed. These metabolites exert the same inhibitory effect on DHFR, TS, and the formlytransferases as MTX or are more active (Allegra *et al.*, 1985b; Whitehead, 1977, Matherly *et al.*, 1983). Because of a higher molecular weight and more hydrophilic properties, the polyglutamate derivatives of MTX are better retained in the cells than the parent compound. The actual concentration of polyglutamates is determined by the balance between folylpolyglutamate synthetase (FPGS) and folylpolyglutamate hydrolase in several tumor cell lines (see Fig. 3). The ability to generate MTX polyglutamate seems to be correlated with the sensitivity to MTX (Samuels *et al.*, 1985; Matherly *et al.*, 1985; Pizzorno *et al.*, 1988). In several resistant cell lines, however, significantly reduced polyglutamylation has been demonstrated (Pizzorno, 1988; Curt and Allegra, 1987). Different mechanisms may explain the decreased polyglutamylation. In some experimental systems, a decreased ability to polyglutamate is demonstrated as a consequence of mutations in FPGS (Bertino, 1990). In other experimental systems, however, reduced polyglutamatation has been observed without qualitative differences in FPGS between resistant cells and the parental cells (Pizzorno *et al.*, 1988; Cowan and Jolivet, 1984). In some of these examples, the mechanism could be increased activity of folylpolyglutamate hydrolase (Rhee *et al.*, 1993) or reduced affinity of Mg^{2+}ATP or glutamate for the enzyme, as suggested by Cook *et al.* (1987).

As previously mentioned, TMTX does not form polyglutamate. In the cases of MTX resistance where decreased polyglutamylation is a major mechanism of resistance, this analog may be able to overcome drug resistance (Cowan and Jolivet, 1984).

8. Altered Binding to DHFR

A number of MTX-resistant cell lines have been found to express DHFR, which demonstrates a reduction in the affinity for MTX that ranges from a few to several thousandfold (Goldie *et al.*, 1981; Dedhar and Goldie, 1983; Melera *et al.*, 1984; Melera *et al.*, 1987; Haber and Schimke, 1981; Blumenthal and Greenberg, 1970; Albrecht *et al.*, 1972; Jackson and Niethammer, 1977; Goldie *et al.*, 1980). In a few studies, cloning and subsequent sequence analysis made it possible to characterize the locus of the mutation. Thus, according to Simonsen (Simonsen and Levinson, 1983), a single point mutation resulted in substitution of arginine with lysine at amino acid 22. In another study, binding of MTX with the purified enzyme revealed that DHFR from the HCT-8 human colon carcinoma cell line had significantly lower affinity for MTX than the parental cell line. In this case, it could be explained by a point mutation, which implied that phenylalanine was replaced by serine at residue 31 in DHFR (Schweitzer *et al.*, 1990). As a result of the multifactorial mechanisms in drug resistance, it is not surprising to find both amplification of DHFR and reduced affinity for MTX to the enzyme in the same cell lines (Haber *et al.*, 1981; Srimatkandada *et al.*, 1989).

9. Decreased Activity of Thymidylate Synthase

As previously described, TS is the only enzyme that oxidizes FH_4 to FH_2, which is the substrate for DHFR. Consequently, reduced TS activity resulting from 5-FU inhibition, depletion of its substrate (deoxyuridylate), or addition of thymidine reduces the effect of MTX.Correspondingly, cell lines with low TS activity demonstrate less sensitivity to MTX (Moran *et al.*, 1973; Ayasawa *et al.*, 1981; Curt *et al.*, 1985). The role of this mechanism of resistance in the development of MTX resistance, however, has not been clarified.

10. Circumvention of MTX Resistance

Since impaired transport is an important factor in MTX resistance, it is possible to overcome this barrier by a considerable increase in the drug concentration. MTX thereby penetrates the membrane by passive diffusion. A selective rescue of normal cells may be obtained by subsequent addition of 5-formyltetrahydrofolate (leucoverin) in concentrations sufficient to reverse normal cells, which are able to transport reduced folate quickly by the reduced carrier transport system. If the concentration of 5-formyltetrahydrofolate is too high, however, the drug will penetrate the membrane of the resistant tumor by diffusion. In this case, the selective toxicity against tumor cells is lost.

Drug resistance may be circumvented by the use of analogs which are not substrates for the known mechanisms of resistance. Trimetrexate, a "non classical" folate antagonist, is such an analog. TMTX binds to DHFR like MTX, but as a result of its high lipophilicity, it probably enters the cells by passive diffusion. Futhermore, unlike MTX, it is unable to formate polyglutamates (Bertino *et al.*, 1987). If resistance to MTX is a result of impaired transport or reduced formation of polyglutamate, there will be no cross-resistance to TMTX—perhaps even collateral sensitivity. If MTX resistance is mostly a result of elevated DHFR level, however, cross-resistance is expected. It is interesting that cross-resistance between TMTX and MDR compounds has been observed (Assaraf *et al.*, 1989). This finding may be explained by a sequential amplification of both DHFR and *mdr1* genes.

B. 5-Fluorouracil

5-Fluorouracil (5-FU or 5-FUra) is one of the most widely used agents in the treatment of human cancer. This fluoropyrimidine was synthesized by Heidelberger *et al.* (1957). It represents one of the few examples of an antineoplastic drug which is synthesized on a rational basis. In this case, it is observed that rat hepatomas utilized more uracil than nonmalignant tissues. In attempts to increase the antitumor activity and reduce toxicity, several analogs have been developed [e.g., Ftorafur 1-(2-tetrahydrofuranyl)-5-fluorouracil and 5′-deoxy-5-flurouridine (5′-dFUrd)].

1. Structure and Mechanism of Action

5-FU represents the simplest derivative of uracil because only a hydrogen atom at the 5 position is replaced by a fluorine atom. 5-FU is not active per se, but requires intracellular conversion to nucleotides. Prior to the intracellular bioactivation, 5-FU has to enter the cell, which takes place by the same transport system as uracil.

The intracellular activation of 5-FU is illustrated in Fig. 4. It appears that cytotoxicity can be a result of three different mechanisms. Probably the most important is the inhibition of TS by 5-fluorouridine-5′-monophosphate (FdUMP). This interaction results in reduced *de novo* synthesis of thymidine, leading to thymineless cell death. TS catalyzes the reductive methylation of dUMP to dTMP. In the presence of the cofactor $N^{5,10}$-CH_2-FH_4 (Glu)n, a covalent ternary complex between FdUMP, TS, and $N^{5,10}$-CH_2-FH_4 (Glu)n is formed. An excess of reduced folate prolongs the enzyme inhibition. On the other hand, in the absence of reduced folate, only an unstable binary complex is formed between FdUMP and TS, resulting in a rather weak inhibition of the enzyme activity.

TS exists as a dimer of two identical subunits, each 36 kDa in size. The level of TS is very low in resting, nondividing cells. The maximum cellular

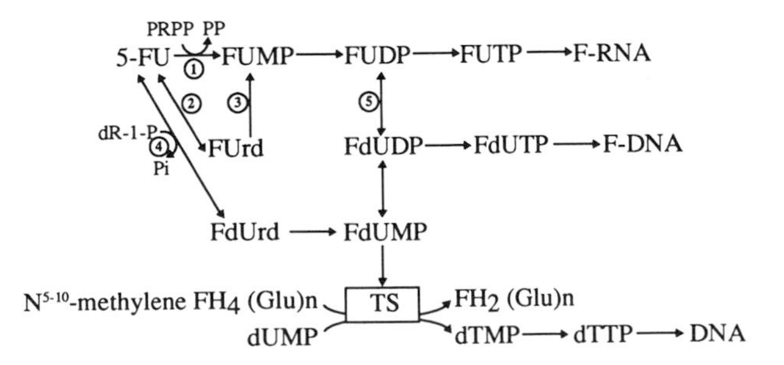

FIG. 4 Metabolism and sites of action of 5-FU. Circled numbers refer to the enzymes: (1) phosphoribosylpyrophosphatetransferase or orotate phosphoriboslytransferase; (2) uridine phosphorylase; (3) uridine kinase; (4) thymidine phosphorylase; (5) ribonucleotide reductase. TS, thymidylate synthetase. PRPP, phosphoribosylpyrophosphate; FdUrd, 2′-deoxy-5-fluorourdine; FUdr, 5-fluoro-2-deoxyuridine; FUMP, 5-fluorouridine-5′-monophosphate; FdUMP, 5-fluoro-2′-deoxyuridine-5′-monophosphate; FUDP, 5-fluorouridine-5′-diphosphate; FUTP, 5-fluoro-uridine-5′-triphosphate; FdUTP, 5′-fluoro-2′-deoxyuridine triphosphate.

activity of TS occurs when cells proceed from the G_1 to the S phase (about a 20-fold increase). Incorporation of FUTP into mRNA may also be a mechanism of action. Finally, the incoporation of FdUTP into DNA is a possible pathway by which 5-FU may produce single-stranded breaks and thereby cytotoxicity.

2. Mechanisms of Resistance

Since TS is the target enzyme of 5-FU, it is not surprising that cellular resistance to 5-FU is most often attributed to altered conditions for forming the ternary complex or to increased activity of TS, both mechanisms leading to diminished TS inactivation.

2A. *Alterations in the Formation of the Ternary Complex*

a. Generally, FdUMP is a very potent inhibitor of TS, with a K_i of about 1 n*M*. In several cases, resistance to 5-FU is accompanied by reduced affinity of FdUMP to TS (Pinedo and Peters, 1988; Bapat *et al.*, 1983; Priest *et al.*, 1980). Thus, in a human lymphocytic leukemia cell line, the dissociation constant for the ternary complex was increased 23-fold after development of resistance to 5-FU (Bapat *et al.*, 1983). The lower affinity of FdUMP for TS may be expressive of a mutation in the TS gene, leading to a variant form of TS (Berger *et al.*, 1988; Peters *et al.*, 1991; Bapat *et al.*, 1983).

b. In a study comparing 5-FU metabolism in human colon mucosa with that in colon tumors, a higher rate of all the activation enzymes was found, suggesting that the selective antitumor effect of 5-FU is related to increased activation of the compound (Peters *et al.*, 1991). Accordingly, resistance to 5-FU has been correlated with reduced activity of the different steps in this pathway. Probably most important is a reduction in orotate phophoribosyltransferase activity (Mulkins and Heidelberger, 1982) because this process represents the primary route for 5-FU activation. Resistance has also been correlated with reduced activity of uridine phosphorylase (Woodman *et al.*, 1980) and uridylate kinase (Ardalan *et al.*, 1980).

c. As described earlier, the availability of reduced folate determines the stability of the binding of FdUMP to TS. Not surprisingly, low folate pools are associated with 5-FU resistance (Houghton *et al.*, 1981; Yin *et al.*, 1983).

d. Reduced affinity for the cofactor $N^{5,10}$-CH_2-FH_4 to TS may also play a role for 5-FU resistance, which is supported by the study by Berger *et al.* (1988).

e. Intracellularly, $N^{5,10}$-CH_2-FH_4 is polyglutamylated as described previously. As demonstrated by Radparvar *et al.* (1989), the affinity of

$N^{5,10}$-CH_2-FH_4 for TS increases with the number of glutamate residues associated with the folate. Consequently, a defective polyglutamylation, which is a common mechanism of resistance to MTX, may also contribute to resistance to 5-FU. The existence of this mechanism of 5-FU resistance has been recently confirmed in a human adenocarcinoma cell line (Wang *et al.*, 1993). This mechanism of resistance may possibly be related to prolonged exposure to low-dose 5-FU (Aschele *et al.*, 1992).

2B. *Increased Activity of Thymidylate Synthase*

Several studies have demonstrated a considerable difference in the level of TS in human colon tumors where the cytotoxicity of 5-FU is probably related to altered activity of TS (Washtien, 1984; Peters *et al.*, 1991; Spears *et al.*, 1982: Washtien, 1982). Correspondingly, an increased level of TS is a common finding in cells resistant to 5-FU (Baskin *et al.*, 1975; Priest *et al.*, 1980; Jenh *et al.*, 1985; Berger *et al.*, 1985; Clark *et al.*, 1987; Spears *et al.*, 1982; Spears *et al.*, 1988; Washtien, 1984; Chu *et al.*, 1990; Priest *et al.*, 1980; Swain *et al.*, 1989). In several cases, the increased activity of TS appears to be a result of amplification of the structural gene encoding for TS (Clark *et al.*, 1987, Jehn *et al.*, 1985; Berger *et al.*, 1985).

3. Modulation and Circumvention of 5-FU Resistance

Although the mechanism of action of 5-FU is complex, the different pathways and the various factors influencing the processes have been extensively mapped. This knowledge makes it possible to perform biochemical modulation in order to increase therapeutic activity or to circumvent drug resistance. Very few other cytostatic compounds have been the subject of so many experimental and clinical trials with the intention of obtaining biochemical modulation of 5-FU.

First of all, many studies have been performed to evaluate the influence of the 5FU dose schedule. The duration and the concentration of 5-FU probably determine the mechanism of action and the mechanism of resistance. As demonstrated by Aschele *et al.* (1992), short-term exposure of cells to high concentrations of 5-FU appears to kill cells by a RNA effect, while prolonged exposure to a low dose appears to selectively inhibit the activity of TS. Interestingly, cells which have become resistant after treatment with intermittent high concentrations of 5-FU retained sensitivity to low dose exposure, whereas the opposite was not the case. The modulating effect of leucoverin has been extensively evaluated both experimentally and clinically (reviewed by Peters *et al.*, 1991; Allegra, 1990; Moran and Keyomarsi, 1987; Rustum *et al.*, 1987). Finally, a series of other compounds (MTX, phosphon-acetyl-L-aspartate, hydroxyurea, allopurinol, uridine, and interferon) are of interest for their ability to circumvent 5-FU resistance or to modulate the activity of 5-FU (Armstrong, 1989; Peters, 1991, Allegra, 1990; Peters and van Groeningen, 1991).

C. Cytosine Arabinoside

1-β-D-Arabinofuranosylcytosine, cytosine arabinoside, or Ara-C is an important drug for the therapy of human acute leukemia.

1. Mechanism of Action

Ara-C is an analog of the physiologic nucleoside, deoxycytidine. It probably enters the cell by the nucleoside carrier transport system. Then it is activated by phosphorylation to Ara-cytidine monophosphate (CMP) by deoxycytidine kinase (the rate-limiting step). Ara-CMP is further phosphorylated to Ara-cytidine diphosphate (CDP) (by dCMP kinase) and to Ara-cytidine triphosphate (CTP) (by nucleoside diphosphokinase), which is the active form of the drug. Ara-CTP inhibits DNA synthesis by competitive inhibition of DNA polymerase α (physiological substrate dCTP). The cytotoxicity, however, is related particularly to the incorporation into DNA. Apart from these mechanisms, Ara-C may also inhibit DNA repair by its effect on DNA polymerases as well as ribonucleotide reductase. Cytidine deaminase (CDD) catalyzes the initial catabolism of Ara-C to Ara-U, which is a much weaker inhibitor than Ara-C.

2. Mechanisms of Resistance

Since Ara-CTP is the active metabolite of Ara-C, the concentration and duration of exposure ($C \times T$) to this compound determine cells killed. Thus, all mechanisms which tend to reduce the intracellular level of Ara-CTP may result in resistance to Ara-C. Accordingly the following mechanisms of resistance may be relevant:

a. A decrease in deoxycytidine kinase is probably the most important mechanism of resistance to Ara-C. Reduced phosphorylation by this initial activating enzyme has been demonstrated in rodent cell lines (Chu and Fischer, 1965; Drahovsky and Kreis, 1970; Dechamps *et al.*, 1974; Young *et al.*, 1985) and in human leukemic cells (Kessel and Wilberding, 1969; Tattersall *et al.*, 1974).

b. There may be a decrease in the cellular uptake of Ara-C. This mechanism of resistance is so far only weakly documented for Ara-C. In a study by Wiley *et al.* (1985), however, a decreased number of nucleoside transport sites has been described in resistant acute leukemia cells.

c. Indications of increases in cytidine deaminase that lead to an accelerated inactivation of Ara-C have been demonstrated by Steuart and Burke (1971).

d. As Ara-C competes with dCTP, drug resistance against Ara-C may also be a consequence of an increased dCTP pool. A high concentration of dCTP in the resistant cells may be a result of increased CTP synthetase activity (de Saint Vincent *et al.*, 1980; Weinberg *et al.*, 1981) or increased ribonucleotide reductase (Meuth *et al.*, 1976).

e. Finally, since Ara-C is a specific inhibitor of cells in the S phase of the cell cycle, drug resistance *in vivo* may be a result of a reduction of the fraction of tumor cells in the S phase. Correspondingly, tumor cells that are localized in phamacological sanctuaries may escape the cytotoxic action of the drug. These phenomenons—kinetic resistance and pharmacological resistance—are reviewed by Pallavicini (1984).

D. Purine Antimetabolites

The clinically relevant purine antimetabolites include 6-mercaptopurine (6-MP) and 6-thioguanine (6-TG), both of which are widely used in the treatment of acute leukemias. 6-MP and 6-TG are substituted analogs of the physiological purines, hypoxanthine and guanine, respectively.

1. Mechanism of Action

The initial step of both compounds is a conversion to their respective ribonucleoside 5-monophosphate derivatives by hypoxanthine-guanine phosphoribosyltransferase (HGPRT). Subsequently, they are further activated to deoxyribonucleotides. In mammalian cells, 6-MP may be converted to 6-thioguanine deoxyribonucleotide. The cytotoxicity is probably correlated with the incorporation of the nucleotide derivatives in DNA, but 6-TG-derivatives may also be incorporated in RNA. 6-MP and 6-TG also inhibit *de novo* purine biosynthesis at several levels.

2. Mechanism of Resistance

An important mechanism of resistance seems to be an inability to perform the initial step of the activation, that is, the formation of the respective ribonucleoside monophosphates by HGPRT. This mechanism has been described in both rodent cell lines (Brockman, 1963; van Diggelen *et al.*, 1979) and human leukemic cells (Wolpert *et al.* 1971). Another important mechanism of resistance is probably an increase in the cellular catabolism of these compounds. The increased degradation may be a result of increased deaminase activity (LePage, 1968; Sartorelli *et al.*, 1958) or increased alkaline phosphatase (Wolpert *et al.*, 1971; Rosman *et al.*, 1974; Scholar and Calbresi, 1979), which inactivates the active mononucleotides. Finally, since phosphoribosyl pyrophosphatetransferase (PRPP) is an important factor both in the purine nucleotide synthesis and in the activation of 6-MP and 6-TG, reduced PRPP synthesis could be a factor contributing to resistance, as suggested by Higuchi *et al.* (1977).

IV. DNA Repair in Resistance to Cisplatin and Alkylating Anticancer Drugs

If it were fully efficient, the repair of DNA damage after insult by chemical and physical agents would result in a cell having no remaining damage, and cell kill would be zero. Viewed in this way, the recognition of cell death as a central goal in cancer chemotherapy provides a unifying theme for successful treatment. The cause of the intended cytotoxicity is often multifactorial and DNA is the intracellular locus of action for a wide variety of chemotherapeutic agents. The antitumor activity may be caused by an indirect or a direct interaction with DNA, or a combination of both.

The modes of binding to DNA are diverse. Simple alkylating agents such as nitrogen mustards and nitrosoureas can form covalent bonds between an electrophile carbon of an alkyl moiety and a nucleophile site in the DNA bases. In general, reaction at the guanine-N^7 and guanine-O^6 positions for agents that can produce positively charged alkylating intermediates correlates well with the nearest-neighbor base effect on the molecular electrostatic potential at the reaction site, resulting in a preferential reaction at runs of guanines. This suggests that the specific biological effects of such compounds may include preferential reaction at G + C-rich genomic locations.

Bifunctional alkylating agents such as *bis*(2-chloroethyl)amines (nitrogen mustards) have two reactive groups on each molecule and can react with two different nucleophilic centers in DNA. If these sites are situated on the same polynucleotide chain of a DNA duplex, the reaction product is referred to as an intrastrand crosslink. If the two sites are on opposite polynucleotide strands, a nitrogen mustard molecule can span the major groove and interstrand crosslinks may occur. Metal derivatives such as *cis*-diamminedichloroplatinum II (cisplatin) can also cross link DNA through the formation of coordination complexes between the N^7 atoms of adjacent purines in d(GpG) and d(ApG) sequences; the predominant cisplatin lesion is intrastrand crosslinks of neighboring guanines. Other agents bind noncovalently to DNA either by intercalating between base pairs (e.g., anthracyclines), or by nonintercalative groove binding (particularly in the minor groove, e.g., distamycin A), employing a combination of hydrophobic, electrostatic, hydrogen-bonding, and dipolar forces.

Little is understood of the basis for the wide differences in sensitivity to cytotoxic drugs shown by tumors of different pathological types. Many mechanisms are likely to be involved, as discussed in this chapter. Susceptibility to DNA damage may be the result of less effective repair when damage occurs at certain regions of susceptible intragenomic regions. The mammalian DNA repair processes appear to be considerably complex. They are regulated by many genes and there are many different repair pathways, depending upon the type of lesion and, as has become evident lately, also upon the location within the genome.

DNA repair may be defined as those cellular responses associated with the restoration of the normal nucleotide sequence and stereochemistry of DNA. It may be activated as a cellular response to spontaneous or induced damage to any of the primary components of the DNA molecule, and to changes in the structural configuration of the DNA. Much of our knowledge of these processes has been obtained in yeast, bacteria, and only more recently, rodents and humans. Yet, the major DNA repair processes appear to have been generally conserved through evolution, having served to counteract a wide spectrum of damage, including chemical, physical, and biological agents.

A. DNA Repair Mechanisms

In general, two fundamental reactions are involved in cellular responses to DNA damage: repair of the damage and tolerance of the damage. The major forms of DNA repair include reversal of damage, excision of damage, and postreplication repair [for a comprehensive text on many aspects of DNA repair, Friedberg (1985) is highly recommended].

1. Reversal of Damage

Direct reversal of DNA damage is in principle the simplest biochemical mechanism in which the cell restores the DNA structure (Fig. 5). In this repair response, a single enzyme catalyzes a single reaction. Photolyases break UV light-induced cyclobutane dimers directly by a reaction which is referred to as direct photoreversal: O^6-methylguanine–DNA methyltransferase, which is "suicidal" in the repair of O^6-methylguanine, reacts as a reagent rather than an enzyme in dealkylating O^6-alkylguanine by transfer of methyl (or alkyl) groups from guanine to a cysteine within an acceptor molecule. Purine insertase, albeit controversial, may alterna-

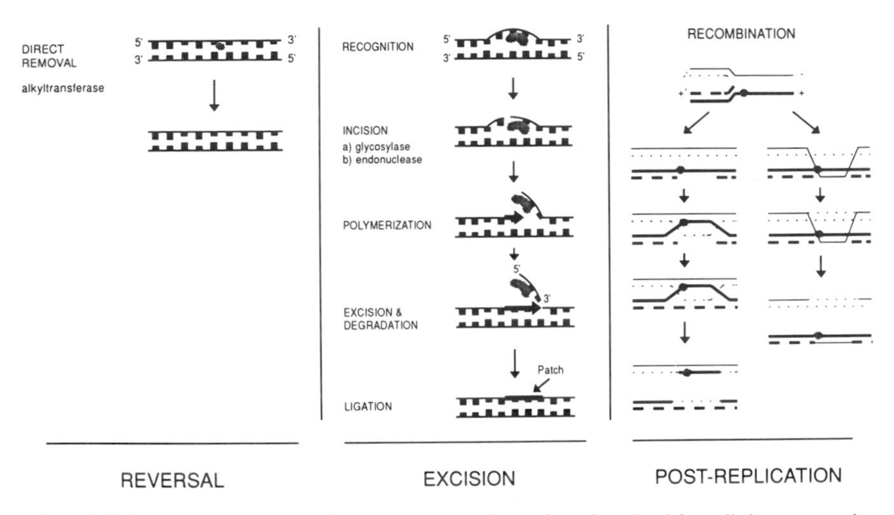

FIG. 5 DNA repair pathways. Three fundamental reactions involved in cellular processing of DNA damage are shown: reversal of damage, excision of damage, and recombination and postreplication.

tively reinsert an appropriate purine into an apurinic site, which may have occurred spontaneously. Polynucleotide ligase (DNA ligase) may be active in simple rejoining of DNA strand interruptions caused by hydrolysis of phosphodiester bonds, an event that occurs frequently upon reaction with several endonucleases, such as DNase I, or after ionizing radiation (Jacobs *et al.*, 1972).

2. Excision of Damage

If we rigorously define "repair" as the removal of lesions from DNA, then with the exception of photoreactivation, purine insertion, and simple DNA strand rejoining, all repair schemes are forms of excision repair (Fig. 5). Two basic types of excision are considered. One system, *base excision repair,* removes smaller types of base damage, such as a number of lesions caused by monofunctional alkylating agents. Another scheme, *nucleotide excision repair,* is a more versatile repair pathway, which operates in cellular responses to a large number of various lesions (e.g., bulky DNA chemical adducts and ionizing radiation) of which the UV light-induced cyclobutane pyrimidine dimer is the prototype.

Unlike the single-step repair of O^6-alkylguanine, excision repair is a multistep process which can be broadly divided into two stages: (1) incision at sites of DNA damage; (2) synthesis of new DNA to replace the damaged nucleotides. In the first stage, a multiprotein system locates a lesion (preincision recognition of DNA damage) and catalyzes enzymatic cleavage of the altered strand by incision of the damaged DNA strand near the defect, followed by excision of the defective site with concomitant degradation of the excised DNA strand. The incision step is achieved by the action of a distinct class of DNA repair enzymes called DNA glycosylases. These enzymes catalyze the hydrolysis of the *N*-glycosylic bond between the sugar and the base in the DNA backbone, leaving apurinic or apyrimidinic sites (AP sites). In the case of *N*-alkylpurines, a glycosylase excises the damaged base, whereas in the case of, for example, UV damage, the glycosylase that specifically recognizes cyclobutane pyrimidine dimers catalyzes the 5′ glycosyl bond in the dimer, leaving the 5′ pyrimidine covalently attached to its 3′ partner through the cyclobutane ring. The AP sites from *N*-alkylpurines and surrounding nucleotides of the lesion-containing strand (e.g., cyclobutane pyrimidine dimers or other DNA distortive lesions) are recognized and excised by AP endonucleases to displace the actual damaged oligonucleotide, leaving a gap. Replicative DNA synthesis proceeds by the action of DNA polymerases to form the gapped segment or a "patch" using the unmodified strand as a template, and repair is completed by ligating the "patch" to the parental strand by the action of a DNA ligase. The size of the repair patch appears to depend on the type of adduct that is being excised (Francis *et al.*, 1981; Huang *et al.*, 1992); the patch size in nucleotide excision repair is much larger (20–100 bases) than in base excision repair (1–5 bases).

3. Postreplication Repair

Excision of more complex damage such as DNA–DNA interstrand cross-links, double-stranded breaks, kinks, bends, or other bulky structural distortion in DNA may require repair of both DNA strands. This poses a special challenge to the repair machinery since the redundant information in the complementary strand will be excised and cannot direct the polymerase in the synthesis of the daughter strand DNA during semiconservative synthesis or the synthesis of a gapped DNA during repair synthesis. It appears that interstrand crosslinks are repaired by a combination of nucleotide excision and recombinational or postreplication repair pathways.

DNA containing physically or chemically induced lesions is a poor template for the DNA replication apparatus. Since DNA polymerase cannot ordinarily read through such damage, a gap is left in the newly synthesized DNA across from the damage site. The gap can be filled in by various postreplication repair mechanisms that utilize information from the sister duplex, such as translesion DNA synthesis or recombination (Fig. 5). Translesion synthesis might involve a copy-choice mechanism in which, upon reaching a block in the template strand, DNA polymerase switches to the undamaged sister duplex, carries out translesion synthesis using the undamaged sister duplex strand as a template, and then switches back to the damaged template strand to resume DNA synthesis beyond the site of the lesion. Alternatively, the gap could be filled in by mutagenic bypass, in which DNA polymerase inserts a wrong base across from the damage site in the template (Prakash *et al.,* 1993). Also, a gap left opposite the lesion in the daughter strand, which is filled in by strand exchange with intact homologous DNA (recombination), is not the result of repaired DNA, since the lesion remains.

In mammalian cells, sister chromatid exchanges can be visualized at the cytogenetic level, and their frequency is increased after DNA damage. Yet this frequency is less than 0.1% of the frequency of damage to the DNA. Recombinational events may be involved in the production of sister chromatid exchange, and disturbances in recombination may be involved in the generation of chromosome abnormalities.

B. The Role of DNA Repair in Determining Cytotoxicity

Along with the fast-growing advances in our knowledge of how cells deal with DNA damage, and parts of the complex biochemistry underlying these events, there has been increasing evidence that DNA repair may play a more important role in drug resistance than previously thought. Processing of DNA damage has evolved as a factor of considerable interest in determining responses to cancer therapy. The biochemical pathways involved in processing alkylating drug damage have been studied extensively. A number of studies with cultured human and rodent cells and human tumor xenografts have shown that the development of cellular resistance to alkylating agents is invariably associated with an enhanced repair activity for the cytotoxic alkyl adducts induced by these drugs (Schold *et al.,* 1989; Kaina *et al.,* 1991). Cells and tumors resistant to 2-haloethyl-*N*-haloethyl-*N*-nitrosourea in particular have a higher repair activity of O^6-alkylguanine, the primary toxic lesion induced by these drugs (Brent *et al.,* 1985). It has been shown, at least in the few cases investigated so far, that the level of O^6-methylguanine–DNA methyltransferase (MGMT) mRNA parallels the MGMT activity (Fornace *et al.,* 1990; Ostrowski *et al.,* 1991; Pieper *et al.,* 1991; Minnick *et al.,* 1993). Assuming that the MGMT message level reflects the rate of its transcription, it is suggested that the development of resistance to these nitrosoureas is due to increased transcription of MGMT genes. The mammalian MGMT is localized in the cell nucleus, except for the nucleolus (Ayi *et al.,* 1992). Transformed human cell lines that have switched off expression of MGMT (Mer^-, methyl excision repair deficient in comparison to Mer^+ cells) are hypersensitive to various genotoxic properties of O^6-methylguanine in DNA, but can be complemented by expression of either bacterial or human MGMT (Kaina *et al.*, 1991; Wu *et al.*, 1992).

Treatment of chloroethylnitrosourea-resistant cells with streptozotozin prior to exposure to 1,3-bis(2-chloroethyl)-*N*-nitrosourea (BCNU), has been shown to result in depleted MGMT activity, increased BCNU-induced interstrand cross-linking, and 2–3 log enhancement of BCNU cytotoxicity *in vitro* (Pieper *et al.*, 1991); streptozotozin is an alkylating agent, which methylates the active site of MGMT and thereby depletes its activity. *In vivo* data from patients with chronic myelogenous leukemia have also demonstrated that a correlation exists between the level of MGMT in peripheral blood lymphocytes and cellular resistance to nitrosoureas (Gerson and Trey, 1988).

Although resistance to nitrosoureas is invariably associated with elevated repair of potentially lethal O^6 alkylguanine residues, some cell lines resistant to haloethyl-nitrosoureas, for example, have, in addition to elevated MGMT levels, a higher repair capacity for 3-methyladenine due to higher levels of 3-methyladenine DNA glycosylase (Matijasevic *et al.*, 1991). Yet, exclusive overexpression of *N*-methylpurine–DNA glycosylase has been reported to result in an increased removal of *N*-methylpurines in DNA without a concomitant increase in resistance to alkylating agents *in vitro* (Ibeanu *et al.*, 1992). This suggests that *N*-methylpurine–DNA glycosylase activity is not limiting in the multistep repair pathway of *N*-alkylpurines.

Expression of MGMT may not be the sole determinant for efficient repair of simple DNA alkylation. A recent report indicated that a lack of either MGMT expression or nucleotide excision repair significantly impaired the ability of human cells to withstand DNA ethylation damage (Bronstein *et al.*, 1992). Evidence from cultured human cells indicates that nucleotide excision repair plays an important role in the repair of ethylated DNA (Maher *et al.*, 1990; Bronstein *et al.*, 1991). Yet, inhibition of MGMT by O^6-benzylguanine in a human cell line derived from xeroderma pigmentosum complementation group A, which expressed MGMT but was deficient in nucleotide excision repair, did not enhance the cytotoxicity of ethylnitrosourea (Bronstein *et al.*, 1992). This suggests that MGMT and nucleotide excision repair cooperate in the removal of DNA ethyl adducts.

The DNA lesions caused by platinum complexes and bifunctional alkylating agents, such as the nitrogen mustards melphalan, chlorambucil, and cyclophosphamide, are more varied, but increased DNA repair has been implicated for some cell lines in resistance to these agents. It is well established that a fraction of platinum–DNA lesions formed *in vivo* are repaired by mammalian cells. Nucleotide excision repair appears to be responsible for alleviating some of the toxic effects of cisplatin since repair-defective mutants are hypersensitive to the drug, as demonstrated by studies of murine leukemia L1210 cells (Sheibani *et al.*, 1989), human xeroderma pigmentosum cells (Poll et al., 1984; Dijt *et al.*, 1988), and human testicular tumor cells *in vitro* (Bedford *et al.*, 1988). Moreover, an increased removal of cisplatin-induced interstrand crosslinks and relatively low induction of unscheduled DNA synthesis in human cervical carcinoma cells resistant to cisplatin *in vitro* (Nishikawa *et al.*, 1992), and an enhanced removal of platinum from DNA in an abnormal cisplatin-sensitive human testicular nonseminomatous germ cell line have been reported (Kelland *et al.*, 1992). In human T lymphocytes with inherently different sensitivity to cisplatin, it was recently suggested that enhanced DNA repair may be a biological compensatory mechanism for cells that cannot prevent cellular uptake of DNA-damaging agents (Dabholkar *et al.*, 1992).

One of the first steps in a given repair pathway must be recognition of damaged DNA. Proteins have now been identified in human tumor cells which can bind to DNA damage produced by ultraviolet radiation or cisplatin (Chu and Chang, 1990; Donahue *et al.*, 1990; Chao *et al.*, 1991). These damage recognition proteins have been shown to be defective in certain xeroderma pigmentosum cell lines (Chu and Chang, 1988), and overexpressed in cisplatin-resistant cell lines with increased DNA repair proficiency. It is therefore believed that these recognition proteins may be involved in the initial steps of the excision repair pathway. However, their role in clinical chemoresistance to platinum compounds has yet to be explored. An increase in damage recognition protein was detected in HeLa cells resistant to cisplatin and UV light (Chao, 1992). On the other hand, repair-defective cells derived from xeroderma pigmentosum patients (defective in UV damage repair) did not express fewer cisplatin damage recognition proteins than repair-competent cells, suggesting that expression of damage recognition proteins may not necessarily be related to DNA repair of ciplatin-mediated DNA lesions. Since the cells resistant to cisplatin consistently expressed high levels of UV-modified DNA damage recognition proteins, these may be different from those that bind to cisplatin-mediated DNA damage. Expression of the human DNA repair gene *ERCC-1* (excision repair cross-complementing group 1) has been implicated in cisplatin resistance because *ERCC-1* confers resistance to cisplatin in UV repair-deficient Chinese hamster ovary cells (van Duin *et al.*, 1986; Bohr *et al.*, 1988; Parker *et al.*, 1990) and is thought to correlate with resistance to cisplatin in human ovarian cancer (Parker *et al.*, 1990).

Historically, nitrogen mustard was the first anticancer agent to demonstrate efficacy for human cancer. This bifunctional akylating agent is still in active use and in particular, has contributed to the cure of Hodgkin's disease. Subsequently, several analogs have been synthesized and have proved useful in the treatment of several malignancies. However, resistance to these agents develops, as is seen with many other chemotherapeutic agents. The mechanism of resistance to nitrogen mustard, however, is still poorly understood. There seems to be substsantial evidence indicating that the overexpression of glutathione and glutathione-dependent enzymes is an important mechanism of acquired drug resistance to bifunctional alkylating agents (Robson *et al.*, 1987b; Lewis *et al.*, 1988; Medh *et al.*, 1991). Also, overexpression of metallothionein (Kelley *et al.*, 1988; Kaina *et al.*, 1990) and cell cycle perturbations (O'Connor *et al.*, 1991; 1992) have been reported to play a role in differential sensitivity to nitrogen mustards.

Whereas numerous reports have suggested an increased repair capacity of cells resistant to cisplatin, the evidence for a role for DNA repair in resistance to nitrogen mustard is far more varied. A breast cancer cell line previously selected for threefold resistance to melphalan had a rate of DNA interstrand crosslink removal two-to threefold higher than in sensitive cells (Batist *et al.*, 1989). Also, in patients with chronic lymphocytic leukemia who were resistant to melphalan, enhanced removal of melphalan-induced DNA crosslinks has been reported (Torres-Garcia *et al.*, 1989). In comparison, melphalan-induced DNA crosslinks persisted in previously untreated patients with chronic lymphocytic leukemia (Torres-Garcia *et al.*, 1989).

C. Modulation of DNA Alkylation Repair

Augmentation of alkylation cytotoxicity by concomitant treatment with various inhibitors of DNA repair may suggest a role for DNA repair in determining the cytotoxicity to these agents. Hence, 1-β-arabinofuranosylcytosine, hydroxyurea, and aphidicolin have been widely used as modulators of DNA repair in order to circumvent drug resistance. The major cytotoxic mechanism of the cytidylic acid analog Ara-C appears to be incorporation into DNA, resulting in an inadequate primer terminal for further chain elongation (Fram and Kufe, 1982; Kufe and Spriggs, (1985). Hydroxyurea specifically inhibits ribonucleotide reductase, which is responsible for the conversion of ribonucleotides to deoxyribonucleotides. Thus, hydroxyurea depleted DNA precursor pools, shutting down DNA synthesis. When Ara-C and hydroxyurea are used together, it has been suggested that these two agents inhibit resynthesis of the repair patch in DNA excision repair in mammalian cells (Collins *et al.*, 1977; Snyder *et al.*, 1981; Fram and Kufe, 1985). This inhibition of the polymerization step is likely due to an Ara-C-mediated inhibition of DNA polymerases.

Combinations of Ara-C and hydroxyurea have been tested in human colon carcinoma cells *in vitro* and showed potentiation of cisplatin-induced cytotoxicity (Swinnen *et al.*, 1989).

Aphidicolin is an inhibitor of DNA polymerase α, which is mainly involved in DNA replication. While aphidicolin has been shown to enhance the sensitivity to cisplatin in a human ovarian cell line (Masuda *et al.*, 1988), another study has reported a negligible effect of this inhibitor in cells with resistance to L-phenylalanine mustard (L-PAM, melphalan) (Frankfurt, 1991). Combinations of Ara-C and hydroxyurea or aphidicolin with Ara-C and hydroxyurea, however, effectively inhibited DNA repair of L-PAM in a human ovarian carcinoma cell line with acquired resistance to L-PAM (Frankfurt, 1991).

Frankfurt *et al.* (1990) generated a monoclonal antibody against nitrogen mustard-treated DNA. They showed that the antibody interacted with single-stranded regions of alkylated DNA, and that the binding of antibody to the cells *in situ* increased in proportion to the decrease in cell viability and was thereby able to distinguish resistant cells from sensitive cells (Frankfurt *et al.*, 1990). Interestingly, by measuring the immunoreactivity with this antibody, cisplatin did enhance the cytotoxicity of L-PAM in a human lung carcinoma cell line with natural resistance to alkylating agents, but not in a human ovarian cell line with acquired resistance to L-PAM (Frankfurt, 1991). Since no reactivity of the antibody could be observed in the lung cells treated with cisplatin alone, this may be interpreted as an inhibition of DNA repair of alkylation damage. The observations also suggest that different DNA repair mechanisms may be active in resistance to nitrogen mustards and cisplatin. While the critical step in cisplatin resistance is likely to be at the stage of preincision recognition of specific lesions, the important repair step in the processing of damage from bifunctional alkylating agents could be a postinsicion event. Furthermore, the potentially lethal interstrand crosslinks induced by cisplatin may be repaired by a combination of nucleotide excision and recombinational or postreplication repair pathways, which may not be active in the removal of alkylation damage. This notion would be supported by the aforementioned report of aphidicolin-mediated sensitization of cisplatin cytotoxicity (Masuda *et al.*, 1988).

Considering DNA repair as a cellular response to restore the stereochemistry of DNA upon distortion of the helical structure, one may regard the action of topoisomerases as part of this process. This has been suggested by studies that demonstrate lower topoisomerase II activity in cells from patients suffering from genetic disorders characterized by reduced DNA repair (Vosberg, 1985; Heartlein *et al.*, 1987). In addition to catalyzing changes in DNA linking number, topoisomerase II has been shown to be a major structural component of the nuclear matrix and chromosome scaffold, and thus helps to maintain chromosomal DNA in an organized looped conformation (Earnshaw *et al.*, 1985; Vosberg, 1985).

Other roles of topoisomerase II have been suggested. They include functions in chromatin assembly, gene replication and transcription, and DNA repair (Vosberg, 1985; Wang, 1985). These indications are in line with the findings of Tan *et al.* (1987b), who reported on the molecular basis of acquired resistance to nitrogen mustard in a human Burkitt lymphoma cell line. They concluded that increased topoisomerase II activity sustained nitrogen mustard resistance, possibly by altering the structure of chromatin so as to increase the accessibility of nitrogen mustard-induced monoadducts to repair enzymes, thus reducing the formation of interstrand crosslinks. This was also hypothesized on the basis of the chromatin in the resistant cells being digested more efficiently by DNase I than be chromatin in the parent cells (Tan *et al.*, 1987a). Moreover, novobiocin, an antibacterial agent that inhibits the topoisomerase II enzyme, has been reported to produce enhanced tumor cytotoxicity of several alkylating agents (Eder *et al.*, 1987). Recently, novobiocin was shown to modulate cyclophosphamide cytotoxicity in refractory cancer patients (Eder *et al.*, 1991). It was suggested that inhibition of topoisomerase II may lead to an increase of the formation of DNA interstrand crosslinks by decreasing the repair of drug-induced monoadducts.

D. Alkylating Anthracyclines

In an effort to dissociate antitumor activity from cardiotoxicity and the problem of multidrug resistance (MDR) frequently encountered with the anthracyclines doxorubicin (DOX) and daunorubicin (DNR), a series of DOX analogs was synthesized by Acton *et al.* (1984, 1988). One of these analogs, cyanomorpholino doxorubicin, was unique in that it was noncardiotoxic at optimal cytotoxic doses (Acton *et al.*, 1984; Sikic *et al.*, 1985), 100- to 1000-fold more potent than DOX (Johnston *et al.*, 1983; Acton *et al.*, 1984; Sikic *et al.*, 1985; Streeter *et al.*, 1985; Wassermann *et al.*, 1986), and active in tumor cells resistant to DOX (Sikic *et al.*, 1985; Streeter *et al.*, 1985). Interestingly, this new analog has been shown to exert its extreme cytotoxic action through an alkylating mechanism by producing DNA interstrand crosslinks (Begleiter and Johnston, 1985; Westendorf *et al.*, 1985; Wassermann *et al.*, 1986, 1990b; Jesson *et al.*, 1987). Recently, a human ovarian carcinoma cell line resistant to cyanomorpholino DOX was established by growing it in increasing concentrations of the drug (Lau *et al.*, 1990). This cell line was also associated with cross-resistance to cross-linking agents such as cisplatin, melphalan, and carmustine, as well as to the DNA strand-breaking agent VP-16, and to ionizing radiation (Lau *et al.,* 1991).

Whereas an increased glutathione content and glutathione S-transferase activity seemed to explain the phenomenon of cross-resistance to the strand-breaking agents, the cells were proficient in repair of cisplatin-induced DNA crosslinks. In case of cyanomorpholino DOX, however, resistance was associated with more apparent DNA cross-linking followed by less DNA strand breakage, as seen in the sensitive parental cell line (Lau *et al.,* 1990). A profound and time-dependent DNA fragmentation following drug removal has been observed in cells sensitive to cyanomorpholino doxorubicin (Wassermann *et al.,* 1986; Lau *et al.,*1990) and in human leukemia cells exposed to *N*-(5,5-diacetoxypentyl) doxorubicin, another DNA cross-linking DOX analog (Zwelling *et al.,* 1991). Neither cyanomorpholino DOX (Wassermann *et al.,* 1990b) nor *N*-(5,5-diacetoxypentyl) DOX (Zwelling *et al.,* 1991) seemed to produce protein-associated DNA cleavage, which suggests that their primary target is not topoisomerase II. Also, 3′,4′-deamino or modified amino anthracyclines have been shown to be inefficient in inhibiting the DNA-joining activity of human replicative DNA ligase (Ciarrocchi *et al.,* 1991). The reported DNA fragmentation was probably due to endonucleolytic cleavage (a feature of apoptosis) and explains the apparent decline in cross-linking following drug removal. Thus, the mechanism of resistance to these DNA cross-linking anthracyclines may be a consequence of increased repair of secondary DNA fragmentation and/or inhibition of specific endonuclease activity.

E. Gene-Specific Repair

Although many reports have demonstrated a role for DNA repair in drug resistance, there are still conflicting observations in several other studies. While several other cellular mechanisms may be active in drug resistance (discussed elsewhere in this chapter), measurements of DNA repair at the level of overall genome may not always correlate with resistance to chemotherapeutic agents. Intragenomic heterogeneity of DNA damage removal has now been reported for various chemically and physically induced DNA lesions, and examination of repair in specific genomic sequences has provided meaningful correlations between DNA repair and biological end points such as survival (Bohr *et al.,* 1986). It was initially demonstrated that repair of UV-induced cyclobutane pyrimidine dimers were much more efficient in the essential gene for DHFR in Chinese hamster ovary cells than the repair in noncoding regions and the overall genome (Bohr *et al.*, 1985). This phenomenon of preferential DNA repair was later extended to several other gene systems and has been confined to the transcribed strand of the gene (Mellon *et al.*, 1987; Mellon and Hanawalt, 1989; May *et al.*, 1993).

Gene-specific repair analysis of a number of different DNA lesions has further revealed preferential repair of nitrogen mustard-induced *N*-alkylpurines in various intragenomic regions (Wassermann *et al.*, 1990a; 1992; 1994) (Fig. 6), intragenomic heterogeneity of the processing of nitrogen mustard-induced interstrand crosslinks (Futcher *et al.*, 1992; Larminat *et al.*, 1993), psoralen adducts (Vos and Hanawalt, 1987), preferential repair of *N*-methyl-*N*-nitrosourea-induced apurinic sites (LeDoux *et al.*, 1990; 1991), and of cisplatin-induced intrastrand crosslinks (Jones *et al.*, 1991). Recently, Zhen *et al.* (1992) reported increased gene-specific repair of cisplatin-mediated lesions in cisplatin-resistant human ovarian cell lines, although no difference was observed at the overall genome level. They demonstrated a more efficient removal of cisplatin-induced interstrand crosslinks from the genes of DHFR, MDR (*mdr1*), and δ-globin genes in

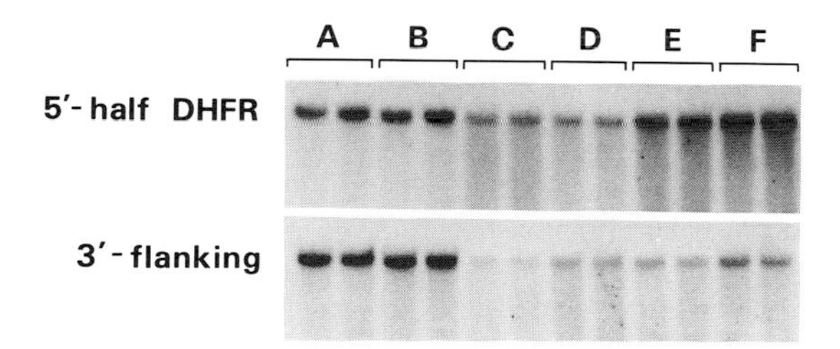

FIG. 6 An example of gene-specific repair of nitrogen mustard-induced *N*-alkylpurines in the hamster dihydrofolate reductase (DHFR) gene. Purified DNA from Chinese hamster ovary cells that had been treated with 150 μM nitrogen mustard for 30 min at 37°C were digested with *Kpn*I and processed as described (Wassermann *et al.*, 1990a). Autoradiography of Southern blots detected hybridization of ^{32}P-labeled genomic probes made from pMB5 to a 14-kb *Kpn*I fragment of the 5′ half of the actively transcribed DHFR gene region and cs-14DO to a 14-kb *Kpn*I fragment of the 3-flanking noncoding region of the DHFR gene. The panels shown are autoradiograms of the Southern blots. Duplicate samples of DNA in lanes A and B were from a nonprocessed control and a processed control, respectively. Duplicate samples in lanes C–F were from cells treated with 150 μM nitrogen mustard and repair-incubated for 0 hr (C), 4 hr (D), 8 hr (E), and 24 hr (F). The increasing intensity of the 14-kb fragment in the 5′-half DHFR gene region with time indicates repair of *N*-alkylpurines. Densitometry analysis revealed that 27%, 52%, and 78% of lesions were removed in 4 hr, 8 hr, and 24 hr, respectively, from the actively transcribed gene region. About 12%, 18%, and 41% of lesions were removed from the noncoding region in 4 hr, 8 hr, and 24 hr, respectively.

the resistant lines than in their parental partners. There was no difference in the removal of intrastrand crosslinks. Reed *et al.* (1988; 1990) have reported a reasonable correlation between the level of cisplatin adducts in leukocyte DNA and therapeutic response in ovarian cancer patients. In a subgroup of these patients, however, a high cisplatin adduct level did not correlate with a favorable response to therapy (Reed *et al.*, 1987). It is conceivable that further studies of gene-specific repair analysis in sensitive and resistant cells would increase the predictive value of chemosensitivity and demonstrate interesting correlations.

V. Glutathione S-Transferase and Alkylation Resistance

The multifunctional isozymes of glutathione S-transferase (GST), which are found in virtually all tissues, are presumed to be involved in the detoxification of not only electrophilic herbicides,insecticides, and carcinogens, but also certain antineoplastic agents and their metabolites (Jakoby, 1978). Recently, studies have indicated that GSTs may play an important role in the resistance of cancer cells to certain anticancer drugs.

Diverse biological functions of the GSTs are mediated by multiple GST enzymes. These enzymes generally exhibit broad and overlapping substrate-specific aspects and functions and are differentially inducible by exposure to diverse drugs and other xenobiotics. Mammalian GSTs are usually grouped into three distinct classes of soluble or cytosolic isozymes, each corresponding to a gene family, on the basis of their immunochemical properties and primary structure. These classes are designated α, μ, and π; there is also a fourth family of membrane-bound GSTs [microsomal GST bound to the membrane of endoplasmatic reticulum; see Mannervik and Danielson (1988) for a review].Unique patterns of tissue-specific gene expression characterize the human GST enzymes.

The multiple isozymes of GST catalyze conjugation of electrophilic compounds to glutathione (GSH), γ-glutamylcysteinglycine, an intracellular cysteine-containing tripeptide at high concentrations, comprising up to 10% of the total soluble cellular protein in certain tissues. The reactions catalyzed by GSTs appear to involve the generation of the thiolate anion of glutathione at the enzyme-active site. The reactive thiolate moiety subsequently participates in the nucleophilic attack on hydrophobic substrates at electrophilic centers such as carbon or oxygen. In some of these reactions, and especially reactions involving alkylating agents such as the nitrogen mustards, electrophilic substrates form stable glutathionyl conjugates that are further metabolized and eliminated as mercapturates and other chemicals (Mannervik, 1985):

$$R\text{-}NR'\text{-}CH_2\text{-}CH_2\text{-}Cl + GSH \rightarrow R\text{-}NR'\text{-}CH_2\text{-}CH_2\text{-}SG + HCl$$

GSH-dependent conjugation and denitrosation of nitrosoureas are preferentially catalyzed by GSTμ subunit 4. In the case of BCNU, this denitrosation leads to an inhibition of DNA crosslink formation and subsequently a decrease in cytotoxicity (Ali-Osman, 1989). Moreover, cisplatin–DNA adducts (Eastman, 1991; Godwin *et al.*, 1992; Satta *et al.*, 1992), and hydroxy peroxides such as reactive organic toxins produced in response to doxorubicin-generated free radicals (Mimnaugh *et al.*, 1989), may be substrates for GST isozymes.

The role of GSTs in the sensitivity of tumor cells to various anticancer agents is frequently based on correlative observations in wild-type versus drug-resistant cells, and experimental approaches to modulation of GSH–GST activity. Several studies have shown that overexpression of GSTs is associated with the emergence of resistance to alkylating agents. Increased GST activity has been demonstrated in cells selected *in vitro* for resistance to chlorambucil and other nitrogen mustards (Robson *et al.*, 1987b). In ovarian carcinoma cells obtained from a patient before and after the appearance of clinical resistance to chlorambucil and cisplatin, elevated expression of GST was observed in the cells obtained during the drug-refractory phase of the patients' illness (Wolf *et al.*, 1987). Other data have also indicated that the complement of isozymes expressed, as well as total GST activity, may influence the pattern of resistance to particular alkylating agents. For instance, GSTμ subunits 3 and 4 possessed a greater catalytic activity toward BCNU than nitrogen mustard in BCNU-resistant brain tumor cells, which were relatively sensitive to nitrogen mustard (Evans *et al.*, 1987; Matsumoto *et al.*, 1989). Resistance to nitrogen mustard in a Chinese hamster ovary cell line has been associated with large overexpression of GSTα (Lewis *et al.*, 1988), although direct evidence for the participation of GST in intracellular drug inactivation was absent. Transfection studies with GSTπ (Moscow *et al.*, 1989), GSTα (Leyland-Jones *et al.*, 1991), or GSTμ and GSTα (Townsend *et al.*, 1992) carried out in MCF7 human breast cancer cells or GSTπ in NIH-3T3 cells (Nakagawa *et al.*, 1990) have failed to confer significant resistance to melphalan, chlorambucil, or cisplatin. Other recent transfection studies with human GSTα, μ, or π class genes, however, provided a modest but significant decrease in sensitivity to nitrogen mustards and chlorambucil (Black *et al.*,1990; Puchalski and Fahl, 1990; Manoharan *et al.*, 1991), supporting a role for certain isozymes in determining the cytotoxicity to these alkylating agents.

Generally, clear evidence for a direct metabolic role of GSTs in tumor cell resistance is still lacking. Interpretation is complicated by the observation that resistant cells selected by prolonged drug exposure often exhibit multiple genetic and biochemical differences from the parental cell lines. As such, P-glycoprotein, cytochrome P-450, selenium-dependent GSH peroxidases, and DNA repair enzymes may all be gene products affected and involved in the selection of subpopulations with acquired drug resistance. Moreover, results obtained with rodent GSTs cannot automatically be extrapolated to human GSTs of the same class, since markedly different substrate specificities can characterize GSTs that exhibit as high as 93% overall amino acid sequence identity (Burgess *et al.,* 1989). Also significant, and of potential relevance for cancer chemotherapy, is a genetic polymorphism. For the human μ class, about 50% of the Caucasian population fail to express the activity in liver or leukocytes (Seidegård *et al.,* 1986). The trait is a genetic polymorphism in both alleles (Seidegård *et al.,* 1988).

Manipulations or pharmacological modulation of intracellular glutathione levels and GST activities have often been used in studies that suggest an influence on the sensitivity of cultured cells to alkylating agents. A reduction in cellular GSH levels can be achieved with the irreversible inhibitor of γ-glutamylcystein synthetase, buthionine sulfoximine, an enzyme required for GSH biosynthesis. Buthionine sulfoximine has been a widely used candidate in the modulation of GSH pools. Thus, it can lead to a substantial reversal of alkylating agent resistance in drug-resistant tumor cells *in vitro* and *in vivo* (Ozols *et al.,* 1988; Medh *et al.,* 1991; Lewis *et al.,* 1992). The diuretic drug ethacrynic acid has also been shown to sensitize human melanoma cells to melphalan (Hansson *et al.,* 1991) and human ovarian carcinoma cells resistant to the DNA cross-linking anthracycline cyanomorpholino DOX (Lewis *et al.,* 1992). Ethacrynic acid interferes with GST activity by being a substrate for GSTμ-mediated GSH conjugation (Hansson *et al.,* 1991). However, experimental procedures using manipulation or modulation strategies may be confounded by the existence of multiple glutathione-dependent reactions, in addition to those involving GSTs. Furthermore, because of the multifactorial nature of cellular resistance to alkylating agents, it is conceivable that in some systems, modulators could have a substantial biochemical effect on GST activity and yet provide only limited reversal of resistance, since the GST–GSH system represents only one of a number of determinants for cellular sensitivity. Although much remains to be learned about the exact biochemistry of the GSH–GST system as it relates to anticancer drug metabolism, numerous studies have provided considerable circumstantial evidence linking alterations in GSH and GST levels (and hence cellular GST expression) to the emergence of resistance of tumor cells to alkylating antineoplastic agents. With gene transfer experiments, we are beginning to assemble confirmatory direct evidence.

VI. Concluding Remarks

During the past 10–15 years a tremendous amount of data on cellular resistance have appeared. This new knowledge, however, has not resulted in clarity or simplification of the field. On the contrary, for each of the compounds, the mechanism of drug resistance appears complex, and within some areas, especially MDR, several hypothesis exist. The mechanisms of resistance as we currently understand them can be summarized as follows:

1. The classic MDR phenomenon, characterized by overexpression of the membrane glycoprotein Pgp.
2. Atypical MDR, which may be related to alterations in the topoisomerases or to overexpression of other membrane proteins, such as MRP. These mechanisms are responsible for cellular resistance to anthracyclines, vinca alkaloids, epipodophyllotoxins, aminoacridines. The exact mechanism by which cells reduce their sensitivity, among other things the function of Pgp, has not been clarified.
3. Increased synthesis or reduced binding affinity of a series of cytoplasmic enzymes. Regarding MTX and 5-FU, the most important enzymes are dihydrofolate reductase and thymidylate synthetase. In the case of the other antimetabolites, cellular resistance is often related to reduced activation and/or increased inactivation.
4. Increased repair of DNA or increased tolerance of DNA damage.
5. Overexpression of the detoxifying enzyme glutathione S-transferase. The two last-mentioned mechanisms are probably involved in resistance to alkylating agents and cisplatin derivatives.
6. Apart from a Pgp-induced redution in drug accumulation, changed membrane transport may play an important role in drug resistance to most of the cytostatics.

Based on this knowledge, an intensive search for compounds that are capable of modulating drug resistance has taken place. Some of these (e.g., verapamil) have been tested in clinical cancers. Circumvention of drug resistance, however, is complicated by both the demand for high specificity of the modulating compound and the general finding of multifactorial drug resistance. At present, however, our knowledge about the mechanisms of resistance may be used to create new drugs on a rational basis.

References

Abdella, B. R. J., and Fisher, J. (1985). *Environ. Health Perspect.* **64,** 3–18.

Ackerman, P., Glover, C. V., and Osheroff, N. (1985). *Proc. Natl. Acad. Sci. USA.***82,** 3164–3168.

Ackerman, P., Glover, C. V., and Osheroff, N. (1988). *J. Biol. Chem.* **263,** 12,653–12,660.

Acton, E. M., Tong, G. L., Mosher, C. W., and Wolgemuth, R. L. (1984). *J. Med. Chem.* **27,** 638–645.

Acton, E. M., Wassermann, K., and Newman, R. A. (1988). *In* "Anthracycline and Anthracenedione-based Anticancer Agents" (J. W. Lown, Ed.), pp. 55–101. Elsevier, New York.

Akman, S. A., Forrest, G., Chu, F. F., Esworthy, S., and Doroshow, J. H. (1990). *Cancer Res.* **50,** 1397–1402.

Alabaster, O., Woods, T., Ortiz-Sanchez, V., and Jahangeer, S. (1989). *Cancer Res.* **49,** 5638–5643.

Albrecht, A. M., Biedler, J. L., and Hutchison, D. J. (1972). *Cancer Res.* **32,** 1539–1546.

Ali-Osman, F. (1989). *Cancer Res.* **49,** 5258–5261.

Allegra, C. J., Drake, J. C., Jolivet, J., and Chabner, B. A. (1985b). *Proc. Natl. Acad. Sci. USA* **82,** 4881–4885.

Allegra, C. J. (1990). *Cancer Chemother.* 110–153.

Alon, N., Busche, R., Tummler, B., and Riordan, J. R. (1991). *In:* "Molecular and Cellular Biology of Multidrug Resistant Tumor Cells" (I. B. Roninson, Ed.), pp. 263–276. Plenum, New York.

Ambudkar, S. V., LeLong, I. S., Zhang, J., Cardarelli, C. O., Gottesman, M. M., and Pastan, I. (1992). *Proc. Natl. Acad. Sci. USA* **89,** 8472–8476.

Ames, G. F. L., Mimura, C. S., and Shyamapa, V. (1990). *FEMS Microbiol.* **6,** 429–446.

Andoh, T., lshii, K., Suzuki, Y., lkegami, Y., Kusunoki, Y., Takemoto, Y., and Okada, K. (1987). *Proc. Natl. Acad. Sci. USA* **84,** 5565–5569.

Anthony, A. C., Kane, M. A., Portillo, R. M., Elwood, P. C., and Kolhouse, J. F. (1985). *J. Biol. Chem.* **260,** 14,911–14,917.

Ardalan, B., Cooney, D. A., Jayaram, H. N., Carrico, C. K., Glazer, R. L., Macdonald, J., and Schein, P. S. (1980). *Cancer Res.* **40,** 1431–1437.

Armstrong, R. D. (1989). *In* "Resistance to Antineoplastic Drugs" (D. Kessel, Ed.), pp. 317–350. CRC Press, Boca Raton, FL.

Aschele, C., Sobrero, A., Faderan, M. A., and Bertino, J. (1992). *Cancer Res.* **52,** 1855–1864.

Assaraf, Y. G., Molina, A., and Schimke, R. T. (1989). *J. Biol. Chem.* **264,** 18,326–18,334.

Austin, C. A., Sng, J.-H., Patel, S., and Fisher, L. M. (1993). *Biochim. Biophys. Acta* **1172,** 283–291.

Ayasawa, D., Koyama, H., and Eono, T. (1981). *Cancer Res.* **41,** 1497–1501.

Ayi, T. C., Loh, K. C., Ali, L. R. B., and Li, B. F. L. (1992). *Cancer Res.* **52,** 6423–6430.

Baas, F., Jongsma, A. P. M., Broxterman, H. J., Lankelma, J., Schoonen, W. G. E. J., van Rijn, J., Pinedo, H. M., and Joenje, H. (1990). *Cancer Res.* **50,** 5392–5398.

Bapat, A. R., Zarov, D., and Danenberg, P. V. (1983). *J. Biol. Chem.* **258,** 4130–4136.

Barrand, M. A., Broxterman, H. J., Wright, K. A., Rhodes, T., and Twentyman, P. R. (1993). *Br. J. Cancer* **65,** suppl. 21.

Baskin, F., Carlin S. C., Kraus, P., Friedkin, M., and Rosenberg, N. (1975). *Mol.Pharmacol.* **11,** 105–117.

Bates, S. E., Mickley, L. A., Chen, Y. N., Richert, N., Rudick, J., Biedler, J. L., and Fojo, A. T. (1989). *Mol. Cell Biol.* **9,** 4337–4344.

Bates, S. E., Currier, S. J., Alvares, M., and Fojo, A. T. (1992). *Biochemistry* **31,** 6366–6372.

Batist, G., Tulpule, A., Sinha, B. K., Katki, A. G., Myers, C. E., and Cowan, K. H. (1986). *J. Biol. Chem.* **261,** 15,544–15,549.
Batist, G., Torres-Garcia, S., Demuys, J.-M., Greene, D., Lehnert, S., Rochon, M., and Panasci, L. (1989). *Mol. Pharmacol.* **36,** 224–230.
Bech-Hansen, N. T., Till, J. E., and Ling, V. (1976). *J. Cell Physiol.* **88,** 23–32.
Beck, W. T., and Cirtain, M. C. (1982). *Cancer Res.* **42,** 184–189.
Beck, W. T. (1987). *Biochem. Pharmacol.* **36,** 2879–2887.
Beck, W. T. (1990). *Cancer Treat. Rev.* **17,** (Suppl. A), 11–20.
Beck, W. T., and Danks, M. K. (1991). *In:* "Molecular and Cellular Biology of Multidrug Resistant Tumor Cells" (I. B. Roninson, Ed.). pp. 3–56. Plenum, New York.
Bedford, P., Fichtinger-Schepman, A. M. J., Shellard, S. A., Walker, M. C., Masters, J. R. W., and Hill, B. T. (1988). *Cancer Res.* **48,** 3019–3024.
Begleiter, A., and Johnston, J. B. (1985). *Biochem. Biophys. Res. Commun.* **131,** 336–338.
Bellamy, W. (1993). *Proc. Am. Assoc. Cancer Res.* **34,** 305.
Belli, J. A., Zhang, Y., and Fritz, P. (1990). *Cancer Res.* **50,** 2191–2197.
Benard, J., Da Silva, J., Teyssier, J. R., and Riou, G. (1989). *Int. J. Cancer* **43,** 471–477.
Berger, S. H., Jenh, C.-H., Johnson, L. F., and Berger, F. G. (1985). *Mol. Pharmacol.* **28,** 461–467.
Berger, S. H., Barbour, K. W., and Berger, F. G. (1988). *Mol. Pharmacol.* **34,** 480–484.
Bertino, J. R., Sobrero, A., Mini, E., Moroson, B. A., and Kashmore, A. (1987). *NCI Monogr.* **5,** 87–91.
Bertino, J. R. (1990). *J. Clin. Pharmacol.* **30,** 291–295.
Bhalla, K., Hindenburg, A., Taub, R. N., and Grant, S. (1985). *Cancer Res.* **45,** 3657–3662.
Biedler, J. L., and Riehm, H. (1970). *Cancer Res.* **30,** 1174–1184.
Bjornsti, M.-A., Benedetti, P., Viglianti, P. A., and Wang, J. C. (1989). *Cancer Res.* **49,** 6318–6323.
Black, S. M., Beggs, J. D., Hayes, J. D., Bartoszek, A., Muramatsu, M., Sakai, M., and Wolf, C. R. (1990). *Biochem. J.* **268,** 309–315.
Blumenthal, G., and Greenberg, D. M. (1970). *Oncology* **24,** 223–229.
Bohr, V. A., Smith, C. A., Okumoto, D. S., and Hanawalt, P. C. (1985). *Cell* **40,** 359–369.
Bohr, V. A., Okumoto, D. S., and Hanawalt, P. C. (1986). *Proc. Natl. Acad. Sci. USA* **83,** 3830–3833.
Bohr, V. A., Chu, E. H. Y., van Duin, M., Hanawalt, P. C., and Okumoto, D. S. (1988). *Nucl. Acids Res.* **16,** 7397–7403.
Boiocchi, M., and Toffoli, G. (1992). *Eur. J. Cancer* **28A,** 1099–1105.
Borst, P. (1991). *Acta Oncol.* **30,** 87–105.
Borst, P., Baas, F., Lincke, C. R., Ouellette, M., Schinkel, A. H., and Smit, J. M. (1992). *In* "Drug Resistance as a Biochemical Target in Cancer Chemotherapy" (T. Tsuruo and M. Ogawa, Eds.), pp. 11–26. Academic Press, New York.
Boscoboinik, D., Gupta, R. S., and Epand, R. M. (1990). *Br. J. Cancer* **61,** 568–572.
Bradley, G., Juranka, P. F., and Ling, V. (1988). *Biochim. Biophys. Acta* **948,** 87–128.
Brent, T. P., Houghton, P. J., and Houghton, J. A. (1985). *Proc. Natl. Acad. Sci USA* **82,** 2985–2989.
Brockman, R. W. (1963). *Cancer Res.* **7,** 129–234.
Bronstein, S. M., Cochrane, J. E., Craft, T. R., Swenberg, J. A., and Skopek, T. R. (1991). *Cancer Res.* **51,** 5188–5197.
Bronstein, S. M., Hooth, M. J., Swenberg, J. A., and Skopek, T. R. (1992). *Cancer Res.* **52,** 3851-3856.
Brox, L. W., Belch, A., Ng, A., and Pollock, E. (1986). *Cancer Chemother. Pharmacol.* **17,** 127–132.
Broxterman, H. J., Pinedo, H. M., Kuiper, C. M., van der Hoeven, J. J. M., de Lange,

P., Quak, J. J., Scheper, R. J., Keizer, H. G., Schuurhuis, G. J., and Lankelma, J. (1989). *Int. J. Cancer* **43,** 340–343.

Bugg, B. Y., Danks, M. K., Beck, W. T., and Suttle, D. P. (1991). *Proc. Natl. Acad. Sci. USA* **88,** 7654–7658.

Burgess, J. R., Chow, N.-W. I., Reddy, C. C., and Tu, C.-P. D. (1989). *Biochem. Biophys, Res. Commun.* **158,** 497–502.

Callen, D. F., Baker, E., Simmers, R. N., Seshadri, R., and Roninson, I. B. (1987). *Hum. Genet.* **77,** 122–126.

Campain, J. A., Padmanabhan, R., Hwang, J., Gottesman, M. M., and Pastan, I. (1993). *J. Cell. Physiol.* **155,** 414–425.

Campos, L., Guyotat, D., Archimbaud, E., Calmard-Oriol, P., Tsuruo, T., Troncy, J., Treille, D., and Fiere, D. (1992). *Blood* **79,** 473–476.

Cano-Gauci, D. F., and Riordan, J. R. (1987). *Biochem. Pharmacol.* **36,** 2115–2123.

Cano-Gauci, D. F., and Riordan, J. R. (1991). *In:* "Molecular and Cellular Biology of Multidrug Resistant Tumor Cells" (I. B. Roninson, Ed.), pp. 337–347. Plenum, New York.

Capranico, G., De Isabella, P., Tinelli, S., Bigioni, M., and Zunino, F. (1993). *Biochemistry* **32,** 3038–3046.

Cardenas, M. E., and Gasser, S. M. (1993). *J. Cell Sci.* **104,** 219–225.

Carlsen, S. A., Till, J. E., and Ling. V. (1976). *Biochim. Biophys. Acta* **455,** 900–912.

Carman, M. D., Schornagel, J. H., Rivert, R., Srimatkandada, S., Portlock, A. S., and Duffy, T. (1984). *J. Clin. Oncol.* **2,** 16–26.

Cass, C. E., Janowska-Wieczorek, A., Lynch, M. A., Sheinin, H., Hindenburg, A. A., and Beck, W. T. (1989). *Cancer Res.* **49,** 5798–5804.

Cazin, J. L., Gosselin, P., Cappelaere, P., Robert, J., and Demaille, A. (1992). *J. Cancer Res. Oncol.* **119,** 76–86.

Chambers, T. C., McAvoy, E. M., Jacobs, J. W., and Eilon, G. (1990). *J. Biol. Chem.* **265,** 7679–7686.

Chambers, T. C., Zheng, B., and Kuo, J. F. (1992). *Molec. Pharmacol.* **41,** 1008–1015.

Chan, H. S. L., Thorner, P. S., Haddad, G., and Ling, V. (1990). *J. Clin. Oncol.* **8,** 689–704.

Chan, H. S. L., Haddad, G., Thorner, P. S., DeBoer, G., Lin, Y. P. Ondrusek, N., Yeger, H., and Ling, V. (1991). *N. Engl. J. Med.* **325,** 1608–1614.

Chan, V. T. W., Ng, S., Eder, J. P., and Schnipper, L. E. (1993). *J. Biol. Chem.* **268,** 2160–2165.

Chao, C. C.-K., Huang, S.-L., Lee, L.-Y., and Lin-Chao, S. (1991). *Biochem. J.* **277,** 875–878.

Chao, C. C.-K. (1992). *Biochem. J.* **282,** 203–207.

Charcosset, J.-Y., Saucier, J.-M., and Jacquemin-Sablon, A. (1988). *Biochem. Pharmacol.* **37,** 2145–2149.

Chaudhary, P. M., and Roninson, I. B. (1992). *Oncol. Res.* **4,** 281–290.

Chauffert, B., Martin, F., Caignard, A., Jeannin, J. F., and Leclerc, A. (1984). *Cancer Chemother. Pharmacol.* **13,** 14–18.

Chauffert, B., Martin, M., Hammann, A., Michel, M. F., and Martin, F. (1986). *Cancer Res.* **46,** 825–830.

Chen, G. L., Yang, C., Rowe, T. C., Halligan, B. D., Tewey, K. M., and Liu, L. F. (1984). *J. Biol. Chem.* **259,** 13,560–13,566.

Chen, C. J., Chin, J. E., Ueda, K., Clark, D. P., Pastan, I., Gottesman, M. M., and Roninson, I. B. (1986). *Cell* **47,** 381–389.

Chen, C. J., Clark, D., Ueda, K., Pastan, I., Gottesman, M. M., and Roninson, I. B. (1990). *J. Biol. Chem.* **265,** 506–514.

Cherif, A., and Farquhar, D. (1992). *J. Med. Chem.* **35,** 3208–3214.

Chevillard, S., Vielh, P., Bastan, G., and Coppey, J. (1992). *Anticancer Res.* **12,** 495–500.
Chin, K. V., Chauhan, S., Pastan, I., and Gottesman, M. M. (1990). *Cell Growth Differ.* **1,** 361–365.
Chin, K. V., Ueda, K., Pastan, I., and Gottesman, M. M. (1992). *Science* **255,** 459–462.
Choi, K., Chen, C. J., Kriegler, M., and Roninson, I. B. (1988). *Cell* **53,** 519–529.
Choi, K., Frommel, T. O., Stern, R. K., Perez, C. F., Kriegler, M., Tsuruo, T., and Roninson, I. B. (1991). *Proc. Natl. Acad. Sci. USA* **88,** 7386–7390.
Chow, K.-C., and Ross, W. E. (1987). *Mol. Cell Biol.* **7,** 3119–3123.
Chu, E., Zinn, S., Boarman, D., and Allegra, C. J. (1990). *Cancer Res.* **50,** 5834–5840.
Chu, G., and Chang, E. (1988). *Science* **242,** 564–567.
Chu, G., and Chang, E. (1990). *Proc. Natl. Acad. Sci. USA* **87,** 3324–3327.
Chu, M. Y., and Fischer, G. A. (1965). *Biochem. Pharmacol.* **14,** 333–341.
Chung, T. D., Drake, F. H., Tan, K. B., Per, S. R., Crooke, S. T., and Mirabelli, C. K. (1989). *Proc. Natl. Acad. Sci. USA* **86,** 9431–9435.
Ciarrocchi, G., Lestingi, M., Fontana, M., Spadari, S., and Montecucco, A. (1991). *Biochem. J.* **279,** 141–146.
Clark, J. L., Berger, S. L., Mittelman, A., and Berger, F. G. (1987). *Cancer Treat. Rep.* **71,** 261–265.
Cohen, D., Piekarz, R. I., Hsu, S. I., DePinho, R. A., Carrasco, N., and Horwitz, S. B. (1991). *J. Biol. Chem.* **266,** 2239–2244.
Cole, S. P. C., Bhardwaj, G., Gerlach, J. H., Mackie, J. E., Grant, C. E., Almquist, K. C., Stewart, A. J., Kurz, E. V., Duncan, A. M. V., and Deeley, R. G. (1992). *Science* **258,** 1650–1653.
Coley, H. M., Workman, P., Twentyman, P. R., and Amos, W. B. (1990). *Br. J. Cancer* **62,** 512.
Coley, H. M, Workman, P., and Twentyman, P. R. (1991). *Br. J. Cancer* **63,** 351–357.
Collett, A. G., and Ralph, R. K. (1992). *Biochim. Biophys. Acta* **1132,** 259–264.
Collins, A. R. S., Schor, S. L., and Johnson, R. T. (1977). *Mutat. Res.* **42,** 413–432.
Cook, J. D., Cichowicz, D. J., George, S., Lawler, A., and Shane, B. (1987). *Biochemistry* **26,** 530–539.
Cardon-Cardo, C., O'Brien, J. P., Casals, D., Rittman-Grauer, L., Biedler, J. L., Melamed, M. R., and Bertino, J. R. (1989). *Proc. Natl. Acad. Sci. USA* **86,** 695–698.
Cornwell, M. M., Gottesman, M. M., and Pastan, I. H. (1986). *J. Biol. Chem.* **261,** 7921–7928.
Cornwell, M. M., Pastan, I., and Gottesman, M. M. (1987). *J. Biol. Chem.* **262,** 2166–2170.
Cornwell, M. M. (1990). *Cell Growth Differ.* **1,** 607–615.
Corrias, M. V., and Tonini, G. P. (1992). *Anticancer Res.* **12,** 1431–1438.
Covey, J. M., Kohn, K. W., Kerrigan, D., Tilchen, E. J., and Pommier, Y. (1988). *Cancer Res.* **48,** 860–865.
Covey, J. M., Jaxel, C., Kohn, K. W., and Pommier, Y. (1989). *Cancer Res.* **49,** 5016–5022.
Cowan, K. H., and Jolivet, J. (1984). *J. Biol. Chem.* **259,** 10,793–10,800.
Cowan, K. H., Goldsmith, M. E., and Ricciardone, M. D. (1986). *Mol. Pharmacol.* **30,** 69–73.
Curt, G. A., Jolivet, J. Carney, D. N., Bailey, B. D., Drake, J. C., Celendin, N. J., and Chabner, B. A. (1985). *J. Clin, Invest.* **76,** 1323–1329.
Curt, G. A., and Allegra, C. J. (1987). *In* "Resistance to Antineoplastic Drugs" (D. Kessel, Ed.), pp. 369–384. CRC Press, Boca Raton, FL.
Cuvier, C., Roblot-Treupel, L., Millot, J. M., Lizard, G., Chevillard, S., Manfait, M., Couvreur, P., and Poupon, M. F. (1992). *Biochem. Pharmacol.* **44,** 509–517.
Dabholkar, M., Parker, R., and Reed, E. (1992). *Mutat. Res.* **274,** 45–56.
Dalton, W. S., Cress, A. E., Alberts, D. S., and Trent, J. M. (1988). *Cancer Res.* **48,** 1882–1888.

Dalton, W. S. (1993). *Proc. Am. Assoc. Cancer Res.* **34,** 305.
Danks, M. K., Yalowich, J. C., and Beck, W. T. (1987). *Cancer Res.* **47,** 1297–1301.
Danks, M. K., Schmidt, C. A., Cirtain, M. C., Suttle, D. P., and Beck, W. T. (1988). *Biochemistry* **27,** 8861–8869.
Danks, M. K., Schmidt, C. A., Deneka, D. A., and Beck, W. T. (1989). *Cancer Commun.* **1,** 101–109.
Danø, K. (1971). *Cancer Chemother. Rep.* **55,** 133–141.
Danø, K. (1972). *Cancer Chemother. Rep.* **56,** 701–708.
Danø, K. (1973). *Biochim. Biophys. Acta* **323,** 466–483.
Darkin, S. J., and Ralph, R. K. (1989). *Biochim. Biophys. Acta* **1007,** 295–300.
Darkin-Rattray, S. J., and Ralph, R. K. (1991). *Biochim. Biophys. Acta* **1088,** 285–291.
DcArpa, P. D., Beardmore, C., and Liu, L. F. (1990). *Cancer Res.* **50,** 6919–6924.
Davies, S. M., Robson, C. N., Davies, S. L., and Hickson, I. D. (1988). *J. Biol. Chem.* **263,** 17,724–17,729.
Dechamps, M., de Saint-Vincent, B. R., Evraad, C., Sassi, M., and Butin, G. (1974). *Cell Res.* **86,** 269–279.
Dedhar, S., and Goldie, J. H. (1983). *Cancer Res.* **43,** 4863–4869.
Dedhar, S., and Goldie, J. H. (1985). *Biochem. Biophys. Res. Commun.* **129,** 536–545.
Deffie, A. M., Alam, T., Seneviratne, C., Beenken, S. W., Batra, J. K., Shea, T. C., Henner, W. D., and Goldenberg, G. J. (1988). *Cancer Res.* **48,** 3595–3602.
Deffie, A. M., Batra, J. K., and Goldenberg, G. J. (1989a). *Cancer Res.* **49,** 58–62.
Deffie, A. M., Bosman, D. J., and Goldenberg, G. J. (1989b). *Cancer Res.* **49,** 6879–6882.
Deffie, A. M., McPherson, J. P., Gupta, R. S., Hedley, D. W., and Goldenberg, G. J. (1992). *Biochem. Cell Biol.* **70,** 354–364.
DeGregorio, M. W., Ford, J. M., Benz, C. C., and Wiebe, V. J. (1989). *J. Clin. Oncol.* **7,** 1359–1364.
de Jong, S., Zijlstra, J. G., de Vries, E. G. E., and Mulder, N. H. (1989). *Proc. Am. Assoc. Cancer Res.* **30,** 526.
Demant, E. J. F., Sehested, M., and Jensen, P. B. (1990). *Biochim. Biophys. Acta* **1055,** 117–125.
Dembo, M., and Sirotnak, F. M. (1976). *Biochem. Biophys. Acta* **448,** 505–516.
DeVore, R. F., Corbett, A. H., and Osheroff, N. (1992). *Cancer Res.* **52,** 2156–2161.
Dijt, F. J., Fichtinger-Schepman, A. M. J., Berends, F., and Reedijk, J. (1988). *Cancer Res.* **48,** 6058–6062.
Doige, C. A., and Sharom, F. J. (1992). *Biochim. Biophys. Acta* **1109,** 161–171.
Dolnick, B. J., Berenson, R. J., Bertino, J. R., Kaufman, R. J., Nunberg, J. H., and Schimke, R. T. (1979). *J. Cell Biol.* **83,** 394–402.
Dolnick, B. J., and Bertino, J. R. (1982). *In* "Tumor Cell Heterogeneity" (A. H. Owens, D. S. Coffey, and S. B. Baylin, Eds.), pp. 169–179. Academic Press, New York.
Domin, B. A., Cheng. Y., and Hakala, M. T. (1982a). *Mol. Pharmacol.* **21,** 231–234.
Domin, B. A., Grill, S. P., and Bastow, K. F. (1982b). *Mol. Pharmacol.* **21,** 478–482.
Donahue, B. A., Augot, M., Bellon, S. F., Treiber, D. K., Toney, J. H., Lippard, S. J., and Essigmann, J. M. (1990). *Biochemistry* **29,** 5872–5880.
Doroshow, J. H., Locker, G. Y., and Myers, C. E. (1980). *J. Clin. Invest.* **65,** 128–135.
Drahovsky, D., and Kreis, V. (1970). *Biochem. Pharmacol.* **19,** 940–944.
Drake, F. H., Zimmerman, J. P., McCabe, F. L., Bartus, H. F., Per, S. R., Sullivan, D. M., Ross, W. E., Mattern, M. R., Johnson, R. K., Crooke, S. T., and Mirabelli, C. K. (1987). *J. Biol. Chem.* **262,** 16,739–16,747.
Duguet, M., Lavenot, C., Harper, F., Mirambeau, G., and DeRecondo, A. M. (1983). *Nucleic Acid Res.* **11,** 1059–1075.

Earnshaw, W. C., Halligan, B., Cooke, C. A., Heck, M. M., and Liu, L. F. (1985). *J. Cell. Biol.* **100,** 1706–1715.

Eastman, A. (1991). *In* "Molecular and Clinical Advances in Anticancer Drug Resistance" (R. F. Ozols, Ed.), pp. 233–249. Kluwer Academic Publishers, Boston.

Eder, J. P., Teicher, B. A., Holden, S. A., Cathcart, K. N. S., and Schnipper, L. E. (1987). *J. Clin. Invest.* **79,** 1524–1528.

Eder, J. P., Wheeler, C. A., Teicher, B. A., and Schnipper, L. E. (1991). *Cancer Res.* **51,** 510–513.

Edwards, C. M., Glisson, B. S., King, C. K., Smallwood-Kentro, S., and Ross, W. E. (1987). *Cancer Chemother. Pharmacol.* **20,** 162–168.

Egorin, M. J., Hildebrand, R. C., Cimino, E. F., and Bachur, N. R. (1974). *Cancer Res.* **34,** 2243–2245.

Endicott, J. A., and Ling, V. (1989). *Annu. Rev. Biochem.* **58,** 137–171.

Eng, W.-K., Faucette, L., Johnson, R. K., and Sternglanz, R. (1988). *Mol. Pharmacol.* **34,** 755–760.

Estey, E., Adlakha, R. C., Hittelman, W. N., and Zwelling, L. A. (1987). *Biochemistry* **26,** 4338–4344.

Evans, C. G., Bodell, W. J., Tokuda, K., Doane-Setzer, P., and Smith, M. T. (1987). *Cancer Res.* **47,** 2525–2530.

Fan, D., Fidler, I. J., Ward, N. E., Seid, C., Earnest, L. E., Housey, G. M., and O'Brian, C. A. (1992). *Anticancer Res.* **12,** 661–668.

Ferguson, P. J., Fisher, M. H., Stephenson, J., Li, D., Zhou, B., and Cheng, Y. (1988). *Cancer Res.* **48,** 5956–5964.

Fernandes, D. J., Danks, M. K., and Beck, W. T. (1990). *Biochemistry* **29,** 4235–4241.

Fernandes, D. J., and Catapano, C. V. (1991). *Cancer Cells* **3,** 134–140.

Fischer, G. A. (1962). *Biochem. Pharmacol.* **11,** 1233–1234.

FitzGerald, D. J., Willingham, M. C., Cardarelli, C. O., Hamada, H., and Tsuruo, T. (1987). *Proc. Natl. Acad. Sci. USA* **84,** 4288–4297.

Foglesong, P. D., Reckord, C., and Swink, S. (1992). *Cancer Chemother. Pharmacol.* **30,** 123–125.

Fojo, A., Akiyama, S. I., Gottesman, M. M., and Pastan, I. (1985). *Cancer Res.* **45,** 3002–3007.

Fojo, A. T., Ueda, K., Slamon, D. J., Poplack, D. G., Gottesman, M. M., and Pastan, I. (1987). *Proc. Natl. Acad. Sci. USA* **84,** 265–269.

Foote, S. J., Thompson, J. K., Cowman, A. F., and Kemp, D. J. (1989). *Cell* **57,** 921–930.

Ford, J. M., and Hait, W. N. (1990). *Pharmacol. Rev.* **42,** 155–199.

Fornace, Jr., A. J., Papathanasiou, M. A., Hollander, M. C., and Yarosh, D. B. (1990). *Cancer Res.* **50,** 7908–7911.

Fram, R. J., and Kufe, D. W. (1982). *Cancer Res.* **42,** 4050–4053.

Fram, R. J., and Kufe, D. W. (1985). *Biochem. Pharmacol.* **34,** 2557–2560.

Francis, A. A., Snyder, R. D., Dunn, W. C., and Regan, J. D. (1981). *Mutat. Res.* **83,** 159–169.

Frankfurt, O. S. (1991). *Int. J. Cancer* **48,** 916–923.

Frankfurt, O. S., Seckinger, D., and Sugarbaker, E. V. (1990). *Cancer Res.* **50,** 4453–4457.

Friche, E., Jensen, P. B., Roed, H., Skovsgaard, T., and Nissen, N. I. (1990). *Biochem. Pharmacol.* **39,** 1721–1726.

Friche, E., Danks, M. K., Schmidt, C. A., and Beck, W. T. (1991). *Cancer Res.* **51,** 4213–4218.

Friedberg, E. C. (1985). "DNA Repair." Freeman, New York.

Fry, D. W., Yalowich, J. C., and Goldman, I. D. (1982). *J. Biol. Chem.* **257,** 1890–1896.

Futcher, B. W., Pieper, R. O., Dalton, W. S., and Erickson, L. C. (1992). *Cell Growth Differ.* **3,** 217–223.
Galivan, J., Johnson, T., Rhee, M., McGuire, J. J., Priest, D., and Kesevan, V. (1987). *Adv. Enzyme Regul.* **26,** 147–155.
Galski, H., Sullivan, M., Willingham, M. C., Chin, K. V., Gottesman, M. M., Pastan, I., and Merlino, G. T. (1989). *Mol. Cell. Biol.* **9,** 4357–4363.
Ganapathi, R., Grabowski, D., Rouse, W., and Riegler, F. (1984). *Cancer Res.* **44,** 5056–5061.
Ganapathi, R., and Grabowski, D. (1988). *Biochem. Pharmacol.* **37,** 185–193.
Gant, T. W., Silverman, J. A., and Thorgeirsson, S. S. (1992). *Nucleic Acids Res.* **20,** 2841–2846.
Garcia-Segura, L. M., Ferragot, J. A., Ferrer-Montiel, A. V., Escriba, P. V., and Gonzales-Ros, J. M. (1990). *Biochim. Biophys. Acta.* **1029,** 191–195.
Gekeler, V., Frese, G., Noller, A., Handgretinger, R., Wilisch, A., Schmidt, H., Muller, C. P., Dopfer, R., Klingebiel, T., Diddens, H., Probst, H., and Niethammer, D. (1992). *Br. J. Cancer* **66,** 507–517.
Gerson, S. L., and Trey, J. E. (1988). *Blood* **71,** 1487–1494.
Gervasoni, Jr., J. E., Fields, S. Z., Krishna, S., Baker, M. A., Rosado, M., Thuraisamy, K., Hindenburg, A. A., and Taub, R. N. (1991). *Cancer Res.* **51,** 4955–4963.
Gill, D. R., Hyde, S. C., Higgins, C. F., Valverde, M. A., Mintenig, G. M., and Sepulveda, F. V. (1992). *Cell* **71,** 23–32.
Giovanella, B. C., Stehlin, J. S., Wall, M. E., Wani, M. C., Nicholas, A. W., Liu, L. F., Silber, R., and Potmesil, M. (1989). *Science* **246,** 1046–1048.
Glisson, B. R., Gupta, R., Smallwood-Kentro, S., and Ross, W. (1986). *Cancer Res.* **46,** 1934–1938.
Godwin, A. K., Meister, A., O'Dwyer, P. J., Huang, C. S., Hamilton, T. C., and Anderson, M. E. (1992). *Proc. Natl. Acad. Sci. USA* **89,** 3070–3074.
Goldie, J. H., and Coldman, A. J. (1979). *Cancer Treat Rep.* **63,** 1727–1733.
Goldie, J. H., Krystal, G., Hartley, D., Andauskas, G., and Dedhar, S. (1980). *Eur. J. Cancer* **19,** 1539–1544.
Goldie, J. H., Dedhar, S., and Krystal, G. (1981). *J. Biol. Chem.* **256,** 11,629–11,634.
Goldman, I. D., Lichtenstein, N. S., and Oliverrio, V. T. (1968). *J. Biol. Chem.* **243,** 5007–5017.
Goldman, I. D. (1969). *J. Biol. Chem.* **244,** 3779–3785.
Goldman, I. D., and Matherly, L. H. (1985). *Pharmacol. Ther.* **28,** 77–102.
Goldstein, L. J., Galski, H., Fojo, A. T., Willingham, M. C., Lai, S. L., Gazdar, A., Pirker, R., Green, A., Crist, W., Brodeur, G., Grant, C., Lieber, M., Cossman, J., Gottesman, M. M., and Pastan, I. (1989). *J. Natl. Cancer Inst.* **81,** 116–124.
Gollapudi, S., McDonald, T., Gardner, P., and Gupta, S. (1992). *Cancer Lett.* **66,** 83–89.
Gosland, M. P., Lum, B. L., and Sikic, B. I. (1989). *Cancer Res.* **49,** 6901–6905.
Gosland, M. P., Brophy, N. A., Duran, G. E., Yahanda, A. M., Adler, K. M., Hardy, R. I., Halsey, J., and Sikis, B. I. (1991). *Proc. Am. Assoc. Cancer Res.* **32,** 426.
Gottesman, M. M. (1993). *Cancer Res.* **53,** 747–754.
Greenberger, L. M., Williams, Y., and Horwitz, S. B. (1987). *J. Biol. Chem.* **262,** 13,685–13,689.
Gromova, I. I., Kjeldsen, E., Svejstrup, J. Q., Alsner, J., Christiansen, K., and Westergaard, O. (1993). *Nucl. Acids Res.* **21,** 593–600.
Gupta, R. S., Gupta, R., Eng, B., Lock, R. B., Ross, W. E., Hertzberg, R. P., Caranfa, M. J., and Johnson, R. K. (1988). *Cancer Res.* **48,** 6404–6410.
Haber, D. A., and Schimke, R. T. (1981). *Cell* **26,** 355–359.
Haber, D. A., Beverly, S. M., Kiely, M. L., and Schimke, R. T. (1981). *J. Biol. Chem.* **256,** 9501–9510.
Hakala, M. T. (1985). *Biochim. Biophys. Acta* **102,** 210–225.

Hamada, H., Hagiwara, K. I., Nakajima, T., and Tsuruo, T. (1987). *Cancer Res* **47,** 2860–2865.

Hamada, H., and Tsuruo, T. (1988). *J. Biol. Chem.* **263,** 1454–1458.

Hamlin, J. L. (1992). *Mutat. Res.* **276,** 179–187.

Hammond, J. R., Johnstone, R. M., and Gros, P. (1989). *Cancer Res.* **49,** 3867–3871.

Hansson, J., Berhane, K., Castro, V. M., Jungnelius, U., Mannervik, B., and Ringborg, U. (1991). *Cancer Res.* **51,** 94–98.

Harker, W. G., Slade, D. L., Drake, F. H., and Parr, R. L. (1991). *Biochemistry* **30,** 9953–9961.

Hasmann, M., Valet, G. K., Tapiero, H., Trevorrow, K., and Lampidis, T. (1989). *Biochem. Pharmacol.* **38,** 305–312.

Heartlein, M. W., Tsuji, H., and Latt, S. A. (1987). *Exp. Cell Res.* **169,** 245–254.

Heck, M. M. S., Hittelman, W. N., and Earnshaw, W. C. (1988). *Proc. Natl. Acad. Sci. USA* **85,** 1086–1090.

Heck, M. M. S., Hittelman, W. N., and Earnshaw, W. C. (1989). *J. Biol.Chem.* **264,** 15,161–15,164.

Heidelberger, C., Chaudhuri, N. K., Danneberg, P., Mooren, D., Griesbach, L., Duschinsky, R., Schnitzer, R. J., Pleven, E., and Scheiner, T. (1957). *Nature (London)* **179,** 663–666.

Henderson, G. B., and Zevely, E. M. (1982). *Biochim. Biophys. Res.* **104,** 474–480.

Henderson, G. B., and Zevely, E. M. (1984). *J. Biol. Chem.* **259,** 1526–1531.

Henderson, G. B., and Zevely, E. M. (1985). *Biochem. Pharmacol.* **34,** 1725–1729.

Henderson, G. B., Tsuji, J. M., and Kumar, H. P. (1986a). *Cancer Res.* **46,** 1633–1638.

Henderson, G. B., Suresh, M. R., Vitols, K. S., and Huennekens, F. M. (1986b). *Cancer Res.* **46,** 1639–1643.

Henderson, G. B., and Tsuji, J. M. (1987). *J. Biol. Chem.* **262,** 13,751–13,778.

Henderson, G. B., and Strauss, B. P. (1990). *Cancer Res.* **50,** 1709–1714.

Herweijer H., Sonneveld, P., Baas, F., and Nooter, K. (1990). *J. Natl. Cancer Inst.* **82,** 1113–1140.

Higgins, C. F. (1992). *Annu. Rev. Cell Biol.* **8,** 67–113.

Higuchi, T., Nakamura, T., Uchino, H., and Wakisaka, G. (1977). *Antimicrob. Agents Chemother.* **12,** 518–522.

Hindenburg, A. A., Baker, M. A., Gleyzer, E., Stewart, V. J., Case, N., and Taub, R. N.(1987). *Cancer Res.* **47,** 1421–1425.

Hindenburg, A. A., Gervasoni, Jr. J. E., Krishna, S., Stewart, V. J., Rosado, M., Lutzky, J., Bhalla, K., Baker, M. A., and Taub, R. N. (1989). *Cancer Res.* **49,** 4607–4614.

Hinds, M., Deisseroth, K., Mayes, J., Altschuler, E., Jansen, R., Ledley, F. D., and Zwelling, L. A. (1991). *Cancer Res.* **51,** 4729–4731.

Hirabayashi, N., Kim, R., Nishiyama, M., Aogi, K., Saegi, S., Toge, T., and Okada, K. (1992). *Proc. Am. Assoc. Cancer Res.* **33,** 436.

Hitchins, R. N., Harman, D. H., Davey, R. A., and Bell, D. R. (1988). *Eur. J. Cancer Clin. Oncol.* **24,** 449–454.

Hofsli, E., and Nissen-Meyer, J. (1990). *Cancer Res.* **50,** 3997–4002.

Holm, C., Covey, J. M., Kerrigan, D., and Pommier, Y. (1989). *Cancer Res.* **49,** 6365–6368.

Horio, M., Gottesman, M. M., and Pastan, I. (1988). *Proc. Natl. Acad. Sci. USA* **85,** 3580–3584.

Houghton, J. A., Maroda, S. J., and Phillips, J. O. (1981). *Cancer Res.* **41,** 144–149.

Hsiang, Y. H., Hertzberg, R., Hecht, S., and Liu, L. F. (1985). *J. Biol. Chem.* **260,** 14,873–14,878.

Hsiang, Y. H., Lihou, M. G., and Liu, L. F. (1989a). *Cancer Res.* **49,** 5077–5082.

Hsiang, Y. H., Liu, L. F., Wall, M. E., Wani, M. C., Nicholas, A. W., Manikumar, G., Kirschenbaum, S., Silber, R., and Potmesil, M. (1989b). *Cancer Res.* **49,** 4385–4389.

Hsu, S. I., Lothstein, L., and Horwitz, S. B. (1989). *J. Biol. Chem.* **264,** 12,053–12,062.
Hsu, S. I., Cohen, D., Kirschner, L. S., Lothstein, L., Hartstein, M., and Horwitz, S. (1990). *Mol. Cell. Biol.* **10,** 3596–3606.
Hu, X. F., Martin, T. J., Bell, D. R., de Luise, M., and Zalcberg, J. R. (1990). *Cancer Res.* **50,** 2953–2957.
Huang, J. C., Svoboda, D. L., Reardon, J. T., and Sancar, A. (1992). *Proc. Natl. Acad. Sci. USA* **89,** 3664–3668.
Huet, S., Schott, B., and Robert, J. (1992). *Br. J. Cancer* **65,** 538–544.
Hyde, S. C., Emsley, P., Hartshorn, M. J., Mimmack, M. M., Gileadi, U., Pearce, S. R., Gallagher, M. P., Gill, D. R., Hubbard, R. E., and Higgins. C. F. (1990). *Nature (London)* **346,** 362–365.
Ibeanu, G., Hartenstein, B., Dunn, W. C., Chang, L.-Y., Hofmann, E., Coquerelle, T., Mitra, S., and Kaina, B. (1992). *Carcinogenesis* **13,** 1989–1995.
Ichikawa, M., Yoshimura, A., Furukawa, T., Sumizawa, T., Nakazima, Y., and Akiyama, S. I. (1991). *Biochim. Biophys. Acta* **1073,** 309–315.
Ikeguchi, M., Teeter, L., Eckersberg, T., Ganapathi, R., and Kuo, M. T. (1991). *DNA Cell Biol.* **10,** 639–649.
Inaba, M., Kobayachi, H., Sakurai, Y., and Johnson, R. R. (1979). *Cancer Res.* **39,** 2200–2203.
Jackson, R. C., Niethammer, D., and Huennekens, F. M. (1975). *Cancer Biochem. Biophys.* **1,** 151–155.
Jackson, R. C., and Niethammer, D. (1977). *Eur. J. Cancer* **13,** 567–575.
Jacobs, A., Bopp, A., and Hagen, U. (1972). *Int. J. Radiat. Biol.* **22,** 431–435.
Jakoby, W. B. (1978). *In* "Advances in Enzymology" (A. Meister, Ed.), 46th ed., pp. 383–414.
Jansen, G., Westerhof, G. R., Kathmann, I., Rademaker, B. C., Rijksen, G., and Schornagel, J. H. (1989). *Cancer Res.* **49,** 2455–2459.
Jaxel, C., Kohn, K. W., Wani, M. C., Wall, M. E., and Pommier, Y. (1989). *Cancer Res.* **49,** 1465–1469.
Jenh, C. H., Geyer, P. K., Baskin, F., and Johnson, L. F. (1985). *Mol. Pharmacol.* **28,** 80–85.
Jenkins, J. R., Ayton, P., Jones, T., Davies, S. L., Simmons, D. L., Harris, A. L., Sheer, D., and Hickson, I. D. (1992). *Nucl. Acids Res.* **20,** 5587–5592.
Jensen, P. B., Roed, H., Sehested, M., Demant, E. J. F., Vindeløv, L., Christensen, I. J., and Hansen, H. H. (1992). *Cancer Chemother. Pharmacol.* **31,** 46–52.
Jesson, M. I., Johnston, J. B., Anhalt, C. D., and Begleiter, A. (1987). *Cancer Res.* **47,** 5935–5938.
Johnson, R. K., McCabe, F. L., Faucette, L. F., Hertzberg, R. P., Kingsbury, W. D., Boehm, J. C., Caranfa, M. J., and Holden, K. G. (1989). *Proc. Am. Assoc. Cancer Res.* **30,** 623.
Johnston, J. B., Habernicht, B., Acton, E. M., and Glazer, R. I. (1983). *Biochem. Pharmacol.* **32,** 3255–3258.
Jones, J. C., Zhen, W., Reed, E., Parker, R. J., Sancar, A., and Bohr. V. A. (1991). *J. Biol. Chem.* **266,** 7101–7107.
Juliano, R. L., and Ling, V. (1976). *Biochim. Biophys. Acta* **455,** 152–162.
Kaina, B., Lohrer, H., Karin, M., and Herrlich, P. (1990). *Proc. Natl. Acad. Sci.USA* **87,** 2710–2714.
Kaina, B., Fritz, G., Mitra, S., and Coquerelle, T. (1991). *Carcinogenesis* **12,** 1857–1867.
Kamen, B. A., and Bertino, J. R. (1980). *Antibiot. Chemother.* **28,** 62–67.
Kanamaru, H., Kakehi, Y., Yoshida, O., Nakanishi, S., Pastan, I., and Gottesman, M. M.(1989). *J. Natl. Cancer Inst.* **81,** 844–849.

Kanda, N., Schreck, R., Alt, F., Bruns, G., Baltimore, D., and Latt, S. (1983). *Proc. Natl. Acad. Sci. USA* **80,** 4069–4073.

Kane, M. A., Portillo, R. M., Elwood, P. C., Antony, A. C., and Kolhouse, J. F. (1986). *J. Biol. Chem.* **261,** 44–51.

Kato, S., Ideguchi, H., Muta, K., Nishimura, J., and Nawata, H. (1990). *Leukemia Res.* **14,** 567–573.

Keizer, H. G., Schuurhuis, G. J., Broxterman, H. J., Lankelma, J., Schoonen, W. G. E. J., van Rijn, J., Pinedo, H. M., and Joenje, H. (1989). *Cancer Res.* **49,**2988–2993.

Keizer, H. G., and Joenje, H. (1989). *J. Natl. Cancer Inst.* **81,** 706–709.

Kelland, L. R., Mistry, P., Abel, G., Freidlos, F., Loh, S. Y., Roberts, J. J., and Harrap, K. R. (1992). *Cancer Res.* **52,** 1710–1716.

Keller, R. P., Altermatt, H. J., Nooter, K., Poschmann, G., Laissue, J. A., Bollinger, P., and Hiestand, P. C. (1992). *Int. J. Cancer* **50,** 593–597.

Kelley, S. L., Basu, A., Teicher, B. A., Hacker, M. P., Hamer, D. H., and Lazo, J. S. (1988). *Science* **241,** 1813–1815.

Kessel, D., Hall, T. C., and Roberts, D. (1965). *Science* **150,** 752–754.

Kessel, D., Hall, T. C., and Rosenthal, D. (1969). *Cancer Res.* **29,** 459–463.

Kessel, D., and Bosmann, H. B. (1970). *Cancer Res.* **30,** 2695–2701.

Kessel, D., and Wilberding, C. (1985). *Cancer Res.* **45,** 1687–1691.

Kjeldsen, E., Bonven, B., Andoh, T., Ishii, K., Okada, K., Bolund, L., and Westergaard, O. (1988). *J. Biol. Chem.* **263,** 3912–3916.

Kleinberger, T., Etkin, S., and Lavi, S. (1986). *Mol. Cell Biol.* **6,** 1958–1963.

Klopman, G., Srivastava, S., Kolossvary, I., Epand, R. F., Ahmed, N., and Epand, R. M.(1992). *Cancer Res.* **52,** 4121–4129.

Kohno, K., Sato, S., Takano, H., Matsuo, K., and Kuwano, M. (1989). *Biochem. Biophys. Res. Commun.* **165,** 1415–1421.

Kramer, R. A., Arceci, R., Weber, T., and Summerhayes, I. (1992). *Proc. Am. Assoc.Cancer Res.* **33,** 476.

Krishnamachary, N., and Center, M. S. (1993). *Cancer Res.* **53,** 3658–3661.

Kufe, D. W., and Spriggs, D. R. (1985). *Semin. Oncol.* **12,** Suppl. 3, 34–48.

Kuiper, C. M., Broxterman, H. J., Baas, F., Schuurhuis, G. J., Haisma, H. J., Schefer, G. L., Lankelma, J., and Pinedo, H. M. (1990). *J. Cell Pharmacol.* **1,** 35–41.

Lankelma, J., Spoelstra, E. C., Dekker, H., and Broxterman, H. J. (1990). *Biochim. Biophys. Acta* **1055,** 217–222.

Larminat, F., Zhen, W., and Bohr, V. A. (1993). *J. Biol. Chem.* **268,** 2649–2654.

Larsson, R., and Nygren, P. (1990). *Cancer Lett.* **54,** 125–131.

Lau, D. H. M., Ross, K. L., and Sikic, B. I. (1990). *Cancer Res.* **50,** 4056–4060.

Lau, D. H. M., Lewis, A. D., Ehsan, M. N., and Sikic, B. I. (1991). *Cancer Res.* **51,** 5181–5187.

LeDoux, S. P., Patton, N. J., Nelson, J. W., Bohr, V. A., and Wilson, G. L. (1990). *J. Biol. Chem.* **265,** 14,875–14,880.

LeDoux, S. P., Thangada, M., Bohr, V. A., and Wilson, G. L. (1991). *Cancer Res.* **51,** 775–779.

Lehnert, M., Dalton, W. S., Roe, D., Emerson, S., and Salmon S. E. (1991). *Blood* **77,** 348–354.

LePage, G. A. (1968). *Can. J. Biochem.* **46,** 655–661.

Lewis, A. D., Hickson, I. D., Robson, C. N., Harris, A. L., Hayes, J. D., Griffiths, S.A., Manson, M. M., Hall, A. E., Moss, J. E., and Wolf, C. R. (1988). *Proc. Natl. Acad. Sci. USA* **85,** 8511–8515.

Lewis, A. D., Duran, G. E., Lau, D. H.-M., and Sikic, B. I. (1992). *Int. J. Radiat. Oncol. Biol. Phys.* **22,** 821–824.

Leyland-Jones, B. R., Townsend, A. J., Tu, C.-P. D., Cowan, K. H., and Goldsmith M. E. (1991). *Cancer Res.* **51,** 587–594.
Li, L. H., Fraser, T. J., Olin, E. J., and Bhuyan, B. K. (1972). *Cancer Res.* **32,** 2643–2650.
Lin, J. T., and Bertino, J. R. (1991). *Cancer Invest.* **9,** 159–172.
Lincke, C. R., van der Bliek, A. M., Schuurhuis, G. J., and van der Velde-Koerts, T. (1990). *Cancer Res.* **50,** 1779–1785.
Lincke, C. R., Smit, J. J. M., van der Velde-Koerts, T., and Borst, P. (1991). *J. Biol. Chem.* **266,** 5303–5310.
List, A. F., Spier, C., Greer, J., Azar, C., Hutter, Q., Wolff, S., Salmon, S., Futscher, B., and Dalton, W. (1992). *Proc. Am. Assoc. Clin. Oncol.* **11,** 264.
Liu, L. F. (1989). *Ann. Rev. Biochem.* **58,** 351–375.
Liu, S. Y., Hwang, B. D., Haruna, M., Imakura, Y., Lee, K. H., and Cheng, Y. C. (1989). *Mol. Pharmacol.* **36,** 78–82.
Long, B. H., Wang, L., Lorico, A., Wang, R. C. C., Brattain, M. G., and Casazza, A. M.(1991). *Cancer Res.* **51,** 5275–5284.
Lothstein, L., and Horwitz, S. B. (1986). *J. Cell Physiol.* **127,** 253–260.
Lönn, U., Lönn, S., and Stenkvist, B. (1992). *Int. J. Cancer* **53,** 574–578.
Lönn, U., Lönn, S., Nylen, U., and Stenkvist, B. (1993). *Int. J. Cancer* **51,** 682–686.
Ma, L., Marquardt, D., Takemoto, L., and Center, M. S. (1991), *J. Biol. Chem.* **266,** 5593–5599.
Maher, V. M., Domoradzki, J., Bhattacharyya, N. P., Tsujimura, T., Corner, R. C., and McCormick, J. J. (1990). *Mutat. Res.* **233,** 235–245.
Madelaine, I., Prost, S., Naudin, A., Riou, G., Lavelle, F., and Riou, J.-F. (1993). *Biochem. Pharmacol.* **45,** 339–348.
Mandelbaum-Shavit, F., and Ramu, A. (1987). *Cell Biol. Int. Rep.* **11,** 389–396.
Mannervik, B. (1985). *In* "Advances in Enzymology" (A. Meister, Ed.), 57th ed., pp. 357–417. Wiley, New York.
Mannervik, B., and Danielson, U. H. (1988). *Crit. Rev. Biochem.* **23,** 283–337.
Manoharan, T. H., Welch, P. J., Gulick, A. M., Puchalski, R. B., Lathrop, A. L., and Fahl, W. E. (1991). *Mol. Pharmacol.* **39,** 461–467.
Mariani, B. D., and Schimke, R. T. (1984). *J. Biol. Chem.* **259,** 1901–1906.
Marino, P. A., Gottesman, M. M., and Pastan, I. (1990). *Cell Growth Differ.* **1,** 57–62.
Markovits, J., Pommier, Y., Kerrigan, D., Covey, J. M., Tilchen, E. J., and Kohn, K. W. (1987). *Cancer Res.* **47,** 2050–2055.
Marsh, W., Sicheri, D., and Center, M. S. (1986). *Cancer Res.* **46,** 4053–4057.
Marquardt, D. McCrone, S., and Center, M. S. (1990). *Cancer Res.* **50,** 1426–1430.
Martinsson, T., and Levan, G. (1987). *Cytogenet. Cell Genet.* **45,** 99–101.
Masuda, H., Ozols, R. F., Lai, G. M., Fojo, A., Rothenberg, M., and Hamilton, T. C. (1988). *Cancer Res.* **48,** 5713–5716.
Matherly, L. H., Fry, D. W., and Goldman, I. D. (1983). *Cancer Res.* **43,** 2694–2699.
Matherly, L. H., Voss, M. K., and Anderson, L. A. (1985). *Cancer Res.* **45,** 1073–1075.
Matijasevic, Z., Bodell, W. J., and Ludlum, D. B. (1991). *Cancer Res.* **51,** 1568–1570.
Matsumoto, A., Vos, J.-M. H., and Hanawalt, P. C. (1989). Mutat. Res. **217,** 185–192.
May, A., Nairn, S., Okumoto, D. S., Wassermann, K., Stevnser, T., Jones, J. C., and Bohr, V. A. (1993). *J. Biol. Chem.* **268,** 1650–1657.
McGuire, J. J., Mini, E., Hseih, P., and Bertino, J. R. (1985). *Cancer Res.* **45,** 6395–6400.
Medh, R. D., Gupta, V., and Awasthi, Y. C. (1991). *Biochem. Pharmacol.* **42,** 439–441.
Melera, P. W., Davide, J. P., Hession, C. A., Hession, C. A., and Scotto, K. W. (1984). *Mol. Cell. Biol.* **4,** 38–48.
Melera, P. W., Davide, J. P., and Oen, H. (1987). *J. Biol. Chem.* **262,** 1978–1990.

Mellon, I., Spivak, G., and Hanawalt, P. C. (1987). *Cell* **51,** 241–249.
Mellon, I., and Hanawalt, P. C. (1989). *Nature (London)* **342,** 95–98.
Meltzer, P. S., Cheng, Y. C., and Trent, J. M. (1985). *Cytogenetics* **17,** 289–293.
Meuth, M., Aufreiter, E., and Reichard, P. (1976). *Eur. J. Biochem.* **71,** 39–43.
Michieli, M., Giacca, M., Fanin, R., Damiani, D., Geromin, A., and Baccarani, M. (1991). *Br. J. Haematol.* **78,** 288–289.
Mickisch, G. H., Pai, L. H., Gottesman, M. M., and Pastan, I. (1992). *Cancer Res.* **52,** 4427–4432.
Mickley, L. A., Bates, S. E., Richert, N. D., Currier, S., Tanaka, S., Foss, F., Rosen, N., and Fojo, A. T. (1989). *J. Biol. Chem.* **264,** 18,031–18,040.
Miller, T. P., Grogan, T. M., Dalton, W. S., Spier, C. M., Scheper, R. J., and Salmon, S. E. (1991). *J. Clin. Oncol.* **9,** 17–24.
Mimnaugh, E. G., Dusre, L., Atwell, J., and Myers. C. E. (1989). *Cancer Res.* **49,** 8–15.
Minnick, D. T., Gerson, S. L., Dumenco, L. L., Veigl, M. L., and Sedwick, W. D. (1993). *Cancer Res.* **53,** 997–1003.
Mirski, S. E. L., Gerlach, J. H., and Cole, S. P. C. (1987). *Cancer Res.* **47,** 2594–2598.
Miskimins, R., Miskimins, W. K., Bernstein, H., and Shimizu, N. (1983). *Exp. Cell Res.* **146,** 53–62.
Miyamoto, K. I., Inoko, K., Ikeda, K., Wakusawa, S., Kajita, S., Hasegawa, T., Takagi, K., and Koyama, M. (1993). *J. Pharm. Pharmacol.* **45,** 43–47.
Moran, R. G., Malkins, M., and Heidelberger, C. (1973). *Proc. Natl. Acad. Sci. USA* **76,** 5924–5928.
Moran, R. G., and Keyomarsi, K. (1987). *NCI Monogr.* **5,** 159–163.
Moscow, J. A., Townsend, A. J., and Cowan, K. H. (1989). *Mol Pharmacol.* **36,** 22–28.
Mulkins, M. A., and Heidelberger, C. (1982). *Cancer Res.* **42,** 956–964.
Myers, Jr., C. E., and Chabner, B. A. (1990). *In* "Cancer Chemotherapy: Principles and Practice" (B. A. Chabner, Jr. and J. M. Collins, Eds.), pp. 356–381. Lippincott, Philadelphia.
Naito, M., Hamada, H., and Tsurou, T. (1988). *J. Biol. Chem.* **263,** 11,887–11,891.
Nakagawa, K., Saijo, N., Tsuchida, S., Sakai, M., Tsunokawa Y., Yokota, J., Muramatsu, M., Sato, K., Terada, M., and Tew, K. D. (1990). *J. Biol. Chem.* **265,** 4196–4301.
Nakagawa, M., Dixon, K. H., Gilbert, L., Goldsmith, M. E., and Cowan, K. H. (1991). *Proc. Am. Assoc. Cancer Res.* **32,** 371.
Nelson, E. M., Tewey, K. M., and Liu, L. F. (1984). *Proc. Natl. Acad. Sci USA* **81,** 1361–1365.
Newman, E. M., Lu, Y., Kashani-Sabet, M., Kesavan, V., and Scanlon, K. J. (1988). *Biochem. Pharmacol.* **36,** 809–811.
Nielsen, D., and Skovsgaard, T. (1992). *Biochim. Biophys. Acta* **1139,** 169–183.
Niethammer, D., and Jackson, R. C. (1975). *Eur. J. Cancer* **11,** 845–854.
Nishikawa, K., Rosenblum, M. G., Newman, R. A., Pandita, T. K., Hittelman, W. N., and Donato, N. J. (1992). *Cancer Res.* **52,** 4758–4765.
Nitiss, J., and Wang, J. C. (1988). *Proc. Natl. Acad. Sci. USA* **85,** 7501–7505.
Nitiss, J. L., Liu, Y.-X., Harbury, P., Jannatipour, M., Wasserman, R., and Wang, J. C. (1992). *Cancer Res.* **52,** 4467–4472.
Noonan, K. E., Beck, C., Holzmayer, T. A., Chin, J. E., Wunder, J. S., Andrulis, I. L., Gazdar, A. F., Willman, C. L., Griffith, B., von Hoff, D. D., and Roninson, I. B. (1990). *Proc. Natl. Acad. Sci. USA* **87,** 7160–7164.
Nooter, K., and Herweijer, H. (1991). *Br. J. Cancer* **63,** 663–669.
Nunberg, J. H., Kaufman, R. J., Chang, A. C. Y., Cohen, S. N., and Schimke, R. T. (1980). *Cell* **19,** 355–364.

Nygren, P., Larsson, R., Gruber, A., Peterson, C., and Bergh, J. (1991). *Br. J. Cancer* **64,** 1011–1018.

O'Connor, P. M., Wasserman, K., Sarang, M., Magrath, I., Bohr, V. A., and Kohn, K. W. (1991). *Cancer Res.* **51,** 6550–6557.

O'Connor, P. M., Ferris, D. K., White, G. A., Pines, J., Hunter, T., Longo, D. L., and Kohn, K. W. (1992). *Cell Growth Differ.* **3,** 43–52.

Osheroff, N. (1989). *Biochemistry* **28,** 6157–6160.

Osheroff, N., Zechiedrich, E. L., and Gale, K. C. (1991). *BioEssays* **13,** 269–275.

Ostrowski, L., von Wronski, M. A., Bigner, S. H., Rasheed, A., Schold, Jr., S. C., Brent, T. P., Mitra, S., and Bigner, D. D. (1991). *Carcinogenesis* **12,** 1739–1744.

Ozols, R. F., Cunnion, R. E., Klecker, Jr., R. W., Hamilton, T. C., Ostchega, Y., Parrillo, J. E., and Young, R. C. (1985). *J. Clin. Oncol.* **5,** 641–647.

Ozols, R. F., Hamilton, T. C., Masuda, H., and Young, R. C. (1988). *In* "Mechanisms of Drug Resistance in Neoplastic Cells" (P. V. Woolley III and K. D. Tew, Eds.), pp. 289–306. Academic Press, New York.

Pallavicini, M. G. (1984). *Pharmacol. Ther.* **25,** 207–238.

Papathanasiou, M. A., and Fornace, Jr., A. J. (1991). *In* "Molecular and Clinical Advances in Anticancer Drug Resistance" (R. F. Ozols, Ed.), pp. 13–36. Kluwer Academic Publishers, Boston.

Parker, R. J., Poirier, M. C., Bostick-Bruton, F., Vionnet, J., Bohr, V. A., and Reed, E. (1990). *In* "DNA Damage and Repair in Human Tissues" (B. M. Sutherland and A. D. Woodhead, Eds.), pp. 251–261. Plenum, New York.

Pastan, I. H., and Gottesman, M. M. (1988). *In* "Important Advances in Oncology" (V. T. De Vita, S. Hellman, and S. A. Rosenberg, Eds.), pp. 3–16. Lippincott, New York.

Patel, S., and Fisher, L. M. (1993). *Br. J. Cancer* **67,** 456–463.

Pearce, H. L., Sata, A. R., Bach, N. J., Winter, M. A., Cirtain, M. C., and Beck, W. T. (1989). *Proc. Natl. Acad. Sci. USA* **86,** 5128–5132.

Peters, G. J., Cees, J. V. G., Laurensse, E. J., and Pinedo, H. M. (1991). *Cancer* **68,** 1903–1909.

Peters, G. J., and van Groeningen, C. J. (1991). *Ann. Oncol.* **2,** 469–480.

Pieper, R. O., Futscher, B. W., Dong, Q., and Erickson, L. C. (1991). *Cancer Res.* **51,** 1581–1585.

Pinedo, H. M., and Peters, G. J. (1988). *J. Clin. Oncol.* **6,** 1653–1664.

Pizzorno, G., Mini, E., and Coronnello, M. (1988). *Cancer Res.* **48,** 2149–2155.

Plumb, J. A., Milroy, R., and Kaye, S. B. (1990). *Biochem. Pharmacol.* **39,** 787–792.

Poll, E. H. A., Abrahams, P. J., Arwert, F., and Eriksson, A. W. (1984). *Mutat. Res.* **132,** 181–187.

Pommier, Y., Mattern, M. R., Schwartz, R. E., and Zwelling, L. A. (1984). *Biochemistry* **23,** 2927–2932.

Pommier, Y., Minford, J. K., Schwartz, R. E., Zwelling, L. A., and Kohn, K. W. (1985). *Biochemistry* **24,** 6410–6416.

Pommier, Y., Kerrigan, D., and Kohn, K. W. (1987). *Natl. Cancer Inst. Monogr.* **4,** 83–87.

Pommier, Y. (1993). *Cancer Chemother. Pharmacol.* **32,** 103–108.

Porter, S. E., and Champoux, J. J. (1989). *Nucl. Acids Res.* **17,** 8521–8532.

Potmesil, M., Hsiang, Y.-H., Liu, L. F., Wu, H.-Y., Traganos, F., Bank, B., and Silber, R. (1987). *Natl. Cancer Inst. Monogr.* **4,** 105–109.

Potmesil, M., Hsiang, Y.-H., Liu, L. F., Bank, B., Grossberg, H., Kirschenbaum, S., Forlenzar, T. J., Penziner, A., Kanganis, D., Knowles, D., Traganos, F., and Silber, R. (1988). *Cancer Res.* **48,** 3537–3543.

Prakash, L., Johnson, R., Sung, P., and Prakash, S. (1993). *In* "DNA Repair Mechanisms"

(V. A. Bohr, K. Wassermann, and K. H. Kraemer, Eds.), pp. 374–381. Munksgaard, Copenhagen.

Presant, G. A., Kennedy, P. S., Wiseman, C., Gala, K., Bouzaglou, S., Wyres, M., and Naessig, V. (1986). *Am. J. Clin. Oncol.* **9,** 355–357.

Priest, D. G., Ledford, S. E., and Doig, M. T. (1980). *Biochem. Pharmacol.* **29,** 1549–1553.

Prosperi, E., Sala, E., Negri, C., Oliani, C., Supino, R., Ricotti, G. B. C. A., and Bottiroli, G. (1992). *Anticancer Res.* **12,** 2093–2100.

Puchalski, R. B., and Fahl, W. E. (1990). *Proc. Natl. Acad. Sci. USA* **87,** 2443–2447.

Radparvar, S., Houghton, P. J., and Houghton, J. A. (1989). *Biochem. Pharmacol.* **38,** 335–342.

Ramu, A., Glaubiger, D., Magrath, I. T., and Joshi, A. (1983). *Cancer Res.* **43,** 5533–5537.

Ramu, A., Spanier, R., Rahamimoff, H., and Fuks, Z. (1984a). *Br. J. Cancer* **50,** 501–507.

Ramu, A., Glaubiger, D., and Fuks, Z. (1984b). *Cancer Res.* **44,** 4392–4395.

Ramu, A., Pollard, H. B., and Rosario, L. M. (1989). *Int. J. Cancer* **44,** 539–547.

Raymond, M., and Gros, P. (1989). *Proc. Natl. Acad. Sci. USA* **86,** 6488–6492.

Raymond, M., and Gros, P. (1990). *Mol. Cell Biol.* **10,** 6036–6040.

Raymond, M., Gros, P., Whiteway, M., and Thomas D. Y. (1992). *Science* **256,** 232–234.

Reed, E., Ozols, R. F., Tarone, R., Yuspa, S. H., and Poirier, M. C. (1987). *Proc. Natl. Acad. Sci. USA* **84,** 5024–5028.

Reed, E., Gupta-Burt, S., Yuspa, S. H., Tarone, R., Ozols, R. F., Katz, D., and Poirier, M. C. (1988). *Clin. Res.* **36,** 499A.

Reed, E., Ostchega, Y., Steinberg, S. M., Yuspa, S. H., Young, R. C., Ozols, R. F., and Poirier, M. C. (1990). *Cancer Res.* **50,** 2256–2260.

Reeve, J. G., Rabbitts, P. H., and Twentyman, P. R. (1990). *Br. J. Cancer* **61,** 851–855.

Rhee, M. S., Wang, Y., Nair, M. G., and Galivan, J. (1993). *Cancer Res.* **53,** 2227–2230.

Richert, N. D., Aldwin, L., Nitecki, D., Gottesman, M. M., and Pastan, I. (1988). *Biochemistry* **2,** 7607–7613.

Riehm, H., and Biedler, J. L. (1972). *Cancer Res.* **32,** 1195–1200.

Riordan, J. R., Rommens, J. M., Kerem, B., Alon, N., Rozmahel, R., Grzelczak, Z., Zielenski, J., Lok, S., Plavsic, N., Chou, J. L., Drumm, M. L., Iannuzzi, M. C., Collins, F. S., and Tsui, L. C. (1989). *Science* **245,** 1066–1073.

Riou, J.-F., Helissey, P., Grondard, L., and Giorgi-Renault, S. (1991). *Mol. Pharmacol.* **40,** 699–706.

Ro, J., Sahin, A., Ro, J. Y., Fritsche, H., Hortobagyi, G., and Blick, M. (1990). *Hum. Pathol.* **21,** 787–791.

Robbie, M. A., Baguley, B. C., Denny, W. A., Gavin, J. B., and Wilson, W. R. (1988). *Cancer Res.* **48,** 310–319.

Robinson, M. J., and Osheroff, N. (1990). *Biochemistry* **29,** 2511–2515.

Robson, C. N., Hoban, P. R., Harris, A. L., and Hickson, I. D. (1987a). *Cancer Res.* **47,** 1560–1565.

Robson, C. N., Lewis, A. D., Wolf, C. R., Hayes, J. D., Hall, A., Proctor, S. J., Harris, A. L., and Hickson, I. D. (1987b). *Cancer Res.* **47,** 6022–6027.

Rodenhuis, S., McGuire, J. J., Narayanan, R., and Bertino, J. R. (1986). *Cancer Res.* **46,** 6513–6515.

Roninson, I. B. (1992). Mutat. Res. **276,** 151–161.

Rosman, M., Lee, M. L., Creasey, W. A., and Sartorelli, A. C. (1974). *Cancer Res.* **34,** 1952–1956.

Rosowsky, A., Wright, J. E., Cucchi, C. A., Flatow, J. L., Tristes, D. H., and Teicher, B. A. (1987). *Cancer Res.* **47,** 5913–5918.

Ross, W. E., Glaubiger, D. L., and Kohn, K. W. (1978). *Biochim. Biophys. Acta* **519,** 23–30.

Ross, W. E. (1985). *Biochem. Pharmacol.* **34,** 4191–4195.

Rothenberg, M. L., Mickley, L. A., Cole, D. E., and Fojo, A. T. (1989). *Blood* **74,** 1388–1395.
Rottmann, M., Schroder, H. C., Gramzow, M., Renneisen, K., Kurelec, B., Dorn, A., Friese, U., and Muller, W. E. G. (1987). *EMBO J.* **6,** 3939–3944.
Rustum, Y. M., Trave, F., Zakrzewski, S. F., Petrelli, N., Herrera, L., Mittellman, A., Arbuch, S. G., and Creaven, P. J. (1987). *NCI Monogr.* **5,** 165–170.
Safa, A. R. (1988). *Proc. Natl. Acad. Sci. USA* **85,** 7187–7191.
de Saint Vincent, B. R., Dechamps, M., and Buttin, G. (1980). *J. Biol. Chem.* **255,** 162–167.
Salmon, S. E., Dalton, W. S., Grogan, T. M., Plezia, P., Lehnert, M., Roe, D. J., and Miller, T. P. (1991). *Blood* **78,** 44–50.
Samuels, L. L., Moccio, D. M., and Sirotnak, F. M. (1985). *Cancer Res.* **45,** 1489–1495.
Sander, M., Nolan, J. M., and Hsieh, T. S. (1984). *Proc. Natl. Acad. Sci. USA* **81,** 6938–6942.
Sartorelli, A. C., LePage, G. A., and Moore, E. C. (1958). *Cancer Res.* **18,** 1232–1239.
Sato, W., Yusa, K., Naito, M., and Tsuruo, T. (1990). *Biochem. Biophys. Res. Commun.* **173,** 1252–1257.
Satta, T., Isobe, K.-I., Yamauchi, M., Nakashima, I., and Takagi, H. (1992). *Cancer* **69,** 941–946.
Schibler, M. J., and Cabral, F. (1986). *J. Cell Biol.* **102,** 1522–1531.
Schilsky, R. L., Baley, B. D., and Chabner, B. A. (1980). *Proc. Natl. Acad. Sci. USA* **77,** 2919–2922.
Schimke, R. T. (1984). *Cell* **37,** 705–713.
Schlemmer, S. R., and Sirotnak, F. M. (1992). *J. Biol. Chem.* **267,** 14,746–14,752.
Schneider, E., Darkin, S. J., Robbie, M. A., Wilson, W. R., and Ralph, R. K. (1988). *Biochim. Biophys. Acta* **949,** 264–272.
Scholar, E. M., and Calabresi, P. (1979). *Biochem. Pharmacol.* **28,** 445–446.
Schold, S. C., Brent, T. P., von Hofe, E., Friedman, H. S., Mitra, S., Bigner, D. D., Swenberg, J. J., and Kleihues, P. (1989). *J. Neurosurg.* **70,** 573–577.
Schuurhuis, G. J., Broxterman, H. J., van der Hoeven, J. J. M., Pinedo, H. M., and Lankelma, J. (1987). *Cancer Chemother. Pharmacol.* **20,** 285–290.
Schuurhuis, G. J., Broxterman, H. J., Cervantes, A., van Heijningen T. H. M., de Lange, J. H. M., Baak, J. P. A., Pinedo, H. M., and Lankelma, J. (1989). *J. Natl. Cancer Inst.* **81,** 1887–1892.
Schweitzer, B. I., Dicker, A. P., and Bertino, J. R. (1990). *FASEB J.* **4,** 2441–2452.
Sehested, M., Skovsgaard, T., van Deurs, B., and Winther-Nielsen, H. (1987). *J. Natl. Cancer Inst.* **78,** 171–179.
Sehested, M., Skovsgaard, T., and Roed, H. (1988). *Biochem. Pharmacol.* **37,** 3305–3310.
Sehested, M., Bindslev, N., Demant, E. J. F., Skovsgaard, T., and Jensen, P. B. (1989). *Biochem. Pharmacol.* **38,** 3017–3027.
Seidegård, J., Perp, R. W., Miller, D. G., and Beattie, E. J. (1986). *Carcinogenesis* **7,** 751–753.
Seidegård, J., Vorachek, W. R., Pero, R. W., and Pearson, W. R. (1988). *Proc. Natl. Acad. Sci. USA* **85,** 7293–7297.
Sengupta, S. K., Anderson, J. E., and Kelley, C. (1982). *J. Med. Chem.* **25,** 1214–1218.
Shen, D. W., Fojo, A., Chin, J. E., Roninson, I. B., Richert, N. Pastan, I., and Gottesman, M. M. (1986). *Science* **232,** 643–645.
Sheibani, N., Jennerwein, M. M., and Eastman, A. (1989). *Biochemistry* **28,** 3120–3124.
Shimabuku, A. M., Saeki, T., Ueda, K., and Kamono, T. (1991). *Agric. Biol. Chem.* **55,** 1075–1080.
Siegfried, J. M., Tritton, T. R., and Sartorelli, A. C. (1983). *Eur. J. Cancer Clin. Oncol.* **19,** 1133–1141.
Sikic, B. I., Ehsan, M. N., Harker, W. G., Friend, N. F., Brown, B. W., Newman, R. A., Hacker, R. A., and Acton, E. M. (1985). *Science* **228,** 1544–1546.

Simonsen, C. C., and Levinson, A. D. (1983). *Proc. Natl. Acad. Sci. USA* **80,** 2495–2499.

Sinha, B. K., Katki, A. G., Batist, G., Cowan, K. H., and Myers, C. E. (1987). *Biochemistry* **26,** 3776–3781.

Sinha, B. K., Mimnaugh, E. G., Rajagopalan, S., and Myers, C. E. (1989). *Cancer Res.* **49,** 3844–3848.

Sinha, B. K., and Eliot, H. M. (1991). *Biochim. Biophys. Acta* **1097,** 111–116.

Sirotnak, F. M., Kurita, S., and Hutchison, D. J. (1968). *Cancer Res.* **28,** 75–80.

Sirotnak, F. M., and Donsback, R. C. (1976). *Cancer Res.* **36,** 1151–1158.

Sirotnak, F. M., Moccio, D. M., Kelleher, L. E., and Goutas, L. J. (1981a). *Cancer Res.* **41,** 4447–4452.

Sirotnak, F. M., Moccio, D. M., and Young, C. W. (1981b). *Cancer Res.* 41, 966–970.

Sirotnak, F. M., Yang, C. H., Mincs, L. S., Oribe, E., and Biedler, J. L. (1986). *J. Cell Physiol.* **126,** 226–274.

Sirotnak, F. M. (1987). *NCI Monogr.* **5,** 27–35.

Skipper, H. E., Schabel, Jr., F. M., and Lloyd, H. H. (1978). *Semin. Hematol.* **15,** 207–219.

Skovsgaard, T. (1978). *Cancer Res.* **38,** 4722–4727.

Skovsgaard, T. (1980). *Cancer Res.* **40,** 1077–1083.

Skovsgaard, T., Danø, K., and Nissen, N. I. (1984). *Cancer Treat. Rev.* **11,** 63–72.

Skovsgaard, T. (1989). *Anticancer Drugs* **191,** 233–244.

Slapak, C. A., Daniel, J. C., and Levy, S. B. (1990). *Cancer Res.* **50,** 7895–7901.

Slapak, C. A., Fracasso, P. M., LeCerf, J. M., and Levy, S. B. (1992). *Proc. Am. Assoc. Cancer Res.* **33,** 457

Slichenmyer, W. J., Rowinsky, E. K., Donehower, R. C., and Kaufmann, S. H. (1993). *J. Natl. Cancer Inst.* **85,** 271–291.

Slovak, M. L., Hoeltge, G. A., Dalton, W. S., and Trent, J. M. (1988). *Cancer Res.* **48,** 2793–2797.

Slovak, M. L., Coccia, M., Melzter, P. S., and Trent, J. M. (1991). *Anticancer Res.* **11,** 423–428.

Snyder, R. D., Carrier, W. L., and Regan, J. D. (1981). *Biophys. J.* **35,** 339–350.

Sonneveld, P., Durie, B. G. M., Lokhorst, H. M, Marie, J.-P., Solbu, G., Sucik, S., Zittoun, R., Löwenberg, B., and Nooter, K. (1992). *Lancet* **340,** 255–259.

Soto, F., Canaves, J. M., Gonzales-Ros, J. M., and Ferragut, J. A. (1992). *FEBS Lett.* **301,** 119–123.

Spears, C. P., Gustavsson, B. G., Berne, M., Frosling, R., Berstein, L., and Hayes, A. A. (1988). *Cancer Res.* **48,** 5894–5900.

Spears, C. P., Shahimian, A. H., Moran, R. G., Heidelberger, C., and Corbett, T. H. (1982). *Cancer Res.* **42,** 450–456.

Spies, T., Cerundolo, V., Colonna, M., Cresswell, P., Townsend, A., and DeMars, R. (1992). *Nature (London)* **255,** 644–646.

Srimatkandada, S., Schweitzer, B. I., Moroson, B. A., Dube, S., and Bertino, J. R. (1989). *J. Biol. Chem.* **264,** 3524–3528.

Stark, G. R. (1986). *Cancer Surv.* **5,** 1–24.

Steuart, C. D., and Burke, P. J. (1971). *Nature New Biol.* **223,** 109–110.

Streeter, D. G., Taylor, D. L., Acton, E. M., and Peters, J. H. (1985). *Cancer Chemother. Pharmacol.* **14,** 160–164.

Sugawara, I., Kataoka, I., Morishita, Y., Hamada, H., Tsuruo, T., Itoyama, S., and Mori, S. (1988). *Cancer Res.* **48,** 1926–1929.

Sugimoto, Y., and Tsuruo, T. (1991). *In:* "Molecular and Cellular Biology of Multidrug Resistant Tumor Cells" (I. B. Roninson, Ed.), pp. 57–72. Plenum, New York.

Sullivan, D. M., Glisson, B. S., Hodges, P. K., Smallwood-Kentro, S., and Ross, W. E. (1986). *Biochemistry* **25,** 2248–2256.

Sullivan, D. M., Latham, M. D., and Ross, W. E. (1987). *Cancer Res.* **47,** 3973–3979.
Sullivan, D. M., Rowe, T. C., Latham, M. D., and Ross, W. E. (1989). *Biochemistry* **28,** 5680–5687.
Sullivan, D. M., and Ross, W. E. (1991). *In* "Molecular and Clinical Advances in Anticancer Drug Resistance" (R. F. Ozols, Ed.), pp. 57–100. Kluwer Academic Publishers, Boston.
Sullivan, D. M., Eskildsen, L. A., Groom, K. R., Webb, C. D., Latham, M. D., Martin, A. W., Wellhausen, S. R., Kroeger, P. E., and Rowe, T. C. (1993). *Mol Pharmacol.* **43,** 207–216.
Swain, S. M., Lippman, M. E., Egan, E. F., Drake, J. C., Steinberg, S. M., and Allegra, D. J. (1989). *J. Clin. Oncol.* **7,** 890–899.
Swinnen, L. J., Barnes, D. M., Fisher, S. G., Albain, K. S., Fisher, R. I., and Erickson, L. C. (1989). *Cancer Res.* **49,** 1383–1389.
Tan, K. B., Mattern, M. R., Boyce, R. A., Hertzberg, R. P., and Schein, P. S. (1987a). *Natl. Cancer Inst. Monogr.* **4,** 95–98.
Tan, K. B., Mattern, M. R., Boyce, R. A., and Schein, P. S. (1987b). *Proc. Natl. Acad. Sci. USA* **84,** 7668–7671.
Tan, K. B., Per, S. R., Boyce, R. A., Mirabelli, C. K., and Crooke, S. T. (1988). *Biochem. Pharmacol.* **37,** 4413–4416.
Tan, K. B., Mattern, M. R., Eng, W.-K., McCabe, F. L., and Johnson, R. K. (1989). *J. Natl. Cancer Inst.* **81,** 1732–1735.
Tan, K. B., Dorman, T. E., Falls, K. M., Chung, T. D. Y., Mirabelli, C. K., Crooke, S. T., and Mao, J. (1992). *Cancer Res.* **52,** 231–234.
Tattersall, M. H. N., Ganeshaguru, K., and Hoffbrand, A. V. (1974). *Br. J. Haematol.* **27,** 39–46.
Taudou, G., Mirambeau, G., Lavenot, C., der Garabedian, A., Vermeersch, J., and Duguet, M. (1984). *FEBS Lett.* **176,** 431–435.
Taylor, C. W., Dalton, W. S., Parrish, P. R., Gleason, M. C., Bellamy, W. T., Thompson, F. H., Roe, D. J., and Trent, J. M. (1991). *Br. J. Cancer* **63,** 923–929.
Teeter, L. D., Atsumi, S., Sen, S., and Kuo, T. (1986). *J. Cell Biol.* **103,** 1159–1166.
Tewey, K. M., Rowe, T. C., Yang, L., Halligan, B. D., and Liu, L. F. (1984). *Science* **226,** 466–468.
Thiebaut, F., Tsuruo, T., Hamada, H., Gottesman, M. M., Pastan, I., and Willingham, M.C. (1987). *Proc. Natl. Acad Sci. USA* **84,** 7735–7738.
Thiebaut, F., Tsuruo, T., Hamada, H. Gottesman, M. M., Pastan, I., and Willingham, M. C. (1989). *J. Histochem. Cytochem.* **37,** 159–164.
Thiebaut, F., Currier, S. J., Whitaker, J., Haugland, R. P., Gottesman, M. M., Pastan, I., and Willingham, M. C. (1990). *J. Histochem. Cytochem.* **38,** 685–690.
Toffoli, G., Viel, A., Tumiotto, L., Biscontin, G., Rossi, C., and Boiocchi, M. (1991). *Br. J. Cancer* **63,** 51–56.
Tokes, Z. A., Rogers, K. E., and Rembaum, A. (1982). *Proc. Natl. Acad. Sci. USA* **79,** 2026–2030.
Torres-Garcia, S. J., Cousineau, L., Caplan, S., and Panasci, L. (1989). *Biochem. Pharmacol.* **38,** 3122–3123.
Townsend, A. J., Tu, C.-P. D., and Cowan, K. H. (1992). *Mol. Pharmacol.* **41,** 230–236.
Trash, C., Voelkel, K., DiNardo, S., and Sternglanz, R. (1984). *J. Biol. Chem.* **259,** 1375–1377.
Trask, D., and Muller, M. (1988). *Proc. Natl. Acad. Sci. USA* **85,** 1417–1421.
Trent, J., Buick, R. N., Olson, S. M., Horns, R. C., and Schimke, R. T. (1984). *J.Clin. Oncol.* **2,** 8–15.
Tritton, T. R., and Yee, G. (1982). *Science* **217,** 248–250.
Tsuruo, T., Iida, H., Tsukagoshi, S., and Sakurai, Y. (1981). *Cancer Res.* **41,** 1967–1972.

Tsuruo, T., Iida, H., Tsukagoshi, S., and Sakurai, Y. (1982). *Cancer Res.* **42,** 4730–4733.

Tsuruo, T., Iida, H., Kitatani, Y., Yokota, K., Tsukagoshi, S., and Yakurai, Y. (1984). *Cancer Res.* **44,** 4303–4307.

Tsuruo, T., Oh-Hara, T., and Saito, H. (1986). *Anticancer Res.* **6,** 637–642.

Tsuruo, T. Matsuzaki, T., Matsushita, M., Saito, H., and Yokokura, T. (1988). *Cancer Chemother. Pharmacol.* **21,** 71–74.

Tsuruo, T., Yusa, K., Sudo, Y., Takamori, R., and Sugimoto, Y. (1989). *Cancer Res.* **49,** 5537–5542.

Twentyman, P. R., Reeve, J. G., Kock, G., and Wright, K. A. (1990). *Br. J. Cancer* **62,** 89–95.

Twentyman, P. R. (1992). *Biochem. Pharmacol.* **43,** 109–117.

Ueda, K., Pastan, I., and Gottesman, M. M. (1987a). *J. Biol. Chem.* **262,** 17,432–17,436.

Ueda, K., Clark, D. P., Chen, C. J., Roninson, I. B., Gotttesman, M. M., and Pastan, I. H. (1987b). *J. Biol. Chem.* **262,** 505–508.

Ueda, K., Okamura, N., Hirai, M., Tanigawara, Y., Saeki, T., Kioka, N., Komano, T., and Hori, R. (1992). *J. Biol. Chem.* **267,** 24,248–24,252.

Vaage, J., Mayhew, E., Lasic, D., and Martin, F. (1992). *Int. J. Cancer* **51,** 942–948.

Valverde, M. A., Diaz, M., Sepulveda, F. V., Gill, D. R., Hyde, S. C., and Higgins, C. F. (1992). *Nature (London)* **355,** 830–833.

van der Bliek, A. M., van der Velde-Koerst, T., Ling, V., and Borst, P. (1986). *Mol. Cell. Biol.* **6,** 1671–1678.

van der Bliek, and Borst, P. (1989). *Adv. Cancer Res.* **52,** 165–203.

van der Zee, A. G. J., Hollema, H., de Jong, S., Boonstra, H., Gouw, A., Willemse, P. H. B., Zijlstra, J. G., and de Vries, E. G. E. (1992). *Cancer Res.* **52,** 5915–5920.

van Diggelen, O. P., Donahue, T. F., and Shin, S. I. (1979). *J. Cell. Physiol.* **98,** 59–71.

van Duin, M., de Wit, J., Odijk, H., Westerveld, A., Yasui, A., Koken, M. H. M., Hoeijmakers, J. H. J., and Bootsma, D. (1986). *Cell* **44,** 913–923.

van Kalken, C. K., Broxterman, H. J., Pinedo, H. M., Feller, N., Dekker, H., Lankelma, J., and Giaccone, G. (1993). *Br. J. Cancer* **67,** 284–289.

Vasanthakumar, G., and Ahmed, N. K. (1989). *Cancer Commun.* **1,** 225–232.

Vayuvegula, B., Slater, L., Meador, J., and Gupta, S. (1988). *Cancer Chemother. Pharmacol.* **22,** 163–168.

Verrelle, P., Meissonnier, F., Fonck, Y., Feillel, V., Dionet, C., Kwiatkowski, F., Plagne, R., and Chassagne, J. (1991). *J. Natl. Cancer Inst.* **83,** 111–116.

Versantvoort, C. H. M., Broxterman, H. J., Pinedo, H. M., de Vries, E. G. E., Feller, N., Kuiper, C. M., and Lankelma, J. (1992a). *Cancer Res* **52,** 17–23.

Versantvoort, C. H. M., Twentyman, P. R., Barrand, M. A., Lankelma, J., Pinedo, H. M., and Broxterman, H. J. (1992b). *Proc. Am. Assoc. Cancer Res.* **33,** 456.

Verweij, J., Herweijer, H., Oosterom, R., van der Burg, M. E., L., Planting A. S. T., Seynaeve, C., Stoter, G., and Nooter, K. (1991). *Br. J. Cancer* **64,** 361–364.

Volm, M., Mattern, J., and Pommerenke, E. W. (1991). *Anticancer Res.* **11,** 579–586.

Vos, J.-M. H., and Hanawalt, P. C. (1987). *Cell* **50,** 789–799.

Vosberg, H.-P. (1985). *In* "Current Topics in Microbiology and Immunology", 114, pp. 19–102. Springer-Verlag, New York/Berlin.

Waltham, M. C., Holland, J. W., Robinson, S. C., Winzor, D. J., and Nixon, P. F. (1988). *Biochem. Pharmacol.* **37,** 535–539.

Wang, F., Aschele, C., Sobrero, A., Cheng, Y., and Bertino, J. R. (1993). *Cancer Res.* **53,** 3677–3680.

Wang, J. C. (1985). *Ann. Rev. Biochem.* **54,** 665–697.

Warren, L., Jardillier, J. C., and Ordentlich, P. (1991). *Cancer Res.* **51,** 1996–2001.

Wassermann, K., Zwelling, L. A., Mullins, T. D., Silberman, L. E., Andersson, B. S., Bakic, M., Acton, E. M., and Newman, R. A. (1986). *Cancer Res.* **46,** 4041–4046.
Wassermann, K., Kohn, K. W., and Bohr, V. A. (1990a). *J. Biol. Chem.* **265,** 13,906–13,913.
Wassermann, K., Markovits, J., Jaxel, C., Capranico, G., Kohn, K. W., and Pommier, Y. (1990b). *Mol. Pharmacol.* **38,** 38–45.
Wassermann, K., Pirsel, M., and Bohr, V. A. (1992). *Cancer Res.* **52,** 6853–6959.
Wassermann, K., and Damgaard, J. (1994). *Cancer Res.* **54,** 175–181.
Washtien, W. L. (1982). *Mol. Pharmacol.* **21,** 723–728.
Washtien, W. L. (1984). *Mol. Pharmacol.* **25,** 171–177.
Watanabe, M., Komeshima, N., Naito, M., Isoe, T., Otake, N., and Tsuruo, T. (1991). *Cancer Res.* **51,** 157–161.
Webb, C. D., Latham, M. D., Lock, R. B., and Sullivan, D. M. (1991). *Cancer Res.* **51,** 6543–6549.
Weinberg, G., Ullman, B., and Martin, D. W. (1981). *Proc. Natl. Acad Sci. USA* **78,** 2447–2451.
Weinstein, R. S., Jakate, S. M., Dominguez, J. M., Lebovitz, M. D., Koukoulis, G. K., Kuszak, J. R., Klusens, L. F., Grogan, T. M., Saclarides, T. J., Roninson, I. B., and Coon, J. S. (1991). *Cancer Res.* **51,** 2720–2726.
Westendorf, J., Groth, G., Steinheider, G., and Marquardt, H. (1985). *Cell Biol. Toxicol.* **1,** 87–101.
Whitehead, V. M. (1977). *Cancer Res.* **37,** 408–412.
Wiley, J. S., Taupin, J., Jamieson, G. P., Snook, M., Sawyer, W. H., and Finch, L. R. (1985). *J. Clin. Invest.* **75,** 632–642.
Willingham, M. C., Cornwell, M. M., Cardarelli, C. O., Gottesman, M. M., and Pastan, I. H. (1986). *Cancer Res.* **46,** 5941–5946.
Wishart, G. C., Plumb, J. A., Going, J. J., McNicol, A. M., McArdle, C. S., Tsuruo, T., and Kaye, S. B. (1990). *Br. J. Cancer* **62,** 758–761.
Woessner, R. D., Chung, T. D. Y., Hofmann, G. A., Mattern, M. R., Mirabelli, C. K., Drake, F. H., and Johnson, R. K. (1990). *Cancer Res.* **50,** 2901–2908.
Woessner, R. D., Eng, W.-K., Hofmann, G. A., Rieman, D. J., McCabe, F. L., Hetzberg, R. P., Mattern, M. R., Tan, K. B., and Johnson, R. K. (1993). *Oncol. Res.* **4,** 481–488.
Wolf, C. R., Hayward, I. P., Lawrie, S. S., Buckton, K., McIntyre, M. A., Adams, D. J., Lewis, A. D., Scott, A. R. R., and Smyth, J. F. (1987). *Int. J. Cancer* **39,** 695–702.
Wolpert, M. K., Damle, S. P., Brown, J. E., Sznycer, E., Agarwal, K. C., and Sartorelli, A. C. (1971). *Cancer Res.* **31,** 1620–1626.
Wolverton, J. S., Danks, M. K., Schmidt, C. A., and Beck, W. T. (1989). *Cancer Res.* **49,** 2422–2426.
Woodman, P. W., Sariff, A. M., and Heidelberger, C. (1980). *Cancer Res.* **40,** 507–511.
Wozniak, A. J., and Ross, W. E. (1983). *Cancer Res.* **43,** 120–124.
Wright, J. A., Smith, H. S., Watt, F. M., Hancock, M. C., Hudson, D. L., and Strak, G. R. (1990). *Proc. Natl. Acad. Sci. USA* **87,** 1791–1795.
Wright, L. C, Dyne, M., Holmes, K. T., and Mountford, C. E. (1985). *Biochem. Biophys. Res. Commun.* **133,** 539–545.
Wu, Z., Chan, C. L., Eastman, A., and Bresnick, E. (1992). *Cancer Res.* **52,** 32–35.
Yamashita, Y., Kawada, S., Fujii, N., and Nakano, H. (1991). *Biochemistry* **30,** 5838–5845.
Yang, C. H., Sirotnak, F. M., and Minez, L. S. (1988). *J. Biol. Chem.* **263,** 9703–9709.
Yang, C. H., Dembo, M., and Sirotnak, F. M. (1983). *J. Membr. Biol.* **75,** 11–20.
Yang, E. H., Dembo, M., and Sirotnak, F. M. (1982). *J. Membr. Biol.* **68,** 19–28.
Yang, L. Y., and Trujillo, J. M. (1989a). *Proc. Am. Assoc. Cancer Res.* **30,** 516.

Yang, C. P. H., DePinho, S. H., Greenberger, L. M., Arceci, R. J., Horwitz, S. B. (1989b). *J. Biol. Chem.* **264,** 782–788.

Yin, M. B., Zakrzewski, S. F., and Hakala, M. T. (1983). *Mol. Pharmacol.* **23,** 190–197.

Yoshida, A., Ueda, T., Wano, Y., and Nakamura, T. (1993). *Jpn. J. Cancer Res.* **84,** 566–573.

Young, I., Young, G. J., Wiley, J. S., and Van Der Weyden, M. B. (1985). *Eur. J. Cancer Clin. Oncol.* **21,** 1077–1082.

Yu, G., Ahmad, S., Aquino, A., Fairchild, C. R., Trepel, J. B., Ohno, S., Suzuki, K., Tsuruo, T., Cowan, K. H., and Glazer, R. I. (1991). *Cancer Commun.* **3,** 181–189.

Yusa, K., and Tsuruo, T. (1989). *Cancer Res.* **49,** 5002–5006.

Zaman, G. J. R., Versantvoort, C. H. M., Smit, J. J. M., Eijdems, E. W. H. M., de Haas, M., Smith, A. J., Broxterman, H. J., Mulder, N. H., de Vries, E. G. E., Baas, F., and Borst, P. (1993). *Cancer Res.* **53,** 1747–1750.

Zamora, J. M., and Beck, W. T. (1986). *Biochem. Pharmacol.* **35,** 4303–4310.

Zamora, J. M., Pearce, H. L., and Beck, W. T. (1988). *Mol. Pharmacol.* **33,** 454–462.

Zhen, W., Link Jr., C. J., O'Connor, P. M., Reed, E., Parker, R., Howell, S. B., and Bohr, V. A. (1992). *Mol. Cell. Biol.* **12,** 3689–3698.

Zijlstra, J. G., de Vries, E. G. E., and Mulder, N. H. (1987). *Cancer Res.* **47,** 1780–1784.

Zunino, F., Gambetta, R., and Di Marco, A. (1975). *Biochem. Pharmacol.* **24,** 309–311.

Zwelling, L. A. (1985). *Cancer Metastasis Rev.* **4,** 263–276.

Zwelling, L. A., Estey, E., Silberman, L., Doyle, S., and Hittelman, W. (1987). *Cancer Res.* **47,** 251–257.

Zwelling, L. A., Hinds, M., Chen, D., Mayes, J., Sie, K. L., Parker, E., Silberman, L., Radcliffe, A., Beran, M., and Blick, M. (1989). *J. Biol. Chem.* **264,** 16,411–16,420.

Zwelling, L. A., Altschuler, E., Cherif, A., and Farquhar, D. (1991). *Cancer Res.* **51,** 6704–6707.

Zwelling, L. A., Mitchell, M. J., Satitpunwaycha, P., Mayes, J., Altschuler, E., Hinds, M., and Baguley, B. C. (1992). *Cancer Res.* **52,** 209–217.

Cellular and Molecular Effects of Thymic Epithelial Cells on Thymocytes during Differentiation and Maturation

Yoshihiro Kinoshita and Fumihiko Hato
Department of Physiology, Osaka City University Medical School, Abeno-ku, Osaka 545, Japan

I. Introduction

Neonatally thymectomized animals show a severe reduction of cellular immune capacity. Transplantation of the thymus to the thymectomized animals restores this capacity (Good *et al.*, 1962; Kinoshita *et al.*, 1970a; Martinez *et al.*, 1962; Miller, 1961, 1962; Sherman *et al.*, 1963; Waksman *et al.*, 1962). In addition to this, intraperitoneal or intravenous injection of thymic lymphocytes (thymocytes) (Law *et al.*, 1964; Yunis *et al.*, 1964) and of the most dense subset of thymocytes (Taub *et al.*, 1965) into neonatally thymectomized animals partially restores the immune capacity. These results suggest that the thymocytes, which differentiate and mature into immunologically competent cells, exist in the thymus, although in small numbers. Furthermore, immunological activity in neonatally thymectomized animals can also be restored by placing a cell-impermeable diffusion chamber containing the thymus into their peritoneal cavity (Levey *et al.*, 1963). In this experimental system, thymic epithelial cells (TECs) adhere mainly to the inner wall surface of the chamber and survive. It is assumed from this study that the hormonal products from TECs pass through the pore of the diffusion chamber, react to undifferentiated lymphoid cells in the thymectomized animals, and endow them with the property of T cells (thymus-dependent lymphocytes).

The preparation of thymus hormonal products (THPs), for example, thymosin (A. L. Goldstein *et al.*, 1966, 1972), thymus humoral factor (Trainin and Small, 1970), and thymin (G. Goldstein, 1974) was a major advance in thymic research. However, whole thymus, in which the amount of TEC in thymic nucleated cells is small, was used as the source. These hormonal products may be mingled with other bioactive factors that origi-

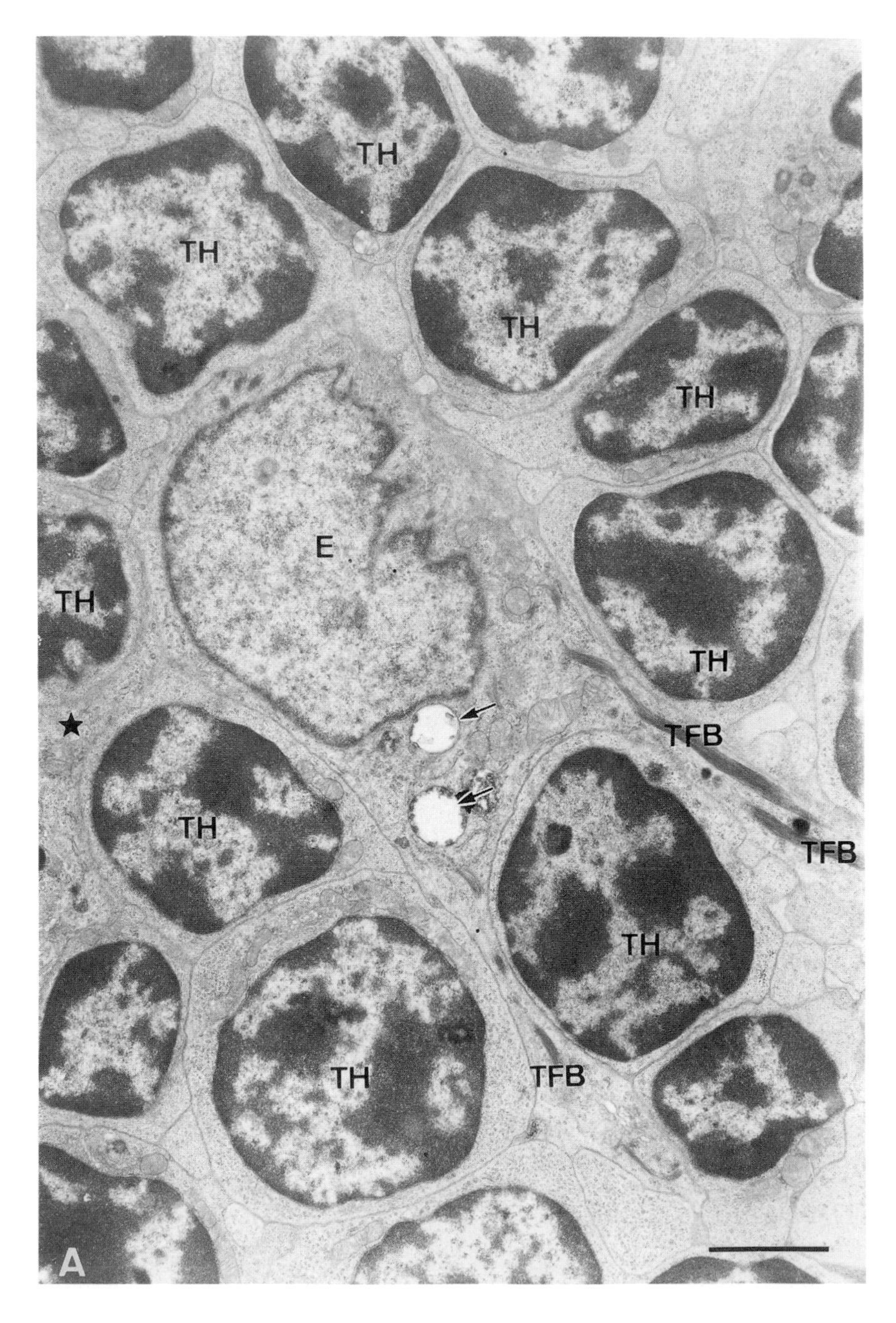
TH
TH
TH
TH
E
TH
TH
TFB
TH
TFB
TH
TH
TFB
A

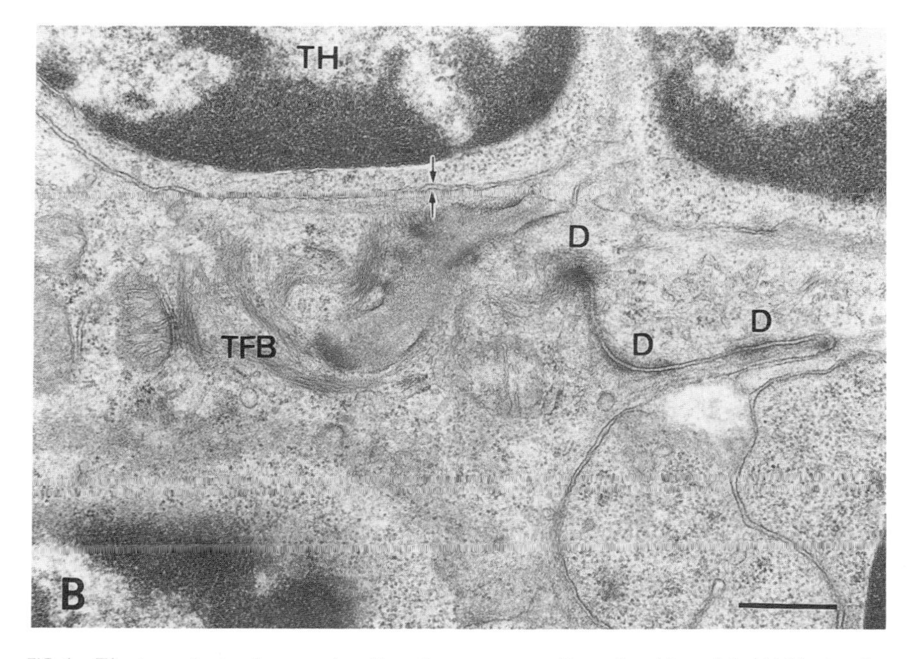

FIG. 1 Electron photomicrographs of rat thymus cortex (6 weeks old, male). (A) Notice the thymocytes (thymic lymphocyte, TH) around thymic epithelial cells (E), which contain two multivesicular bodies (arrows) and tonofilament bundles (TFB). Bar = 2 μm. The cell process (*) is magnified and presented in B. (B) Typical desmosomes (D) are observed between epithelial cells. Arrows indicate a close contact between epithelial cell and thymocyte. Bar = 0.5 μm. (Photography courtesy of Dr. K. P. Takahashi, Department of Anatomy, Osaka City University Medical School.)

nate from other thymic cells (dendritic cells, interdigitating cells, thymocytes, etc.). The best way to prepare THPs is to use a source that has TEC in a highly pure state. Precursors of T cells from bone marrow migrate into thymic parenchyma through the blood. The cells, under the influence of TECs, proliferate, differentiate, and mature into relatively mature T cells which migrate into the peripheral lymphoid organs (Morioka *et al.*, 1989; Mosier and Cantor, 1971; Taniyama *et al.*, 1989). However, the detailed pathway of differentiation and maturation of thymocytes as well as the cellular and molecular relation between thymocytes and TECs remains to be solved. Moreover, TECs have a heterogeneous constitution from the standpoint of embryology (Cordier and Haumont, 1980), immunohistology (Van Ewijk, 1984; Haynes, 1990), surface antigenicity (Hirokawa *et al.*, 1986; Kaneshima *et al.*, 1987), and electron microscopy (Kendal *et al.*, 1985). Accordingly, the population is assumed to consist of several subpopulations which mutually release different hormonal products. It is still not known to which step each hormonal product reacts in the differentiation and maturation pathway.

In this chapter, the following items are reviewed: (1) the cellular and molecular relationship between thymocytes and TECs, (2) partial purification of THPs from TEC culture supernatant, (3) the parameters reflecting thymocyte differentiation and maturation, (4) the thymocyte subset which is sensitive for each THP, and (5) regulation of TEC function from a viewpoint of endocrine organs and the autonomic nervous system.

II. Close Adhesion between Thymocytes and Thymic Epithelial Cells

Electron microscopy has shown that the surface membrane of TECs adheres closely to that of thymocytes (see Fig. 1). A recent review (Haynes, 1990) suggested that a number of factors are essential for close contact between the cell types. TECs express adhesion molecules such as intercellular adhesion molecule-1, (ICAM-1) (Dustin *et al.,* 1986; Marlin and Springer, 1987; Rothlein *et al.,* 1986), lymphocyte function-associated antigen-3 (LFA-3) (Krensky *et al.,* 1983), major histocompatibility complex (MHC) class I (Norment *et al.,* 1988), and class II (Doyle and Strominger, 1987) on their surface membranes. On the other hand, thymocytes express LFA-1 (Arnaout, 1990; Krensky *et al.,* 1983; Marlin and Springer, 1987), cluster of differentiation 2 (CD2, Kato *et al.,* 1992), CD4 (Brideau *et al.,* 1980; Hale *et al.,* 1987), and CD8 (Brideau *et al.,* 1980; Hale *et al.,* 1987), and so on. These adhesion molecules participate in close contact between TECs and thymocytes. It is assumed that the signals to induce thymocyte differentiation and maturation are provided by the binding between the adhesion molecules on thymocytes and those on TECs.

III. Thymocyte Differentiation and Maturation

A. Hormonal Products of Thymic Epithelial Cells That Affect Thymocytes

One of the ways that TECs signal to thymocytes is thought to be through thymus hormonal products. The preparation of THP from TEC is key in solving the complicated questions of TEC function. The ideal way to prepare THP is to use a source that contains highly pure TEC, as described earlier. Several authors have reported that the supernatant from TEC primary culture enhances the thymocyte response to mitogenic and allogenic cell stimulation (Hensen *et al.,* 1978; Kruisbeek *et al.,* 1977, 1978;

Oosterom *et al.*, 1979; Oosterom and Kater, 1980). However, our work using high-performance liquid chromatography (HPLC) and a gel-filtration column showed that a medium from a primary culture of TEC also contained cytotoxic polypeptides (estimated molecular weight, 4.7 kDa) which reduced the percentage of viable thymocytes and suppressed the response to mitogenic stimulation (Oshitani *et al.*, 1989). Furthermore, contamination of TEC primary cultures by other thymic stromal cells is unavoidable; these cells release other cytokines which affect thymocyte function. As a result, the authors decided that a supernatant of the cultured TEC line should be used as a source for THP and that the hormonal products must be separated from the cytotoxic factors mingled in the supernatant (Kinoshita *et al.*, 1990; Morioka *et al.*, 1989; Taniyama *et al.*, 1989). Nieburgs *et al.* (1985) reported earlier that a culture supernatant of a TEC line contained two different mediators capable of enhancing and suppressing the thymocyte mitogenic response; the estimated molecular weights are 9.7 kDa and 1–5 kDa, respectively. The following paragraphs describe our method for preparing THP from a TEC culture supernatant and the problems encountered in purifying THP.

In our department, two thymus epithelial cell lines—IT-45R1 (Itoh *et al.*, 1981) and TAD3 (Kinoshita *et al.*, 1990; Masuda *et al.*, 1985)—are maintained. This is necessary in order to confirm whether successively cultured cell lines keep the characteristics of epithelial cells. As shown in Fig. 2, intracellular keratin distribution of TECs is examined by indirect immunofluorescence when TEC lines are used for every experiment (Hashimura *et al.*, 1987; Taniyama *et al.*, 1989; Toyokawa *et al.*, 1987). It is generally accepted that protein synthesis by cultured cells is more active in the confluent state, whereas DNA synthesis is more active in the preconfluent state. Moreover, the THPs released from the cultured TEC line—which induce proliferation, differentiation, and maturation of thymocytes—appear to be polypeptides or proteins. On the basis of this information, a large amount of the culture supernatant is collected after incubating TEC for 24 hr in the confluent state with fetal calf serum (FCS)-free balanced salt solution (PBS or Earle's balanced salt solution). The reason for using serum-free solution is that serum proteins might disturb the separation of THPs from the culture supernatant. It is important to concentrate the culture supernatant in preparing THP. We think that ultrafiltration or dialysis of the culture supernatant results in a low THP recovery rate because we have found (Kinoshita *et al.*, 1990; Morioka *et al.*, 1989; Taniyama *et al.*, 1989) that the molecular weight of THPs is equal or less than 10 kDa and such low-molecular-weight products presumably pass through the ultrafiltration apparatus or dialysis membrane.

Next, we use lyophilization to concentrate the THP. The lyophilized material of the culture supernatant is dissolved in double-distilled water and subjected to HPLC with gel filtration. The eluent is monitored by

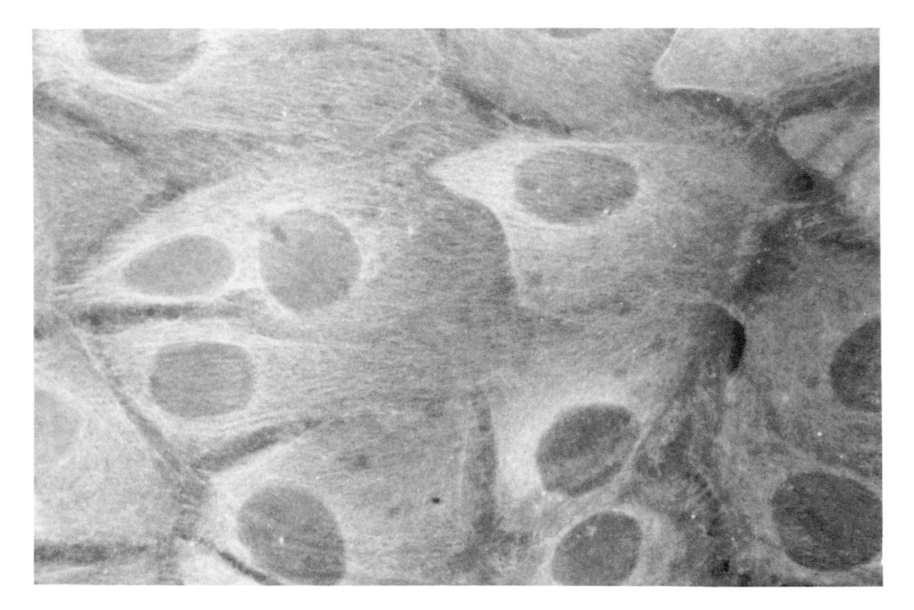

FIG. 2 Photomicrograph of a thymic epithelial cell line stained with antihuman keratin antibody by indirect immunofluorescence method. ×298. Reprinted from *Cell. Mol. Biol.* **36,** Kinoshita, Y., Toyokawa, T., Hato, F., Masuda, A., and Matsuyama, M., New thymocyte growth factor from thymic epithelial cell line, 429–438, copyright 1990, with kind permission from Elsevier Science Ltd., The Boulevard, Langford Lane, Kidlington 0X5 1GB, UK.

absorbance at 280 or 214 nm and the part corresponding to each peak is fractionated (Fig. 3). Each fraction is again lyophilized for further purification of THP in the collected fraction. Thus, lyophilization is carried out twice. When lyophilization is done three or more times, the physiological activity of THP in the collected fraction is markedly reduced. The elution buffer for the gel filtration is 50 m*M* sodium phosphate (pH 6.7), so the lyophilized material of the polypeptide fraction, which can induce thymocyte differentiation and maturation, contains not only THP but also sodium phosphate salt. This material, dissolved in double-distilled water, is used to assay thymus hormonal activity. The biological activity of thymocytes treated *in vitro* with the lyophilized and buffered material is depressed by higher osmotic pressure, which is caused by the addition of THP fractions containing relatively large amounts of salt. This might possibly lead to an erroneous conclusion that the activity of the polypeptide fraction prepared by the above method is too weak.

Further purification of THP is necessary to study the effects of TECs on thymocyte differentiation and maturation. Therefore, the fractions containing THP isolated by gel-filtration column are further separated by reverse-phase chromatography. Generally, reverse-phase chromatogra-

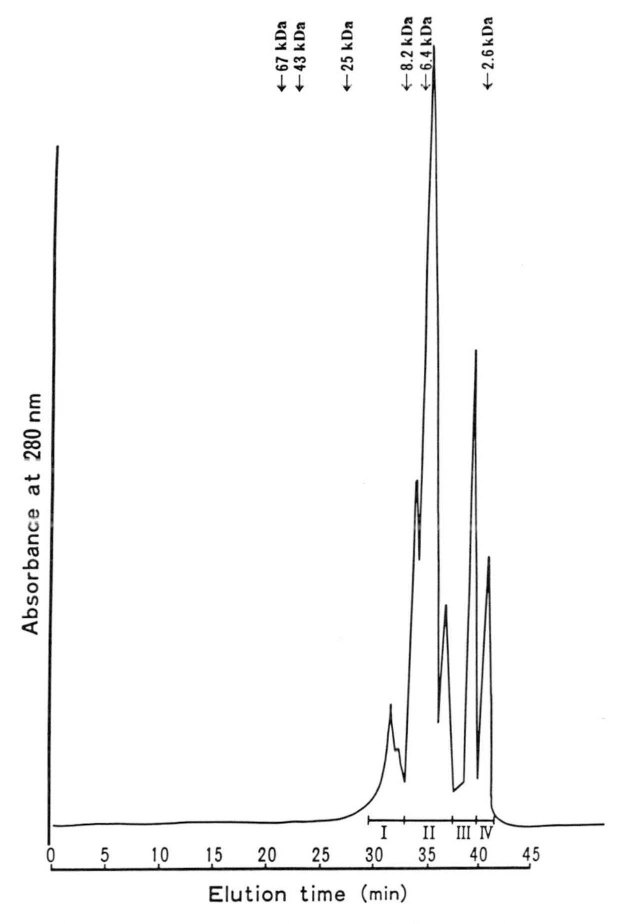

FIG. 3 Elution pattern of polypeptides in thymus epithelial cell culture supernatant as determined by high-performance liquid chromatography and gel-filtration column. Standard materials for estimating molecular weight: bovine serum albumin, 67 kDa; ovalbumin, 43 kDa; chymotrypsinogen, 25 kDa; whale sperm myoglobin products partially cleaved by cyanogen bromide, 8.2 kDa, 6.4 kDa, and 2.6 kDa, I, II, III, and IV: each peak, eluted at 280 nm in the order of the highest to the lowest molecular weight (10 kDa, 7 kDa, 3 kDa, and 2.5 kDa). Reprinted from *Cell. Mol. Biol.* **35,** Morioka, K., Kinoshita, Y., Tominaga, K., Hato, F., and Itoh, T., Separation of polypeptides which induce a mitogen response of thymocytes from the culture supernatant of a thymic epithelial cell line, 523–534, copyright 1989, with kind permission from Elsevier Science Ltd., The Boulevard, Langford Lane, Kidlington 0X5 1GB, UK.

phy is performed by using a gradient of organic solvent under acidic conditions. The organic solvent and acidic substance (trifluoracetic acid) are removed by lyophilization (Morioka *et al.,* 1989; Oshitani *et al.,* 1989; Taniyama *et al.,* 1989). However, it is known that organic solvents and acidic substances reduce thymus hormonal activity. Therefore, it is necessary to shorten the separation time when reverse-phase chromatography is used as the further purification step. Combining gel-filtration chromatography with reverse-phase chromatography provides good results (Morioka *et al.*, 1989). Briefly, one of the THP fractions (estimated molecular weight, 2.5 kDa) separated by gel-filtration chromatography possesses two kinds of activities which enhance the concanavalin A (Con A) and phytohemagglutinin responses of thymocytes.

Next, the THP fraction is further separated by reverse-phase chromatography to see whether the fraction contains two or more different THPs (Fig. 4). The enhanced concanavalin A response of thymocytes is observed in the hydrophobic subfraction, whereas an enhanced phytohemagglutinin response is detected in the hydrophilic subfraction (Fig. 5). The former

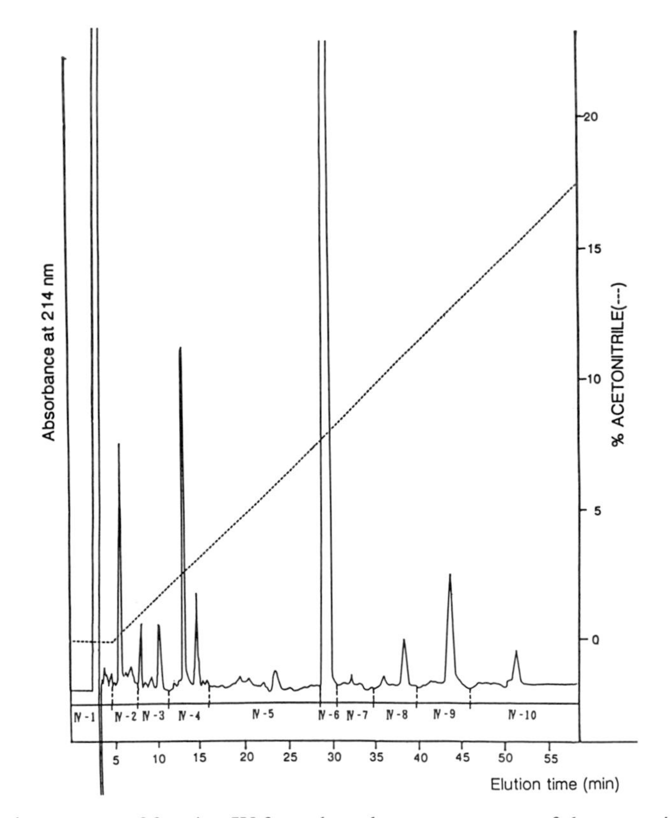

FIG. 4 Elution pattern of fraction IV from the culture supernatant of thymus epithelial cell line (IT-45R1) using HPLC with reverse-phase column. Ordinate: absorbance at 214 nm (left axis) and concentration of acetonitrile (right axis). Abscissa: subfractions from fraction IV and elution time (minutes). Reprinted from *Cell. Mol. Biol.* **35,** Morioka, K., Kinoshita, Y., Tominaga, K., Hato, F., and Itoh, T., Separation of polypeptides which induce a mitogen response of thymocytes from the culture supernatant of a thymic epithelial cell line, 523–534, copyright 1989, with kind permission from Elsevier Science Ltd., The Boulevard, Langford Lane, Kidlington 0X5 1GB, UK.

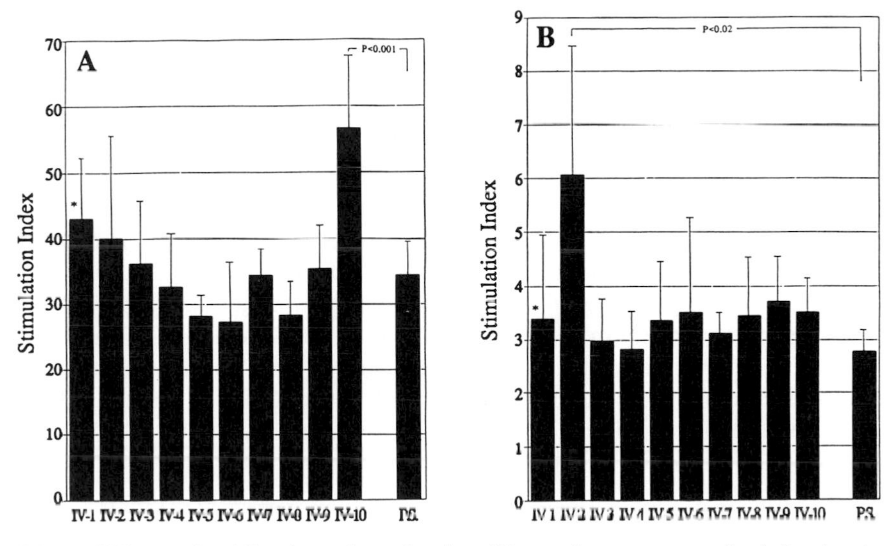

FIG. 5 Effects of subfractions from fraction IV on the response of whole thymic lymphocytes to T-cell mitogen stimulation. (A) Response to concanavalin A stimulation. (B) Response to phytohemagglutinin stimulation. Ordinate: Stimulation index =

$$\frac{\text{DNA synthetic activity of whole thymocytes stimulated with the mitogen}}{\text{DNA synthetic activity of whole thymocytes without the mitogen}}$$

Abscissa: Subfractions from fraction IV from TEC culture supernatant. P.S., physiological saline. Asterisk indicates mean and standard deviation calculated from three or more independent experiments. Reprinted from *Cell. Mol. Biol.* **35,** Morioka, K., Kinoshita, Y., Tominaga, K., Hato, F., and Itoh, T., Separation of polypeptides which induce a mitogen response of thymocytes from the culture supernatant of a thymic epithelial cell line, 523–534, copyright 1989, with kind permission from Elsevier Science Ltd., The Boulevard, Langford Lane, Kidlington 0X5 1GB, UK.

activity is the highest in the lighter, small thymocyte subset and the latter is the highest in the heavier, small thymocyte subset. These small thymocyte subsets are separated by a combination of discontinuous density-gradient centrifugation with an adhesion column (Kinoshita *et al.,* 1974a, 1976, 1979). These results suggest that two kinds of THPs from the 2.5-kDa polypeptide fraction are separated by reverse-phase chromatography and that the molecular structure and target cells of both THPs are different.

B. Parameters Reflecting Thymocyte Differentiation and Maturation

1. Density of Thy 1 Antigen on Thymocyte Surface

The surface density of Thy 1 antigen (Raff, 1969; Reif and Allen, 1963) on T-cell precursors immediately before entering the thymus via blood from bone marrow is extremely low. After the cells penetrate into thymic parenchyma, the density rises markedly with their differentiation to T cells under the influence of thymic stromal cells. The cells continue to acquire the T cell property with further maturation and move from the cortex to the medulla, in which the surface density of Thy 1 antigen on the thymocytes is low. Komuro and Boyse (1973a, b) reported that thymosin, a thymic extract (Goldstein *et al.,* 1966, 1972), enhances *in vitro* Thy 1 antigen density on the cells (possibly T-cell precursors) of the less dense fractions, which are separated from bone marrow by density-gradient centrifugation using bovine serum albumin media. Furthermore, the conditioned medium from TEC culture, which is capable of inducing thymocyte maturation, decreases the density of Thy 1 antigen of thymocytes, especially the peanut agglutinin (PNA)-agglutinating subset, which appears to consist of immature thymocytes (Kruisbeek, 1979; Kruisbeek and Astaldi, 1979).

2. Changes in Cellular Specific Gravity

Thymic small lymphocytes (small thymocytes) are separated into three subsets by combining discontinuous density-gradient centrifugation using gum acacia media with an adhesion column stuffed with cotton or nylon fiber; then antiserum against each subset is prepared (Kinoshita and Kimura, 1971; Kinoshita *et al.,* 1974a, c). The following results are obtained with these antisera. Surface antigenicity of the least dense subset seems to be similar to that of the cells in the least dense and nonadhesive bone marrow cells, which are rich in stem cells or T-cell precursors (Kinoshita *et al.,* 1974a; Okamoto *et al.,* 1986). It appears that some of small thymocytes in the least dense fraction, possibly the most immature cells ($CD4^{-}8^{-}$ cells; see Section III, B, 8) transform into blastoid cells (Fig. 6) under the effect of growth factors (Haynes, 1990; Kinoshita *et al.,* 1990, 1992) from TECs; divide successively; and then change into smaller thymocytes which are separated into more dense fractions. The surface antigenic property of small thymocytes in the most dense fraction coincides with that of lymphocytes in thymus-dependent areas (Parrott *et al.,* 1966) of lymph nodes and spleen (Kinoshita *et al.,* 1974a). It is assumed that the cellular specific gravity of small lymphocytes in thymus increases as the lymphocytes mature (Kinoshita *et al.,* 1974a, 1976, 1979). This increase with their maturation might possibly depend upon an increase in intracellular protein level and a decrease in water content with development (Baba *et al.,* 1979).

3. Sensitivity and Resistance to Glucocorticoid

Cortical thymocytes appear to be sensitive to glucocorticoid exposure and to be destroyed, while medullary cells are not (Blomgren and Anders-

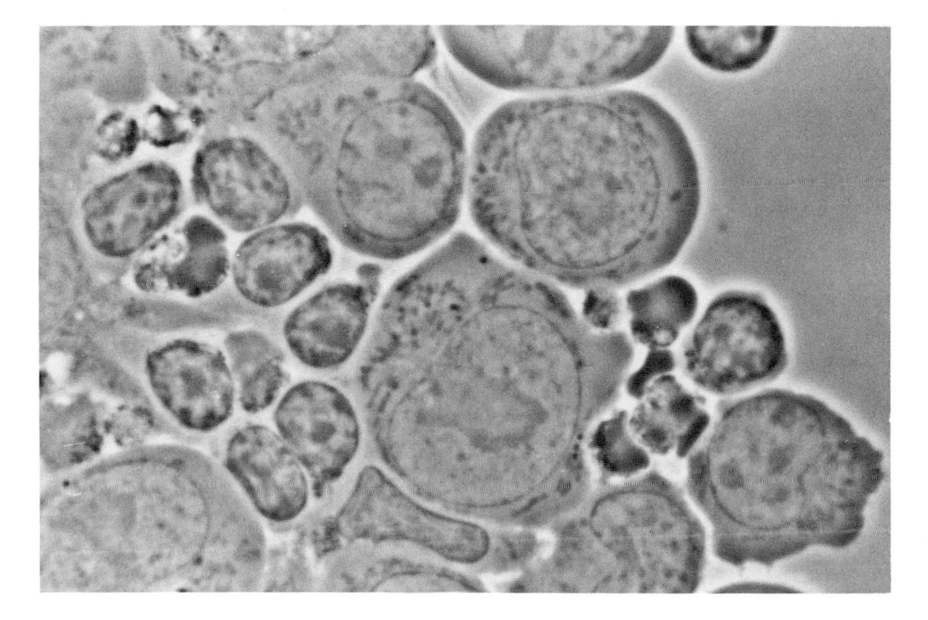

FIG. 6 Phase-contrast photomicrograph of blastoid cells found in the culture of nonrosette-forming thymocytes stimulated with a 10-kDa fraction separated from thymus epithelial cell culture supernatant. ×930. Reprinted from *Cell. Mol. Biol.* **36,** Kinoshita, Y., Toyokawa, T., Hato, F., Masuda, A., and Matsuyama, M., New thymocyte growth factor from thymic epithelial cell line, 429–438, copyright 1990, with kind permission from Elsevier Science Ltd., The Boulevard, Langford Lane, Kidlington 0X5 1GB, UK.

son, 1969; Kinoshita *et al.,* 1974b). Accordingly, the cells collected at 48–72 hr after cortisol administration are presumed to be derived from the medulla and used as medullary thymocytes. Since approximately 95% of the thymocytes in normal murine animals exist in the cortical area, a suspension of whole thymocytes from normal animals for practical purposes is considered to consist of cortical thymocytes (Kinoshita *et al.,* 1974b). It is generally accepted that medullary thymocytes are more mature than cortical ones (Fig. 7). Furthermore, our experiments have shown that rat thymocytes, which form rosettes with guinea pig erythrocytes in the presence of nonheated fetal calf serum (Elfenbein and Winkelstein, 1978; Oka *et al.,* 1984a), are double positive ($CD4^+8^+$; see Section III, B, 8) cells (Hato *et al.,* 1991) and the most sensitive to glucocorticoid (Tohji *et al.,* 1991). Therefore, rosette-forming thymocytes, separated from nonrosette-forming cells by one-step density-gradient centrifugation (Figs. 8, 9) (Kinoshita *et al.,* 1990; Parish *et al.,* 1974), are a valuable material for studying apoptosis (programmed cell death) of thymocytes (von Boehmer *et al.,* 1989; Jenkinson *et al.,* 1989; Kerr *et al.,* 1972) and further maturation to single positive ($CD4^+8^-$ or $CD4^-8^+$) cells.

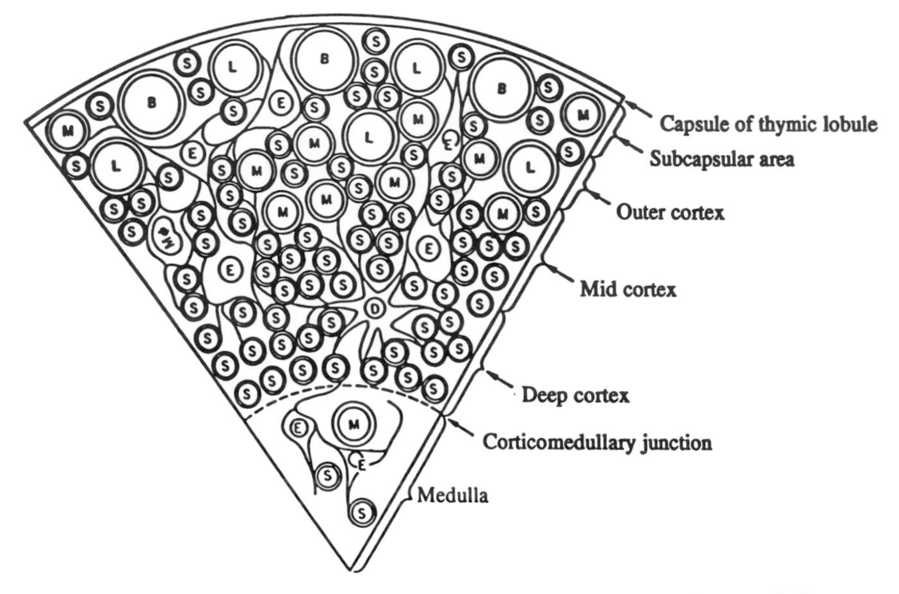

FIG. 7 Scheme for the histological structure of a thymic lobule. Each arrow indicates a portion of the thymic lobule. B, blastoid cell; L, large lymphocyte; M, medium lymphocyte; S, small lymphocyte; E, epithelial cell; Mø, macrophage; D, dendritic cell.

4. Response to Stimulation of T-cell Mitogens or Allogenic Cells

Phytohemagglutinin (PHA) and concanavalin A, which cause blastoid transformation of thymus-dependent lymphocytes, are designated as T-cell mitogens (Powell and Leon, 1970; Stobo, 1972; Stobo *et al.*, 1972).

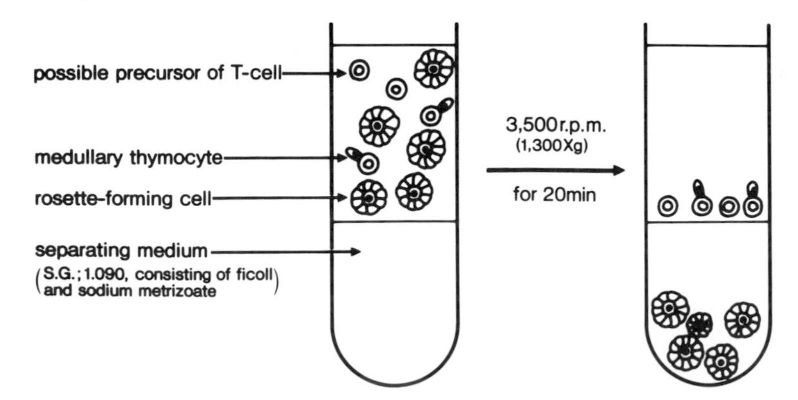

FIG. 8 Scheme for separation of rosette-forming and nonrosette-forming cell populations by one-step density-gradient centrifugation using Ficoll–Hypaque solution as the separating medium. Left. Before centrifugation. Right. After centrifugation.

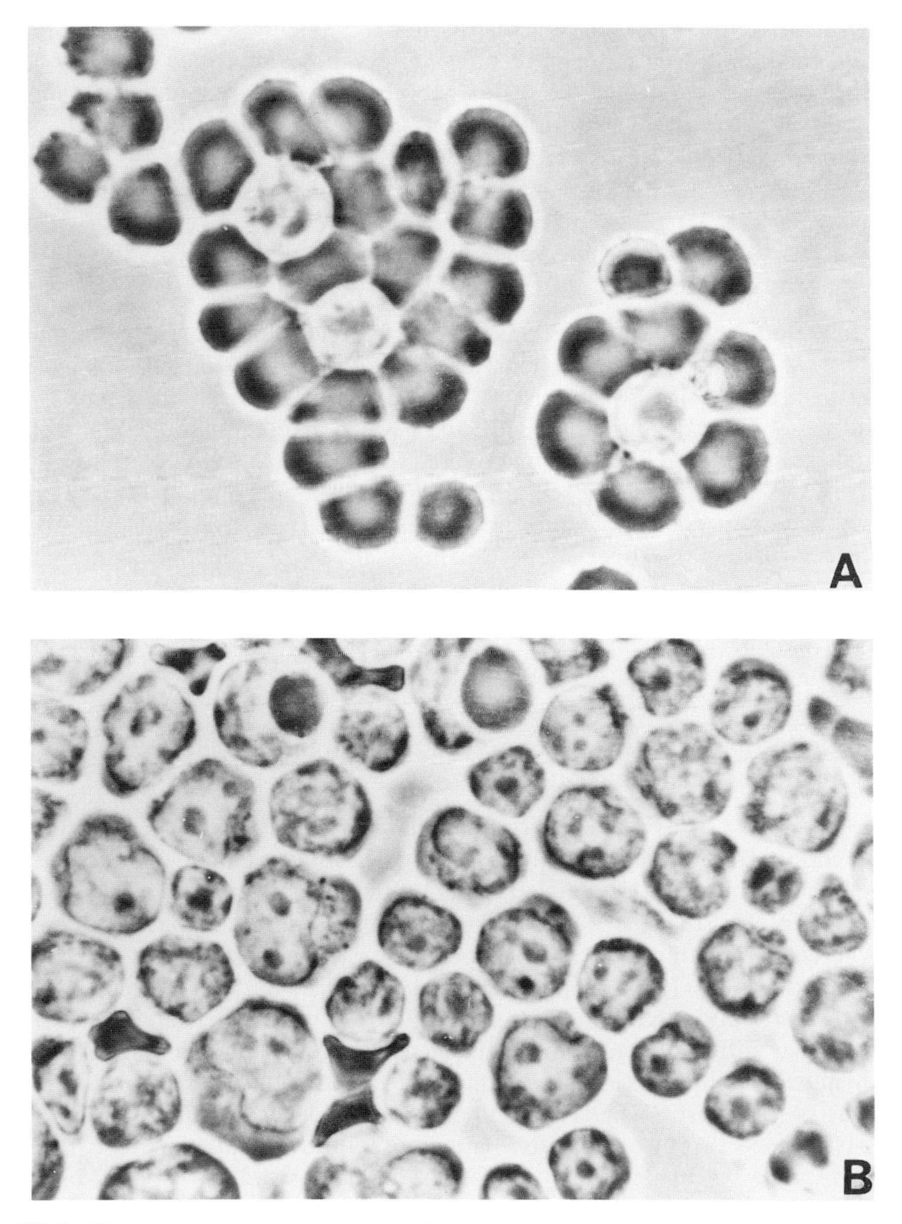

FIG. 9 Phase-contrast photomicrographs of rosette-forming cells (A) and nonrosette-forming cells (B) separated respectively by one-step density-gradient centrifugation. ×930.

Medullary thymocytes harvested after cortisol administration are capable of responding to PHA, Con A, and allogenic stimulation, while cortical thymocytes, generally speaking, are not sensitive to this stimulation

(Kruisbeek, 1979). The cortical thymocytes, being PNA-agglutinating cells, acquire T-cell mitogen responsiveness in the presence of TEC culture supernatant (Kruisbeek, 1979). Moreover, Con A-responding thymocytes seem to be less mature than PHA-responding cells (Jacobsson and Blomgren, 1974; Morioka *et al.*, 1989).

5. Expression and Disappearance of Peanut Agglutinin Receptor

Mouse cortical thymocytes possess the receptor for peanut agglutinin, whereas medullary ones do not (Reisner *et al.*, 1976). Mouse thymocytes suspended in an isotonic solution containing PNA are divided into two subsets—PNA agglutinating cells and PNA nonagglutinating cells. Both subsets suspended in 20% of inactivated fetal calf serum (FCS) are respectively separated by a 1-*g* sedimentation (Miller and Phillips, 1969). PNA agglutinating cells sink and form sediments, whereas PNA nonagglutinating cells float in the supernatant (Reisner *et al.*, 1976). The PNA agglutinating subset, which is less mature than the nonagglutinating one, is reported to be a target for THP in the TEC culture supernatant (Kruisbeek, 1979; Kruisbeek and Astaldi, 1979).

6. Expression and Disappearance of Responsiveness against Soybean Lectin Stimulation

Soybean lectin (SBL) (Liener and Pallansch, 1952; Lis *et al.*, 1966), a glycoprotein, binds specifically to *N*-acetyl-D-galactosamine (Lotan *et al.*, 1974). Phytohemagglutinin also binds to the sugar (Borberg *et al.*, 1966). It is of interest that PHA causes blastoid transformation in mature T cells, whereas SBL does not. Our previous papers (Kinoshita *et al.*, 1986b, c; Okamoto *et al.*, 1986; Okimoto *et al.*, 1988) presented data showing that SBL markedly enhances DNA synthesis by the least dense subset containing undifferentiated lymphoid cells from bone marrow, tonsil, and thymus. It is therefore assumed that the target cells for SBL are pre-T-cells in the subset. This assumption is supported by the finding that the least dense subset treated with thymosin fraction 5 (Goldstein *et al.*, 1966, 1972) acquires the property of T cells (concanavalin A responsiveness and E rosette-forming capacity), whereas the subset with the thymic product loses its responsiveness to SBL stimulation (Kinoshita *et al.*, 1986b, c). These results suggest that the pre-T cells in the least dense subset differentiate into a T-cell family by an *in vitro* exposure of thymosin fraction 5. Furthermore, this experimental system would be a valuable way to identify the pre-T-cell step and to analyze the mechanism by which the cells differentiate into a T-cell family.

7. Acquisition and Loss of Rosette-Forming Capacity

A majority of rat cortical thymocytes bind to guinea pig erythrocytes in the presence of nonheated (FCS) and thus form rosettes. On the other hand, thymocytes in the medulla and subcapsular cortex do not bind to the erythrocytes. Rat whole thymocytes subjected to rosette-formation techniques (Elfenbein and Winkelstein, 1978; Oka *et al.*, 1984a) are composed of rosette-forming and nonrosette-forming subsets. The former are separated from the latter by one-step density-gradient centrifugation using Ficoll–Hypaque medium (Fig. 8) (Hato *et al.*, 1991; Kinoshita *et al.*, 1990; Parish *et al.*, 1974). Flow cytometry using monoclonal antibodies against differentiation antigens, CD4 and CD8 (Lanier *et al.*, 1983; Reichert *et al.*, 1986; Tohji *et al.*, 1991), shows that the rosette-forming subset consists of double positive ($CD4^+8^+$) cells, as mentioned earlier, and the nonrosette-forming subset consists of double negative ($CD4^-8^-$) and single positive ($CD4^+8^-$ and $CD4^-8^+$) cells (Fig. 10) (Hato *et al.*, 1991).

Taniyama *et al.* (1989) demonstrated that the 3-kDa polypeptide fraction (Fig. 3) from TEC culture supernatant, and thymosin α_1 (Low *et al.*, 1979) prepared from thymosin fraction 5 endow the nonrosette-forming cells with rosette-forming capacity. Since the hormonal products do not endow the medullary thymocytes that are single positive cells with this capacity

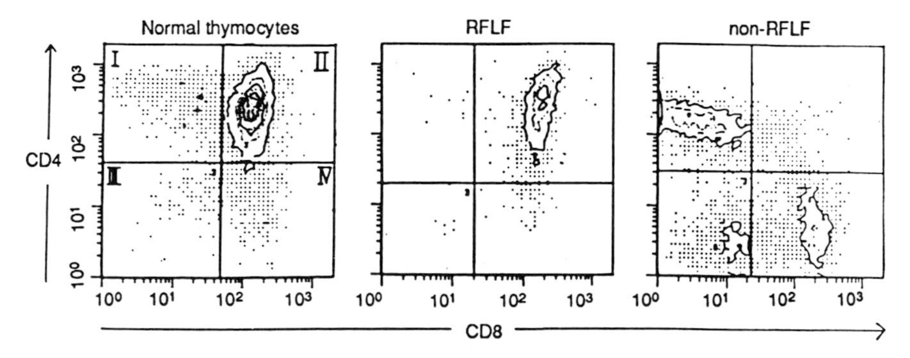

FIG. 10 Flow cytometric analysis of each cell material by CD4 and CD8 antigens. Expression of CD4 and CD8 was detected by labeling anti-CD4 and -CD8 monoclonal antibodies with phycoerythrin and fluorescein-isothiocyanate, respectively. RFLF, rosette-forming lymphocyte fraction; non-RFLF, nonrosette-forming lymphocyte fraction. Ordinate: Fluorescence intensity of phycoerythrin. Abscissa: Fluorescence intensity of fluorescein-isothiocyanate. Fractions I, II, III, and IV represent the area for CD4 single positive cells, double positive cells, double negative cells, and CD8 single positive cells, respectively. Reprinted from *Cell. Mol. Biol.* **37,** Hato, F., Tohji, Y., Kinoshita, Y., and Terano, Y., Surface antigenic analysis of rosette-forming cells in rat thymus using monoclonal antibodies against T cell surface markers, 839–845, copyright 1991, with kind permission from Elsevier Science Ltd., The Boulevard Langford Lane, Kidlington 0X5 1GB, UK.

(Kinoshita *et al.* 1986a), the target cells for the products would be double negative cells in the subcapsular cortex. A small number of double positive cells with rosette-forming capacity appear to change their surface differentiation markers (CD4, CD8) and to mature into single positive cells which lose rosette-forming capacity. In this maturation pathway, some of the receptors for rosette formation might possibly be masked by sialic acid, since the percentage of rosette-forming cells in medullary thymocytes is significantly enhanced by *in vitro* treatment with neuraminidase (Oka *et al.*, 1984b).

8. Expression and Disappearance of Differentiation Molecules (CD3, CD4, and CD8) on Thymocytes

The thymocyte suspension is treated with monoclonal antibodies against CD4 and CD8, which are respectively labeled with different fluorescent dyes, and then the cell surface antigenicity is analyzed by flow cytometry (Lanier *et al.*, 1983; Reichert *et al.*, 1986; Tohji *et al.*, 1991). In this analysis, the thymocytes are divided into three subsets according to their level of maturation—double negative ($CD4^{-}8^{-}$), double positive ($CD4^{+}8^{+}$), and single positive ($CD4^{+}8^{-}$ or $CD4^{-}8^{+}$) cells. Raulet (1985) presented data showing that about 30% of the cells in the least mature (double negative) adult thymocyte subset express the receptors for interleukin-2 (IL-2) without *in vitro* induction and that DNA synthesis by the least mature subset is vigorously enhanced by adding IL-2 with a costimulating mitogen.

An experimental attempt to recolonize the thymus by intrathymically transferred double negative (DN) thymocytes suggested that the IL-2R positive DN thymocytes are more differentiated than the IL-2R negative DN ones (Shimonkevitz *et al.*, 1987). Other authors (MacDonald *et al.*, 1988) have reported that there is an intermediate stage between double negative cells and double positive cells; the surface marker of this stage is $CD3^{-}4^{-}8^{+}$, which differs in CD3 antigenicity from that of $CD3^{+}4^{-}8^{+}$, one of single positive cells. The target cells for the polypeptide that induces rosette formation (3-kDa fraction) and thymocyte growth factor (10-kDa fraction) from TEC culture supernatant are suggested to be a double negative subset (Kinoshita *et al.*, 1990, 1992; Taniyama *et al.*, 1989). It is possible that the 3-kDa polypeptide fraction induces differentiation of double negative cells (nonrosette-forming cells) to double positive cells (rosette-forming cells). Antigenically distinct T-cell subsets are separated by a fluorescence-activated cell sorter or panning method using monoclonal antibodies against the different antigens on T-cell surface membranes. Functional analysis of the T-cell subsets showed that CD4 positive and CD8 negative ($CD4^{+}8^{-}$) cells consist of helper T and inducer T cells, and $CD4^-8^+$ cells are killer T and suppressor T cells (Ledbetter *et al.*, 1981).

IV. Thymocyte Subset, the Target for Thymus Hormonal Products

1. Mouse cortical thymocytes possessing the receptors for peanut agglutinin agglutinate mutually through binding of PNA to the receptors, whereas medullary cells do not (Reisner *et al.*, 1976). Accordingly, mouse thymocytes consist of PNA-agglutinating and PNA-nonagglutinating subsets which are separated by a 1-*g* sedimentation method (Miller and Phillips, 1969) using 20% v/v inactivated fetal calf serum in an isotonic solution (Kruisbeck and Astaldi, 1979; Reisner *et al.*, 1976). Since the agglutination seems to be induced by an interaction between PNA agglutinin and galactose residue on thymocyte surface receptors, PNA-agglutinating cells are individually dissociated in an isotonic D-galactose solution (Reisner *et al.*, 1976). PNA-nonagglutinating thymocytes are capable of responding to the stimulation of phytohemagglutinin, concanavalin A, and allogenic cells (Kruisbeek, 1979; Kruisbeek and Astaldi, 1979), while PNA-agglutinating cells respond only to concanavalin A (Reisner *et al.*, 1976). This result suggests that the PNA-agglutinating subset is less mature than PNA-nonagglutinating cells.

The conditioned medium from TEC primary culture enhances mitogen and allogenic cell responsiveness in the PNA-agglutinating subset, whereas the medium does not do so in the PNA-nonagglutinating subset (Kruisbeek, 1979; Kruisbeek and Astaldi, 1979). It is assumed from these results that TEC culture supernatant contains the THPs that induce a functional maturation of thymocytes and that the target cells for THPs are PNA-agglutinating cells.

2. Nonheated fetal calf serum contains a heat-stable factor (HSF) and a heat-labile factor (HLF) (Elfenbein and Winkelstein, 1978). HSF binds to the receptors on rat thymocytes and HLF binds to the resulting complex of thymocytes and HSF. Guinea pig erythrocytes bind to the sequential complex of thymocytes and the two factors (Elfenbein and Winkelstein, 1978; Oka *et al.*, 1984a). Rat thymocytes in the mid- and deep cortex possess the receptor for HSF, bind to guinea pig red blood cells, and thus form rosettes. On the other hand, thymocytes in the subcapsular cortex do not appear to possess the HSF receptors, while the receptors on the medullary thymocytes seem to be masked by sialic acid residues (Oka *et al.*, 1984b). These thymocytes could not form rosettes (Elfenbein and Santos, 1978; Hashimura *et al.*, 1987; Oka *et al.*, 1984a; Taniyama *et*

al., 1989). Accordingly, rat thymocytes are composed of rosette- and nonrosette-forming subsets which can be separated by one-step density-gradient centrifugation using Ficoll–Hypaque medium (Hato *et al.*, 1991; Kinoshita *et al.*, 1990; Parish *et al.*, 1974). Rat thymocyte suspension treated with a rosette formation technique is gently layered on the top of the separating medium (specific gravity, 1.090) and then centrifuged at $1300 \times g$ for 20 min. The rosette-forming cells sink to the bottom, while nonrosette-forming cells are separated into the boundary between supernatant and separating medium (Fig. 8). The binding erythrocytes on the surface of rosette-forming cells (Fig. 9A) are eliminated by hypotonic shock (Hato *et al.*, 1991).

A flow cytometric analysis using anti-CD4 and anti-CD8 monoclonal antibodies respectively conjugated with different fluorescent dyes disclosed that the rosette-forming subset consists of $CD4^{+}8^{+}$ thymocytes (Fig. 10), while the nonrosette-forming subset (Fig. 9B) is composed of $CD4^{-}8^{-}$, $CD4^{+}8^{-}$, and $CD4^{-}8^{+}$ cells (Fig. 10) (Hato *et al.*, 1991).

The ratio of the single positive cell number to the total thymocyte number in the rat thymus is remarkably enhanced at 3 days after administration of cortisol compared with that before administration, while the ratio of double positive cells is strikingly decreased (Tohji *et al.*, 1991). This result clearly shows that the thymocytes suffering from cytolysis after glucocorticoid treatment are double positive cells. The 3-kDa polypeptide fraction from TEC culture supernatant endows nonrosette-forming thymocytes with rosette-forming capacity, while the fraction does not endow medullary thymocytes with this capacity (Taniyama *et al.*, 1989). It is possible to conclude from these data that double negative cells in the subcapsular area (Kim *et al.*, 1990; Takacs *et al.*, 1984) differentiate into cells expressing the receptor for HSF and then acquire rosette-forming capacity (Taniyama *et al.*, 1989). When the 3-kDa fraction is pretreated with antithymosin α_1 antibody (Hato *et al.*, 1992; Hirokawa *et al.*, 1982), the activity inducing the rosette formation is markedly depressed (Fig. 11) (Taniyama *et al.*, 1989). It is thus assumed that the 3-kDa fraction from TEC culture supernatant contains several polypeptide molecules which are identical with thymosin α_1 or express an epitope closely similar to the thymic agent (Taniyama *et al.*, 1989).

3. Murine thymocytes are separated into three fractions by discontinuous density-gradient centrifugation using gum acacia solution as the separating medium (Fig. 12). Physicochemical properties (pH, osmotic pressure, etc.) of the solution are physiologically adjusted (Kimura *et al.*, 1960; Kinoshita *et al.*, 1970b, 1974a, 1976, 1979, 1986b). The first fraction is composed of large, medium, and small lymphocytes. The second fraction consists of medium and small lymphocytes. The third is formed of small lymphocytes (upper portion of Fig. 13). Small lymphocytes are found in

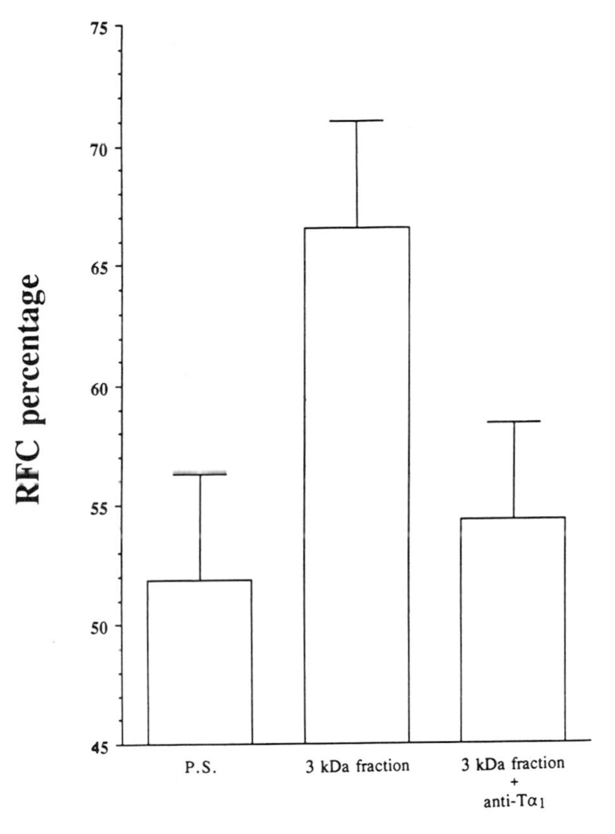

FIG. 11 Inhibitory effect of antithymosin α_1 antiserum on the ability of a 3-kDa fraction from the culture supernatant to endow nonrosette-forming cells with rosette-forming capacity. Ordinate: Percentage of rosette-forming cells (RFC). Abscissa: Treatment of non-RFC-rich population with various agents. P.S., physiological saline; anti-Tα_1, antithymosin α_1 antiserum.

all fractions. It is thus shown that the small lymphocyte group consists of cells that are heterogeneous in their specific gravity (Kinoshita *et al.*, 1974a).

By removing larger thymocytes from each fraction with an adhesion column stuffed with cotton or nylon fibers, three kinds of small lymphocyte subsets can be harvested (lower portion of Fig. 13). The surface antigenicity of the most dense small lymphocytes corresponds with that of mature T cells in the thymus-dependent area (Parrott *et al.*, 1966) of peripheral lymphoid tissues (Tables I, II) (Kinoshita *et al.*, 1974a). On the other hand, the antigenicity of the least dense coincides with that of undifferentiated lymphoid cells in bone marrow (Kinoshita *et al.*, 1974a, 1976).

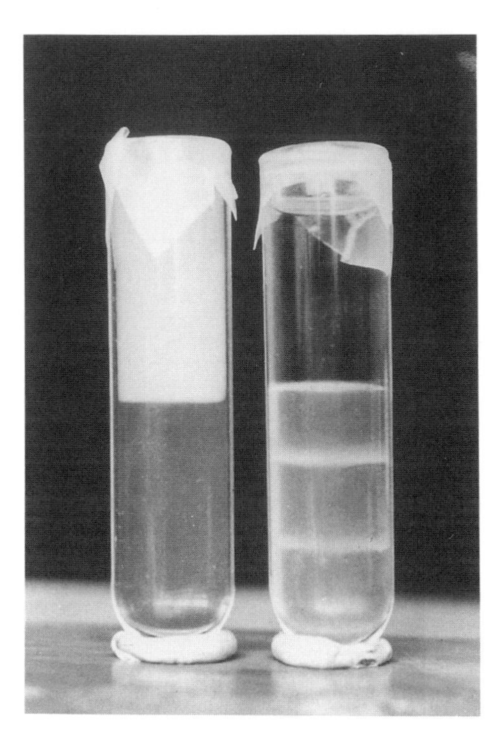

FIG. 12 Fractions of thymic lymphocytes separated by discontinuous density-gradient centrifugation using gum acacia solution as separating media. Left. Before centrifugation. Three kinds of separating media (specific gravity; 1.064, 1.068, and 1.072) are used. Right. After centrifugation. Whole thymocytes are separated into three thymocyte fractions (fraction I, between supernatant and separating medium with S.G. 1.064; fraction II, between separating media with S.G. 1.064 and that with S.G. 1.068; fraction III, between separating media with S.G. 1.068 and that with S.G. 1.072).

Next, three subsets are used to estimate the subset sensitive for each THP from the TEC culture supernatant. The 2.5-kDa polypeptide fraction isolated by HPLC with a gel-filtration column (Fig. 3) possesses the potent capacity of inducing Con A responsiveness in the lighter, small lymphocyte (LSL) subset and of inducing PHA responsiveness in the heavier, small lymphocyte (HSL) subset. This result suggests that there are functionally different polypeptides in the fraction. One induces the PHA response, while the other induces the Con A response. Further purification of the 2.5-kDa fraction by HPLC with a reverse-phase column (Fig. 4) showed that the hydrophobic subfraction endows the LSL subset with Con A responsiveness, whereas the hydrophilic subfraction endows the HSL subset with PHA responsiveness (Fig. 5) (Morioka *et al.*, 1989).

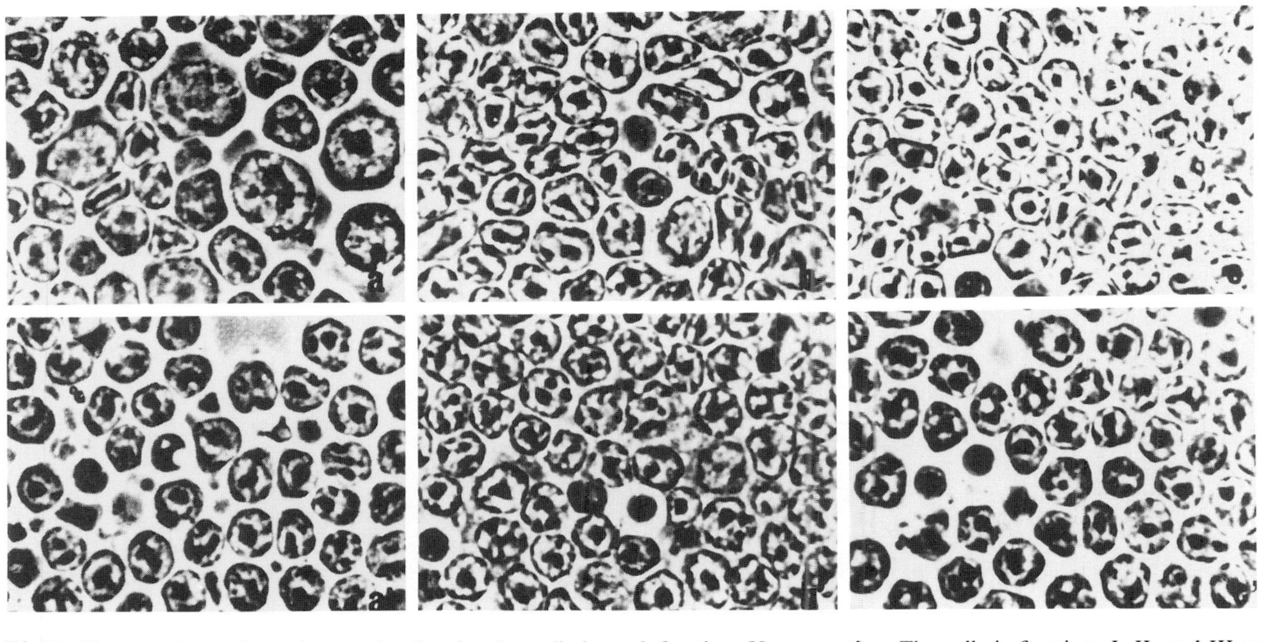

FIG. 13 Phase-contrast photomicrographs showing the cells in each fraction. Upper portion: The cells in fractions I, II, and III as mentioned in Fig. 12 from left to right. Lower portion: Photomicrographs of the cells in the fractions shown on top treated with the adhesion column. One can notice that larger lymphocytes are removed by the adhesion column method and then small lymphocytes are harvested. ×1000.

TABLE I

Effects of Various Sera on the Lymphocytes of Thymus-Dependent Areas in Peripheral Lymphoid Organs[a]

Areas of selective depletion of lymphocytes	Experimental group treated with		
	AHSLAS[b]	ALSLAS[c]	NRC[d]
Surrounding zones of central arterioles of splenic follicles	10/10[e]	0/10	0/10
Paracortical areas of lymph nodes	9/10	0/10	0/10

[a] From Kinoshita *et al.* (1974a).
[b] Rabbit antiserum against heavier small lymphocyte subpopulation separated from thymuses of Wistar strain rats.
[c] Rabbit antiserum against thymic lighter small lymphocyte subpopulation.
[d] Normal rabbit serum.
[e] No. of animals depleted of lymphocytes in the thymus-dependent areas / No. of animals used in each group

V. Regulators of Thymic Epithelial Cell Function

A. Endocrine Organs

There is a close relation between the development of the living body and thymus growth (Boyd, 1932). Since growth hormone (GH) is one of main effectors of development, it has been thought to affect thymus growth.

TABLE II

Effects of AHSLAS Absorbed with the Thymic HSL or LSL Subpopulation Respectively on Lymphocytes of Thymus-Dependent Areas[a]

Area of selective depletion of lymphocytes	Experimental group treated with AHSLAS absorbed with		
	Non	Thymic HSL[b] fraction	Thymic LSL[c] fraction
Surrounding zones of central arterioles of splenic follicles	10/10[d]	0/5	5/5
Paracortical areas of lymph nodes	9/10	1/5	4/5

[a] From Kinoshita *et al.* (1974a).
[b] Heavier small lymphocytes.
[c] Lighter small lymphocytes.
[d] No. of animals depleted of lymphocytes in the thymus-dependent areas / No. of animals used in each group

Thymic atrophy is detected in congenital hypopituitarism (Fabris *et al.*, 1971) and hypophysectomized animals (Berczi, 1986). Moreover, physiological thymic involution, which occurs during aging, can be reversed by injecting GH (Maggiano *et al.*, 1992). On the other hand, the thymus falls into atrophy when antiserum against GH is injected into mice (Maggiano *et al.*, 1992). It has been reported that there are receptors for GH in thymocytes (Arrenbrecht, 1974).

It is naturally assumed that GH participates directly in thymocyte proliferation. Indeed, the hormone induced a significant increase in DNA synthesis by the thymocytes cultured *in vitro*, but the grade of increment was weak (Yamada *et al.*, 1994). The GH receptor in thymocytes might possibly play other roles related to induction of thymocyte maturation in a T-cell series, such as endowment with T-cell surface markers. Recently, we obtained data showing that insulin-like growth factor-1 (IGF-1), one of the mediators brought about by GH stimulation (Chung *et al.*, 1985; Turner *et al.*, 1988), significantly enhances thymocyte DNA synthetic activity (Yamada *et al.*, 1994). Ban *et al.* (1991) reported that the specific receptors for GH are also detected in TECs. On the other hand, protein synthesis by TECs cultured in the confluent state is significantly enhanced by addition of the hormone (Tominaga *et al.*, 1993). This enhancement is assumed to be linked to the binding of GH to the receptors of the confluent TECs and the production and release of IGF-1 in the cells. The authors measured IGF-1 level in the TEC culture supernatant with and without GH stimulation by radioimmunoassay using anti-IGF-1 antibody (Furlanetto and Marino, 1987). They found that TECs treated with GH released significantly more IGF-1 into the supernatant than the cells without the hormone. It is thus presumed that this IGF-1 causes more active proliferation of thymocytes than IGF-1 released without the stimulation, and that GH participates indirectly in thymocyte proliferation. This increase in IGF-1 gene expression in GH-stimulated TECs is being studied.

Adrenalectomy brings about thymic hypertrophy (Shortman and Jackson, 1974). Stress to the living body, which raises blood glucocorticoid level, also causes thymic atrophy (Munck *et al.*, 1984). Intramuscular injection of cortisol induces cytolysis of cortical thymocytes and then markedly reduces thymus weight (Fig. 14) (Blomgren and Andersson, 1969; Kinoshita *et al.*, 1974b). These results suggest that glucocorticoid takes part in the suppression of thymic development. This phenomenon appears to be one of the causes inducing apoptosis (Jenkinson *et al.*, 1989; Kerr *et al.*, 1972) and negative selection of thymocytes (von Boehmer *et al.*, 1989; Swat *et al.*, 1991). According to flow cytometric analysis for thymocytes using monoclonal antibodies against CD4 and CD8 antigens, the target cells for the steroid hormone seem to be double positive ($CD4^{+}8^{+}$) thymocytes (Fig. 15) (Hato *et al.*, 1991; Tohji *et al.*, 1991).

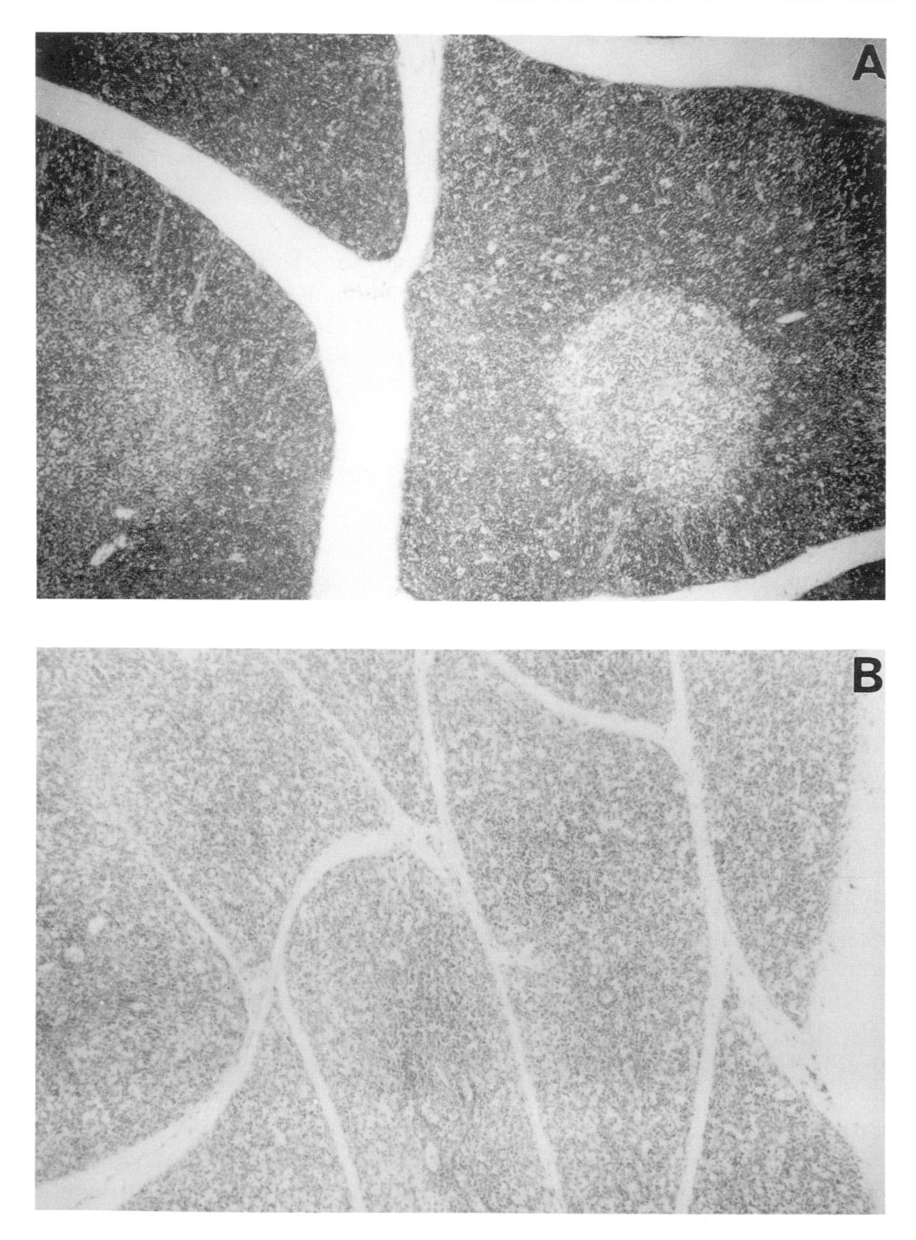

FIG. 14 Histological photomicrographs of rat thymic lobuli before and 3 days after glucocorticoid (GC) administration. (A) Before treatment. (B) Three days after GC administration. Reprinted from *Cell. Mol. Biol.* **36,** Kim, T. J., Kinoshita, Y., Hato, F., and Kimura, S. Rat thymus reconstitution after thymocyte destruction by glucocorticoid treatment. From the view of endogenous DNA synthesis and soybean lectin responsiveness in thymocytes, 705–716, copyright 1990, with kind permission from Elsevier Science Ltd., The Boulevard, Langford Lane, Kidlington 0X5 1GB, UK.

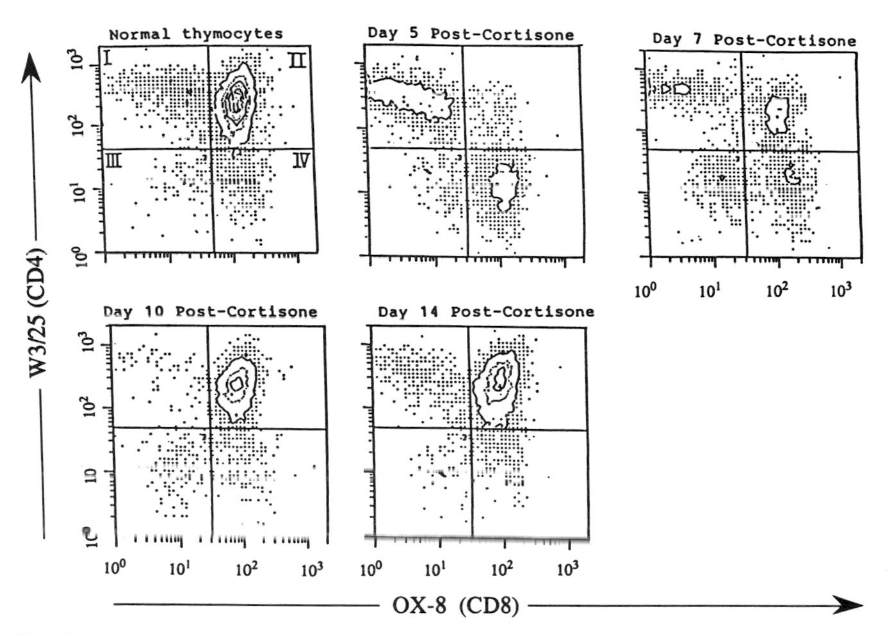

FIG. 15 Two-color analysis of CD4 and CD8 expression in normal rat thymocytes and 5, 7, 10, and 14 days after GC injection. The flow cytometric pattern on day 5 postcortisone shows that there are very few cells in area II. It is revealed from this pattern that the target thymocytes for glucocorticoid are double positive cells. Reprinted from *Cell. Mol. Biol.* **37,** Tohji, Y., Hato, F., Kinoshita, Y., and Terano, Y., Surface characteristics of thymocytes in glucocorticoid-treated rats using rosette-formation technique and surface marker analysis, 713–721, copyright 1991, with kind permission from Elsevier Science Ltd., The Boulevard, Langford Lane, Kidlington 0X5 1GB, UK.

The surviving cells after cortisol administration are mainly single positive ($CD4^+8^-$ or $CD4^-8^+$) thymocytes located in the medullary area (Fig. 15).

A severe disturbance of the intracellular metabolism of nucleic acids and proteins, accompanied by impaired glucose and amino acid incorporation and reduced RNA polymerase activity, has been proposed as a possible mechanism for cortical thymocytolysis by glucocorticoid (Claman, 1972; Makman *et al.,* 1968; Nakagawa and White, 1970). Alnemri and Litwack (1989) have reported that the steroid hormone activates a constitutive endogenous endonuclease and then induces internucleosomal DNA fragmentation. The regeneration process in cortical thymus atrophy begins 3 days after cortisol administration. The process is thought to be a useful way to estimate the pathway of thymocyte differentiation and maturation in adult animals. Briefly, it is difficult in the normal postnatal thymus to study the effect of epithelial cells on this pathway, since cortical thymocytes overwhelmingly outnumber epithelial cells. On the other hand, the marked decrease of thymocytes after glucocorticoid administration reduces the ratio of thymocytes to epithelial cells and thus makes it relatively

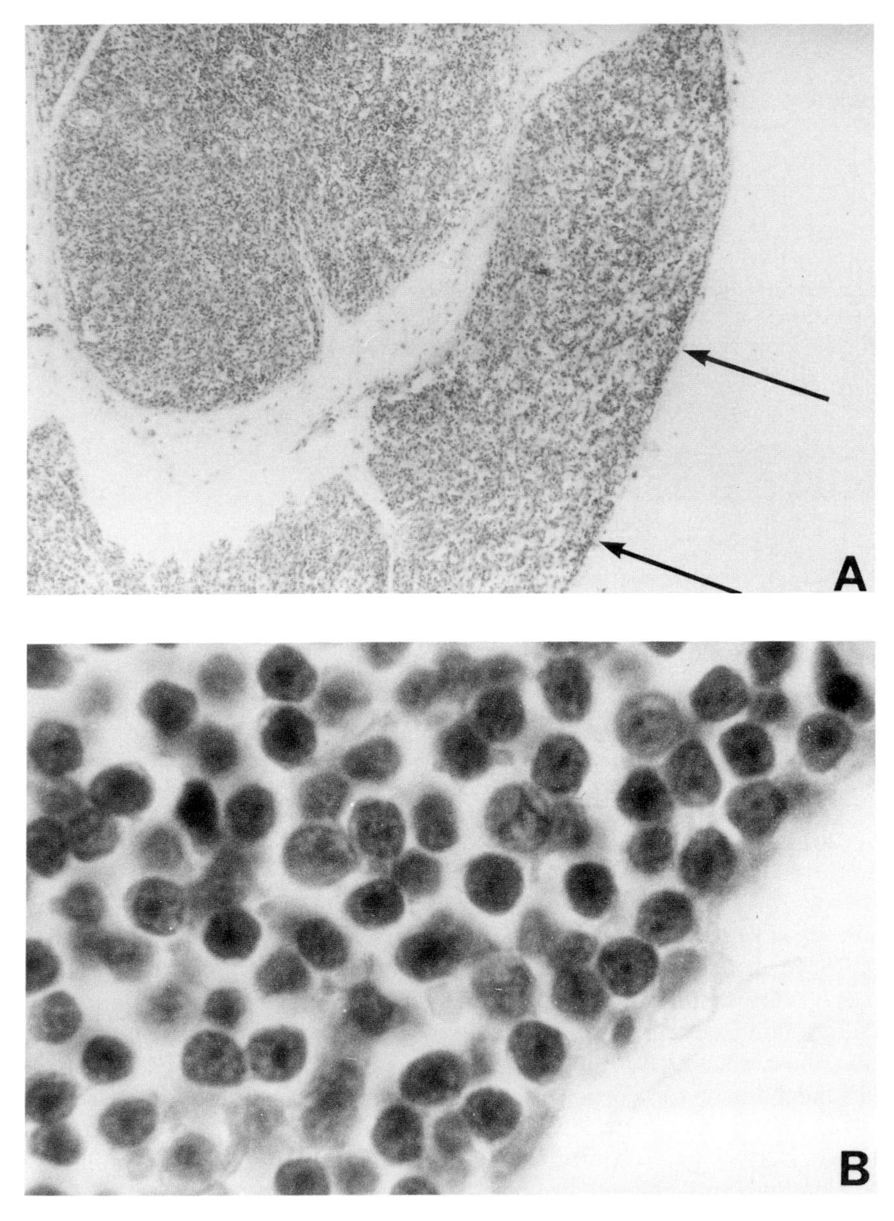

FIG. 16 Histological photomicrographs of rat thymic lobuli 4 days after GC treatment. (A) The arrows indicate the clusters of small round cells in the subcapsular area. ×37. (B) Higher magnification of the same part as A. ×930. Reprinted from *Cell. Mol. Biol.* **36,** Kim, T. J., Kinoshita, Y., Hato, F., and Kimura, S., Rat thymus reconstitution after thymocyte destruction by glucocorticoid treatment. From the view of endogenous DNA synthesis and soybean lectin responsiveness in thymocytes, 705–716, copyright 1990, with kind permission from Elsevier Science Ltd., The Boulevard, Langford Lane, Kidlington 0X5 1GB, UK.

easier to examine the effect of TECs on the pathway. On the fourth day after steroid injection, the small round cells were visible, gathered in a tiny mass in the subcapsular area of the thymic lobule (Fig. 16A). When the same specimen was observed at a higher magnification, clusters of small round cells, presumably small lymphocytes without mitotic activity, were found (Fig. 16B).

A photomicrograph taken at a higher magnification on the sixth day revealed blastoid cells with mitotic activity among the cells gathering in the subcapsular area (Fig. 17). These histological findings (Kim *et al.*, 1990) suggest that the small round cells are transformed into blastoid cells by exposure to growth factors from TECs, for example, various interleukins (Haynes, 1990), a 10-kDa polypeptide fraction (Kinoshita *et al.*, 1990; Toyokawa *et al.*, 1987), and insulin-like growth factor-1 (Yamada *et al.*, 1994). At 6–8 days after glucocorticoid administration, a significant increase in endogenous DNA synthesis in thymocytes is observed (Kim *et al.*, 1990). Subsequently, continuous division from blastoid cells to large, medium, and small lymphocytes is observed in the cortical area.

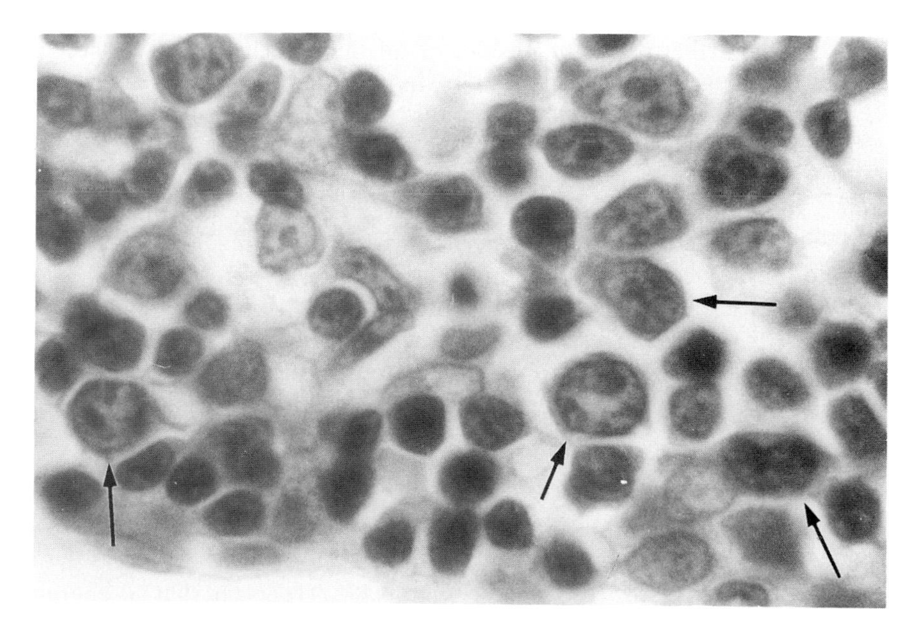

FIG. 17 A histological photomicrograph of rat thymic lobule 6 days after GC injection. The arrows indicate the blastoid cells found in almost the same subcapsular area as in Fig. 16. ×930. Reprinted from *Cell. Mol. Biol.* **36,** Kim, T. J., Kinoshita, Y., Hato, F., and Kimura, S., Rat thymus reconstitution after thymocyte destruction by glucocorticoid treatment. From the view of endogenous DNA synthesis and soybean lectin responsiveness in thymocytes, 705–716, copyright 1990, with kind permission from Elsevier Science Ltd., The Boulevard, Langford Lane, Kidlington 0X5 1GB, UK.

These results might possibly be related to the restoration process in the thymus after cortisol administration (Tohji *et al.*, 1991).

The next important step would be to study the expression of genes for the production of polypeptidic growth factors in the thymus after cortisol injection. However, the relationship between induction of cortical thymocytolysis by cortisol and that of gene expression of the polypeptidic growth factors in TECs remains unclear. A thorough investigation of this relationship appears to be a most important project.

Sex steroid hormones are also capable of causing thymic involution (Barr *et al.*, 1984; Dougherty, 1952). On the other hand, castration in experimental animals induces thymic hypertrophy in both sexes (Chiodi, 1940; Eidinger and Garrett, 1972; Grossman, 1985). These findings suggest that the steroid hormones also participate in the regulation of T-cell production in the thymus. However, the restoration mechanism after thymic atrophy caused by administration of the sex steroids remains to be clarified. Immunohistochemical analysis using monoclonal antibodies against estrogen and progestin receptors disclosed that receptors exist in TECs which are identified by polyclonal antikeratin antibody (Kawashima *et al.*, 1991). Furthermore, the results of an immunoelectron-microscopic technique using polyclonal antibody to thymulin (Bach and Carnaud, 1976; Bach *et al.*, 1977) suggest that the thymic hormone appears to be secreted from TECs (Kawashima *et al.*, 1991). According to Stimson and Crilly (1981), the primary culture supernatant of TECs stimulated with estradiol and testosterone contains more immunoregulatory factors than that of cells without the steroids. It is assumed from these results that the sex steroid hormones modulate the production and release of THPs through binding of the hormone to the receptors in TECs and thus participate indirectly in the system regulating thymocyte proliferation, differentiation, and maturation. The next problem is to analyze the mechanism for gene expression of polypeptidic hormonal factors in TECs stimulated with sex steroid hormones.

B. Autonomic Nervous System

Bulloch and Pomerantz (1984) and Bulloch (1987) reported that adrenergic nerves in the thymus are derived from sympathetic ganglions (nodose, stellate, and superior cervical ganglions), while cholinergic nerves in the thymus originate in vagus, recurrent laryngeal, and phrenic nerves. The techniques for identifying autonomic nerve fibers (ANF) have been much improved. Catecholaminergic nerve fibers (Ca-NFs) can be examined by a glyoxylic acid-induced fluorescent histochemical method (Bulloch and Pomerantz, 1984; de la Torre and Surgeon, 1976; Felten *et al.*, 1985). Cholinergic nerve fibers (Ch-NFs) can be scrutinized by histochemical staining for acetylcholinesterase using acetylthiocholine iodide as the substrate (Bulloch and Pomerantz, 1984; Felten *et al.*, 1985), and by an immunocytochemical method using monoclonal antibody against choline acetyltransferase (Fatani *et al.*, 1986). These techniques have clarified the innervation of autonomous nerve fibers in the thymic parenchyma.

Ca-NFs penetrate the thymus lobuli along the blood vessels, the capsule, and the interlobular septa (Bulloch and Pomerantz, 1984; Felten *et al.*, 1985). The Ca-NF plexuses in the subcapsular area move to the outer zone of the thymic cortex. Then, the nerve fibers extend into the cortical and medullary parenchyma (Fig. 7). Perivascular nerve plexuses are detected in the interlobular septa and corticomedullary boundary, whereas nonvascular-associated free fibers are found in the cortex and medulla (Bulloch and Pomerantz, 1984).

Ch-NFs penetrate through the capsule and trabeculae of thymic lobuli and extend to the mid and deep cortex, corticomedullary junction, and medulla (Fig. 7) (Fatani *et al.*, 1986). The most dense distribution of Ch-NFs is found in the corticomedullary boundary (Felten *et al.*, 1985). Ch-NF innervation takes place mostly in close association with vessels, but a few nerve fibers are also observed within the thymic parenchyma (Fatani *et al.*, 1986).

Autonomic nerve fibers seem to end among thymocytes and thymic stromal cells. However, the exact terminations of ANFs in the thymus gland remain to be further examined (Fatani *et al.*, 1986). Although the functional significance of thymic ANF innervation remains unclear, this innervation appears to be related to control of the following phenomena: (1) migration of pre-T cells from bone marrow into thymic parenchyma via the blood, (2) regulation of thymocyte differentiation and maturation by modulating the function of TECs which produce and release thymus hormones, and (3) migration of mature thymocytes into peripheral lymphoid organs.

Engel *et al.* (1977) reported earlier that the existence of acetylcholine receptors (AChRs) on TECs was defined by a histochemical method using α-bungarotoxin, a competitive inhibitor for the receptors. However, the functional roles of AChRs on the TECs have not been clarified. We found that stimulation with cholinergic agonists to TECs cultured in the preconfluent state significantly enhanced DNA synthetic and mitotic activities (Fig. 18), while stimulation of cells in the confluent state enhanced their synthesis of protein (Fig. 19). Moreover, these enhancements are specifically depressed by α-bungarotoxin (Figs. 18, 19) (Tominaga *et al.*, 1989). These results suggest that binding of the agonist to AChRs on the TECs activates their production system for thymic hormonal polypeptides (Kinoshita *et al.*, 1990; Morioka *et al.*, 1989; Taniyama *et al.*, 1989) and then promotes thymocyte proliferation, differentiation, and maturation. On the other hand, application of catecholaminergic agonists to this TEC culture

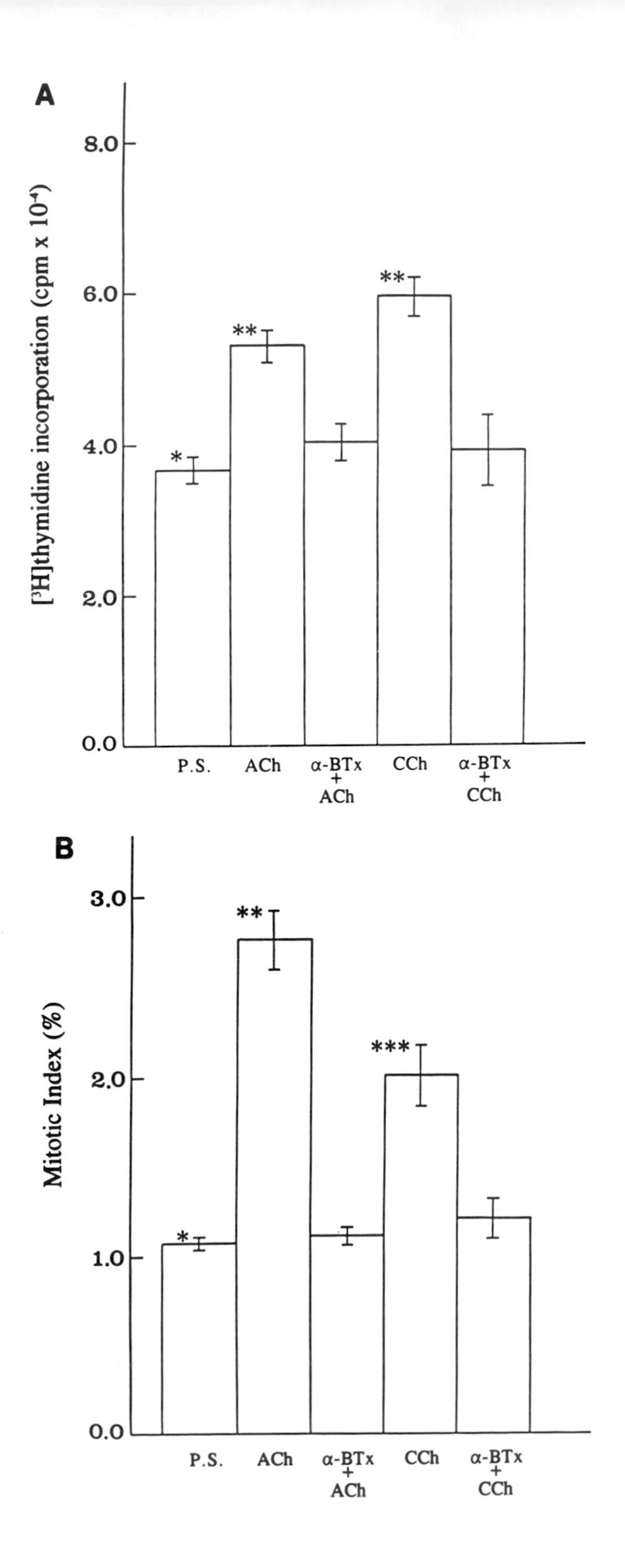
A
8.0
6.0
4.0
2.0
0.0
[3H]thymidine incorporation (cpm x 10-4)
*
**
**
P.S.
ACh
α-BTx
+
ACh
CCh
α-BTx
+
CCh
B
3.0
2.0
1.0
0.0
Mitotic Index (%)
*
**
***
P.S.
ACh
α-BTx
+
ACh
CCh
α-BTx
+
CCh

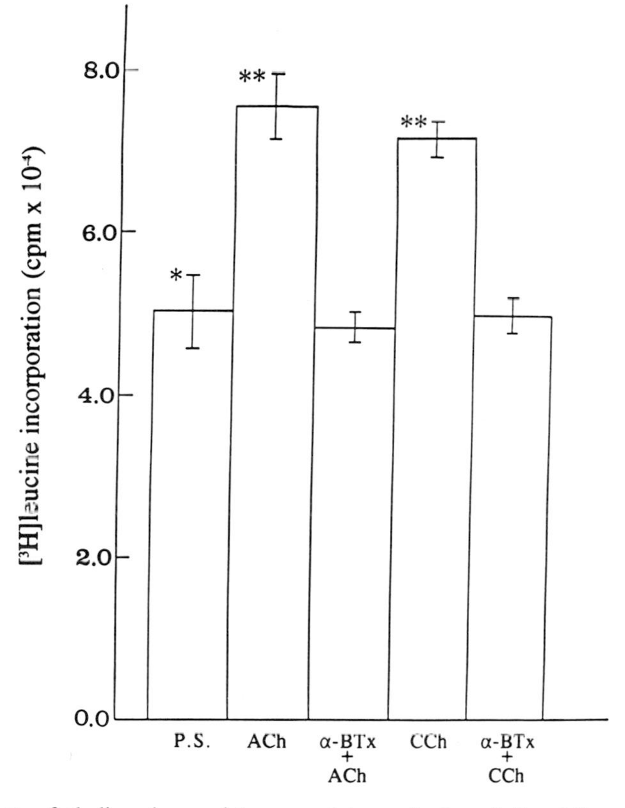

FIG. 19 Effects of cholinergic agonists on protein synthetic activity of thymus epithelial cell line (TAD3) in the confluent state. Ordinate: Protein synthetic activity [^{3}H]leucine incorporation; (cpm × 10^{-4}). P.S., physiological saline; ACh, acetylcholine; α-BTx, α-bungarotoxin; CCh, carbamylcholine. Reprinted from *Cell. Mol. Biol.* **35,** Tominaga, K., Kinoshita, Y., Hato, F., Masuda, A., and Matsuyama, M., Effects of cholinergic agonists on the proliferation and protein synthesis in a cultured thymic epithelial cell line, 679–686, copyright 1989, with kind permission from Elsevier Science Ltd., The Boulevard, Langford Lane, Kidlington 0X5 1GB, UK.

FIG. 18 Effects of cholinergic agonists on DNA synthetic activity (A) and mitotic index (B) of a thymus epithelial cell line (TAD3) in the preconfluent state. P.S., physiological saline; ACh, acetylcholine; α-BTx, α-bungarotoxin; CCh, carbamylcholine. Ordinate: DNA synthetic activity [^{3}H] thymidine incorporation, (cpm, × 10^{-4}) and mitotic index (%). *; mean ± S.D.; **, $p < 0.01$; ***, $p < 0.05$. Reprinted from *Cell. Mol. Biol.* **35,** Tominaga, K., Kinoshita, Y., Hato, F., Masuda, A., and Matsuyama, M., Effects of cholinergic agonists on the proliferation and protein synthesis in a cultured thymic epithelial cell line, 679–686, copyright 1989, with kind permission from Elsevier Science Ltd., The Boulevard, Langford Lane, Kidlington 0X5 1GB, UK.

system could not induce the same macromolecular modulation found with cholinergic stimulation (Tominaga *et al.*, 1993).

According to a report by Singh (1979, 1980), in an organ culture of fetal thymus, phenylephrine, one of the α-adrenergic agonists, induces a significant enhancement of [^{125}I]2-L-deoxyuridine ([^{125}I]dU) incorporation and cell yields compared with the untreated control. The α-agonist might possibly cause an intracellular increase in cyclic guanosine monophosphate in undifferentiated lymphoid cells and then make the cells enter DNA synthetic and mitotic phases (Singh, 1979, 1980). In contrast, isoproterenol, one of the β-adrenergic agonists, shows the opposite effect on thymocyte proliferation. Namely, cell yields and [^{125}I]dU incorporation in a fetal thymic organ culture stimulated with the β-agonist are slightly lower than those in the unstimulated control (Singh, 1979, 1980). The β-agonist presumably enhances the intracellular cyclic AMP (cAMP) level for a longer period and then induces the arrest of the cells in the G_1 phase.

The functional significance of intracellular enhancement of the cAMP level with a β-agonist remains unclear. The agonist might possibly play a role in endowing the undifferentiated thymocytes with surface differentiation molecules. It is also of interest that acetylcholine added to a fetal thymic organ culture causes the same increase in [^{125}I]dU incorporation and cell yields as the α-adrenergic agonist (Singh, 1979, 1980). It is emphasized that ultrastructurally marked changes in thymocytes and thymic stromal cells have not been observed in the neurotransmitter-stimulated organ culture, and that the transmitters from ANF capable of enhancing or suppressing thymic lymphopoiesis would act directly on the thymocytes and not on the thymic stromal cells. This does not, however, eliminate the possibility that the TECs stimulated with the neurotransmitters release thymic hormonal products without markedly changing their morphological characteristics, which act on the nearby thymocytes. Furthermore, active lymphopoiesis appeared to be detected in the thymus-implanted anterior eye chamber of the cervical sympathectomized side of syngenic mice, while it did not in the sham-operated side (Singh, 1985). The results mentioned earlier suggest that adrenergic nerve innervation can inhibit thymic lymphopoiesis, whereas cholinergic nerve innervation can stimulate lymphopoiesis (Fatani *et al.*, 1986). It was also reported that the number of sympathetic nerve fibers in the thymic parenchyma continues to increase and reaches a maximum in the adult (Singh, 1984).

It is possible that the development of sympathetic nerve innervation provides an increased inhibitory input to thymocyte proliferation and plays a role in the induction of thymus involution (Fatani *et al.*, 1986). The following research project would be one way to discover the functional role of the autonomous nervous system in the thymic parenchyma. Adrenergic and cholinergic nerve fibers taken from a point just before they enter the thymic capsule could be prepared for viewing on a stereoscopic microscope. Then, the fibers could be stimulated with chemical or electrical means and the cellular and molecular changes in thymocytes and TECs could be analyzed. This experimental system might possibly give us new information on how thymus epithelial cells affect thymocyte differentiation and maturation.

VI. Conclusion

It has been shown that thymic epithelial cells release thymus hormonal products which participate in T-cell production by affecting the proliferation, differentiation, and maturation of thymocytes. Several hormones from endocrine organs and neurotransmitters from the autonomic nervous system innervated into thymic parenchyma appear to regulate the function of thymic epithelial cells. The new information presented here clearly indicates that there is neuro-endocrine-immunological coordination in the thymic microenvironment.

References

Alnemri, E. S., and Litwack, G. (1989). Glucocorticoid induced lymphocytolysis is not mediated by an induced endonuclease. *J. Biol. Chem.* **264,** 4104–4111.

Arnaout, M. A. (1990). Structure and function of the leukocyte adhesion molecules CD11/CD18. *Blood* **75,** 1037–1050.

Arrenbrecht, S. (1974). Specific binding of growth hormone to thymocytes. *Nature (London)* **252,** 255–257.

Baba, T., Kinoshita, Y., and Kimura, E. (1979). Cyclic AMP phosphodiesterase activities of small lymphocytes isolated from rat thymuses and lymph nodes. *Folia Biol. (Kraków)* **27,** 293–298.

Bach, J.-F., and Carnaud, C. (1976). Thymic factors. *Prog. Allergy* **21,** 342–408.

Bach, J.-F., Dardenne, M., and Pleau, J.-M. (1977). Biochemical characterization of a serum thymic factor. *Nature (London)* **266,** 55–56.

Ban, E., Gagnerault, M.-C., Jammes, H., Postel-Vinay, M.-C., Haour, F., and Dardenne, M. (1991). Specific binding sites of growth hormone in cultured mouse thymic epithelial cells. *Life Sci.* **48,** 2141–2148.

Barr, I. G., Pyke, K. W., Pearce, P., Toh, B.-H., and Funder, J. W. (1984). Thymic sensitivity to sex hormones develops post-natally; an in vivo and in vitro study. *J. Immunol.* **132,** 1095–1099.

Berczi, I. (1986). Immunoregulation by pituitary hormones. *In* "Pituitary Function and Immunity" (I. Berczi, Ed.), pp. 227–240. CRC Press, Boca Raton, FL.

Blomgren, H., and Andersson, B. (1969). Evidence for a small pool of immunocompetent cells in the mouse thymus. *Exp. Cell Res.* **57,** 185–192.

Borberg, H., Woodruff, J., Hirschhorn, R., Gesner, B., Miescher, P., and Silber, R. (1966). Phytohemagglutinin: Inhibition of the agglutinating activity by N-acetyl-D-galactosamine. *Science* **154,** 1019–1020.

Boyd, E. (1932). The weight of the thymus gland in health and disease. *Am. J. Dis. Child.* **43,** 1162–1214.

Brideau, R. J., Carter, P. B., McMaster, W. R., Mason, D. W., and Williams, A. F. (1980). Two subsets of rat T lymphocytes defined with monoclonal antibodies. *Eur. J. Immunol.* **10,** 609–615.

Bulloch, K. (1987). The innervation of immune system. Tissues and organs. *In* "The Neuro-Immuno-Endocrine Connection" (C. W. Cotman, Ed.), pp. 33–47. Raven Press, New York.

Bulloch, K., and Pomerantz, W. (1984). Autonomic nervous system innervation of thymic-related lymphoid tissue in wildtype and nude mice. *J. Comp. Neurol.* **228,** 57–68.

Chiodi, H. (1940). Relationship between thymus and sexual organs. *Endocrinology (Baltimore)* **26,** 107–116.

Chung, C. S., Etherton, T. D., and Wiggins, J. P. (1985). Stimulation of swine growth by porcine growth hormone. *J. Anim. Sci.* **60,** 118–130.

Claman, H. N. (1972). Glucocorticoids and lymphoid cells. *N. Engl. J. Med.* **287,** 388–397.

Cordier, A. C., and Haumont, S. M. (1980). Development of thymus, parathyroids, and ultimo-branchial bodies in NMRI and nude mice. *Am. J. Anat.* **157,** 227–263.

de la Torre, J. C., and Surgeon, J. W. (1976). A methodological approach to rapid and sensitive monoamine histofluorescence using a modified acid technique: The SPG method. *Histochemistry* **49,** 81–93.

Dougherty, T. F. (1952). Effect of hormones on lymphatic tissue. *Physiol. Rev.* **32,** 379–401.

Doyle, C., and Strominger, J. L. (1987). Interaction between CD4 and class II MHC molecules mediated cell adhesion. *Nature (London)* **330,** 256–259.

Dustin, M. L., Rothlein, R., Bhan, A. K., Dinarello, C. A., and Springer, T. A. (1986). Induction by IL 1 and interferon-γ: Tissue distribution, biochemistry, and function of a natural adherence molecule (ICAM-1). *J. Immunol.* **137,** 245–254.

Eidinger, D., and Garrett, T. J. (1972). Studies of the regulatory effects of the sex hormones on antibody formation and stem cell differentiation. *J. Exp. Med.* **136,** 1098–1116.

Elfenbein, G. J., and Santos, G. W. (1978). Rosette formation between rat thymocytes and guinea pig erythrocytes requires "active" fetal calf serum. II. Characterization of the receptor-bearing thymocytes. *Cell. Immunol.* **37,** 199–208.

Elfenbein, G. J., and Winkelstein, J. A. (1978). Rosette formation between rat thymocytes and guinea pig erythrocytes requires "active" fetal calf serum. I. Characteristics of "active" serum factors and receptors on thymocytes and erythrocytes. *Cell. Immunol.* **37,** 188–198.

Engel, W. K., Trotter, J. L., McFarlin, D. E., and McIntosh, C. L. (1977). Thymic epithelial cell contains acetylcholine receptor. *Lancet* **2,** 1310–1311.

Fabris, N., Pierpaoli, W., and Sorkin, E. (1971). Hormones and immunological capacity. 3. The immunodeficiency disease of the hypopituitary Snell-Bagg dwarf mouse. *Clin. Exp. Immunol.* **9,** 209–225.

Fatani, J. A., Qayyum, M. A., Mehta, L., and Singh, U. (1986). Parasympathetic innervation of the thymus: a histochemical and immunocytochemical study. *J. Anat.* **147,** 115–119.

Felten, D. L., Felten, S. Y., Carlson, S. L., Olschowka, J. A., and Livnat, S. (1985). Noradrenergic and peptidergic innervation of lymphoid tissue. *J. Immunol.* **135,** 755S–765S.

Furlanetto, R. W., and Marino, J. M. (1987). Radioimmunoassay of somatomedin C/insulin-like growth factor. I. *In* "Methods in Enzymology" (D. Barnes and D. Sirbasku, Eds.), Vol. 146, pp. 216–226. Academic Press, Orlando, FL.

Goldstein, A. L., Slater, F. D., and White, A. (1966). Preparation, assay, and partial purification of a thymic lymphopoietic factor (thymosin). *Proc. Natl. Acad. Sci. USA* **56,** 1010–1017.

Goldstein, A. L., Guha, A., Zatz, M. M., Hardy, M. A., and White, A. (1972). Purification and biological activity of thymosin, a hormone of the thymus gland. *Proc. Natl. Acad. Sci. USA* **69,** 1800–1803.

Goldstein, G. (1974). Isolation of bovine thymin: A polypeptide hormone of the thymus. *Nature (London)* **247,** 11–14.

Good, R. A., Dalmasso, A. P., Martinez, C., Archer, O. K., Pierce, J. C., and Papermaster, B. W. (1962). The role of the thymus in development of immunologic capacity in rabbits and mice. *J. Exp. Med.* **116,** 773–796.

Grossman, C. J. (1985). Interactions between the gonadal steroids and the immune system. *Science* **227,** 257–261.

Hale, M. L., Greiner, D. L., and McCarthy, K. F. (1987). Characterization of rat prothymocyte with monoclonal antibodies recognizing rat lymphocyte membrane antigenic determinants. *Cell. Immunol.* **107,** 188–200.

Hashimura, H., Toyokawa, T., Hato, F., Oshitani, N., Kimura, S., and Kinoshita, Y. (1987). Separation of biologically active polypeptides from rat thymus epithelial-cell-culture-supernatant by high-performance liquid chromatography. *Cell. Mol. Biol.* **33,** 375–386.

Hato, F., Tohji, Y., Kinoshita, Y., and Terano, Y. (1991). Surface antigenic analysis of rosette-forming cells in rat thymus using monoclonal antibodies against T cell surface markers. *Cell. Mol. Biol.* **37,** 839–845.

Hato, F., Tominaga, K., Kinoshita, Y., and Terano, Y. (1992). Preparation of monoclonal antibody against synthetic thymosin α_1 and its effect on the biological activity of thymosin α_1. *Recent Adv. Cell. Mol. Biol.* **1,** 35–41.

Haynes, B. F. (1990). Human thymic epithelium and T cell development: Current issues and future directions. *Thymus* **16,** 143–157.

Hensen, E. J., Hoefsmit, E. C. M., and Tweel, V. D. (1978). Augmentation of mitogen responsiveness in human lymphocytes by a humoral factor obtained from thymic epithelial cultures. *Clin. Exp. Immunol.* **32,** 309–317.

Hirokawa, K., McClure, J. E., and Goldstein, A. L. (1982). Age-related changes in localization of thymosin in the human thymus. *Thymus* **4,** 19–29.

Hirokawa, K., Utsuyama, M., Moriizumi, E., and Handa, S. (1986). Analysis of the thymic microenvironment by monoclonal antibodies with special reference to thymic nurse cells. *Thymus* **8,** 349–360.

Itoh, T., Azumi, S., Kasahara, S., and Mori, T. (1981). Establishment of a functioning epithelial cell line from the rat thymus. A cell line that induces the differentiation of rat bone marrow cells into T cell lineage. *Biomed. Res.* **2,** 11–19.

Jacobsson, H., and Blomgren, H. (1974). Responses of mouse thymic cells to mitogens. A comparison between phytohemagglutinin and concanavalin A. *Cell. Immunol.* **11,** 427–441.

Jenkinson, E. J., Kingston, R., Smith, C. A., Williams, G. T., and Owen, J. J. T. (1989). Antigen-induced apoptosis in developing T cells: A mechanism for negative selection of the T cell receptor repertoire. *Eur. J. Immunol.* **19,** 2175–2177.

Kaneshima, H., Ito, M., Asai, J., Taguchi, O., and Hiai, H. (1987). Thymic epithelial reticular cell subpopulations in mice defined by monoclonal antibodies. *Lab. Invest.* **56,** 372–380.

Kato, K., Koyanagi, M., Okada, H., Takanashi, T., Wong, Y. W., Williams, A. F., Okumura, K., and Yagita, H. (1992). CD48 is a counter-receptor for mouse CD2 and is involved in T cell activation. *J. Exp. Med.* **176,** 1241–1249.

Kawashima, I., Sakabe, K., Seiki, K., Fujii-Hanamoto, H., Akatsuka, A., and Tsukamoto, H. (1991). Localization of sex steroid receptor cells, with special reference to thymulin (FTS)-producing cells in female rat thymus. *Thymus* **18,** 79–93.

Kendal, M. D., Van de Wijngaert, F. P., Schuurman, H.-J., Rademakers, L. H. P. M., and Kater, L. (1985). Heterogeneity of the human thymus epithelial microenvironment at the ultrastructural level. In microenvironments in the lymphoid system. *Adv. Exp. Med. Biol.* **186,** 289–297.

Kerr, J. F. R., Wyllie, A. H., and Currie, A. R. (1972). Apoptosis: A basic biological phenomenon with wide-ranging implications in tissue kinetics. *Br. J. Cancer* **26,** 239–257.

Kim, T. J., Kinoshita, Y., Hato, F., and Kimura, S. (1990). Rat thymus reconstitution after thymocyte destruction by glucocorticoid treatment. From the view of endogenous DNA synthesis and soybean lectin responsiveness in thymocytes. *Cell. Mol. Biol.* **36,** 705–716.

Kimura, E., Suzuki, T., and Kinoshita, Y. (1960). Separation of reticulocytes by means of multi-layer centrifugation. *Nature (London)* **188,** 1201–1202.

Kinoshita, Y., and Kimura, S. (1971). Remarkable enhancement of transplantable tumor growth by antiserum prepared from the heavier lymphocyte fraction. *Exp. Cell Res.* **68,** 471–476.

Kinoshita, Y., Sherman, J. D., and Dameshek, W. (1970a). Effect of thymectomy on DNA synthesis of lymph node cells. *Arch. Pathol.* **89,** 266–270.

Kinoshita, Y., Kimura, S., Takeshita, T., Kimura, E., Yukioka, M., and Morisawa, S. (1970b). Isolation of heterogeneous lymphocytes according to their cellular densities by multilayer centrifugation and detection of the separated fraction containing the immunological memory cells. *Exp. Cell Res.* **59,** 299–306.

Kinoshita, Y., Kimura, S., Fukamizu, M., and Nagasawa, T. (1974a). Separation of two antigenically different subpopulations of small lymphocytes from rat thymus. *Exp. Cell Res.* **86,** 136–142.

Kinoshita, Y., Kimura, S., and Fukamizu, M. (1974b). Cytolytic effects of glucocorticoid on thymus-medullary lymphocytes incubated in vitro. *Exp. Cell Res.* **87,** 387–392.

Kinoshita, Y., Kimura, S., Komano, Y., Koshikawa, M., and Inamoto, E. (1974c). Small-lymphocyte subpopulation rich in effector cells against Walker carcinosarcoma 256. *J. Natl. Cancer Inst. (U.S.)* **53,** 867–874.

Kinoshita, Y., Kimura, S., and Fukamizu, A. (1976). Changes in the stimulatory capability of mouse small thymocytes to give a proliferation of allogeneic spleen cells during differentiation. *Cell Differ.* **5,** 217–224.

Kinoshita, Y., Baba, T., Kimura, S., and Kimura, E. (1979). Difference in cAMP phosphodiesterase activity among subpopulations of small lymphocytes isolated from rat thymuses. *Folia Biol. (Kraków)* **27,** 299–303.

Kinoshita, Y., Kimura, S., Hashimura, H., Toyokawa, T., and Goldstein, A. L. (1986a). New system examining biological activity of thymus hormone using rosette formation assay of rat thymocytes. *Cell. Mol. Biol.* **32,** 391–397.

Kinoshita, Y., Yamaga, K., Kimura, S., and Nakai, Y. (1986b). Soybean lectin-responding cells in human tonsil. I. Separation of soybean lectin-responding cells and an alteration of the responsiveness of the separated cells by thymosin. *Cell. Mol. Biol.* **32,** 399–406.

Kinoshita, Y., Yamaga, K., Kimura, S., and Nakai, Y. (1986c). Soybean lectin-responding cells in human tonsil. II. An induction of concanavalin A-responsiveness and E-rosette-forming capacity to soybean lectin-responding cells by thymosin. *Cell. Mol. Biol.* **32,** 407–410.

Kinoshita, Y., Toyokawa, T., Hato, F., Masuda, A., and Matsuyama, M. (1990). New thymocyte growth factor from thymic epithelial cell line. *Cell. Mol. Biol.* **36,** 429–438.

Kinoshita, Y., Hato, F., Tominaga, K., and Terano, Y. (1992). Target cells for new thymocyte growth factor derived from the rat thymus epithelial cell line (TAD3). *Recent Adv. Cell. Mol. Biol.* **1,** 19–24.

Komuro, K., and Boyse, E. A. (1973a). In vitro demonstration of thymic hormone in the mouse by conversion of precursor cells into lymphocytes. *Lancet* **1,** 740–743.

Komuro, K., and Boyse, E. A. (1973b). Induction of T lymphocytes from precursor cells in vitro by a product of the thymus. *J. Exp. Med.* **138,** 479–482.

Krensky, A. M., Sanchez-Madrid, F., Robins, E., Nagy, J. A., Springer, T. A., and Burakoff, S. J. (1983). The functional significance, distribution, and structure of LFA-1, LFA-2, and LFA-3: Cell surface antigens associated with CTL-target interactions. *J. Immunol.* **131,** 611–616.

Kruisbeek, A. M. (1979). Thymic factors and T cell maturation in vitro: A comparison of the effects of thymic epithelial cultures with thymic extracts and thymus dependent serum factors. *Thymus* **1,** 163–185.

Kruisbeek, A. M., and Astaldi, G. C. B. (1979). Distinct effects of thymic epithelial culture supernatants on T cell properties of mouse thymocytes separated by the use of peanut agglutinin. *J. Immunol.* **123,** 984–991.

Kruisbeek, A. M., Kröse, T. C. J. M., and Zijlstra, J. J. (1977). Increase in T cell mitogen responsiveness in rat thymocytes by a humoral factor from thymic epithelial culture supernatant. *Eur. J. Immunol.* **7,** 375–381.

Kruisbeek, A. M., Astaldi, G. C. B., Blankwater, M.-J., Zijlstra, J. J., Levert, L. A., and Astaldi, A. (1978). The in vitro effect of a thymic epithelial culture supernatant on mixed lymphocyte reactivity and on antibody production to SRBC by Nu/Nu spleen cells. *Cell. Immunol.* **35,** 134–147.

Lanier, L. L., Le, A. M., Phillips, J. H., Warner, N. L., and Babcock, G. F. (1983). Subpopulations of human natural killer cells defined by expression of the Leu-7 (HNK-1) and Leu-11 (NK-15) antigens. *J. Immunol.* **131,** 1789–1796.

Law, L. W., Dunn, T. B., Trainin, N., and Levey, R. H. (1964). Studies of thymic function. *In* "The Thymus" (V. Defendi and D. Metcalf, eds.), pp. 105–117. Wistar Institute Press, Philadelphia.

Ledbetter, J. A., Evans, R. L., Lipinski, M., Cunningham-Rundles, C., Good, R. A., and Herzenberg, L. A. (1981). Evolutionary conservation of surface molecules that distinct T lymphocyte helper/inducer and cytotoxic/suppressor subpopulations in mouse and man. *J. Exp. Med.* **153,** 310–323.

Levey, R. H., Trainin, N., and Law, L. W. (1963). Evidence for function of thymic tissue in diffusion chambers implanted in neonatally thymectomized mice. *J. Natl. Cancer Inst. (U.S.)* **31,** 199–217.

Liener, I. E., and Pallansch, M. J. (1952). Purification of a toxic substance from defatted soybean flour. *J. Biol. Chem.* **197,** 29–36.

Lis, H., Sharon, N., and Katchalski, E. (1966). Soybean hemagglutinin, a plant glycoprotein. *J. Biol. Chem.* **241,** 684–689.

Lotan, R., Siegelman, H. W., Lis, H., and Sharon, N. (1974). Subunit structure of soybean agglutinin. *J. Biol. Chem.* **249,** 1219–1224.

Low, T. L. K., Thurman, G. B., McAdoo, M., McClure, J., Rossio, J. L., Naylor, P. H., and Goldstein, A. L. (1979). The chemistry and biology of thymosin. I. Isolation, characterization and biological activities of thymosin $\alpha1$ and polypeptide $\beta1$ from calf thymus. *J. Biol. Chem.* **254,** 981–986.

MacDonald, H. R., Budd, R. C., and Howe, R. C. (1988). A $CD3^-$ subset of $CD4^-8^+$ thymocytes: A rapidly cycling intermediate in the generation of $CD4^+8^+$ cells. *Eur. J. Immunol.* **18,** 519–523.

Maggiano, N., Ricci, R., Larroca, L. M., Lauriola, L., Piantelli, M., Capelli, A., and Ranelletti, F. O. (1992). Growth hormone (GH) and somatomedin C (SM-C) immunoreactive cells in human thymus. Effects of GH and SM-C on thymocyte proliferation and phenotype. *Recent Adv. Cell. Mol. Biol.* **1,** 99–106.

Makman, M. H., Dvorkin, B., and White, A. (1968). Influence of cortisol on the utilization of precursors of nucleic acids and protein by lymphoid cells in vitro. *J. Biol. Chem.* **243,** 1485–1497.

Marlin, S. D., and Springer, T. A. (1987). Purified intercellular adhesion molecule-1 (ICAM-1) is a ligand for lymphocyte function-associated antigen 1 (LFA-1). *Cell (Cambridge, Mass.)* **51,** 813–819.

Martinez, C., Dalmasso, A., and Good, R. A. (1962). Acceptance of tumour homografts by thymectomized mice. *Nature (London)* **194,** 1289–1290.

Masuda, A., Nishimoto, Y., Morita, T., and Matsuyama, M. (1985). Dexamethasone-induced changes in morphology and keratin organization of rat thymic epithelial cells in primary culture. *Exp. Cell Res.* **160,** 343–355.

Miller, J. F. A. P. (1961). Immunological function of the thymus. *Lancet* **1,** 748–749.

Miller, J. F. A. P. (1962). Role of the thymus in transplantation immunity. *Ann. N.Y. Acad. Sci.* **99,** 340–354.

Miller, R. G., and Phillips, R. A. (1969). Separation of cells by velocity sedimentation. *J. Cell. Physiol.* **73,** 191–202.

Morioka, K., Kinoshita, Y., Tominaga, K., Hato, F., and Itoh, T. (1989). Separation of polypeptides which induce a mitogen response of thymocytes from the culture supernatant of a thymic epithelial cell line. *Cell. Mol. Biol.* **35,** 523–534.

Mosier, D. E., and Cantor, H. (1971). Functional maturation of mouse thymic lymphocytes. *Eur. J. Immunol.* **1,** 459–461.

Munck, A., Guyre, P. M., and Holbrook, N. J. (1984). Physiological functions of glucocorticoids in stress and their relation to pharmacological actions. *Endocr. Rev.* **5,** 25–44.

Nakagawa, S., and White, A. (1970). Properties of an aggregate ribonucleic acid polymerase from rat thymus and its response to cortisol injection. *J. Biol. Chem.* **245,** 1448–1457.

Nieburgs, A. C., Korn, J. H., Picciano, P., and Cohen, S. (1985). The production of regulatory cytokines for thymocyte proliferation by murine thymic epithelium in vitro. *Cell. Immunol.* **90,** 426–438.

Norment, A. N., Salter, R. D., Parham, P., Engelhard, V. H., and Littman, D. R. (1988). Cell-cell adhesion mediated by CD8 and MHC class I molecules. *Nature (London)* **336,** 79–81.

Oka, H., Kimura, S., and Kinoshita, Y. (1984a). Sugar specificity of the receptor of rat thymocytes in rosette formation with guinea pig erythrocytes. *Cell. Immunol.* **89,** 235–241.

Oka, H., Kimura, S., and Kinoshita, Y. (1984b). Effect of neuraminidase treatment of rat thymocytes on rosette formation with guinea pig erythrocytes. *Jpn. J. Physiol.* **34,** 1141–1145.

Okamoto, M., Kimura, S., Matsuya, H., and Kinoshita, Y. (1986). Separation of soybean lectin-responding cells from rat bone marrow by combination of the adhesion column method with discontinuous density gradient centrifugation. *Cell. Mol. Biol.* **32,** 63–72.

Okimoto, T., Kinoshita, Y., Hato, F., Toyokawa, T., Kimura, S., and Kinoshita, H. (1988). Effects of thymosin on soybean lectin responsiveness of thymic small lymphocyte subsets from streptozotocin-induced diabetic rats. *Cell. Mol. Biol.* **34,** 465–472.

Oosterom, R., and Kater, L. (1980). Effect of human thymic epithelial conditioned medium on in vitro and in vivo alloantigen-induced lymphocyte activation in the mouse. *Clin. Immunol. Immunopathol.* **17,** 173–182.

Oosterom, R., Kater, L., and Oosterom, J. (1979). Effects of human thymic epithelial-conditioned medium on mitogen responsiveness of human and mouse lymphocytes. *Clin. Immunol. Immunopathol.* **12,** 460–470.

Oshitani, N., Hato, F., and Kinoshita, Y. (1989). Functional diversity of polypeptides in primary culture supernatant of thymus epithelial cells. *Cell. Mol. Biol.* **35,** 657–668.

Parish, C. R., Kirov, S. M., Bowern, N., and Blanden, R. V. (1974). A one-step procedure for separating mouse T and B lymphocytes. *Eur. J. Immunol.* **4,** 808–815.

Parrott, D. M. V., DeSousa, M. A. B., and East, J. (1966). Thymus-dependent areas in the lymphoid organs of neonatally thymectomized mice. *J. Exp. Med.* **123,** 191–204.

Powell, A. E., and Leon, M. A. (1970). Reversible interaction of human lymphocytes with the mitogen concanavalin A. *Exp. Cell Res.* **62,** 315–325.

Raff, M. C. (1969). Theta isoantigen as a marker of thymus-derived lymphocytes in mice. *Nature (London)* **224,** 378–379.

Raulet, D. H. (1985). Expression and function of interleukin-2 receptors on immature thymocytes. *Nature (London)* **314,** 101–103.

Reichert, R. A., Weissman, I. L., and Bucher, E. C. (1986). Phenotypic analysis of thymocytes that express homing receptors for peripheral lymph nodes. *J. Immunol.* **136,** 3521–3528.

Reif, A. E., and Allen, J. M. V. (1963). Specificity of isoantisera against leukemia and thymic lymphocytes. *Nature (London)* **200,** 1332–1333.

Reisner, Y., Linker-Israeli, M., and Sharon, N. (1976). Separation of mouse thymocytes into two subpopulations by the use of peanut agglutinin. *Cell. Immunol.* **25,** 129–134.

Rothlein, R., Dustin, M. L., Marlin, S. D., and Springer, T. A. (1986). A human intercellular adhesion molecule (ICAM-1) distinct from LFA-1. *J. Immunol.* **137,** 1270–1274.

Sherman, J. D., Adner, M. M., and Dameshek, W. (1963). Effect of thymectomy on the golden hamster (*Mesocricetus auratus*). I. Wasting disease. *Blood* **22,** 252–271.

Shimonkevitz, R. P., Husmann, L. A., Bevan, M. J., and Crispe, I. N. (1987). Transient expression of IL-2 receptor precedes the differentiation of immature thymocytes. *Nature (London)* **329,** 157–159.

Shortman, K., and Jackson, H. (1974). The differentiation of T lymphocytes. I. proliferation kinetics and interrelationships of subpopulations of mouse thymus cells. *Cell. Immunol.* **12,** 230–246.

Singh, U. (1979). Effect of catecholamines on lymphopoiesis in fetal mouse thymic explants. *J. Anat.* **129,** 279–292.

Singh, U. (1980). In vitro lymphopoiesis in foetal thymic organ cultures: Effect of various agents. *Clin. Exp. Immunol.* **41,** 150–155.

Singh, U. (1984). Sympathetic innervation of fetal mouse thymus. *Eur. J. Immunol.* **14,** 757–759.

Singh, U. (1985). Lymphopoiesis in the nude fetal thymus following sympathectomy. *Cell. Immunol.* **93,** 222–228.

Stimson, W. H., and Crilly, P. J. (1981). Effects of steroids on the secretion of immunoregulatory factors by thymic epithelial cell cultures. *Immunology* **44,** 401–407.

Stobo, J. D. (1972). Phytohemagglutinin and concanavalin A: Probes for murine T cell activation and differentiation. *Transplant. Rev.* **11,** 60–86.

Stobo, J. D., Rosenthal, A. S., and Paul, W. E. (1972). Functional heterogeneity of murine lymphoid cells. I. responsiveness to and surface binding of concanavalin A and phytohemagglutinin. *J. Immunol.* **108,** 1–17.

Swat, W., Ignatowicz, L., von Boehmer, H., and Kisielow, P. (1991). Clonal deletion of immature $CD4^{+}8^{+}$ thymocytes in suspension culture by extrathymic antigen-presenting cells. *Nature (London)* **351,** 150–153.

Takacs, L., Osawa, H., and Diamantstein, T. (1984). Detection and localization by the monoclonal anti-interleukin 2 receptor antibody AMT-13 of IL 2 receptor-bearing cells in the developing thymus of the mouse embryo and in the thymus of cortisone-treated mice. *Eur. J. Immunol.* **14,** 1152–1156.

Taniyama, T., Kinoshita, Y., Hato, F., Tominaga, K., Kimura, S., and Itoh, T. (1989). Separation of rosette formation-inducing polypeptides from the supernatant of cultures of rat thymic epithelial cell line. *Cell. Mol. Biol.* **35,** 535–545.

Taub, R. N., Wong, F. M., Kinoshita, Y., Sherman, J. D., and Dameshek, W. (1965). Effect of separated thymic cell fractions on thymectomized hamsters. *Fed. Am. Soc. Exp. Biol.* **128,** 681.

Tohji, Y., Hato, F., Kinoshita, Y., and Terano, Y. (1991). Surface characteristics of thymocytes in glucocorticoid-treated rats using rosette-formation technique and surface marker analysis. *Cell. Mol. Biol.* **37,** 713–721.

Tominaga, K., Kinoshita, Y., Hato, F., Masuda, A., and Matsuyama, M. (1989). Effects of cholinergic agonists on the proliferation and protein synthesis in a cultured thymic epithelial cell line. *Cell. Mol. Biol.* **35,** 679–686.

Tominaga, K., Yamada, M., Hato, F., Tsuji, Y., Tominaga, M., and Kinoshita, Y. (1993).

Study on physiological significance of growth hormone in thymus. *Jpn. J. Physiol.* **43(Suppl.),** 61.

Toyokawa, T., Hato, F., Mizoguchi, S., Kimura, S., Matsuyama, M., and Kinoshita, Y. (1987). Remarkable changes in mitogen responsiveness and rosette-forming capacity of thymocytes with aging due to possible disorders of thymic hormone secretion from thymus epithelial cells of spontaneous thymoma Buffalo/Mna rats. *Cell. Mol. Biol.* **33,** 363–374.

Trainin, N., and Small, M. (1970). Studies on some physicochemical properties of a thymus humoral factor conferring immunocompetence on lymphoid cells. *J. Exp. Med.* **132,** 885–897.

Turner, J. D., Rotwein, P., Novakofski, J., and Bechtel, P. J. (1988). Induction of mRNA for IGF-I and -II during growth hormone-stimulated muscle hypertrophy. *Am. J. Physiol.* **255,** E513–E517.

Van Ewijk, W. (1984). Immunohistology of lymphoid and non-lymphoid cells in thymus in relation to T lymphocyte differentiation. *Am. J. Anat.* **170,** 311–330.

von Boehmer, H., Teh, H. S., and Kisielow, P. (1989). The thymus selects the useful, neglects the useless and destroys the harmful. *Immunol. Today* **10,** 57–61.

Waksman, B. H., Arnason, B. G., and Janković, B. D. (1962). Role of the thymus in immune reactions in rats. *J. Exp. Med.* **116,** 187–206.

Yamada, M., Hato, F., Kinoshita, Y., Tominaga, K., and Tsuji, Y. (1994). The indirect participation of growth hormone in the thymocyte proliferation system. *Cell. Mol. Biol.* **40,** 111–121.

Yunis, E. J., Hilgard, H., Sjodin, K., Martinez, C., and Good, R. A. (1964). Immunological reconstitution of thymectomized mice by injections of isolated thymocytes. *Nature* (*London*) **201,** 784–786.

Actin-Binding Proteins in Cell Motility

Sadashi Hatano[1]
Department of Molecular Biology, School of Science, Nagoya University
Chikusa-ku Nagoya 464-01, Japan

I. Introduction

Many actin-binding proteins (ABP) have been isolated from nonmuscle cells and are classified into several groups based on specific properties which are characterized *in vitro* (Schliwa, 1981; Weeds, 1982; Korn, 1982; Craig and Pollard, 1982; Hatano *et al.*, 1983; Stossel *et al.*, 1985; Mabuchi, 1986; Pollard and Cooper, 1986; Hartwig and Kwiatkowski, 1991). It is generally considered that these ABPs play key roles in the dynamic behavior of the actin cytoskeleton in nonmuscle cells. On the other hand, two types of actin-myosin-based cell motility have been studied for many years.

1. Streaming of the cytoplasmic sol on the inner surface of the cortical gel layer. The typical cytoplasmic streaming is seen in internodal cells of Characeae. The sol continuously slides on bundles of actin filaments located on the cortical gel layer. These bundles of actin filaments are stable structures and show no appreciable structural changes during cytoplasmic streaming. The problem is determining how the contractile apparatuses work in the cells.

2. Passive flow of the cytoplasmic sol within the cortical gel layer. This type of cytoplasmic streaming is seen in plasmodium of an acellular slime mold, *Physarum polycephalum* (Kamiya and Kuroda, 1958), and in a giant ameba, *Chaos chaos* (Allen and Roslansky, 1959). The movement of amebas, leukocytes, and tissue-cultured cells could be classed with the latter type of movement (Kamiya, 1959; Allen, 1961; Komnick *et al.*, 1973; Taylor and Condeelis, 1979; Kessler, 1982; Stockem and Brix, 1994). The contractile apparatuses are constructed in cell regions when they contract, and disintegrate after the contraction. Therefore it is a complex problem to determine where and when the contractile apparatuses are

[1] Present address: Chiyogaoka 1-107-1001, Chikusa-ku, Nagoya 464, Japan.

constructed, in addition to how they work in these cells. The ABPs are the key to solving these problems.

Cytoplasmic streaming in characean cells as well as in other plant cells has been recently reviewed by Kuroda (1990), Higashi-Fujime (1991), and Nagai (1993). In this chapter, I would discuss passive cytoplasmic streaming in *Physarum* plasmodium. Ameboid movements of other cells are also included in the discussion.

II. Cytoplasmic Streaming and Actin Cytoskeleton in *Physarum* Plasmodium

Plasmodia of *Physarum polycephalum* can be cultivated on the surfaces of filter papers by supplying oats each day. They grow into large plasmodia anywhere from several to several tens of centimeters in size. The plasmodium has no cell wall, but is surrounded by a slime layer more than 2 μm thick. A plasmodium advancing on a substratum, such as an agar plate, assumes a fan-like shape (1 × 5 cm in size in a Petri dish; see Fig. 1). The anterior region of the plasmodium consists of a sheet of cytoplasm. The frontal region of the plasmodial sheet repeats contraction and relaxation every 1.5–3 min at room temperature. The streaming of the cytoplasmic sol (endoplasm) originates in the frontal region and develops into microscopic streams in the intermediate region. The posterior region is constituted of a network of plasmodial strands (50 μm–2 mm in diameter). Thus plasmodium is a huge mass of cytoplasm composed of a cortical gel layer (ectopolasm) and cytoplasmic sol (endoplasm).

Invagination systems of the cell membrane are extensively differentiated in the ectoplasm (Hoffmann *et al.*, 1981; Achenbach and Wohlfarth-Bottermann, 1981). The rate of cytoplasmic streaming reaches more than 1 mm/sec at maximum and the direction of the streaming reverses every 1.5–3 min, which is the same time taken by the contraction cycle of the cortical gel layer. This type of cytoplasmic streaming is called shuttle streaming (Kamiya, 1959). The plasmodium moves on a substratum at a rate of 2–3 cm/hr in a definite direction, for example, in the direction of food. The structure and contractility of *Physarum* plasmodium have been reviewed by Komnick *et al.* (1973), Kessler (1982), and recently by Stockem and Brix (1994).

A. Passive Cytoplasmic Streaming in Plasmodium

Kamiya and Kuroda (1958) analyzed the velocity distribution of cytoplasmic streaming in plasmodial strands. The rate of the streaming is greatest and most constant in the region near the central axis of the streaming, but it decreases to zero at the inner surface of the gel wall, giving a flattened truncated parabola for the profile of the rate distribution. Thus, the cytoplasm flows like a bloc of endoplasm in the center of the streaming region. This indicates that the flow of endoplasm is non-Newtonian or that the endoplasm shows a structural viscosity. The non-Newtonian nature of endoplasm was also shown in cytoplasmic streaming in a giant ameba, *Chaos chaos* (Allen and Roslansky, 1959) and in an artificially induced flow of endoplasm isolated from internodal cells of *Nitella* (Kamiya and Kuroda, 1965). The viscosity coefficient of endoplasm varies according to the shearing force (force applied/area of moving plane). There is a yield value which is the minimum shearing force causing the flow. In the center of the streaming region, the shearing forces are less than the yield value of the endoplasm.

Kamiya and Kuroda (1958) applied a pressure difference (up to 15 mm, H_2O) between the two terminal ends of a plasmodial strand, when the cytoplasmic streaming in the strand ceased. The difference induced a passive flow of endoplasm in the strand. It was shown that the profile of the rate distribution of the flow is a truncated parabola which is essentially identical to that of the natural cytoplasmic streaming in the strand. Kamiya and Kuroda concluded that the cytoplasmic streaming in the plasmodial strand is caused passively by a pressure difference between the anterior and posterior ends of the strand.

B. Cyclic Contraction of Plasmodial Strand

When a segment of plasmodial strand is excised from the network of plasmodial strands and suspended by a fine glass hook in a moist chamber, it begins to contract and relax after 10 to 20 min. The cycle of contraction and relaxation repeats every 1.5–3.0 min, which is the same time as the shuttle cytoplasmic streaming in the plasmodium. The contraction is around one-tenth of the total length (Kamiya *et al.*, 1972). The isometric or isotonic tension of the plasmodial segment can be measured by high-sensitivity tension meters (Kamiya *et al.*, 1972; Kamiya and Yoshimoto, 1972; Kamiya, 1979; Wohlfarth-Bottermann, 1975, 1977). When the length of the segment is kept constant (isometric contraction), the tension oscillates between 6 and 13 mg (i.e., between 18 g/cm^2 and 35 g/cm^2 for a strand 220 μm in diameter in this case; Kamiya *et al.*, 1972). When the strand is stretched quickly to 10–20% of its total length, the isometric tension suddenly increases to a higher level and then decreases to the original level (Nagai *et al.*, 1978; Aschenbach, 1982).

Matsumura *et al.* (1980) reconstituted actomyosin threads from purified

Physarum F-actin and myosin in the molar ratio of 1 : 1 (protein concentration, 15 mg/ml). The thread contracts on addition of Mg-ATP. The extent of the contraction is dependent on the ATP concentration in the medium, and the contraction is reversible in relation to the ATP concentration. The maximum and half-maximum isometric tensions are generated at the ATP concentrations of 10 μM and 2.4 μM, respectively, under the physiological conditions of plasmodium (30 mM KCl, 5 mM $MgCl_2$, pH 7.0). The maximum isometric tension of the actomyosin thread is as high as 10 g/cm^2 at 10 μM of ATP. This value is comparable to the isometric tension of plasmodial strands. The sudden increase and decrease in isometric tension can be seen in quick stretching up to 2.5% of the total thread length.

These facts demonstrate that the tension involved in the contraction of the plasmodial strand is generated by the interaction of actin and myosin in the presence of Mg-ATP. There appears to be a periodic cycle in the interaction of myosin heads with the actin filaments. The sudden increase in isometric tension by quick stretching of the strand indicates strong associations of myosin heads with actin filaments which resist the stretching. The decrease in tension followed by the sudden increase in tension reflects a fast kinetic cycle of association-dissociation of myosin heads with actin filaments by which the rearrangement of myosin heads on actin filaments occurs within a short time (see the scheme in Fig. 8). Yoshimoto and Kamiya (1978a) showed that the isometric tension of a segment of posterior plasmodial strand gradually increases to a higher level yet reveals no significant oscillation within 10–20 min. The isometric tension of the strand begins to oscillate around the tension, which is higher than the initial level 30 min after the excision. Next, Yoshimoto and Kamiya (1978b) excised a rectangular piece from the anterior plasmodial sheet. The anterior plasmodial piece began to contract and relax without a lag phase after the excision. Yoshimoto and Kamiya (1978b) concluded that the posterior plasmodial strands maintain a stable internal pressure. The internal pressure in the anterior region becomes higher or lower than the posterior region according to the contraction or relaxation of the frontal region of plasmodium. The pressure difference induces a passive streaming of the endoplasm from the anterior to the posterior region or vice versa.

C. Actin Cytoskeleton in Plasmodium

Actin filaments are organized into two kinds of actin cytoskeleton—cytoplasmic fibrils and a cortical actin layer. Both structures are located in the ectoplasm of plasmodium. It has been shown that the cytoplasm isolated from the endoplasm of plasmodium contracts in artificial media

(Kuroda, 1979), suggesting that an actin cytoskeleton which is contractile is also present in the endoplasm (Section II,C,4). Salles-Passador *et al.* (1991) recently demonstrated networks of microtubules in plasmodium, but the domains of these microtubule networks are distinct from the microfilamentous domains.

1. Cytoplasmic Fibrils

Wohlfarth-Bottermann (1962) first observed bundles of microfilaments in *Physarum* plasmodium by electron microscopy. The bundles of microfilaments differentiate into cytoplasmic fibrils of light microscopic dimension (Wohlfarth-Bottermann, 1963). The microfilaments are specifically decorated with muscle heavy meromyosin (HMM) to form arrowhead-like structures, indicating that the microfilaments are actin filaments (Allera *et al.*, 1971). Localization of myosin molecules on the cytoplasmic fibrils was shown by immunofluorescence using antibodies against myosins (Osborn *et al.*, 1983). Thus, the cytoplasmic fibrils in plasmodium were found to be actomyosin fibrils which consist of bundles of actin filaments decorated with myosin molecules. However, long myosin filaments resembling thick filaments in striated muscle are not found in the cytoplasmic fibrils (Nagai and Kato, 1975; Fleischer and Wohlfarth-Bottermann, 1975; Osborn *et al.*, 1983).

Cytoplasmic fibrils are absent in the endoplasm, but differentiate in accordance with the transformation from the endoplasm to the ectoplasm (Achenbach and Wohlfarth-Bottermann, 1981). The cytoplasmic fibrils are associated with membrane invaginations which develop extensively in the ectoplasm of plasmodial strands and the border between the ectoplasm and the endoplasm in the anterior plasmodial sheet (Wohlfarth-Bottermann, 1974).

a. Contractility of Cytoplasmic Fibrils The contractility of cytoplasmic fibrils has been shown in cryosections of plasmodial strands stained with nitrobenzooxadiazol (NBD)-labeled phalloidin (Pies and Wohlfarth-Bottermann, 1984). Phalloidin is a reagent which specifically binds actin filaments. Ishigami and Hatano (1986) prepared Triton cell models of thinly spread plasmodia either in the contraction or relaxation phase. Many cytoplasmic fibrils differentiate in the contraction phase of plasmodia and they are well preserved in the cell models (Fig. 1). The cytoplasmic fibrils contract rapidly with Mg-ATP, resulting in aggregates at several points of the fibril. However, the contraction of the entire cell model is slow and the extent of the contraction is small (about 10% in length). On the other hand, few if any cytoplasmic fibrils are present in the plasmodial relaxation phase. The cell models prepared from its relaxation phase show almost

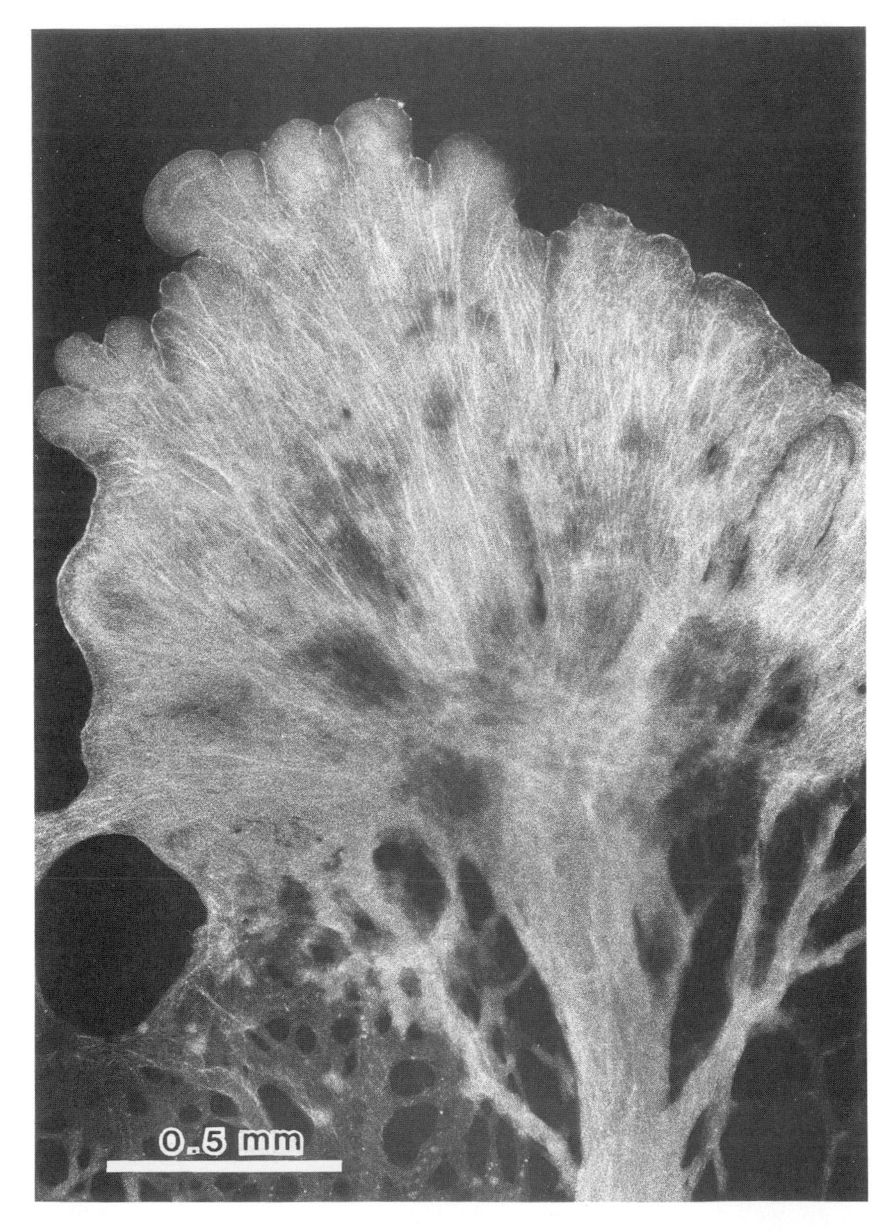

FIG. 1 A fluorescence micrograph showing cytoplasmic fibrils in the contraction phase of *Physarum* plasmodium. Many fibrils differentiate in the contraction phase. However, a few short and slender fibrils are seen in the relaxation phase. (Photo courtesy of M. Ishigami.)

no contraction on addition of Mg-ATP. Thus, the contractility of the cell models is regulated in accordance with the cycle of contraction and relaxation of plasmodium.

Based on the biochemical results, two possibilities may be considered for the regulation of contractility of the cell models. Ogihara *et al.* (1983) prepared two kinds of actomyosin (natural actomyosin), both active and inactive, from *Physarum* plasmodium and found that the heavy chain of myosin is phosphorylated in active actomyosin. Active and inactive actomyosins convert each other by phosphorylation and dephosphorylation of myosin heavy chains (Section III,B). Sugino and Matsumura (1983) demonstrated that fragmin (a Ca^{2+}-sensitive F-actin-severing protein from *Physarum* plasmodium; Section IV, B,2) reduces the isometric tension of *Physarum* actomyosin threads to 15–30% of the control tension at Ca^{2+} concentrations of more than 10^{-6} *M*. Electron micrographs of the threads showed that actin filaments in the threads become shorter in the presence of Ca^{2+}. It is thought that actin filaments are severed by fragmin into short pieces in the relaxation phase. However, there is no evidence at present to indicate the dephosphorylation of myosin heavy chains or the fragmentation of actin filaments in the relaxation phase of plasmodium.

b. Spatial Localization of Cytoplasmic Fibrils Spatial localization of cytoplasmic fibrils in plasmodia migrating unidirectionally was shown by immunofluorescence using antibodies raised in rabbits against sodium dodecyl sulfate (SDS)-denatured chicken gizzard actin (Naib-Majani *et al.*, 1982, 1984), chicken-gizzard myosin heavy chain, *Physarum* myosin and fragmin (Osborn *et al.*, 1983), and by NBD-labeled phallacidin (Ishigami, 1986) (Fig. 1). Phallacidin is a reagent which specifically binds actin filaments.

Naib-Majani *et al.* (1982) showed that the spatial localization of cytoplasmic fibrils continuously changes in migrating plasmodia. Some fibrils are oriented parallel to the cell membrane in the frontal region of a plasmodial sheet. In the intermediate region of a plasmodial sheet, fibrils form polygonal networks in which the fibrils are interconnected with focal nodes which are not stained with actin antibody. Fibrils are aligned parallel to the anterior–posterior axis in the posterior region. In plasmodial strands, helically twisted fibrils are located in the gel tubes (see Fig. 1). The cytoplasmic fibrils are specifically stained with antibodies against myosin and fragmin (Osborn *et al.*, 1983). As mentioned later, stress fibers in tissue-cultured cells are stained periodically with antibodies of the actin-associated proteins such as myosin, tropomyosin, and α-actinin (Section II, C, 5). However, cytoplasmic fibrils in plasmodium are stained continuously with antibodies of myosin and fragmin (Osborn *et al.*, 1983).

c. Differentiation of Cytoplasmic Fibrils Wohlfarth-Bottermann (1964, 1965) initially demonstrated that cytoplasmic fibrils are abundant when plasmodial strands contract, but scarce when the strands expand. In the 1970s, electron microscopy made it clear that cytoplasmic fibrils differentiate in the contraction phase of plasmodial strands under isometric conditions (Wohlfarth-Bottermann and Fleischer, 1976; Nagai *et al.,* 1978), whereas cytoplasmic fibrils transform into other states, dense networks (Nagai *et al.,* 1978), or plaques (Fleischer and Wohlfarth-Bottermann, 1975; Wohlfarth-Bottermann and Fleischer, 1976) in the contraction phase under isotonic conditions.

Cytoplasmic fibrils in thinly spread plasmodia have been found to show birefringence. The birefringent fibrils appear and disappear according to the cycle of contraction and relaxation of plasmodia (Nakajima and Allen, 1965; Kamiya, 1973; Ogihara and Kuroda, 1979; Ishigami *et al.,* 1981; Naib-Majani *et al.,* 1984). This birefringence is caused by oriented actin filaments in the fibrils. The birefringence is positive in relation to the longitudinal direction of the fibril. The coefficient of birefringence, (n_e ^-n_o), the measured retardation divided by the thickness of the fibril, is around 2.3×10^{-3}, which is comparable to birefringence of the I-band in a myofibril (Sato *et al.,* 1981). Unfortunately, it is difficult to observe birefringent fibrils in a tension meter because the strands are too thick.

Ishigami (1986) prepared thin-spread plasmodia. The cytoplasmic fibrils were visualized by staining with NBD-phallacidin after the plasmodia were fixed and permeabilized. It was possible to confirm the preservation of cytoplasmic fibrils by a polarizing microscope after plasmodia fixation. The same investigator showed that slender and flabby fibrils develop from the inside of the cell membrane and nodes in the early stage of the contraction. They develop into thicker and straight cytoplasmic fibrils in the contraction stage. The straight fibrils form the polygonal networks and parallel arrays which Naib-Majani *et al.* (1982) demonstrated by immunofluorescence. These straight fibrils grow up to several hundreds of microns in length and 10 μm in diameter. Most cytoplasmic fibrils are located in the dorsal cortical layer, although a few are present in the ventral cortical layer. The cytoplasmic fibrils degenerate gradually and eventually disappear almost completely, remaining only around the nodes. These observations give the impression that the inside of the membrane and the nodes work as the nucleation sites for the development of cytoplasmic fibrils. Ishigami also reported that the fibril pattern in plasmodial strands in the posterior region does not change, an observation consistent with that of Yoshimoto and Kamiya (1978b), who noted that the posterior plasmodial strands do not show the periodic contraction *in situ.*

To determine the relation between the differentiation of the cytoplasmic fibrils and the cycle of contraction and relaxation of plasmodium in detail, Ishigami *et al.* (1987) prepared thin, minute, plasmodia of discoid or ellipsoid shape, 100–200 μm in size. There is little cytoplasmic streaming in most of these small plasmodia, but the plasmodia inhibit a contraction and relaxation which can be monitored by measuring changes of the areas of plasmodia spread between two thin agar plates. The areas are about 9% smaller during the contraction. The patterns of birefringent fibrils in the contraction and relaxation cycle were recorded on micrographs and analyzed by an image processor. The change in birefringence was monitored by measuring the total area of birefringent fibrils in the plasmodia. It was revealed that cyclic change of the area of a plasmodium and that of the birefringence in the plasmodium proceed in phase. The phase difference is mostly within ± one-sixteenth of a cycle. Since the birefringence is caused by oriented actin filaments in the cytoplasmic fibril, it is concluded that the contraction of plasmodium is coupled with the parallel alignment of actin filaments in the cytoplasmic fibrils.

Ishigami and Hatano (1986) indicated that about 60% of the total actin of plasmodia is Triton-insoluble and preserved in the cell models. In contrast, virtually all myosin (about 90%) is preserved in the cell models. There were no appreciable differences in the amounts of actin and myosin between the cell models prepared from the contraction and relaxation phases of plasmodia. However, the cell models included the posterior regions where the birefringence does not change in accordance with the contraction and relaxation cycle of plasmodium (Ishigami, 1986), so that these investigators could not exclude the possibility that the polymerization and depolymerization of actin are involved in the contraction and relaxation cycle of plasmodium (Isenberg and Wohlfarth-Bottermann, 1976). It is safe to say at present that most of the actin filaments are oriented in parallel to form a bundle of actin filaments in the contraction phase and then disintegrate in the dispersed state of the relaxation phase.

Some birefringent fibrils differentiate into large fibrils 10 μm in diameter and 500 μm in length, as mentioned earlier. The density of actin filaments in the cytoplasmic fibril is estimated to be about $1.5 \times 10^3/\mu m^2$ (Ishigami *et al.*, 1981). If the length of actin filaments is assumed to be 1 μm, these birefringent fibrils are composed of 5.5×10^6 actin filaments. It is surprising that such an enormous number of actin filaments are aligned to form single fibrils and then dispersed every 1.5 to 3 min.

2. Cortical Actin Layer

Naib-Majani *et al.* (1982) showed by immunofluorescence that a continuous actin layer, several microns in diameter, is closely associated with the membrane in *Physarum* plasmodium. Brix *et al.* (1987) further revealed by electron microscopy that the cortical actin layer consists of a rather irregular meshwork of actin filaments and is continuous along the entire cell surface, including all invaginations. They pointed out that the cortical actin layer in plasmodium is strikingly similar to the cortical layer in amebas, although cytoplasmic fibrils are not present in amebas. In plasmodium, cytoplasmic fibrils extend through the cytoplasmic matrix and often terminate at the membrane in focal spikes resembling the focal contacts in tissue-cultured cells. Brix *et al.* (1987) discussed the possibility that the cytoplasmic fibrils help to fasten the plasmodium to a substratum and the cortical actin layer is responsible for generating motive force. This possibility has also been discussed in the case of tissue-cultured cells (Kreis and Birchmeier, 1980; Burridge, 1981).

3. Birefringent Structures in Caffeine Drops

Caffeine drops are droplets of plasmodium 10–200 μm in diameter which can be obtained by treating plasmodium with a 5–10 m*M* caffeine solution at 25°C (Hatano, 1970). Treatment with adenosine or adenine produces similar plasmodial droplets (Nachmias, 1979; Nachmias and Meyers, 1980). In caffeine drops, the granular cytoplasm separates from the membrane, and the hyaline layer appears between the membrane and the granular cytoplasm. The granular cytoplasm shows active cytoplasmic streaming, such as a reversal fountain type of streaming (Kuroda, 1979), and/or repeats the cycle of the contraction and relaxation that accompanies the formation of birefringent structures (Hatano, 1970; Sato *et al.*, 1981) (Fig. 2). The mean value of the coefficient of birefringence ($n_e - n_o$), the measured retardation divided by the thickness of the birefringent structure, is 2.3×10^{-3}, which is the same order as that of the cytoplasmic fibrils in plasmodium. Electron micrographs of the birefringent structures revealed parallel alignments of microfilaments in the structures.

Achenbach *et al.* (1979) reported that 5 m*M* caffeine inhibits the formation of the membrane invaginations in plasmodium. The cytoplasmic fibrils are associated with membrane invaginations in plasmodium that is not treated with caffeine (Wohlfarth-Bottermann, 1974). Achenbach *et al.* assumed that caffeine interrupts the association of cytoplasmic fibrils with the cell membrane, so that the formation of the membrane invaginations is inhibited.

Caffeine also interrupts the association of the cortical actin layer with the cell membrane. Kukulies *et al.* (1984) and Kukulies and Stockem

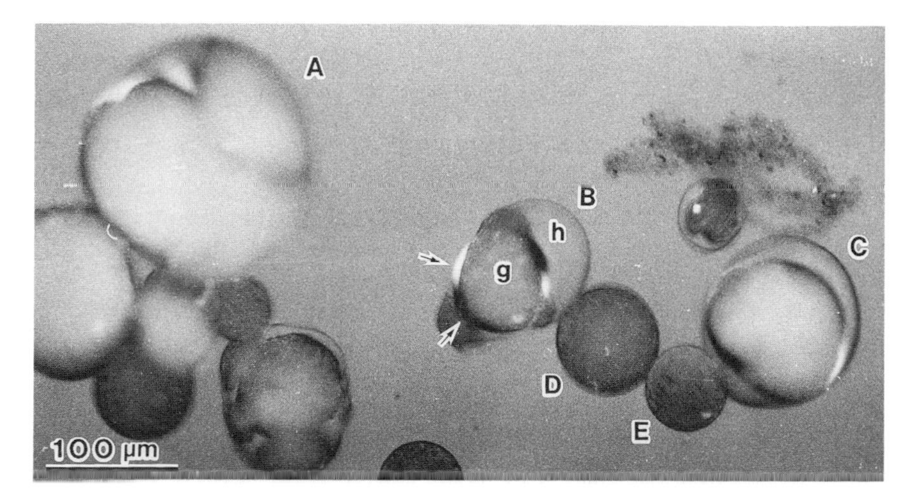

FIG. 2 Birefringent structures in plasmodial caffeine drops. In caffeine drop A, the upper left part of the granular cytoplasm is contracting. In caffeine drops B and C, the entire granular cytoplasm contracts. The cap-shaped or circular birefringent structure appears in between the hyaline layer and the granular cytoplasm. Birefringent structures disappear after contraction of granular cytoplasm. In caffeine drops D and E, the granular cytoplasm relaxes and expands in the entire drop. No birefringent structure is appreciable in the relaxed caffeine drops. h, hyaline layer; g, granular cytoplasm; arrows, birefringent structures. (Reprinted from Sato *et al.*, 1981, by permission of Springer-Verlag.)

(1985) microinjected fluorescein-labeled G-actin into caffeine drops. The injected G-actin was first evenly distributed in the whole caffeine drop, but soon incorporated into the thin actin layer beneath the membrane. When the granular cytoplasm began to contract, the thin actin layer which enveloped the granular cytoplasm detached from the membrane. Both the intensity of the fluorescence of the thin actin layer and the latter's thickness increased during the contraction of the granular cytoplasm. After the contraction, the actin layer disappeared and a new actin thin layer appeared beneath the membrane during relaxation of the granular cytoplasm. Electron micrographs revealed that the thin actin layer is composed of meshworks of actin filaments. Some actin filaments are oriented parallel to one another. Thus, the actin thin layer features coincide structurally with those of the cortical actin layer of plasmodium (Kukulies *et al.* 1984; Kukulies and Stockem, 1985).

It should be mentioned that when a high concentration of phalloidin is microinjected into caffeine drops, the drops show normal dynamic activities for more than 20 hr (Kukulies *et al.*, 1984), indicating that the actin layer is repeatedly constructed from phalloidin-loaded actin filaments and disintegrates into actin filaments in the contraction and relaxation cycle of the granular cytoplasm. As mentioned earlier, phalloidin specifically binds actin filaments to inhibit the depolymerization of the filaments.

On the other hand, birefringence appears on the surface of the granular cytoplasm after it begins to contract. No birefringence can be seen beneath the membrane when the granular cytoplasm expands throughout the entire caffeine drop (Sato *et al.*, 1981) (Fig. 2). Taking into account the observation of birefringence in caffeine drops together with the results of the microinjection experiments (Kukulies *et al.*, 1984; Kukulies and Stockem, 1985), it is likely that actin filaments incorporate into the actin layer beneath the membrane to form three-dimensional meshworks of actin filaments. When the contraction of the actin layer begins, the filaments are oriented parallel to the surface of the granular cytoplasm. The parallel alignment of actin filaments proceeds in accordance with the contraction of the actin layer. The dynamics of these actin filaments are seen in superprecipitation of *Physarum* actomyosin *in vitro* (see Fig. 7) and are discussed in Section V,B.

4. Granular Cytoplasm

It has been shown that granular cytoplasm itself contracts in plasmodial caffeine drops (Kuroda, 1979) and in naked (membraneless) cytoplasmic masses isolated from amebas (*Chaos carolinensis*) (Taylor *et al.*, 1973) and *Physarum* plasmodium (Kuroda, 1979). It is thought that the contraction of granular cytoplasm is a kind of syneresis, because it has been shown in caffeine drops that water is squeezed out from the granular cytoplasm to the hyaline layer during the contraction of the granular cytoplasm (Kuroda, 1979).

Taylor and his colleagues (Taylor *et al.*, 1973) isolated membraneless masses of cytoplasm from amebas (*Chaos carolinensis*). The cytoplasmic masses are *in vitro* cell models of the granular cytoplasm because the hyaloplasm has been diffused out from the cytoplasmic masses, so that the granular cytoplasm is suspended in artificial solutions. The same authors examined the contractility of the cytoplasmic masses in solutions which mimic the conditions for contraction, relaxation, and rigor of striated muscle. The cytoplasm contracts in the contraction solution containing Ca^{2+} of more than 7.0×10^{-7} *M*. In the solution containing the threshold concentration of Ca^{2+} (7.0×10^{-7} *M*), the cytoplasm flows outward and returns to the cytoplasmic mass to form many loops on the surface of the mass in the solution. The protein concentration of the cytoplasm increases at the tip of the loop from the level of cytoplasmic sol to that of cytoplasmic gel in ameba cells. This means that granular cytoplasm contracts at the tip of the loop to pull the cytoplasm from the cytoplasmic mass. Taylor *et al.* (1973) emphasized that the consistency is maintained even when the cytoplasm is in the state of sol in the solution containing the threshold concentration of Ca^{2+}, and the contraction of the cytoplasm is coupled with the sol-gel transformation of the cytoplasm (Section IV,B,3).

Kuroda (1979) also isolated membraneless cytoplasmic masses from the endoplasm of plasmodium. The cytoplasmic mass radiates numerous moving lobes in a solution containing 1 m*M* Mg-ATP and 55 m*M* EGTA. The cytoplasm flows outward and returns to the cytoplasmic mass in the lobes. When the cytoplasm reaches the top of the lobe, filaments appear intermittently within a period of 10 sec. Filaments contract to form bundles at the base of the lobe. Electron microscopic studies revealed that there are many actin filaments in the cytoplasm of the lobe and they aggregate progressively into bundles at the base of the lobe. The same author speculated that the contraction of cytoplasmic fibrils pulls the cytoplasm toward the base of the lobe. It should be noted that the granular cytoplasm contracts in the artificial media and bundles of actin filaments are formed during the contraction. The dynamics of these actin filaments are discussed in Section V,B.

5. Comparison with Actin Cytoskeletion in Tissue-Cultured Cells

I mentioned in the introduction that the ameboid movement of tissue-cultured cells could be classified in the same category of cell motility as cytoplasmic streaming in *Physarum* plasmodium. This idea comes from studies on actin cytoskeletons in tissue-cultured cells, which have made rapid progress in the past 20 years. The following paragraphs summarize the studies on actin cytoskeletons of tissue-cultured cells which are necessary for the discussion in this chapter.

The actin cytoskeleton of tissue-cultured cells is composed of cytoplasmic fibers termed stress fibers and a cortical actin layer (Buckley, 1981; Byers *et al.,* 1984, for review). Stress fibers exhibit birefringence (Goldman *et al.,* 1975), suggesting parallel alignment of subfilaments in the fibers. It was shown by electron microscope that stress fibers are bundles of microfilaments about 7.5 nm in diameter (Buckley and Porter, 1967). The microfilaments are decorated with muscle heavy meromyosin (HMM) to form structures on the filaments that resemble arrowheads, indicating that microfilaments are actin filaments (Ishikawa *et al.,* 1969). In spread cells, stress fibers localize in the dorsal and ventral cell cortices and run in diverse directions, focusing at the cell poles (Buckley and Porter, 1967). The stress fibers together with less parallel arrays of actin filaments form a sheath of actin filaments in the cortex that envelops the subcortical cytoplasm (Buckley and Porter, 1967; Spooner *et al.*, 1971).

It has been shown that stress fibers are not directly attached to the cell membrane but rather to electron-dense plaques (focal contacts or adhesion plaques) immediately beneath the cell membrane (see Geiger *et al.*, 1987 and Burridge *et al.*, 1987, for review). It can be said that the actin cytoskeleton in tissue-cultured cells is identical to that of *Physarum* plasmodium. The presence of focal contacts in plasmodium is hypothesized (Brix *et al.*, 1987), but the structure and the chemical components of the focal contacts in plasmodium remain unknown.

The stress fibers are stained continuously by indirect immunofluorescence using antibodies against actin (Lazarides and Weber, 1974; Owaribe and Hatano, 1975). In contrast, immunofluorescence of myosin (Weber and Groeschel-Stewart, 1974; Fujiwara and Pollard, 1976), tropomyosin (Lazarides, 1975), and α-actinin (Lazarides and Burridge, 1975) show a periodical localization of these proteins in the stress fibers. Double staining for these proteins confirms that myosin and tropomyosin localize in the same periodic units, while α-actinin localizes in other units. The two units (myosin units and α-actinin units) are arranged alternately in the stress fiber.

The polarity of actin filaments in the stress fibers was demonstrated by labeling with HMM by Begg *et al.* (1978), and Sanger and Sanger (1980). Actin filaments which attach to the cell membrane show uniform polarity in that the arrowheads always point away from the membrane. When cells are fixed with tannic acid together with glutaraldehyde for electron microscopy, the stress fibers show a banded pattern of alternating electron-dense and electron-lucid bands, each measuring up to 0.16 μm in length. HMM-labeled actin filaments are arranged in parallel and antiparallel orientations in stress fibers, and individual actin filaments extend over a band pattern. Sanger and Sanger (1980) presented a model for the arrangements of actin filaments based on the banded pattern of stress fibers and the polarity of the actin filaments.

No myosin filaments similar to the thick filaments in striated muscle have been observed in stress fibers. Langanger *et al.* (1986) demonstrated with double immunogold labeling that myosin is localized in the electron-lucid bands, while α-actinin is localized in the intervening electron-dense bands. Antibodies which are reactive with the head portion of nonmuscle myosin and light chains labeled two opposite bands adjacent to the band of α-actinin (the electron-dense band), while antibodies to light meromyosin (the tail of myosin molecule) labeled a single central zone of the electron-lucid band. Langanger *et al.* concluded that myosin (myosin-II) is organized into short bipolar filaments with a central bare zone in the electron-lucid bands of stress fibers.

The contractility of stress fibers has been shown in glycerinated cell models (Isenberg *et al.*, 1976), and in digitonin- or Triton-treated cell models (Kreis and Birchmeier, 1980; Masuda *et al.*, 1983). Some Triton cell models show a Ca^{2+}-dependent contraction on addition of Mg-ATP. It was found that the contraction is regulated through Ca^{2+}-calmodulin-dependent phosphorylation of the myosin 20-kDa light chain (Masuda *et al.*, 1984).

Stress fibers are abundant in well-spread cells on substratum, but there are few, if any, in locomoting cells (Herman *et al.*, 1981). In addition, Kreis and Birchmeier (1980) demonstrated that microinjected fluorescein-labeled α-actinin incorporates into stress fibers, giving the periodical banded pattern along the stress fibers. When the cells are pretreated with glycerin or digitonin, the stress fibers contract up to 25% on addition of Mg-ATP. Shortening proceeds in the bands which are not stained with α-actinin. However, the contraction of the stress fibers could not be induced by microinjection of Ca^{2+} or ATP into living cells. These facts demonstrated that stress fibers are contractile, but they generate isometric tensions between focal contacts where the cells adhere to the substratum (Burridge, 1981).

It should be noted that cell models further shrink to less than 50% of the original length after the contraction of the stress fibers (Kreis and Birchmeier, 1980). As mentioned above, the meshwork structure of actin filaments in the cortical actin layer envelops the subcortical cytoplasm. This observation suggests that the cortical actin layer is also contractile and responsible for the contraction of the cell models.

Thus, the structure and contractility of the actin cytoskeleton in tissue-cultured cells are essentially identical to those of the actin cytoskeleton in *Physarum* plasmodium. These facts strongly suggest that there is a common molecular mechanism underlying the amoboid movement of tissue-cultured cells and the cytoplasmic streaming in plasmodium.

III. Actin and Myosin

A. Actin

1. Structure of G-Actin

Actin in nonmuscle cells was first isolated from plasmodium of *Physarum polycephalum* in the 1960s (Hatano and Oosawa, 1966; Adelman and Taylor, 1969; Pollard, 1981). We isolated actin from crude extracts of plasmodium using its specific binding to rabbit striated myosin and obtained partially purified *Physarum* actin (Hatano and Oosawa, 1966). The *Physarum* G-actin polymerizes to filaments on addition of 0.1 *M* KCl. It was shown by electron microscope that the actin filaments are microfilaments 6 nm in diameter having two-stranded helical structures (Hatano *et al.,* 1967; Totsuka and Hatano, 1970).

The amino acid sequences of muscle and nonmuscle actins are very similar (Vandekerckhove and Weber, 1978a, 1978b). For example, *Physarum* actin (375 amino acid residues) differs from rabbit skeletal muscle actin in 35 residues (9.3%), and from mammalian cytoplasmic γ-actin in only 17 residues (4.5%). Thus actin is a very conservative protein. In 1990, Kabsch *et al.* succeeded in determining the three-dimensional structure of rabbit striated muscle G-actin by X-ray analysis of the complex of G-actin and bovine pancreatic deoxyribonuclease (DNase I) at an effective resolution of 0.28 nm. The G-actin molecule has a large and a small domain, each of which has two subdomains. There is a cleft between the large and small domains, and one molecule of ATP and one molecule of Ca^{2+} are located in the cleft.

2. Binding Sites of Actin-Binding Proteins

The binding sites of the actin-binding proteins on the actin sequence have been examined by chemical cross-linking methods. As far as we can tell, many types of actin-binding proteins bind a limited surface of the actin molecule. The amino-terminal (N-terminal) and carboxyl-terminal (C-terminal) segments participate in the binding of many actin-binding proteins, including myosin (Sutoh, 1982), profilin (Vandekerckhove *et al.,* 1989), depactin (Sutoh and Mabuchi, 1984), cofilin (Muneyuki *et al.,* 1985), fragmin (Sutoh and Hatano, 1986), gelsolin (Doi *et al.,* 1987; Sutoh and Yin, 1989), α-actinin, and actinogelin (Mimura and Asano, 1987). Both segments are located in the subdomain 1 of the small domain. It has been shown that fragmin caps the barbed end of the actin filament. Fragmin cannot sever actin filaments when myosin is associated with the filaments (Hasegawa *et al.,* 1980). These results coincide with the fact that fragmin and myosin have common binding sites on the surface of subdomain 1, which is located at the barbed end of the actin filament (Holmes *et al.,* 1990).

3. Polymerization of G-Actin and Structure of F-Actin

Oosawa and coworkers first demonstrated that the polymerization of G-actin is a kind of condensation process. After polymerization, the critical concentration of G-actin coexists with F-actin (Kasai *et al.,* 1962). The polymerization process consists of at least two processes, nucleation and elongation. Nucleation is the rate-limiting process for actin polymerization (Kasai *et al.*, 1962; Kasai, 1969; Oosawa and Asakura, 1975). It is now well recognized that these physical properties of the actin polymerization are very important for understanding the roles of actin–binding proteins in actin polymerization *in vitro* and *in vivo* (Section IV, B).

Actin filaments have a two-stranded helical structure (Hanson and Lowy, 1963). They are decorated with muscle HMM to form arrowhead structures with pointed and barbed ends. Thus, each actin filament has a polarity (Huxley, 1963). These filaments elongate from the two terminal ends, but the elongation from the barbed ends is much faster than from the pointed ends (Woodrum *et al.*, 1975; Kondo and Ishiwata, 1976). Bonder *et al.* (1983) determined the elongation rates and the critical concentrations at the barbed ends and pointed ends under physiological conditions (75 m*M* KCl and 5 m*M* $MgCl_2$, pH 7.2, 22°C). They used bundles of actin filaments isolated from *Limulus* sperm as nuclei and measured the elongation of actin filaments from the two ends of the actin bundles by electron microscopy. The critical concentration was determined from the concentration of G-actin below which the actin filaments do not elongate from the barbed or pointed ends of the bundles.

According to Bonder *et al.* (1983), when the G-actin concentration is 2.8 μM, actin filaments elongate at the rates of 0.073 and 0.0083 μm sec^{-1} at the barbed end and the pointed end, respectively. This means that it takes about 12 sec for actin filaments to elongate 1 micron. This would explain the elongation of actin filaments in the cells. They calculated the rate constants at the two ends: k^+(B) and k^+(P) are 12.3 and 1.5 [molecules sec^{-1} μM^{-1} (μM; concentration of G-actin)], and k^-(B) and k^-(P) are 2.0 and 0.7 (molecules sec^{-1}), respectively. k(B) and k(P) are the respective rate constants at the barbed end and the pointed end. The critical concentrations of the barbed end and pointed ends are 0.1 μM and 0.5 μM, respectively. In a similar way, Korn and co-workers (Korn *et al.*, 1987) obtained the values of 0.1 μM and 4 μM for the critical concentrations of the barbed end and the pointed end, respectively (see Wegner and Isenberg, 1983; Pollard and Mooseker, 1981; Pollard, 1986).

There must be an equilibrium between G-actin and F-actin in cells. It is necessary to shift the equilibrium to induce the polymerization of G-actin or the depolymerization of F-actin in the cells. Two possible ways to induce the polymerization or depolymerization in the cells can be considered at present. (1) G-actin-sequestering proteins inactivate G-actin. This could control the concentration of the active state of G-actin. (2) Capping proteins cap the barbed ends of actin filaments and this induces the change in the critical concentration of G-actin (Section IV, B,1). Relatively recent studies on the kinetics of actin polymerization have been reviewed by Korn (1982), Neuhaus *et al.* (1983), Frieden (1985), Pollard and Cooper (1986), and Korn *et al.* (1987).

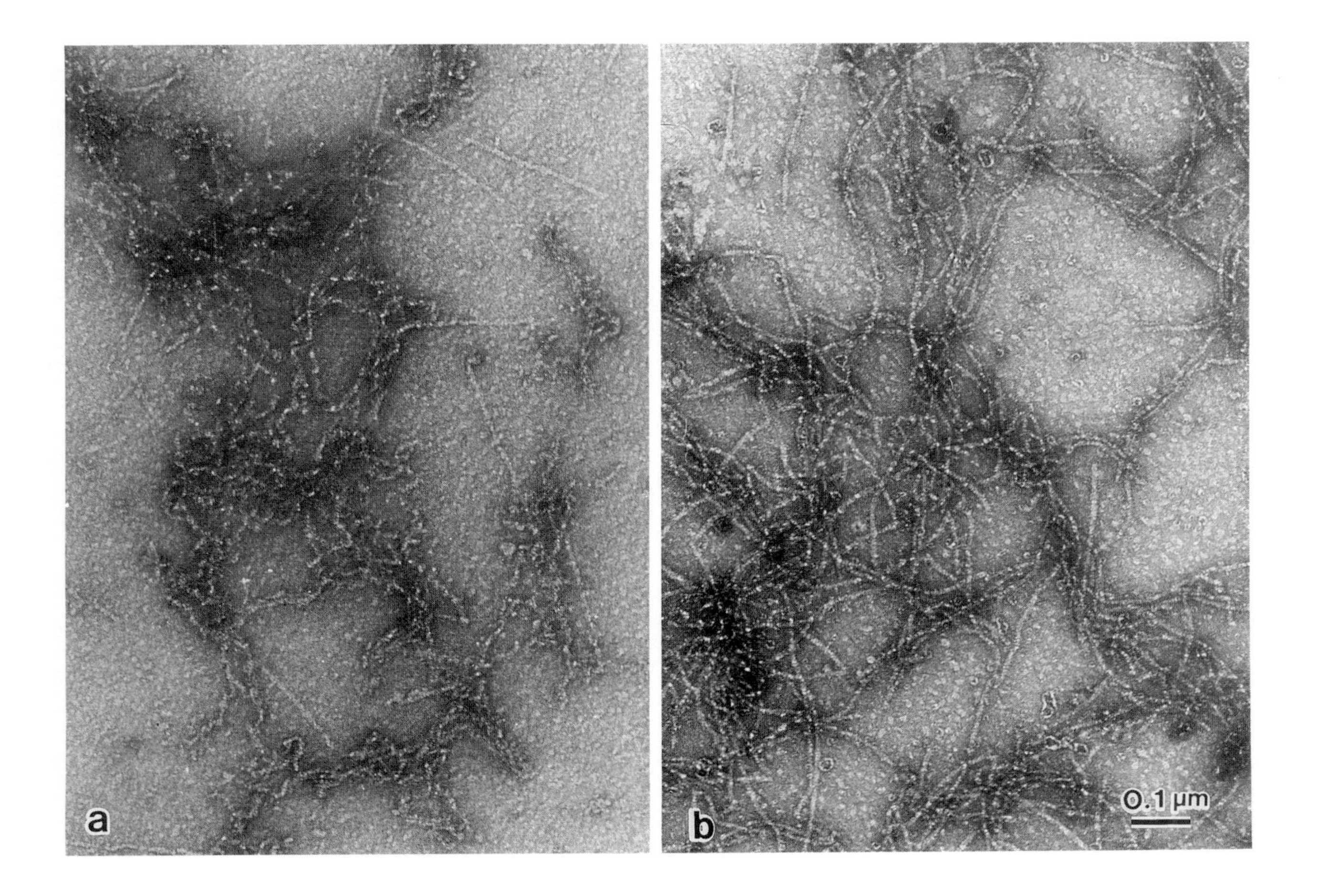
a
b
0.1 μm

4. Mg-Polymer of *Physarum* Actin

Hatano and Oosawa used the specific binding of actin to muscle myosin to isolate *Physarum* actin from the crude extract in 1966. Actin was extracted from acetone-dried actomyosin, which is the complex of *Physarum* actin and muscle myosin. However, actin extracted from acetone-dried actomyosin does not polymerize to F-actin, even when 0.1 *M* KCl and 2 m*M* $MgCl_2$ are added to the extract. It seems that G-actin polymerizes to globular polymers with sedimentation coefficients of around 30S (cf. 3.1S for G-actin, 100–200S for F-actin). We further purified the actin by isoelectric precipitation followed by fractionation with ammonium sulfate. The partially purified G-actin polymerizes to F-actin ("KCl-F-actin") on addition of 0.1 *M* KCl, the viscosity of which is a little lower than that of striated muscle F-actin (Hatano and Oosawa, 1966). However, the G-actin was found to polymerize to another state of polymer (Mg-polymer) on addition of 2 m*M* $MgCl_2$. Mg-polymer has a much lower viscosity than F-actin polymerized by 0.1 *M* KCl and shows ATPase activity at room temperature (Hatano *et al.*, 1967; Hatano and Totsuka, 1972). Fujime and Hatano (1972) demonstrated by quasi-elastic scattering of laser light that Mg-polymer is more flexible than KCl–F-actin. The flexibility of Mg-polymer changes according to the concentration of ATP (0.1–0.6 m*M*) in the medium containing 2 m*M* $MgCl_2$ and 0.1 *M* KCl. The higher the ATP concentration, the less flexible the Mg-polymer. The transformation of Mg-polymer is reversible as to the ATP concentration. Figure 3 shows electron micrographs of Mg-polymers in the 0.05 m*M* ATP solution (Fig. 3a). The Mg-polymers are transformed into less flexible polymers which are very similar in appearance to KCl–F-actins after addition of 2 m*M* ATP (Fig. 3b) (Hatano, 1972).

Oosawa and co-workers suggested that in F-actin each monomer is bound to four neighboring monomers by at least two different kinds of bonds. If some of the bonds are broken or weakened under appropriate conditions, a partially linear polymer having a higher flexibility will be

FIG. 3 Electron micrographs of Mg-polymer showing the conformational change after addition of ATP. (a) Before addition of ATP. (b) After addition of 2 m*M* ATP in the presence of 2 m*M* $MgCl_2$ and 0.1 *M* KCl. *Physarum* actin was partially purified using specific binding of *Physarum* actin to muscle myosin: G-actin was extracted from actomyosin which was synthesized from *Physarum* actin and muscle myosin in crude extracts from plasmodia and fractionated by ammonium sulfate. Partially purified G-actin polymerized to Mg-polymer on addition of 2 m*M* $MgCl_2$, but it polymerized to F-actin on addition of 0.1 *M* KCl. Note that flexible Mg-polymers (a) were transformed into less flexible polymers (b) very similar in appearance to F-actin after addition of 2 m*M* ATP. (Reprinted from Hatano, 1972, by permission of Gordon and Breach Science Publishers.)

formed (Oosawa and Kasai, 1962; Oosawa and Asakura, 1975). Hatano (1972) presented a model of Mg-polymer in which one of the bonds is broken, so that the Mg-polymer is more flexible than F-actin.

Hatano and Owaribe (1976, 1977) further purified *Physarum* actin. The purified G-actin polymerized to F-actin on addition of 0.1 *M* KCl or 2 m*M* $MgCl_2$. The viscosity of F-actin is nearly identical to that of F-actin purified from striated muscle. No Mg-polymer is formed on addition of 2 m*M* $MgCl_2$ to purified G-actin. These results indicated that any regulatory factors for the actin polymerization are present in the partially purified actin preparation. Maruyama and coworkers verified that purified muscle G-actin polymerized to Mg-polymer, when β-actinin, a capping protein isolated from striated muscle, is added to purified muscle G-actin (Kamiya *et al.,* 1972; Maruyama, 1973; Shimaoka *et al.,* 1991).

Actin filaments form meshworks and bundles of actin filaments in the cortical gel layer of plasmodium (Section II,C). However, no actin filaments have been observed in the endoplasm in spite of extensive studies by electron microscope (Achenbach and Wohlfarth-Bottermann, 1981). In the case of tissue-cultured cells, meshworks of short actin filaments have been observed in the subcortical cytoplasm (the endoplasm) (Buckley, 1981). If actin is present as flexible or globular polymers in the endoplasm of plasmodium, it would be very difficult to observe them by electron microscopy because the endoplasm contains many granules and vacuoles. Condeelis and Taylor (1977) observed aggregates of short actin filaments in extracts of *Dictyostelium* amebas, when the extracts were warmed under relaxation conditions. The aggregates of short actin filaments appear very similar to those of Mg-polymer. Taylor and Condeelis (1979) suggested that flexible polymers of actin might be formed from G-actin by warming the extracts (Section IV, B,3).

B. Myosin

Myosin is the most important actin-binding protein whose binding is specifically regulated by ATP. Myosin shows very low Mg-ATPase at low ionic strengths, but the Mg-ATPase is highly activated by F-actin. The liberated energy of ATP is converted into the mechanical energy for cell motility. Myosin has been isolated from a variety of eukaryotic cells from ameba to tissue cells of mammals (see Warrick and Spudich, 1987; Korn and Hammer, 1988; Kiehart, 1990, for review). Two kinds of myosin have been identified in nonmuscle cells. One is the conventional myosin which has two globular heads with a long rod-like tail, as is seen in striated muscle myosin. Another is the minimyosin which has one head with a short tail. These myosins are now called two-headed myosin (or myosin-

II) and single-headed myosin (or myosin-I). The ATPase and the actin-binding sites are located in the head regions in both myosins. The two-headed myosins form bipolar filaments at low ionic strengths, whereas the single-headed myosin is monomeric at low ionic strengths. The two myosins seem to have different roles in cell motility.

1. *Physarum* Myosin

Myosin was isolated from *Physarum* plasmodium and characterized by Hatano and Tazawa (1968), Hatano and Ohnuma (1970), and Adelman and Taylor (1969). This was the first isolation of myosin (myosin-II) from nonmuscle cells. *Physarum* myosin has two globular heads with a long rod-like tail (Hatano and Takahashi, 1971). The tail (170 nm) is longer than striated muscle myosin (150 nm). This is consistent with the fact that the molecular weight of the rodlike tail (140 kDa) of *Physarum* myosin is larger than that of the tail (120 kDa) of striated muscle myosin (Kohama, 1987). The myosin molecule consists of pairs of heavy chains (230 kDa), phosphorylatable light chains (P-Lc, 18 kDa), and Ca-binding light chains (Ca-Lc, 14 kDa) (Kohama, 1987).

a. Formation of Thick Filaments *Physarum* myosin is unique in its solubility in the salt solution. It is soluble and forms only small aggregates with the sedimentation coefficients of up to 15S at low ionic strengths (0.03–0.15 *M* KCl, pH 7.0), including the physiological salt concentration of plasmodium (0.03 *M* K^+) (Hatano and Takahashi, 1971). Nachmias (1972) further purified *Physarum* myosin by a gel filtration in the presence of KI. Purified myosin forms short bipolar filaments up to 0.45 μm long in the presence of milimolar concentrations of $CaCl_2$ or $MgCl_2$. Hinssen *et al.* (1978) and D'Haese and Hinssen (1979) observed another type of *Physarum* myosin filament measuring 0.4 to 4 μm in length in the presence of 2 to 10 m*M* $MgCl_2$ at low ionic strengths. Myosin heads arrange along both sides of the entire filament with 14 nm periodicity. The cross bridges of both sides point in opposite directions. At the two terminal ends of the filament, bare zones about 0.2 μm long are observed. This side polar arrangement of the myosin heads has been observed in the assembly of smooth muscle myosin (Sobieszek, 1972).

Matsumura and Hatano (1978) found that thick filaments of *Physarum* myosin which have been formed at low ionic strengths in the presence of 6 m*M* $MgCl_2$ are immediately broken by addition of *Physarum* F-actin. Myosin binds actin filaments to form actin filaments decorated with myosin molecules. This clearly indicates that the affinity among myosin molecules in the thick filament is much weaker than that of myosin molecules for actin filaments. When ATP is added to the mixture solution, myosin molecules separate from the actin filaments. Short bipolar filaments 0.2 to 0.4 μm in length appear in the clearing phase of superprecipitation. This phase corresponds to the relaxation phase in the contraction and relaxation cycle of plasmodium, so I think that myosin exists as short bipolar filaments in the relaxation phase of plasmodium.

b. Regulation of Filament Formation by Phosphorylation Ogihara *et al.* (1983) prepared two kinds of actomyosin (natural actomyosin), inactive and active, from *Physarum* plasmodium. Inactive actomyosin has low superprecipitation activity and the superprecipitation is Ca^{2+}-insensitive. Active actomyosin, on the other hand, exhibits a very high superprecipitation activity in the absence of Ca^{2+}, and its activity is reduced by addition of Ca^{2+}. Active actomyosin converts to the inactive type when incubated with potato acid phosphatase. Inactive actomyosin converts to the active type by incubation with ATP-γ-S. The heavy chain of active myosin is phosphorylated in the ratio of 1 mole Pi/mole of heavy chain. In the same study, Ogihara *et al.* isolated phosphorylated and dephosphorylated myosins from these actomyosins. The Mg-ATPase of phosphorylated myosin is activated seven to eightfold by F-actin in the Ca^{2+}-sensitive way, whereas dephosphorylated myosin is not activated by F-actin. Ogihara *et al.* found that phosphorylated myosin forms bipolar filaments 0.4 to 0.5 μm long at low ionic strength in the presence of 5 m*M* $MgCl_2$, while dephosphorylated myosin does not form filaments. It should be noted that the effect of phosphorylation of heavy chains of *Physarum* myosin is the inverse of the cases of *Acanthamoeba* and *Dictyostelium* myosins (myosin-II; Section III,B,2 and 3).

c. Ca^{2+}-Sensitivity of Actomyosin Actomyosin, which was synthesized from *Physarum* myosin purified by Hatano and Ohnuma (1970) and F-actin, shows no Ca^{2+} sensitivity; it superprecipitates in the presence and absence of Ca^{2+}. In caffeine drops, the granular cytoplasm displays the active movement described above. When Ca^{2+} is removed from the caffeine solution, the movement ceases and the granular cytoplasm expands in the entire caffeine drop. The threshold concentration of Ca^{2+} was determined in caffeine solutions containing a Ca^{2+}-EGTA buffer and found to be about 10^{-6} *M*, which is on the same order as that for muscle contraction (Hatano, 1970; Sato *et al.* 1981). This suggests that some factors which give Ca^{2+} sensitivity to actomyosin are present in plasmodium.

Kohama *et al.* (1980) succeeded in preparing Ca^{2+}-sensitive actomyosin (natural actomyosin) from *Physarum* plasmodium. They found that a reducing reagent such as mercaptoethanol should be included in the solutions during the preparation. Purified actomyosin (natural actomyosin) shows superprecipitation more slowly in the presence of Ca^{2+} above 10^{-6} *M* than in the absence of Ca^{2+}. Thus, Ca^{2+} inhibits superprecipitation. Kohama and Kendrick-Jones (1986) isolated myosin from Ca^{2+}-sensitive actomyosin. According to Kohama (1987), purified *Physarum* myosin is phosphorylated in the molar ratio of 4.0–6.8. The major phosphorylation sites are in the heavy chain, although P-Lc is also phosphorylated. Mg-ATPase of the myosin is activated 30-fold by either purified *Physarum* or striated muscle F-actin in the absence of Ca^{2+}, but this actin-activated ATPase is reduced to around one fifth on addition of 50 μM Ca^{2+} (Kohama and Kendrick-Jones, 1986).

Kohama *et al.* (1987) found that *Physarum* myosin binds 2 moles of Ca^{2+}/1 mole in a solution containing 3.5 mM Mg^{2+}, 0.03 M KCl, EGTA–Ca buffer (pCa 5.5–8), pH 7.5. The K_D is about 3 μM, which is comparable with that of scallop myosin. Purified Ca-Lc binds Ca^{2+}. Kobayashi *et al.* (1988) determined the amino acid sequence of Ca-Lc and found two E-F hand consensus sequences in the sequence.

Actin-activated ATPase of *Physarum* myosin is inhibited by micromolar levels of Ca^{2+} half maximally under the physiological conditions of *Physarum* plasmodium (0.2–0.5 mM ATP, 1 mM Mg^{2+}, and 30 mM KCl at pH 6.9; Kohama *et al.*, 1984). Kohama concluded that Ca^{2+} regulation of *Physarum* actomyosin is mediated mainly through the binding of Ca^{2+} to Ca-Lc of myosin, and this causes the inhibition of actomyosin ATPase (Kohama, 1985, 1987; Kohama *et al.*, 1992). This inhibitory effect of Ca^{2+} was unexpected, because Hatano (1970) and Sato *et al.* (1981) had demonstrated that Ca^{2+} activates the movement of the granular cytoplasm in caffeine drops. This seeming contradiction of Ca^{2+} effects has not been clarified.

2. *Acanthamoeba* Myosin

A soil ameba, *Acanthamoeba castellanii,* and a cellular slime mold, *Dictyostelium discoideum,* are also useful organisms for biochemical studies, since both amebas can be easily cultured in laboratories, and large amounts of them can be used as starting materials for isolation of the proteins. Important results on nonmuscle myosins have been obtained from studies on myosins isolated from these amebas.

a. Single-Headed Myosin (Myosin-I) Pollard and Korn (1973a) isolated an enzyme with K^+-EDTA ATPase from *Acanthamoeba castellanii.* The enzyme was considered to be *Acanthamoeba* myosin, because purified enzyme binds F-actin in an ATP dependent way, and the Mg-ATPase is activated by F-actin. It has a single head with a short tail. The length of the tail is about 10 nm (Pollard *et al.*, 1991). The molecule consists of a single heavy chain of 140 kDa and two light chains of 16 kDa and 14 kDa, giving the molecular weight of native myosin as 180 kDa. It is soluble and monomeric at high and low ionic strengths (Pollard and Korn, 1973a, 1973b; Maruta *et al.*, 1979). For activation of Mg-ATPase by F-actin, a cofactor is necessary (Pollard and Korn, 1973b), and it was identified as a heavy-chain kinase (Maruta and Korn, 1977a; Hammer *et al.*, 1983). Thus, the actin-activated ATPase of the myosin is regulated by phosphorylation of the heavy chain (Albanesi *et al.*, 1983; Brzeska *et al.*, 1989).

This single-headed myosin is called myosin-I to distinguish it from the conventional two-headed myosin (myosin-II), which was also discovered in *Acanthamoeba* (see later discussion). Recent reports indicate that single-headed myosins are widely distributed in eukaryotic cells (see Pollard *et al.*, 1991, for review).

There are at least three or four isomers in *Acanthamoeba,* and they have been termed myosin IA, myosin IB, myosin IC, etc. (Maruta *et al.*, 1979; Lynch *et al.*, 1989). The amino acid sequences of the heavy chains of myosins IB and IC have been determined. The amino acid sequences of the N-terminal 80-kDa segments of myosin-I are similar to the subfragment 1 of muscle myosin. About 55% of the amino acid sequence of each myosin-I is identical to the corresponding sequence of the muscle myosin subfragment 1 (myosin IA, Jung *et al.*, 1987; myosin IB, Jung *et al.*, 1989; Brzeska *et al.*, 1989). These sequences are also retained in the heads of heavy chains of myosin-II (Warrick and Spudich, 1987; Pollard *et al.*, 1991). In other words, myosin-I consists of minimyosin, which is equivalent to the head piece (subfragment 1) of conventional myosin (myosin-II). The ATPase site and the ATP-sensitive actin-binding site are located in the head regions of the heavy chains of all myosins.

Myosin-I has a C-terminal extension with a membrane binding site (Doberstein and Pollard, 1989). *Acanthamoeba* myosin-I binds membranes isolated from *Acanthamoeba* and pure lipid vesicles. Vesicles isolated from *Acanthamoeba* move along actin cables *in vitro,* and the movement is inhibited with anti-myosin-I (Adams and Pollard, 1986). It is considered that myosin-I is a motor protein for the transportation of vesicles and granules within the cells. Interestingly, *Acanthamoeba* myosin-I isozymes have an ATP-insensitive actin-binding site which is located in a C-terminal segment of the heavy chains (Lynch *et al.*, 1986; Brzeska *et al.*, 1988), allowing them to cross-link actin filaments and induce superprecipitation of F-actin (Fujisaki *et al.*, 1985).

b. Two-Headed Myosin (Myosin-II) Maruta and Korn (1977b) and Pollard *et al.* (1978) isolated a myosin from *Acanthamoeba* that is similar to myosin from rabbit striated muscle, and termed this conventional myosin "myosin-II." It has two globular heads with a tail 90 nm in length. The molecule of myosin-II consists of two heavy chains of 175 kDa and four light chains (two each of 17.5 kDa and 16.5 kDa) that give a native molecular weight of 400 kDa (Maruta and Korn, 1977b; Pollard *et al.*, 1978).

Each heavy chain of *Acanthamoeba* myosin-II contains three phosphorylatable sites located near the tail end of the heavy chain. When four to six sites are phosphorylated *in vitro* and *in vivo,* myosin shows no actin-activated ATPase, while dephosphorylated myosin has the maximum ATPase activity (Côté *et al.*, 1981; Collins *et al.*, 1982; Collins and Korn, 1980, 1981). Dephosphorylated myosin-II forms bipolar filaments up to 230 nm in length at low ionic strengths (10–100 m*M* KCl). The filament is composed of eight myosin-II molecules. In the presence of milimolar concentrations of divalent cations or at acidic pH, the bipolar filaments associate laterally to form thicker filaments. However, no long filaments resembling thick filaments of muscle myosin are formed (Pollard, 1982; Sinard *et al.*, 1989; Sinard and Pollard, 1990). Phosphorylated myosin II also forms bipolar filaments, the sedimentation coefficient of which is smaller than that of bipolar filaments of dephosphorylated myosin. For the lateral association of the bipolar filaments, more concentrated divalent cations are necessary (Sinard and Pollard, 1989; Kuznicki *et al.*, 1983).

3. *Dictyostelium* Myosin

Clarke and Spudich (1974) isolated and characterized two headed myosin from *Dictyostelium discoideum. Dictyostelium* myosin-II is composed of two heavy chains of 210 kDa and pairs of two light chains of 18 kDa and 16 kDa (Clarke and Spudich, 1974).

The actin-activated ATPase is increased when the heavy chain is dephosphorylated by a specific phosphatase isolated from the amobas. When the heavy chain is phosphorylated again by a specific kinase isolated from the amebas, both ATPase and filament formation are inhibited (Kuczmarski and Spudich, 1980). Dephosphorylated myosin-II forms long thread-like filaments. Phosphorylated myosin-II forms bipolar filaments of uniform length (0.43 μm) when the pH is lowered or the magnesium concentration is increased to 10 m*M* (Kuczmarski *et al.*, 1987).

In myosin-II isolated from *Dictyostelium* amebas, the heavy chains are phosphorylated in the molar ratio of 0.3 mole Pi/mole of 210 kDa heavy chain. The 18-kDa light chains are also phosphorylated in the ratio of 1 mole Pi/1 mole of the light chain (Kuczmarski and Spudich, 1980). The 18-kDa light chain can be phosphorylated and dephosphorylated by myosin light chain kinase and myosin light chain phosphatase purified from *Dictyostelium* amebas, respectively. The actin-activated ATPase is increased when the light chain is phosphorylated. Thus, the actin-activated ATPase is directly related to the phosphorylation of the light chain. However, the formation of myosin filament is independent of the extent of the phosphorylation of the 18-kDa light chain (Griffith *et al.*, 1987). The activity of myosin light chain kinase is not Ca^{2+}–calmodulin dependent in the case of *Dictyostelium* myosin, and it is not known how the light chain kinase is regulated (Griffith *et al.*, 1987).

The amino acid sequences of the heavy chains of *Dictyostelium* and *Acanthamoeba* myosin IIs have been determined from the DNA sequence analysis of the cloned genes and compared with those of nematode and mammalian muscle myosins (Warrick *et al.*, 1986; Hammer *et al.*, 1987). Specific regions of the heads are highly conserved. However, there are some differences in the sequence of the tail between *Dictyostelium* myosin-II and muscle myosins. This might reflect the difference in the formation of the thick filaments in both myosins (Warrick *et al.*, 1986; Hammer *et al.*, 1987).

4. Roles of Single- and Two-Headed Myosin

a. Single-Headed Myosin (Myosin-I) The physiological roles of myosin-I are not clear at present. One possibility is that myosin-Is are motor proteins which transport vesicles and granules in the cells. This is supported by *in vitro* motility assays indicating that myosin-I-bound lipid vesicles and myosin-I-coated plastic beads move unidirectionally along the bundles of actin filaments of *Nitella* cell in the presence of Mg-ATP (Adams and Pollard, 1986; Albanesi *et al.*, 1985).

Fukui *et al.* (1989) applied antibodies specific to myosin-I and myosin-II for the immunofluorescence of *Dictyostelium* amebas. Fukui *et al.* revealed that myosin-I localizes near the membrane in the front region of the migrating ameba. When the cell engulfs bacteria, myosin-I together with actin filaments concentrate beneath the phagocytotic vesicles. On the other hand, myosin-II distributes in the rear region of the ameba where the endoplasm contracts. Myosin-I binds lipid vesicles and purified cell membranes (Adams and Pollard, 1989; Miyata *et al.*, 1989) so it is assumed that myosin-I is responsible for the pseudopod extension and endocytosis in the front region of amebas (Pollard *et al.*, 1991).

b. Two-Headed Myosin (Myosin-II) Kiehart *et al.* (1982) microinjected anti-*Asterias* egg myosin (myosin-II) into one of the blastomeres at the two-cell stage of development. The microinjection of antimyosin does not affect spindle elongation and the anaphase chromosome movement in the blastomere. However, the contractile ring is not formed, so the cytokinesis of the injected blastomere is inhibited. De Lozanne and Spudich (1987) and Knecht and Loomis (1987) succeeded in preparing strains of *Dictyostelium* which were deficient in myosin-II. Surprisingly, the amebas formed pseudopods and moved on the substratum. Despite the fact that the nuclear division occurred normally, cytokinesis was completely inhibited. These experiments clearly demonstrated that myosin-II is responsible for the formation of the contractile ring for cytokinesis.

As mentioned above, myosin-II forms bipolar filaments *in vitro*. The formation of myosin-II filaments is regulated by the phosphorylation of heavy chains, suggesting that myosin-II shows dynamic behavior in cells. Many workers have attempted to observe myosin filaments in nonmuscle cells by immunofluorescence and electron microscopy. However, myosin filaments are hardly observed. These facts probably suggest that myosin filaments, if present, are so labile that the filaments are broken during the preparation of specimens for immunofluorescence and electron microscopy. Yumura and Fukui (1985) succeeded in staining myosin filaments in amebas of *Dictyostelium* by an improved method of immunofluorescence. The filament dimensions are similar to those of myosin filaments formed *in vitro*. Reines and Clarke (1985) succeeded in preparing monoclonal antibodies which specifically reacted with monomeric and polymeric *Dictyostelium* myosin-IIs, respectively. They quantified the amounts of filamentous myosin in Triton-extracted cytoskeletons of *Dictyostelium* amebas. It was shown that all myosin exists in filamentous form in the cytoskeletons of amebas.

IV. Actin-Binding Proteins

A. Isolation and Characterization of Actin-Binding Proteins

Crude extracts from nonmuscle cells contain a variety of actin-binding proteins which inhibit actin polymerization. These proteins can be fractionated by anion-exchange chromatography (Fig. 4). Many actin-binding proteins have been purified from crude extracts of nonmuscle cells and characterized *in vitro*. Actin-binding proteins could be classified into several groups based on the characteristic properties examined *in vitro* (see Schliwa, 1981; Weeds, 1982; Korn, 1982; Craig and Pollard, 1982; Hatano *et al.*, 1983; Stossel *et al.*, 1985; Mabuchi, 1986; Pollard and Cooper, 1986; Hartwig and Kwiatkowski, 1991, for review). However, there are many actin-binding proteins which have not been identified in the crude extract from nonmuscle cells (see Fig. 4). Therefore, systematic studies on isolation of actin-binding proteins are still necessary to determine the wide variety of roles for actin-binding proteins *in vivo*.

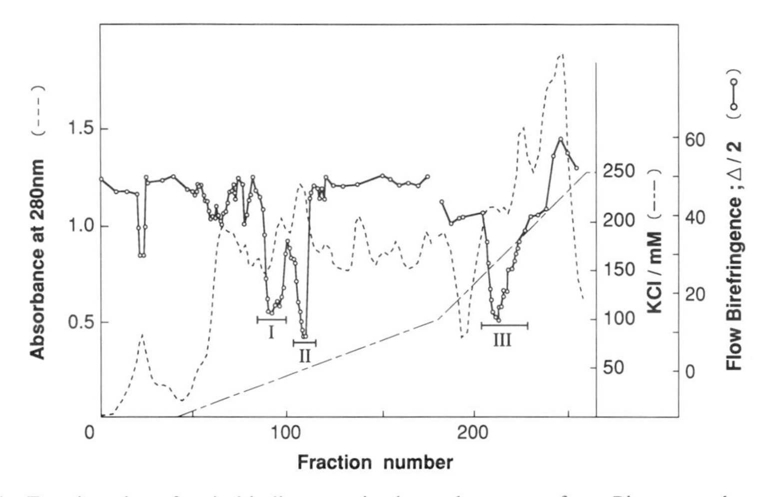

FIG. 4 Fractionation of actin-binding proteins in crude extracts from *Physarum* plasmodia. Actin-binding proteins were fractionated by DEAE-cellulose column chromatography. Their activity in each fraction was expressed by flow birefringence, which was measured after muscle G-actin was mixed in each fraction (final G-actin concentration, about 1 mg/ml) and polymerization was induced on addition of 0.1 *M* KCl and 2 m*M* $MgCl_2$. When the fraction contained any capping or F-actin-severing proteins, the flow birefringence of the mixture solution became smaller than that of the control F-actin solution. When the fraction contained G-actin-sequestering proteins such as profilin, the flow birefringence after a short incubation (e.g., 3–5 min) became smaller because the proteins in particular inhibited the nucleation process for actin polymerization. Fractions I, II, and III contained fragmin 60 (60 kDa), fragmin (42 kDa), and the complex of fragmin and G-actin, respectively. Profilin was eluted just before fraction II, and F-actin bundling protein was eluted in the passthrough fractions. Note that unidentified actin-binding proteins reduced the flow birefringence of the fractions. (From Furuhashi and Hatano, 1989.)

B. Roles of Actin-Binding Proteins in Nonmuscle Cells

In this section the structures and functions of several groups of actin-binding proteins are summarized and their possible roles in a variety of cell functions are discussed.

1. Nonpolymerized Actin

It has been shown that more than 50% of the total actin of nonmuscle cells is extracted by a dilute nonionic detergent such as 0.1–1% of Triton X-100 solution. It is speculated that the Triton-soluble actin exists in unpolymerized forms in the cells. Protein factors which inhibit actin polymerization and depolymerize F-actin have been isolated and characterized. DNAase I (Hitchcock *et al.*, 1976; Mannherz *et al.*, 1976), vitamin

D-binding protein (Baelen *et al.*, 1980), profilin (Carlsson *et al.*, 1977), and thymosin β_4 (Safer *et al.*, 1990, 1991) are G-actin-sequestering proteins. The proteins bind monomeric actin to form 1 : 1 complexes which cannot polymerize. On the other hand, depactin (Mabuchi, 1981, 1983), actin depolymerizing factor (ADF) (Bamburg *et al.*, 1980; Giuliano *et al.*, 1988), and cofilin (Nishida *et al.*, 1984; Yonczawa *et al.*, 1985) are F-actin-sequestering proteins. The molecular weight of these proteins ranges from 17.5 kDa to 18.5 kDa. The proteins bind F-actin and depolymerize F-actin much faster than G-actin-sequestering proteins.

It has been shown that phosphorylation of G-actin by kinases (Grazi and Magri, 1979, Sonobe *et al.*, 1986) and ADP ribosylation of G-actin by a bacterial toxin (Aktories *et al.*, 1986) inhibit actin polymerization. It is thought that the critical concentration of G-actin, the complexes of G-actin with G-actin- and F-actin-sequestering proteins, phosphorylated and ADP ribosylated G actin are possibly unpolymerized actin in nonmuscle cells. The concentrations of profilin and/or thymosin β_4 are much higher than other protein levels in nonmuscle cells. Therefore, at present, these proteins are considered to be the main regulators of actin polymerization in nonmuscle cells (Safer *et al.*, 1991).

a. Profilin Profilin was first isolated from calf spleen by Carlsson *et al.* (1977) and its presence has been confirmed in eukaryotic cells from ameba to mammalian cells (see Haarer and Brown, 1990, for review). Ozaki *et al.* (1983) isolated profilin from *Physarum* plasmodium. Takagi *et al.* (1990) determined the amino acid sequence of *Physarum* profilin. *Physarum* profilin is composed of 125 amino acid residues (13,132 Da). The amino acid sequence is similar to *Acanthamoeba* profilin isomers (51% identical) and to yeast profilin (42% identical), but is less similar to calf spleen and human platelet profilins. The C-terminal region is similar to a segment of *Physarum* fragmin (Ampe and Vanderckhove, 1987). It contains a pentapetid sequence LADYL, which is a variant of a motif LTDYL. The variants of LTDYL sequence have been shown to be common among many actin-binding proteins, including gelsolin, fragmin, etc. (Tellam *et al.*, 1989).

Ozaki *et al.* (1983) showed that profilin isolated from plasmodium reduces the rates of nucleation and elongation of the polymerization of *Physarum* G-actin and reduces the extent of polymerization as well. The apparent critical concentration for the actin polymerization increases with the addition of profilin. These results can be explained by a simple G-actin sequestration model (Tobacman and Korn, 1982; Tseng and Pollard, 1982; Ozaki *et al.*, 1983) in which *Physarum* profilin binds *Physarum* G-actin to form a 1 : 1 complex which cannot polymerize. The G-actin concentration available for actin polymerization decreases. In the steady state, equilibria exist among F-actin, G-actin (of the critical concentration), profilin, and the G-actin–profilin complex:

$$\text{F-actin} \rightleftharpoons \text{G-actin} + \text{profilin} \rightleftharpoons K_D \text{ G-actin–profilin complex}$$

The dissociation constant (K_D) for the profilin–G-actin complex can be calculated from the initial rate of actin polymerization, the rate of elongation of F-actin nuclei, and the apparent critical concentration in the presence of various concentrations of *Physarum* profilin under the physiological conditions of *Physarum* plasmodium, that is, in a solution containing 50 m*M* KCl or 50 m*M* KCl and 1 m*M* $MgCl_2$ (pH 7.0). The dissociation constant was estimated to be in a range of 1.4–3.7 μM. Tobacman *et al.* (1983) reported that the dissociation constant for the complex of *Acanthamoeba* profilin and *Acanthamoeba* actin is between 4 and 9 μM. It is possible to calculate the F-actin concentration in equilibria. The total concentration of profilin in plasmodium is on the same order as that of actin (the actin concentration, 50 μM; Nachmias, 1979; 220 μM; D'Haese and Hinssen, 1979) and the critical concentration of *Physarum* G-actin under physiological ionic conditions (50 m*M* KCl and 1 m*M* $MgCl_2$) is 0.6 μM (Ozaki *et al.*, 1983). If we assume that the total concentrations of actin and profilin are 50 μM, respectively, the F-actin concentration is calculated to be 37 μM (74% of total actin).

Tobacman and Korn (1982) pointed out that a small change in critical concentration might cause a large change in F-actin concentration. For example, actin filaments have polarity, and the critical concentrations at two ends of actin filaments are different. The critical concentrations of the barbed end and the pointed end of *Physarum* actin filaments are about 0.1μM and 1.5 μM, respectively (Ozaki *et al.*, 1983) (Section III, A,3). If the barbed ends of actin filaments are capped by any capping protein, the critical concentration increases to 1.5 μM, thereby causing a decrease in F-actin concentration to 28 μM (56% of total actin).

Pollard and Cooper (1984) presented another model to explain the effects of *Acanthamoeba* profilin on the polymerization of *Acanthamoeba* G-actin. *Acanthamoeba* profilin shows little effect on the extent of actin polymerization under physiological ionic strength (1 m*M* $MgCl_2$ and 50 m*M* KCl). The dissociation constant estimated from the apparent critical concentration and the steady-state actin monomer concentration are 12.5 μM and 55 μM, respectively. Assuming that the actin and profilin concentrations are 100 μM in *Acanthamoeba* (Tobacman and Korn, 1982) and the dissociation constant is 55 μM, the F-actin concentrations are estimated to be 99 μM and 96 μM at the critical concentrations of 0.18 μM and 1.5 μM, respectively, so that *Acanthamoeba* profilin shows almost no effects on the F-actin concentration in the steady state of *Acanthamoeba* actin polymerization, even when the critical concentration changes. The same investigators also showed that *Acanthamoeba* profilin inhibits nucleation more strongly than the elongation of actin filaments. The inhibition of elongation occurs far more often at the pointed end than at the barbed end. They explained these results by a complex model which includes two different dissociation constants, one for profilin binding the actin monomer and one for profilin binding the barbed end of the actin filament. The affinity for the actin filament was much lower (K_D, 50–100 μM) than the affinity for the actin monomer (K_D, 5 μM).

b. Thymosin β_4 Safer *et al.* (1990) reported that the bulk of the unpolymerized actin of resting human platelets is a complex with a small peptide (5 kDa). They subsequently purified the peptide from calf thymocyte and deduced it to be thymosin β_4, which is a thymic hormone (Safer *et al.*, 1991). Thymosin β_4 binds muscle G-actin and forms 1 : 1 complexes which cannot polymerize. The dissociation constant of the complex is estimated to be 2 to 3 μM, which is on the same order as the G-actin–profilin complex. The platelet contains 600 μM actin, and 50% of the total actin is estimated to be unpolymerized. Platelets contain 600 μM of thymosin β_4 at six times the concentration of profilin (100 μM). The authors concluded that thymosin β_4 serves as a major G-actin-sequestering protein in platelets. The other mammalian tissues contain relatively high concentrations of thymosin β_4. The concentrations of thymosin β_4 are much higher than the other G-actin-sequestering protein factors such as profilin.

c. Phosphorylated G-Actin Grazi and Magri (1979) reported that G-actin as well as F-actin is phosphorylated by isolated liver plasma membranes. Phosphorylated G-actin does not polymerize. On the other hand, Ohta *et al.* (1987) found that G-actin is phosphorylated by protein kinase C and cAMP-dependent kinase. Phosphorylated G-actin polymerizes to F-actin, but G-actin phosphorylated by protein kinase C polymerizes faster than unphosphorylated G-actin, whereas G-actin phosphorylated by cAMP-dependent kinase polymerizes slower than unphosphorylated G-actin.

Sonobe *et al.* (1986) found that actin is phosphorylated in crude extracts from *Amoeba proteus* in a Ca^{2+}-dependent way; that is, actin is phosphorylated in 10^{-4} M Ca^{2+}, but not in 10^{-6} M and 10^{-8} M Ca^{2+}. They demonstrated that G-actin purified from *Amoeba proteus* is phosphorylated by a kinase(s) partially purified from crude extracts. Phosphorylated G-actin does not polymerize. The kinase does not phosphorylate *Amoeba* F-actin, indicating that the phosphorylation site(s) of *Amoeba* G-actin is masked in F-actin. The kinase phosphorylates about 34% of G-actin even in a dilute G-actin solution (0.4 μM). This is due to the rapid polymerization of G-actin in the reaction solutions, which contain 0.4–2 mM $MgCl_2$.

Sonobe *et al.* demonstrated that when profilin purified from *Amoeba proteus* is added to the reaction solutions, more than 80% of G-actin is phosphorylated. Phosphorylated G-actin can be separated from profilin by chromatography, showing that the phosphorylated G-actin remains in the monomeric state.

Kuroda and Sonobe (1981) and Sonobe *et al.* (1985) prepared interesting cell models of *Amoeba proteus* by extraction of amebas with a 50% buffered glycerol solution at −20°C. The cell models show the contraction of the cytoplasm and/or the cytoplasmic streaming in a Ca^{2+}-dependent way on addition of 1 m*M* Mg-ATP. In 10^{-8} *M* Ca^{2+}, the entire cytoplasm contracts, while in 10^{-6} *M* Ca^{2+}, the cytoplasm shows active streaming. In 10^{-4} *M* Ca^{2+}, no contraction or streaming of the cytoplasm is observed. Electron micrographs of the cell models revealed that actin filaments in the cytoplasm disappear in the presence of 10^{-4} *M* Ca^{2+}. In the crude extracts from *Amoeba proteus, Amoeba* actin is phosphorylated in the presence of 10^{-4} *M* Ca^{2+}. Thus, Sonobe *et al.* (1985) suggested that the disappearance of actin filaments in 10^{-4} *M* Ca^{2+} could possibly be caused by the phosphorylation of *Amoeba* actin by *Amoeba* G-actin kinase. However, it is not known at present how much *Amoeba* actin may be phosphorylated in the cell models or living amebas.

Schweiger *et al.* (1992) reported that tyrosine and serine residues of *Dictyostelium* actin are transiently phosphorylated shortly after starving amebas are transferred into a nutrient medium. An inhibitor of phosphotyrosine phosphatase induces alterations of the actin cytoskeletons of amebas. These results suggest that phosphorylation of actin is involved in the organization of the actin cytoskeleton in amebas.

2. Fragmentation of Actin Filaments

The most interesting protein which we have isolated from *Physarum* is fragmin (Hasegawa *et al.*, 1980). Fragmin severs actin filaments into short pieces immediately after its addition to an F-actin solution. The activity of fragmin is regulated by physiological concentrations of Ca^{2+}. It has been shown that F-actin-severing proteins are widely distributed in cells and tissues of eukaryotic cells. There are two kinds of F-actin-severing proteins, the invertebrate and vertebrate types. Invertebrate-type proteins have been isolated from *Physarum* (fragmin: Hasegawa *et al.*, 1980; actin modulating protein: Hinssen, 1981a, 1981b), *Dictyostelium* (severin: Brown *et al.*, 1982; Yamamoto *et al.*, 1982), and sea urchin eggs (Wang and Spudich, 1984), while the vertebrate type has been isolated from rabbit macrophage (gelsolin: Yin and Stossel, 1979, 1980), human plasma (actin depolymerizing factor: Norberg *et al.*, 1979; brevin: Harris and Schwartz, 1981), bovine thyroid (Kobayashi *et al.*, 1983), bovine adrenal medulla (Ashino *et al.*, 1987), smooth muscle (Hinssen *et al.*, 1984), and chicken intestinal brush border (villin: Bretscher and Weber, 1979, 1980a).

Ca^{2+} is the second messenger of cell signals, so it is assumed that the dynamic behavior of the actin cytoskeleton in nonmuscle cells could be modulated by Ca^{2+}-sensitive, F-actin-severing proteins. The actin cytoskeletons would disintegrate and the polymerization of G-actin would be initiated by Ca^{2+}-sensitive F-actin-severing proteins when the Ca^{2+} concentration in the cells increases to more than 10^{-6} *M*.

a. Fragmin Hasegawa *et al.* (1980) isolated and purified F-actin-severing protein factor from *Physarum* plasmodium and termed it fragmin. The molecular weight is nearly identical to that of *Physarum* actin (42 kDa). The activity of fragmin is regulated by physiological concentrations of Ca^{2+} in plasmodium (10^{-7}–10^{-6} *M;* Kuroda *et al.*, 1988). It enhances the polymerization of G-actin and severs actin filaments in the presence of Ca^{2+} of more than 10^{-6} *M*. In the absence of Ca^{2+}, fragmin shows no effects on actin polymerization and actin filaments. Sugino and Hatano (1982) demonstrated that fragmin forms a complex of 1 fragmin and 1 G-actin and the complex becomes a nucleus for actin polymerization. Many nuclei are formed on addition of fragmin to a G-actin solution, thereby greatly enhancing the polymerization of G-actin. Fragmin severs actin filaments into short fragments; it caps the barbed ends of the short actin filaments and inhibits annealing of the short actin filaments. However, the complex of fragmin and G-actin is very stable. Even when the Ca^{2+} concentration is reduced to less than 10^{-6} *M,* fragmin is hardly separated from the complex.

An isomer of fragmin with the molecular weight of 60 kDa was isolated from *Physarum* plasmodium and was termed fragmin 60 (Furuhashi and Hatano, 1989). It shows properties identical to those of fragmin of 42 kDa based on *in vitro* examination. However, the tryptic peptide map, and the cross-reactivity with polyclonal antifragmin antibodies are clearly different in the two isoforms of fragmin. *Physarum* myxameba contains a fragmin-like protein very similar to, but different from, plasmodial fragmin (Uyeda *et al.*, 1988a). Antibodies raised against plasmodial and myxamebal fragmins scarcely cross-react with each other. Myxamebal fragmin is specifically present in myxamebas, while plasmodial fragmin is found in plasmodia (Uyeda *et al.*, 1988a). Thus, there are three forms of F-actin-severing proteins in *Physarum polycephalum;* fragmin, fragmin 60, and myxamebal fragmin. The expressions of these fragmins are regulated in the life cycle of *Physarum*.

Ampe and Vandekerckhove (1987) determined approximately 88% of the amino acid sequence of fragmin. The authors found that the available sequence of fragmin is more than 36% homologous with the N-terminal half of human gelsolin (Kwiatkowski *et al.*, 1986) (see Section IV, B,2,c).

b. Disintegration of Backbone in* Physarum *Flagellate The *Physarum* ameba (myxameba), about 12–15 μm in diameter, is another distinct form in the life cycle of *Physarum polycephalum*. The myxameba contains a single nucleus and increases like soil amebas. Myxamebas transform to flagellates with two flagella when suspended in a liquid solution. The transformation is reversible; flagellates transform to myxamebas again when spread on an agar plate (see Goodman, 1982).

Uyeda and Furuya (1985) showed that the concentration of actin filaments is especially high in pseudopods and filopods of myxambas. During the transformation from myxameba to flagellate, the actin filaments are organized into a bundle of actin filaments which run along the dorsal axis of the flagellate. This bundle is called the "backbone," because the spindle shape of the flagellate is maintained by the backbone. The structure termed a "ridge," which consists of a meshwork of actin filaments, is tentatively formed during the transformation (Pagh and Adelman, 1982). Uyeda and Furuya (1986) found that rapid cooling of flagellates on ice induces the breakdown of the backbones in the flagellates. At the same time, the elongated spindle-like shape of the flagellates becomes spherical. When the cells are warmed to 25°C, the backbone is reconstructed within 1 min and the cells regain their spindle shapes (see Fig. 5, b, d, and f). Uyeda *et al.* (1988b) further demonstrated that when flagellates are first treated with a Triton X-100 solution containing 1 m*M* EGTA, that is, in the absence of Ca^{2+} at 25°C, the flagellates are not stained by antimyxamebal fragmin (Fig. 5, a). Thus, fragmin is soluble in the cells at room temperature, but they become evenly stained by the antibody after cold treatment (Fig. 5c). Similarly, they are evenly stained with fluorescein-labeled phalloidin, which specifically stains actin filaments (Fig. 5d). These observations indicate that fragmin is associated with Triton-insoluble actin filaments after cold treatment. Uyeda *et al.* further demonstrated that when the flagellates are treated with Triton X-100 solutions containing various concentrations of Ca^{2+} (10 μM–1 m*M*) at room temperature, the backbones disintegrate, depending on the Ca^{2+} concentration in the solutions. Similarly, the amount of fragmin which binds actin filaments increases, depending on the Ca^{2+} concentration. The disintegration of the backbones and the binding of myxamebal fragmin to actin filaments show a similar dependence on the Ca^{2+} level in the solutions. These results strongly suggest that myxamebal fragmin-Ca^{2+} binds actin filaments of the backbones and severs the actin filaments by cold treatment. This causes the disintegration of the backbones. It is safe to assume that myxamebal

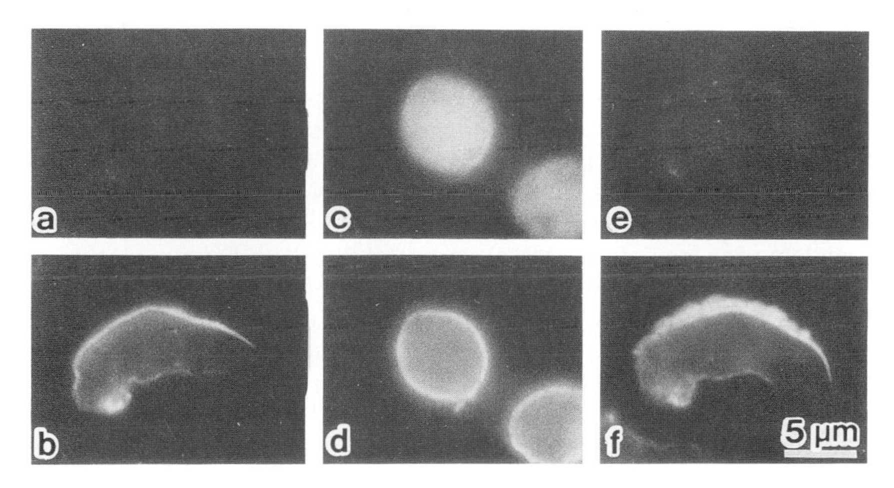

FIG. 5 Fluorescence micrographs of cold-treated *Physarum* flagellates. Suspensions of flagellates at 25°C were incubated in ice water for 2 min and then in a water bath at 25°C for 2 min. Cells were fixed, permeabilized, and double-stained with FITC-antimyxamebal fragmin (a, c, e) and rhodamine–phalloidin (b, d, f). (a) and (b) Before cold treatment. (c) and (d) After cold treatment. (e) and (f) After rewarming. (Reprinted from Uyeda *et al.*, 1988b, by permission of Wiley-Liss, A Division of John Wiley & Sons, Inc. © 1988.)

fragmin-Ca^{2+} participates in the disintegration of the backbone during the transformation of the flagellate into the myxameba.

*c. **Gelsolin*** Yin and Stossel (1979, 1980) isolated a protein factor from rabbit macrophages which induces solation of a gel composed of F-actin and high-molecular-weight actin-binding protein (an F-actin-gelling protein isolated from rabbit macrophage; Section IV,B,5) in a Ca^{2+}-dependent manner. Called gelsolin, the factor severs actin filaments and enhances actin polymerization like fragmin. The molecular weight of gelsolin is around 90 kDa, approximately twofold that of F-actin-severing proteins of the invertebrate cells, fragmin and severin (molecular weight, 40–42 kDa). Gelsolin is universally found in vertebrate tissues (Kwiatkowski *et al.*, 1988). A gelsolin-like protein was also isolated from human plasma. The protein was called actin depolymerizing factor (Norberg *et al.*, 1979) or brevin (Harris and Schwartz, 1981). Its molecular weight of 93 kDa is a little more than that of gelsolin (Yin *et al.*, 1984). The protein factor also goes by the name of plasma gelsolin now. Cytoplasmic gelsolin and plasma gelsolin are produced from a single gene by alternative splicing (Kwiatkowski *et al.*, 1988). Plasma gelsolin is frequently used for biochemical experiments because it can be easily prepared from human or pig plasma (Harris and Schwartz, 1981; Harris and Weeds, 1983).

Gelsolin binds two G-actins to form a ternary complex of 1 gelsolin and 2 G-actins (Yin and Stossel, 1980; Bryan and Kurth, 1984). One bound G-actin easily separates from the complex when the Ca^{2+} concentration is reduced to less than 10^{-6} *M,* while another one does not separate from gelsolin even in the presence of 1 m*M* EGTA (Bryan and Kurth, 1984; Kurth and Bryan, 1984).

The amino acid sequence of plasma gelsolin was deduced from the complete sequence of the cDNA clone and revealed that plasma gelsolin is identical to cytoplasmic gelsolin except that plasma gelsolin has a short extension at the N-terminal end (25 amino acid residues in the case of human plasma gelsolin and 9 residues in the case of pig plasma gelsolin; Kwiatkowski *et al.,* 1986; Way and Weeds, 1988). There is a tandem repeat in the sequence, that is, the N-terminal half is very similar to the C-terminal half in the sequence. Fragmin and severin, whose molecular weights are approximately one half that of gelsolin, are more closely related to the N-terminal half of gelsolin (Ampe and Vandekerckhove, 1987; Andre *et al.,* 1988). There is a further repeat of a shorter sequence (130–150 amino acids, up to 15 kDa). The smaller repeat occurs six times in gelsolin and three times in fragmin and severin. The N-terminal and C-terminal halves of gelsolin are connected with a short peptide (18 amino acids). Each half has three domains (domains 1–3 and domains 4–6, respectively. Fragmin and severin have three domains which correspond to domains 1–3 of gelsolin. It is speculated that gelsolin evolved from a protein similar to fragmin or severin by gene duplication (Andre *et al.,* 1988; Way and Weeds, 1988; Yin *et al.,* 1990).

Gelsolin is rapidly cleaved by proteinases into the N-terminal and the C-terminal halves, the molecular weights of which are 41–45 kDa (Chaponnier *et al.,* 1986; Bryan and Hwo, 1986; Janmey and Matsudaira, 1988). The N-terminal half binds F-actin and retains F-actin-severing activity. However, the severing activity is not regulated by Ca^{2+}. On the other hand, the C-terminal half binds only G-actin. This binding is regulated by Ca^{2+}. The N-terminal half of gelsolin is further cleaved into two peptides of 14 kDa and 24 kDa. The 14 kDa and 24 kDa peptides correspond to domain 1 and domains 2–3, respectively (Yin *et al.,* 1988; Bryan, 1988). Domain 1 can bind only G-actin and caps F-actin. Domains 2 and 3 together bind F-actin and can sever F-actin. However, the severing activity is much lower than in the N-terminal half (domains 1, 2, and 3). Therefore domains 1, 2, and 3 are necessary for the full severing activity of gelsolin.

Thus, there are three actin-binding domains in gelsolin, domain 1, domains 2–3, and domains 4–6. Domain 1 and domains 4–6 bind the same region on subdomain 1 of G-actin (Pope *et al.,* 1991). Way *et al.* (1989) and Pope *et al.* (1991) proposed a model to show the association of gelsolin with F-actin. The actin-binding sites of three domains are masked in the absence of Ca^{2+}. Ca^{2+} induces a conformational change of gelsolin and the three actin-binding sites become free to bind G-actins. According to the model, a gelsolin molecule binds three adjacent G-actins (A, B, and C) on F-actin. G-actins A and B are adjacent on the one-start helix of F-actin and A and C on the two-start helix. Domain 1 of gelsolin binds the subdomain of G-actin A; domains 4–6 bind the equivalent domain of G-actin B; and domains 2–3 bind at the interface of G-actins B and C.

The immunofluorescence of gelsolin showed that gelsolin resides in the cortical cytoplasm and is diffusely distributed in the cytoplasm (Yin *et al.*, 1981; Carron *et al.*, 1986; Cooper *et al.*, 1988). Microinjection of gelsolin into living fibroblasts and macrophages does not affect the shape, actin distribution, deformability, or ruffling activity of the cells. However, microinjection of the proteolytic fragment of the N-terminal half of gelsolin shows marked effects; the cells round up, stop ruffling, and stress fibers disappear (Cooper *et al.*, 1987), somewhat as with the microinjection of a capping protein (Füchtbauer *et al.*, 1983). The capping protein caps the barbed ends of actin filaments, but cannot sever them. It seems likely that the dramatic effects by the N-terminal half of gelsolin are caused by capping of actin filaments in the injected cells. Gelsolin might be inactive in the resting cells, where the Ca^{2+} concentration is maintained at less than 10^{-6} *M*.

Hartwig *et al.* (1989) investigated ultrastructural localization of gelsolin in macrophages and platelets by immunoelectron microscopy. They showed that gelsolin molecules localize both in the cytoplasm and near the plasma membrane. Gelsolin molecules associate with the one-terminal ends of short actin filaments on the membrane. Upon stimulation of platelets with thrombin, cytoplasmic gelsolin moves to the membrane up to 15 sec after stimulation and returns to the cytoplasm after around 20 min. Hartwig *et al.* (1989) speculated that the Ca^{2+} concentration in the cytoplasm increases to more than 0.1 μM by stimulation with thrombin. Gelsolin molecules activated by Ca^{2+} bind actin at the one-terminal ends of short actin filaments. The short actin filaments diffuse and attach to the membrane through the binding site for polyphosphoinositides on gelsolin molecules (Yin *et al.*, 1988).

d. Villin The apical surface of intestinal epithelial cells is covered with brush borders, which consist of packed microvilli. Within a microvillus is a core bundle of 20–30 actin filaments. These filaments are connected to the terminal web that is a meshwork of actin filaments beneath the microvilli (Ishikawa *et al.*, 1969; Tilney and Mooseker, 1971; Mooseker and Tilney, 1975). Myosin is not present in a microvillus, but localizes in the terminal web (Drenckhahn *et al.*, 1980; Herman and Pollard, 1981; Hirokawa *et al.*, 1982). Demembraned brush borders contract on addition of Mg-ATP in the presence of Ca^{2+}. The contraction consists of a retraction of the core actin filaments into and through the terminal web (Mooseker, 1976; Rodewald *et al.*, 1976; see Mooseker, 1985, for review).

Villin, an F-actin-severing protein, was isolated from demembraned microvilli or brush borders by affinity chromatography of G-actin bound to immobilized DNAase I. Villin is released from the villin–G-actin complex on the DNAase I when the Ca^{2+} concentration is less than 10^{-6} *M*. The molecular weight of villin is around 95 kDa, which is a little larger than gelsolin (Bretscher and Weber, 1979, 1980a). Villin binds G-actin to form a 1 villin : 2 G-actin complex in the presence of Ca^{2+} of more than 10^{-6} *M*. Villin shows nucleating and F-actin-severing activities in the presence of Ca^{2+}, just as do fragmin and gelsolin. In addition, villin cross-links actin filaments to form an actin gel or bundles of actin filaments in the presence and absence of Ca^{2+}.

Villin is cleaved into the core peptide of the N-terminal side (90 kDa) and the head piece of the C-terminal side (8.5 kDa) by mild proteolysis. The core peptide retains the Ca^{2+}-dependent nucleating and severing activities. The head piece bundles actin filaments irrespective of Ca^{2+} (Glenney and Weber, 1981; Glenney *et al.*, 1981). The amino acid sequence of villin was deduced from the cDNA clone (Bazari *et al.*, 1988; Arpin *et al.*, 1988). The villin core contains the conserved sequence of F-actin-severing proteins that is repeated six times. It is interesting that the head piece has sequences in common with synapsin I, an F-actin-bundling protein purified from synaptic vesicles (Bazari *et al.*, 1988). The unique bundling activity of villin seems to derive from the head piece, which is the extension of 76 amino acids from the C-terminal end of the core peptide.

Friederich *et al.* (1989) found that large amounts of villin are synthesized in transfecting fibroblasts with cloned human cDNAs encoding villin. The high level of villin synthesis results in the reorganization of stress fibers and induces the growth of numerous long microvilli on the cell surfaces. These facts clearly demonstrate that villin plays a key role in the morphogenesis of microvilli.

3. Gelation and Contraction of Cytoplasmic Fraction

Thompson and Wolpert (1963) first prepared motile cytoplasmic fractions from *Amoeba proteus*. They found that cold supernatants from low-speed centrifugation of ameba homogenates become gels and contract when warmed to 20°C in the presence of ATP. Cytoplasmic streaming resembling cytoplasmic streaming in amebas is also observed in the supernatants. The supernatants of *Amoeba* homogenates appear to be motile cell models without cell membranes. The cell models are simply described here as cytoplasmic fractions.

Pollard and Ito (1970) studied the gelation and contraction of cytoplasmic fractions from *Amoeba proteus* in detail and showed that numerous microfilaments appear in the cytoplasmic fraction in the initial phase of gelation and contraction. The microfilaments form meshworks incorporating clusters of thick filaments 16 nm in diameter and 0.5 μm in length. In the second phase, the microfilaments aggregate to form birefringent fibrils. In the end stage, large pseudocrystalline aggregates of microfilaments and thick filaments form. The microfilaments are identified as actin filaments (Pollard and Korn, 1971) and the thick filaments are assumed to be myosin filaments. Both thin and thick filaments have been observed in *Amoeba proteus* (Pollard and Ito, 1970) and *Chaos chaos* (Nachmias, 1964, 1968). It was speculated that the contraction of the cytoplasmic fraction is caused by the interaction of actin and myosin filaments in the presence of ATP (Pollard and Ito, 1970).

It is impossible to follow the process of gelation with the high-shear viscometers (more than 100 sec^{-1}) used to measure the viscosity of an F-actin solution because the gel is broken and precipitates under the high shear rates. MacLean-Fletcher and Pollard (1980) developed a low-shear falling ball device (below 10 sec^{-1}) to measure the apparent viscosity of *Acanthamoeba* extracts. They succeeded in describing in detail the process of gelation upon warming cold cytoplasmic fractions of *Acanthamoeba* (supernatants of ameba homogenates obtained by centrifugation at 100,000 × *g*, 1 hr, 4°C). There is a short lag phase (1–2 min) before gelation of the fraction. The gel is not formed when the fraction is diluted to protein concentrations less than 5–6 mg/ml (actin concentration, 0.7–0.8 mg/ml). Thus, there is a critical protein concentration for gelation. Gelation is reversible at low temperatures. The same authors prepared desalted cytoplasmic fractions and found that Mg^{2+} and ATP are necessary for gelation. The optimal concentrations of ATP and Mg^{2+} are 1 m*M* and 2 m*M*, respectively, in physiological concentrations (0–20 m*M*) of KCl solution. The micromolar range of Ca^{2+} inhibits gelation. Preincubation of the cytoplasmic fraction with 1 m*M* ATP and 2 m*M* $MgCl_2$ considerably shortens the lag phase for the gelation. In the same study, actin gels were reconstituted from purified F-actin and partially purified F-actin-gelling proteins. The reconstituted gels no longer require m*M* Mg-ATP, or no longer show Ca^{2+} and cold sensitivities.

These experiments indicate that at least two processes are involved in gelation of the cytoplasmic fractions. In the lag phase of gelation, actin filaments are formed, a process which depends on the temperature and Mg-ATP concentration. In the second process, the actin filaments are cross-linked by gelling proteins. The latter process does not depend on the temperature and Mg-ATP concentration. When gelling proteins are Ca^{2+}-sensitive, the gelation process becomes Ca^{2+}-sensitive. Therefore the second process can be simply explained by cross-linking of actin filaments by gelling proteins. However, the first process, the formation of actin filaments, is not so simple. F-Actin has a known tendency to depolymerize into G-actin when the temperature of the F-actin solution is decreased from room temperature to 0–4°C. This takes a long time (hours), and the extent of the depolymerization is not complete under physiological conditions of nonmuscle cells (for example, 50 m*M* KCl, 2 m*M* $MgCl_2$, pH 7.0).

On the other hand, the cytoplasmic fractions form actin gels on warming from 0 to 4°C to 20–30°C, and the gels are reversibly liquefied by cooling the fractions. MacLean-Fletcher and Pollard (1980) further demonstrated that preincubation of the fraction with Mg-ATP shortens the lag phase of the gelation. Several peptides are phosphorylated by incubation of the fraction with ATP, suggesting that phosphorylation of some proteins is involved in the formation of actin filaments when the cytoplasmic fraction is warmed.

Hatano (1972) and Fujime and Hatano (1972) have shown that flexible actin polymers (Mg-polymers) are transformed into actin filaments shortly upon addition of 0.5–1.0 m*M* ATP in the presence of 2 m*M* $MgCl_2$. The transformation is reversible, depending on the ATP concentration in the medium. Condeelis and Taylor (1977) observed aggregates of short actin filaments in extracts from *Dictyostelium* under a relaxation condition. The aggregates look very similar to those of Mg-polymers. The transformation of Mg-polymer might be involved in the formation of actin filaments in the cytoplasmic fractions.

The relationships between the gelation and contraction of cytoplasmic fractions from homogenized ameba cells (*Amoeba proteus, Chaos carolinenis, Dictyostelium discoideum*) were extensively studied by Taylor and colleagues (see Taylor and Condeelis, 1979, for review). The gelation and contraction of cytoplasmic fractions (supernatants of *Dictyostelium* ameba homogenates obtained by centrifugation at 3000x*g*, 45,000x*g*, and 100,000 x*g*) are influenced by Ca^{2+}, pH, ionic strength, Mg^{2+}, and ATP. The optimal gelation is induced in a relaxation solution containing 2.5 m*M* EGTA, 1 m*M* Mg-ATP, and 20 m*M* KCl (pH 7.0). However, no contraction and cytoplasmic streaming occur in the relaxation solution, showing that gelation and contraction are two different processes. Micromolar ranges of Ca^{2+} inhibit gelation; the gelation of the cytoplasmic fraction is therefore Ca^{2+} sensitive. In the presence of micromolar levels of Ca^{2+} and 1 m*M* Mg-ATP, the cytoplasmic fraction is solated but contracts rapidly (Condeelis and Taylor, 1977). Condeelis and Taylor (1977) emphasized that solation is necessary for the cytoplasmic fraction to contract. Based on these and other results (see Section II, C,4), Hellewell and Taylor (1979) and Taylor and Condeelis (1979) presented a solation-contraction coupling hypothesis for ameboid movements according to which actin filaments are weakly cross-linked in the endoplasm; the endoplasm has a moderate consistency to contract, and it is transformed into gel after the contraction.

4. Bundling of Actin Filaments

a. Fascin Kane (1975, 1976) found that cold cytoplasmic fractions (supernatants of sea urchin egg homogenates obtained by centrifugation at 100,000 × *g*, 1 hr, 2°C) form dense actin gels on warming to 35–40°C. Bryan and Kane (1978) isolated an F-actin-bundling protein (58 kDa) from the gel and termed it fascin (Otto *et al.*, 1980).

Fascin cross-links actin filaments to form microscopic needles, which are composed of parallel aggregates of actin filaments with a banding pattern showing 11 nm periodicity (Kane, 1975, 1976; Bryan and Kane, 1978). Otto *et al.* (1979, 1980) and Otto and Bryan (1981) described the roles of fascin in the formation of filopodia of *Echinoderm* coelomocytes and the formation of microvilli of the sea urchin egg cortex. Coelomocytes change their shape from a flattened lamellipodial form to a filopodial form by hypotonic shock. Fascin distributed in the circumferential lamellipodia is incorporated into the actin cores (the bundles of actin filaments) of the filopodia after the transformation. Isolated filopodial actin cores show the characteristic banding pattern of actin filaments bundled by fascin (Otto *et al.*, 1979; Otto and Bryan, 1981). Unfertilized sea urchin eggs have short microvilli on the egg cortex which elongate immediately after fertilization. The isolated, elongated microvilli show a banding pattern with 12 nm periodicity in the cores of actin filaments. Immunofluorescence using antifascin reveals the localization of fascin in the elongated fertilized microvilli (Otto *et al.*, 1980). Otto *et al.* (1980) suggested that fascin organizes actin cores by cross-linking actin filaments.

b. Fimbrin Microvilli of intestinal epithelial cells contain core bundles of actin filaments. The isolated actin cores contain four major actin-binding proteins of 110 kDa, 95 kDa (villin), 68 kDa (fimbrin), and 16.5 kDa (calmodulin) (Matsudaira and Burgess, 1979; Bretscher and Weber, 1980a, 1980b; Howe *et al.*, 1980). Bretscher (1981) isolated and purified fimbrin from brush borders of chicken intestines. Fimbrin is a monomeric protein of 68 kDa. It cross-links actin filaments into straight bundles in a 30-m*M* KCl solution. These properties of fimbrin are very similar to those of fascin isolated from sea urchin eggs. Immunofluorescence using antibody against fimbrin revealed fimbrin in a wide variety of cells; it localizes in straight, highly organized microfilament bundles in microvilli, microspikes, and stereocilia (Bretscher and Weber, 1980b).

c. **Physarum *52-kDa Protein*** Itano and Hatano (1991) isolated an F-actin-bundling protein of 52 kDa from *Physarum* plasmodium. This protein also cross-links microtubules to form bundles. In this way, it cross-links actin filaments and microtubules together to form cobundles of actin filaments and microtubules. The 52-kDa protein cross-reacts with a monoclonal antibody raised against an HeLa F-actin-bundling protein of 55 kDa (Yamashiro-Matsumura and Matsumura, 1985). Toriyama *et al.* (1988) then isolated an F-actin-bundling protein of 51 kDa from sea urchin eggs, and showed by immunofluorescence that the 51-kDa protein locates in the centrosome. Ohta *et al.* (1990) discovered that the amino acid composition of the 51-kDa protein is very similar to that of yeast polypeptide elongation factor-1α (EF-1α). The 51-kDa protein was found to cross-react with antibody raised against yeast EF-1α. Itano and Hatano (1991) revealed that the 52-kDa protein from *Physarum* plasmodium cross-reacts with antibody against yeast EF-1α. Yang *et al.* (1990) cloned and sequenced ABP 50, an F-actin bundling protein isolated from *Dictyostelium discoideum* and identified ABP 50 as *D. discoideum* EF-1α. Therefore, it seems likely that there is a superfamily of F-actin-bundling proteins that belong to EF-1α, which is involved in protein synthesis in the cells. Yang *et al.* thus suggested that the binding of EF-1α to actin may affect EF-1α activity in protein synthesis within the cells.

5. F-Actin Gel

a. High Molecular Weight of Actin-Binding Protein Hartwig and Stossel (1975) and Stossel and Hartwig (1976) isolated an F-actin-gelling protein from homogenates of rabbit lung macrophages and termed it high-molecular-weight actin-binding protein (HMWABP). It cross-links actin filaments to form an actin gel. F-actin-bundling proteins and F-actin-gelling proteins have been isolated from many organisms and tissues of eukaryotes at present (Stossel *et al.*, 1985). Hartwig and Stossel (1981) reported that the HMWABP molecule is an asymmetrical dimer composed of 270 kDa subunits at physiological ionic strengths. Negatively stained HMWABP molecules in electron micrographs appear to be very flexible strands 162 nm in length. One dimer of HMWABP binds 14 actin monomers of actin filaments with a K_D value of 0.5×10^{-6} *M*, indicating the weak association of ABP with actin filaments. Electron micrographs of HMWABP which cross-links two actin filaments show that the dimers are bipolar; two monomers are connected, head to head, and the free ends of monomers are closely associated with the actin filaments.

Sutoh *et al.* (1984) purified a high-molecular-weight actin-binding protein from *Physarum* plasmodia. The molecular weight of the subunit is estimated to be 230–270 kDa and the subunits form dimers at physiological ionic strength (50 m*M* KCl, pH 7.0). Electron micrographs show that the subunit is a flexible rod 41 nm in length. Its dimer has a dumbbell-shaped structure 60 nm in length which is composed of two globular domains 14 nm in diameter connected with a thin, flexible strand. A similar high-molecular-weight actin-binding protein termed filamin was isolated from smooth muscle (Wang *et al.*, 1975; Wang, 1977). These high-molecular-weight actin-binding proteins of nonmuscle and muscle cells are homodimers with a subunit that has a molecular weight of 250 to 270 kDa. Electron micrographs of these homodimers revealed that they are flexible strands 160 nm in length (Tyler *et al.*, 1980). They have no Ca^{2+} sensitivity and do not bind calmodulin.

b. Spectrin There is another family of high-molecular-weight actin-binding proteins that cross-link actin filaments to form actin gels. This group includes fodrin from brain (Levine and Willard, 1981) and TW-260/240 from brush borders of epithelial cells (Glenney *et al.*, 1982). They are very similar to spectrin, which is the main component of the membrane cytoskeleton of mammalian erythrocyte, so the term nonerythroid spectrins is generally used for these proteins. The proteins of this family are highly asymmetric tetramers composed of two nonidentical subunits (α and β) (see Coleman *et al.*, 1989, for review). It is interesting that spectrin-like proteins are also isolated from *Acanthamoeba* (Pollard, 1984), *Dictyostelium* (Bennet and Condeelis, 1988), and sea urchin eggs (Kuramochi *et al.*, 1986), reflecting the wide distribution of nonerythroid spectrins in eukaryotic cells.

6. Gel–Sol Transformation

The gel–sol and sol–gel transformations of cytoplasm are clearly visible in locomoting amebas under a light microscope. The ectoplasm is transformed into the endoplasm in the tail regions of amebas, and the endoplasm is converted back into the ectoplasm when it reaches the advancing tip regions. The gel–sol and sol–gel transformations of cytoplasm occur not only in ameboid cells but also in other eukaryote cells in which cytoplasm differentiates into ectoplasm (or cortical gel layer) and endoplasm (or subcortical cytoplasm). Gelling proteins together with other actin-binding proteins are considered to play key roles in the gel–sol transformation in the eukaryotic cells. The following two molecular mechanisms of the cytoplasmic gel–sol transformation may be considered at present.

a. Ca^{2+}-Sensitive Gelling Protein α-Actinin was first isolated from striated muscle as a factor promoting superprecipitation of muscle actomyosin (Ebashi *et al.*, 1964). α-Actinin was soon found to cross-link actin filaments to form an actin gel (Maruyama and Ebashi, 1965). The molecule of α-actinin is composed of two identical subunits with a molecular weight of 100 kDa (Suzuki *et al.*, 1976). Mimura and Asano (1979) isolated an F-actin-gelling protein (actinogelin) from Ehrlich tumor cells and indicated that the gelling activity of actinogelin is inhibited by the micromolar range of Ca^{2+}. Hence, actinogelin is a Ca^{2+}-sensitive, F-actin-gelling protein. Actinogelin is now classified as nonmuscle α-actinin based on its immunological cross-reactivity (Mimura and Asano, 1986), similarity in amino acid composition, and molecular shape (Ohtaki *et al.*, 1985).

Burridge and Feramisco (1981) isolated an α-actinin-like protein from HeLa cells which cross-links actin filaments to form an actin gel. In the micromolar range, Ca^{2+} inhibits F-actin gelation by HeLa α-actinin, whereas Ca^{2+} does not inhibit F-actin gelation by muscle α-actinin. α-Actinin is a homodimer of identical subunits with a molecular weight of about 100 kDa. The subunit molecule has three distinct domains; an N-terminal domain containing an actin-binding site, four spectrin-like repeats, and a C-terminal domain containing EF-hand Ca^{2+}-binding motifs (Blanchard *et al.*, 1989). Waites *et al.* (1992) determined the complete sequence of chicken brain α-actinin (892 amino acids, 107644 Da) and compared it with that of chicken smooth α-actinin. The sequences of both α-actinins, nonmuscle and muscle α-actinin, differ only in the region of the first EF-hand motif. The authors suggested that the difference in the Ca^{2+}-sensitivity of both α-actinins could be explained by the difference in their EF-hand motifs.

The regulation of gel-sol transformation of the actin gel by Ca^{2+} appears to be very simple as far as it is observed *in vitro*. However, Ohtaki *et al.* (1985) reported that the binding of actinogelin to actin filaments is strongly inhibited by skeletal muscle tropomyosin and partially inhibited by smooth muscle filamin. Therefore the other actin-binding proteins also might regulate the gelation of actin filaments by actinogelin in the cells.

b. Solation of Actin Gel by Gelsolin Control of the gel–sol transformation of actin gels by the gelsolin-Ca^{2+} complex was shown by Yin and Stossel (1979) and Yin *et al.* (1980). Yin *et al.* (1980) demonstrated that the critical concentration of HMWABP to induce the gelation of actin filaments is inversely proportional to the length of actin filaments. This means that when actin filaments are severed into shorter filaments by gelsolin, the critical concentration of HMWABP increases and higher concentrations of HMWABP are required for gelation of shorter actin filaments. Ca^{2+} regulates gelsolin activity and thus the gel–sol transformation of the actin gel containing gelsolin. The effect of Ca^{2+} is very rapid. However, the retransformation from the solated actin gel into the actin gel after removal of Ca^{2+} takes some time (3–19 hr). As shown in the next section (IV,B,7), phosphatidylinositol 4,5-*bis*phosphate (PIP_2), an intermediate of the phosphoinositide cycle, accelerates the dissociation of gelsolin from the actin–gelsolin complex (Lind *et al.*, 1987; Chaponnier *et al.*, 1987). PIP_2 may well accelerate the reverse transformation from the solated actin gel to the actin gel in the cells.

7. Regulation of Actin-Binding Proteins

To have physiological roles within the cells, ABPs would have to be located in the proper regions in the cells where they operate, and activated at the correct times when the functions initiate. In this sense, it is important to know how actin-binding proteins are regulated *in vitro*. As mentioned above, F-actin-severing proteins and Ca^{2+}-sensitive, F-actin-gelling proteins have Ca^{2+}-binding sites. Moreover Ca^{2+}, the second messenger of the signal transduction of the cells, directly regulates the activities of these actin-binding proteins. It has been shown that ABPs are regulated by Ca^{2+}-dependent phosphorylation and by intermediates of the phosphoinositide cascade of cell signaling. Such regulation of ABPs will be the key process in inducing the dynamic behavior of the actin cytoskeleton in cells, when the cells are stimulated by chemical substances such as hormones.

a. Phosphorylation More than 50% of fragmin is extracted from *Physarum* plasmodium in the form of a complex of 1 fragmin and 1 G-actin. The complex can be separated from the free state of fragmin by DEAE chromatography (Fig. 4). Furuhashi and Hatano (1990) found that when the extract is incubated with ATP in the presence of Ca^{2+}, the nucleating activity of the complex fractions decreases remarkably. However, the activity recovers when the extract is further incubated with ATP in the absence of Ca^{2+}. On the other hand, the nucleating and severing activities of the fragmin fractions were not affected by these incubations. It was shown that the factors are kinases with molecular weights of 78 kDa and 80 kDa. Purified kinases specifically phosphorylate G-actin of the complex and are termed actin kinase(s) (Furuhashi and Hatano, 1992). Actin kinases phosphorylate Thr-201 and probably Thr-202 and/or Thr-203 of *Physarum* G-actin with 1 mole of phosphate distributed among them (Furuhashi *et al.*, 1992).

Maruta *et al.* (1983) have isolated a protein complex which caps the barbed end of actin filament from *Physarum* plasmodium and termed it cap 42(a + b). Capping activity is regulated by phosphorylation of cap 42b at a threonine residue by an endogenous kinase. Cap 42(a + b) was later identified as a complex of fragmin (cap 42a) and *Physarum* G-actin (cap 42b) (Ampe and Vandekerckhove, 1987).

Gettemans *et al.* (1992) isolated and purified a kinase with the molecular weight of 80 kDa and termed it actin-fragmin kinase. The kinase specifically phosphorylates G-actin of the fragmin and G-actin complex. The major phosphorylation site is Thr-203 and a second minor site is Thr-202. The kinases isolated from *Physarum* plasmodium by Furuhashi and Hatano (1992) and Gettemans *et al.* (1993) are presumably the same kinases. The barbed end domain of G-actin is capped with fragmin in the complex. The Thr-201 to 203 residues are positioned on the pointed end domain of G-actin (Kabsch *et al.*, 1990; Holmes *et al.*, 1990). Thus, the elongation of actin filament from the pointed end side of the G-actin of the complex is inhibited by the phosphorylation of the threonine residues (Furuhashi and Hatano, 1992; Gettemans *et al.*, 1992). The inactivation of the complex in the crude extract is Ca^{2+}-sensitive. However, purified actin kinases exhibit no Ca^{2+}-sensitivity. Some factors providing Ca^{2+} sensitivity to actin kinases are considered to be lost during purification, or some phosphatases with Ca^{2+} sensitivity are present in the extract (Furuhashi and Hatano, 1992). Gettemans *et al.* (1992) reported that the fragmin-actin kinase activity is inhibited by micromolar Ca^{2+} concentrations when extra G-actin is added to the kinase solution.

b. Intermediates of Phosphoinositide Cycle It has been shown that the affinity of profilin to G-actin is very high in the case of profilactin, which is directly isolated from calf spleen. It does not dissociate into profilin and G-actin in physiological saline solution (Carlsson *et al.*, 1977). However, when platelets are activated by thrombin, profilin and G-actin are no longer associated. This induces the polymerization of actin in platelets (Carlsson *et al.*, 1979; Markey *et al.*, 1981). Lind *et al.* (1987) reported that the extract from resting platelets contains nearly no high-affinity complexes of profilin–G-actin and gelsolin–G-actin. Thrombin stimulation induces the formation of high-affinity complexes of profilin and gelsolin with G-actin, but these complexes are not present 5 min after stimulation. Thus, profilin and gelsolin are associated and dissociated with G-actin within short periods of time in platelets.

In 1985 Lassing and Lindberg found that phosphatidylinositol 4,5-*bis*phosphate (PIP_2), an intermediate of the phosphoinositide cycle, induces a rapid dissociation of profilactin accompanying the polymerization of G-actin. In 1988 the same investigators presented a model linking transmembrane signaling to actin polymerization at the plasma membrane through the phosphatidylinositol cycle. Chaponnier *et al.* and Lind *et al.* demonstrated in 1987 that PIP_2 dissociates the EGTA-resistant complex of gelsolin and G-actin. It has been shown that PIP_2 and phosphatidylinositol monophosphate inhibit the severing activity of gelsolin (Janmey and Stossel, 1987; Janmey *et al.*, 1987). PIP_2 inhibits binding of CT28N to actin filaments. CT28N is a proteolytic fragment of the N-terminal side of gelsolin which has a domain binding the actin filament (Yin *et al.*, 1988) (Section IV,B,2). Profilin specifically binds PIP_2 vesicles with a high affinity and it inhibits PIP_2 hydrolysis by phospholipase C through the competitive interaction of phospholipase C with PIP_2 (Goldschmidt-Clermont *et al.*, 1990). Similarly, gelsolin inhibits phospholipase C, but less so with the addition of large amounts of PIP_2 (Banno *et al.*, 1992). These results indicate that profilin and gelsolin are involved in the transmembrane signalings of the cells.

8. Consistency of Cytoplasm

Living cytoplasm exists in the form of viscoelastic fluid (endoplasm) or a solid (ectoplasm). In either state, it exhibits contractility (Section II). It was the various studies on the physical properties of cytoplasm which led to the concept of "cytoplasmic consistency" (Allen, 1961; Taylor and Condeelis, 1979). It is possible now to discuss cytoplasmic consistency based on the three-dimensional networks of cytoplasmic filaments which constitute cytoskeletons in the cytoplasm. Four kinds of cytoplasmic filaments have been identified in eukaryotic cells; microfilaments (actin filaments), microtubules, intermediate filaments, and elastic filaments (connectin or titin filaments). Buckley (1975, 1981) demonstrated by electron microscopy that cytoplasmic filaments are interconnected by fine filaments 30–60 nm in length and 1.5–2.0 nm in diameter to form a three-dimensional network in tissue-cultured cells. The network extends to all parts of the cytoplasm, and cell organelles and polyribosomes are included within the network structure. This cytoplasmic network has been termed microtrabeculae (Wolosewick and Porter, 1976).

It is reasonable to assume that such fine filaments consist of F-actin-gelling proteins and/or F-actin-bundling proteins, since it has been shown that these proteins cross-link actin filaments to form actin gels or bundles of actin filaments *in vitro*. On the other hand, elastic proteins which form lattices of superthin filaments 2–3 nm in diameter have been isolated from muscle and nonmuscle cells. Kimura *et al.* (1984) demonstrated that the superthin filaments are associated with actin and myosin filaments. Thus, the superthin filaments of elastic proteins are also quite possibly the fine filaments which interlink cytoskeletons in nonmuscle cells.

Connectin or Titin Elastic protein was first isolated from striated muscle and termed connectin (Maruyama *et al.*, 1976a) or titin (Wang *et al.*, 1979). It has been shown to localize in a lattice of fine filaments connecting thick (myosin) filaments with Z-lines in sarcomers (Maruyama *et al.*, 1985; Früst *et al.*, 1988). There are two types of connectin, α-connectin

(2,800 kDa) which is barely soluble, and β-connectin (2,100 kDa), which is soluble in buffer solutions (Maruyama *et al.*, 1984). Electron micrographs of purified muscle titin molecules revealed a beaded structure for most of their length (Trinick *et al.*, 1984; Whiting *et al.*, 1989). Single molecules span from Z- to M-lines (about 1 μm in length) in a sarcomere of striated muscle (Früst *et al.*, 1988; Whiting *et al.*, 1989). Labeit *et al.* (1992) isolated cDNAs encoding regions of titin located in the A-band and determined 1000 kDa of the amino acid sequence of the titin molecule. The A-band sequence consists of 100 amino acid repeats.

These investigators also showed that the expressed titin fragments bind the light meromyosin part of myosin and C-protein from striated muscle, showing the interconnection of titin molecules with thick filaments and molecules in the A-band. Connectin- or titin-like proteins have been isolated from nonmuscle cells; *Physarum* plasmodium (Ozaki and Maruyama, 1980; Gassner *et al.*, 1985), erythrocyte cell membrane (Maruyama *et al.*, 1977), and the cortical gel layer of sea urchin eggs (Maruyama *et al.*, 1976b). It is thus likely that connectin- or titin-like elastic proteins are widely distributed in eukaryotic cells.

When plasmodium is homogenized and extensively washed with a solution containing 0.5 *M* KI, 6 *M* urea, 1 *M* acetic acid, and 1% sodium dodecyl sulfate, some residues still remain in the precipitate. Ozaki and Maruyama (1980) found that the amino acid composition of the insoluble proteins is very similar to that of muscle connectin. Gassner *et al.* (1985) further purified *Physarum* titin (connectin) by solubilization in a hot (90°C) buffer solution containing 10% SDS and 40 m*M* dithiothreitol (DTT) and separating it by an agarose gel filtration. *Physarum* titin is a huge aggregate of molecules even in SDS solutions. Electron micrographs revealed that *Physarum* titin forms networks of superthin filaments 2–3 nm in diameter. Gassner *et al.* further showed that Triton- and SDS-treated ghosts of endoplasm consist of networks of superthin filaments with beaded structures which are characteristic of titin filaments. The tension of Triton- and KI-extracted ghosts of plasmodial strands increases when the ghosts are stretched (the elasticity) and then gradually decreases to a definite level (the viscosity). Similar tension curves which indicate the viscoelasticity of living plasmodial strands have been shown by stretching the strands (Kamiya and Yoshimoto, 1972). The viscoelasticity of plasmodial strands could be explained at least qualitatively by the viscoelastic nature of networks of superthin filaments in the cytoplasm.

Native connectin has been isolated from *Physarum* plasmodium by Kimura *et al.* (1984). Native connectin was extracted and purified in a solution containing 1 m*M* $NaHCO_3$ without the denaturation reagents. The molecular weight of native *Physarum* connectin is larger than that of native muscle connectin (2–3 million Da). Kimura *et al.* (1984) showed that native *Physarum* connectin binds actin and myosin filaments. Therefore it is thought that connectin or titin filaments cross-link actin and myosin filaments to form network structures in the cytoplasm. The diameter of connectin or titin filaments is on the same order (2–3 nm) as that of fine filaments which have been observed in tissue-cultured cells (Buckley, 1975; Wolosewick and Porter, 1976), although there is no evidence at present to show that the fine filaments which interlink actin filaments in tissue-cultured cells are connectin or titin filaments.

V. Mechanism of Cytoplasm Contraction

Finally, I would like to discuss the molecular mechanism of cytoplasmic streaming of *Physarum* plasmodium based on results reviewed in this chapter: (1) The cytoplasmic streaming of the endoplasm is a passive flow (Section II, A). (2) This passive flow is caused by the contraction of the ectoplasm (Section II,B). 3) The ectoplasm contracts in accordance with the formation of cytoplasmic fibrils in the ectoplasm (Section II, C). (4) The cytoplasmic fibrils are bundles of actin filaments decorated with myosin molecules (Section II, C).

A. Superprecipitation of *Physarum* Actomyosin

Superprecipitation of actomyosin is considered to be a model of muscle contraction *in vitro*. Matsumura and Hatano (1978) found that superprecipitation of actomyosin synthesized from *Physarum* F-actin and *Physarum* myosin is reversible. The extent of superprecipitation which is measured by the turbidity of actomyosin solution is dependent on the ATP concentration. The turbidity decreases with the increase of the ATP concentration in the solution. On the other hand, the actin-activated Mg-ATPase increases with the ATP concentration (Fig. 6). A reciprocal relation between turbidity and the actin-activated Mg-ATPase activity has been shown in the case of muscle F-acto-heavy meromyosin, which is the complex of F-actin and heavy meromyosin from muscle (Sekiya *et al.*, 1967). Tonomura (1972) proposed a kinetic model for the actin-activated ATPase of myosin in which there are two states in the binding of myosin to F-actin. At low ATP concentrations, myosin binds F-actin strongly.

Superprecipitation of *Physarum* actomyosin is reversible, and the whole process could be followed under an electron microscope (Matsumura and

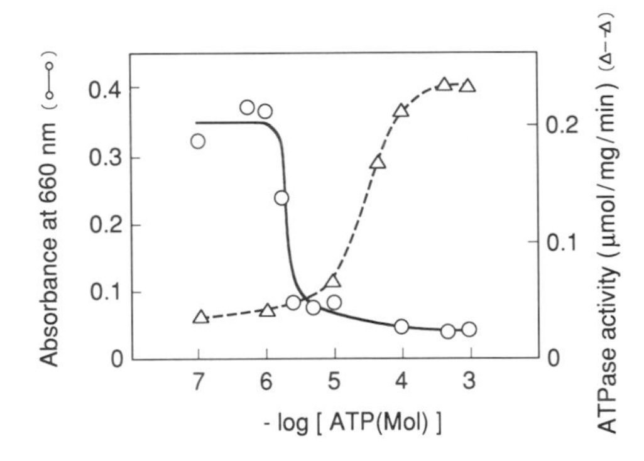

FIG. 6 Dependence of superprecipitation and ATPase activity of *Physarum* actomyosin on ATP concentration. Superprecipitation of *Physarum* actomyosin was reversible as to the ATP concentration in the solution. The extent of superprecipitation and the actin-activated Mg-ATP of actomyosin solutions were measured at various ATP concentrations. (Reprinted from Matsumura and Hatano, 1978, by permission of Elsevier Science.)

Hatano, 1978) (Fig. 7). Extracted *Physarum* myosin is phosphorylated (Kohama *et al.*, 1987). It forms long bipolar filaments up to 2.5 μm in 0.05 *M* KCl solution (pH 7.0) containing 5 m*M* Mg^{2+}. When F-actin is added to thick filaments in the absence of ATP, the thick filaments are immediately disintegrated and actin filaments in the solution are decorated with myosin molecules. When ATP is added to the solution, myosin molecules separate from actin filaments and form short myosin filaments 0.1 to 0.4 μm in length (Fig. 7a). It seems likely that myosin is present in the form of similar bipolar filaments in plasmodium. In the initial phase of superprecipitation, actin filaments tend to aggregate side by side and finally form bundles of actin filaments decorated with myosin molecules (actomyosin fibrils) (Fig. 7, b, c). The structure of actomyosin fibrils is essentially identical to that of cytoplasmic fibrils (birefringent fibrils) which are formed in accordance with the contraction of plasmodium (Ishigami *et al.*, 1987). It is reasonable to conclude that the contraction of plasmodial ectoplasm is caused by superprecipitation of actomyosin *in vivo*.

B. Molecular Mechanism of Superprecipitation

A hypothetical scheme for superprecipitation of *Physarum* actomyosin is presented in Fig. 8. Superprecipitation consists of two processes. In the first, there is the bundling of actin filaments by bipolar myosin filaments [Fig. 8 (1)], and in the second, the contraction of the bundle of actin filaments by actin-activated myosin ATPase [Fig. 8 (2)].

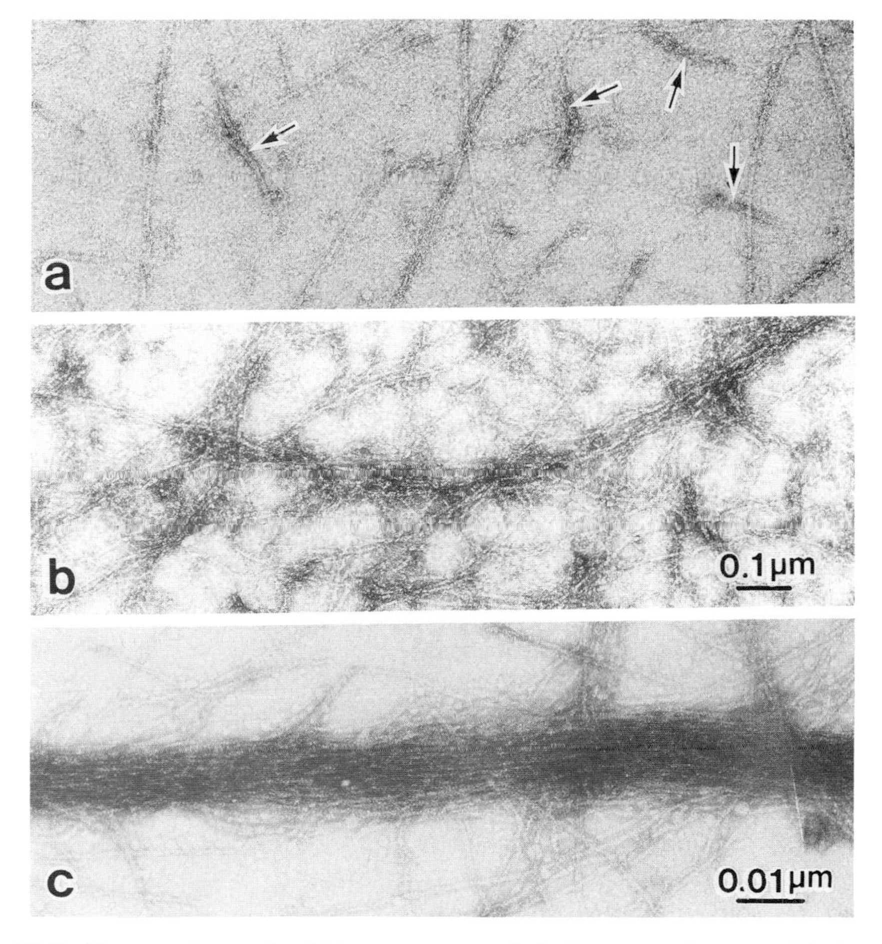

FIG. 7 Electron micrographs of *Physarum* actomyosin in the process of superprecipitation. (a) Clearing phase after addition of ATP to actomyosin. Arrows indicate short myosin filaments. (b) Initial phase. Lateral aggregation of actin filaments began. (c) Superprecipitated actomyosin. The bundle of actin filaments was formed. The magnification is the same in a and b. (Reprinted from Matsumura and Hatano, 1978, by permission of Elsevier Science.)

1. Bundling of Actin Filaments

The initial phase of superprecipitation is characterized by the side-by-side aggregation of actin filaments in this scheme [Fig. 8 (1)]. Myosin must be either long or short bipolar filaments in order to bundle actin filaments. Heavy meromyosin or subfragment 1 (S1, a fragment of muscle myosin molecule with a single head) slides on actin filaments in the presence of

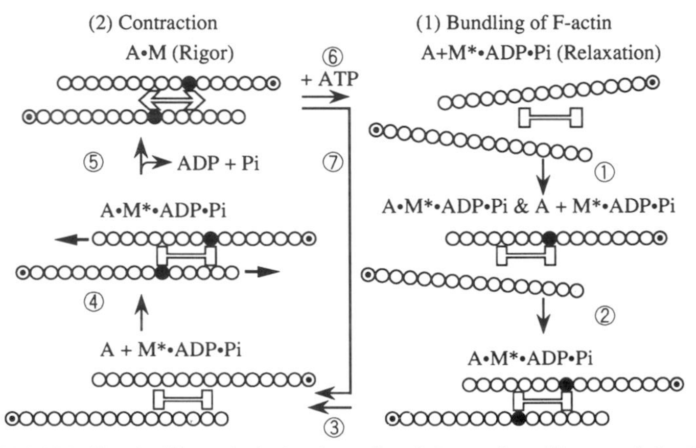

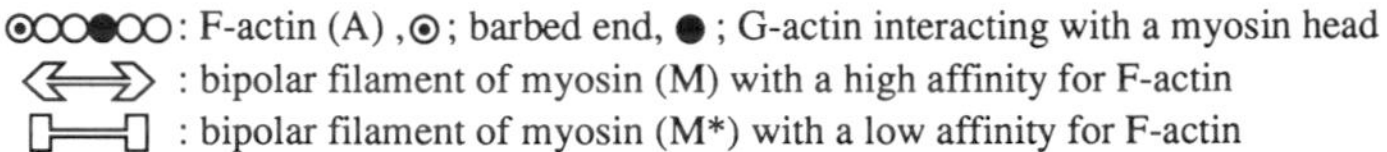

FIG. 8 Scheme for superprecipitation of *Physarum* actomyosin. It is assumed in this scheme that the process of superprecipitation consists of (1) bundling of actin filaments and (2) contraction of the bundle. The bipolar myosin filament works as a bundling factor in the process (1). Two actin filaments slide along each other in opposite directions in the presence of Mg-ATP, when the bipolar myosin filament is located between the actin filaments. This causes shortening of the bundle of actin filaments [process (2)].

Mg-ATP. However, these fragments of myosin cannot induce superprecipitation, because they cannot bundle actin filaments under physiological conditions. In my view, bundling of actin filaments is essential to initiate superprecipitation. *Acanthamoeba* myosin-I isomers are single-headed myosins, as mentioned earlier. However, they have two actin-binding sites (Lynch *et al.,* 1986; Brzeska *et al.,* 1988) and can bundle actin filaments and induce superprecipitation of actin filaments (Fujisaki *et al.,* 1985).

Actin filaments in I-bands of myofibrils are bundled by Z membranes so that myofibrils are ready to contract by sliding between thin (actin) and thick (myosin) filaments (Huxley and Niedergarke, 1954; Huxley and Hanson, 1954). Takiguchi (1991) succeeded in demonstrating that bundles of actin filaments several tens of microns in length are formed when methylcellulose is added to an F-actin solution. The bundles of actin filaments contract to about one-third the initial length when muscle heavy meromyosin and Mg-ATP are added to the solution. Thus, heavy meromyosin has no actin filament bundling activity, but it causes the contraction of bundles of actin filaments in the presence of Mg-ATP. De Lozanne and Spudich (1987) and Knecht and Loomis (1987) succeeded in preparing *Dictyostelium* amebas in which myosin-II is deleted. In the amebas the contractile ring for cytokinesis cannot be constructed, so cytokinesis is inhibited.

Tada and Tonomura (1967) reported important results indicating that there is a critical concentration in superprecipitation of muscle actomyosin. Superprecipitation does not occur below the critical protein concentration. When small amounts of superprecipitated actomyosin which have been fragmented by sonic vibration are added in the initial phase of superprecipitation, the superprecipitation is much accelerated according to the amounts of fragmented actomyosin added. These results thus demonstrate that superprecipitation is a kind of condensation process, and its initial phase is nucleation, which is the rate-limiting process of superprecipitation. The fragments of superprecipitated actomyosin serve as nuclei for superprecipitation. Tada and Tonomura (1967) emphasized that the actin-activated Mg-ATPase is not activated by the addition of small amounts of fragmented suprcprecipitated actomyosin. The acceleration of superprecipitation by actomyosin fragments is not accompanied by activation of actomyosin ATPase. One may readily infer from these results that the bundling of actin filaments by either short bipolar nonmuscle myosin filaments or long bipolar muscle myosin filaments is the nucleation process of superprecipitation. Actomyosin ATPase is not necessary for bundling of actin filaments, for myosin filaments work as F-actin-bundling proteins with multiple actin-binding sites [Fig. 8 (1)].

The affinity of each myosin head with an actin filament is weak in the initial phase of superprecipitation because relatively high concentrations of ATP are present in the solution (Fig. 6). According to Pollard (1987), the dissociation constant for the complex of myosin and F-actin in the presence of ATP is 10^{-5}-10^{-4} *M*, while that for the complex in the absence of ATP is less than 10^{-9} *M*. Myosins showing weak and strong affinity with F-actin are expressed as M* and M, respectively, in Fig. 8. However, the affinity of myosin filaments for actin filaments in the presence of ATP (K_D, 10^{-5}–10^{-4} *M*) will be multiplied by the number of myosin heads of the filaments, so that myosin filaments could work as actin bundling proteins which have stronger bundling activity.

It has been shown that the polarity of actin filaments in the stress fiber is not uniform, but random (Sanger and Sanger, 1980; Section II, C,5). This would be true in the case of actomyosin fibrils in *Physarum* plasmodium. Let us consider the polarity of actin filaments bundled by a single bipolar myosin filament [Fig. 8 (1)]. Myosin heads at either polar side have the same polarity, so they bind actin filaments with uniform polarity. Since the myosin filament is bipolar, the polarity of actin filaments bundled by the single myosin filament should be a mixture of opposing polarities. I suspect that actomyosin fibrils such as cytoplasmic fibrils in plasmodium and stress fibers in tissue-cultured cells are constructed through the bundling of actin filaments by bipolar myosin filaments in the initial phase of actomyosin superprecipitation. This would be the reason why the actin filaments in the actomyosin fibrils have such a mixture of opposite polarities.

2. Contraction of Bundle of Actin Filaments

The second process following bundling of actin filaments is the contraction of those bundles in which bipolar myosin filaments are located [Fig. 8 (2)]. Cytoplasmic fibrils in plasmodium and stress fibers in tissue-cultured cells contract longitudinally to generate tension (Section II, C). The best models are actomyosin threads in which actin filaments are arranged parallel to the long axis of the threads. The threads contract on addition of Mg-ATP (Section II,B). Unfortunately, at present there is no reasonable explanation for the contraction of actomyosin fibrils or actomyosin threads.

It is generally thought that actin filaments slide along each other by actin-activated Mg-ATPase activity of myosin filaments and this induces the shortening of fibrils or threads. Let us consider a case in which two actin filaments with opposite polarity (two actin filaments arranged in antiparallel) are connected by a single bipolar myosin filament [Fig. 8 (2)]. Each actin filament slides on the myosin filament in the direction of the pointed end, so that two actin filaments slide in opposite directions. This assumption is reasonable considering that myosin molecules or myosin filaments slide in the direction of the barbed ends of actin filaments in *Characean* cells (Kersey *et al.*, 1976; Sheetz and Spudich, 1983; Shimmen and Yano, 1984). When myosin filaments are fixed, actin filaments slide in the direction of their pointed ends. However, this does not always cause shortening of the total length of two actin filaments, because the total length decreases or increases depending on the overlapping of the two actin filaments.

Where two actin filaments with the same polarity are connected by a single myosin filament, the myosin filament only slides on the actin filaments in the direction of the barbed ends of the filaments. It appears that the shortening of actomyosin fibrils could not be explained by a simple sliding movement between actin filaments and myosin filaments. In addition to the sliding mechanism, there must be some mechanism to stop the sliding movement when there is maximum overlap of actin filaments. This is the most difficult problem and one which must be solved in the future. In the scheme for superprecipitation (Fig. 8), I assumed that generation of the sliding force was coupled with ATP hydrolysis by myosin ATPase, based on the models of Lymn and Taylor (1971), Tonomura (1986), and Tawada and Sekimoto (1991); however, I greatly simplified the scheme.

My tentative speculation as to the contraction of the cytoplasmic fibril is as follows: When actin filaments are bundled by bipolar myosin filaments, Mg-ATPase of myosin is activated. This activation will be very effective because each myosin head could interact with the actin filaments surrounding the myosin filaments. The activated Mg-ATPase generates a mechanical force. The polarity of actin filaments in the bundle is random, and short myosin filaments are bipolar, as mentioned above. Therefore they could not induce the unidirectional movement of myosin filaments within the bundle. At one moment a myosin filament moves in one direction, but in the next it moves in the opposite direction. This bidirectional movement induces a mechanical agitation of actin filaments within the bundle so as to cause their rearrangement. The interaction of myosin filaments with actin filaments continues until the free energy of the state reaches the minimum level, in which point myosin heads may interact with actin filaments at the maximum, or the overlap between myosin and actin filaments becomes maximum. This is achieved when the actin filaments reach the stage of maximum overlap with one another in the bundle so that all myosin heads can interact with the actin filaments.

VI. Concluding Remarks

A. Cytoplasmic Streaming and Actin Cytoskeleton in *Physarum* Plasmodium

The experimental results reviewed in this chapter could be summarized as follows. The actin cytoskeleton of *Physarum* plasmodium consists of cytoplasmic fibrils and a cortical actin layer. The cytoplasmic fibrils are bundles of actin filaments (Allera *et al.*, 1971). It is assumed that myosin exists as short bipolar filaments up to 0.5 μm in length in the bundles of actin filaments. On the other hand, the cortical actin layer consists of meshworks of actin filaments. The cortical actin layer together with the cytoplasmic fibrils envelops the entire cortical gel layer of the plasmodium (Naib-Majani *et al.*, 1982; Brix *et al.*, 1987). The plasmodial actin cytoskeleton has been found to be essentially identical to the actin cytoskeleton of tissue-cultured cells and that of amebas, although cytoplasmic fibrils are absent in amebas (Brix *et al.*, 1987). Most cytoplasmic fibrils are located in the dorsal cortical layer of plasmodium and generate the isotonic tension for plasmodial contraction (Ishigami, 1986). On the other hand, the cytoplasmic fibrils located in the ventral cortical layer serve to attach the plasmodium to the substratum (Brix *et al.*, 1987). It is shown by *in vivo* and *in vitro* cell models of plasmodium that the cortical actin layer and the granular cytoplasm each contracts by itself (Kuroda, 1979; Sato *et al.*, 1981; Kukulies *et al.*, 1984; Kukulies and Stockem, 1985). The contraction of these actin cytoskeletons will be responsible for the ameboid movement of amebas (Taylor *et al.*, 1973; Taylor and Condeelis, 1979). The amounts of actin which is not extracted with a Triton solution do not change in the contraction and relaxation cycle for plasmodium (Ishigami and Hatano, 1986). Phalloidin does not inhibit the contraction and relaxation of cytoplasm in caffeine drops (Kukulies *et al.*, 1985). It seems likely that most actin filaments aggregate into bundles of actin filaments in the contraction phase and they disperse to single actin filaments in the relaxation phase.

Here is a brief hypothetical scheme for the contraction of plasmodium caused by superprecipitation of actomyosin. Superprecipitation consists of two processes, the bundling of actin filaments and the contraction of the bundle of such filaments. Both processes are mediated by myosin filaments. Myosin may be taken to be bipolar filaments. In this sense, the activity of myosin (two-headed myosin) is regulated by a double process in which filament formation is regulated by the phosphorylation of the heavy chains (Ogihara *et al.*, 1983) and the actin-activated Mg-ATPase is regulated by the binding of Ca^{2+} to the light chains (Kohama *et al.*, 1987). It is not known, however, which process is dominant in the regulation of myosin activity in plasmodium. In any case, myosin would be activated in regions where the cytoplasmic fibrils are formed and contract. Myosin kinases presumably play key roles in the activation of myosin.

B. Roles of Actin-Binding Proteins in Nonmuscle Cells

The structures and functions of actin-binding proteins are briefly summarized and their possible roles in a variety of cell functions are discussed based on their specific properties, which have been characterized *in vitro*. More than 50% of actin is present in the unpolymerized state in nonmuscle cells. The unpolymerized actin can be explained by the inactivation of G-actin by G-actin- and F-actin-sequestering proteins. Concentrations of profilin and/or thymosin β_4 are much higher than other proteins in the cells. Therefore profilin and/or thymosin β_4 are now considered to be the main regulators of actin polymerization in nonmuscle cells (Safer *et al.*, 1991). The disintegration of the backbones of *Physarum* flagellates by rapid cooling is caused by severing actin filaments of the backbone by myxamebal fragmin. This process is reversible, since the backbone is reconstructed when the flagellates are warmed (Uyeda *et al.*, 1988b).

The high level of villin synthesis results in the growth of numerous microvilli on the surface of fibroblasts with cloned cDNA encoding villin

(Friederich *et al.*, 1989). Similarly, fascin organizes actin filaments into actin cores of filopodia of *Echinoderm* coelomocyte (Otto *et al.*, 1979; Otto and Bryan, 1981) and microvilli on the surface of the sea urchin egg (Otto *et al.*, 1980). F-actin-gelling proteins cross-link actin filaments to form actin gels. The actin gel is solated by Ca^{2+} when the gelling protein is Ca^{2+} sensitive (Mimura and Asano, 1979; Burridge and Feramisco, 1981). The actin gel which is cross-linked with Ca^{2+}-insensitive gelling protein is solated by the Ca^{2+}–gelsolin complex (Yin *et al.*, 1980). The nucleation activity of the fragmin–G-actin complex is lost when G-actin of the complex is phosphorylated by actin kinases (Furuhashi and Hatano, 1992) or actin-fragmin kinase Gettemans *et al.*, 1992). It is also shown that intermediates of the phosphoinositol cycle inhibit the severing activity of gelsolin (Janmey and Stossel, 1987; Janmey *et al.*, 1987). Conversely, profilin and gelsolin inhibit the hydrolysis of PIP_2 in the phosphoinositol cycle (Goldschmidt-Clermont *et al.*, 1990; Banno *et al.*, 1992), indicating that profilin and gelsolin are also regulators of cell signals. The consistency of the cytoplasm is maintained by the three-dimensional structures of the cytoskeletons in the cytoplasm. The viscoelasticity of the plasmodial strand can be explained at least qualitatively by the lattice structures of superthin filaments of connectin or titin (Kamiya and Yoshimoto, 1972; Gassner *et al.*, 1985).

When I first attempted to isolate actin and myosin from *Physarum* plasmodium in the 1960s, I could never have imagined such rapid progress in studies on the actin and myosin system in nonmuscle cells. However, the molecular mechanism of cytoplasmic streaming in plasmodium is not yet clear. The migration of plasmodium on a substratum is considered to be a kind of ameboid movement, and I believe there must be a general molecular mechanism underlying such movement. It is hoped that the hypothetical scheme for superprecipitation I have suggested will help in developing future experimental plans to understand the molecular mechanism of cytoplasmic streaming in *Physarum* plasmodium as well as ameboid movement of amebas and tissue-cultured cells.

Acknowledgments

I thank Dr. Mitsuo Ishigami, who provided the photograph in Fig. 1. Useful comments from Dr. Koshin Mihashi are also acknowledged.

References

Achenbach, U. (1982). The influence of mechanical stimuli on the oscillatory contraction activity of *Physarum polycephalum*. *Cell Biol. Int. Rep.* **6,** 1005–1011.

Achenbach, F., Achenbach, U., and Wohlfarth-Bottermann, K. E. (1979). Plasmalemma invaginations, contraction and locomotion in normal and caffeine-treated protoplasmic drops of *Physarum*. *Eur. J. Cell Biol.* **20,** 12–23.

Achenbach, F., and Wohlfarth-Bottermann, K. E. (1981). Morphogenesis and disassembly of the circular plasmalemma invagination system in *Physarum polycephalum*. *Differentiation* **19,** 179–188.

Adams, R. J., and Pollard, T. D. (1986). Propulsion of organelles isolated from *Acanthamoeba* along actin filaments by myosin-I. *Nature (London)* **322,** 754–756.

Adams, R. J., and Pollard, T. D. (1989). Binding of myosin I to membrane lipids. *Nature (London)* **340,** 565–568.

Adelman, M. R., and Taylor, E. W. (1969). Further purification and characterization of slime mold myosin and slime mold actin. *Biochemistry* **8,** 4976–4988.

Aktories, K., Bärmann, M., Ohishi, I., Tsuyama, S., Jakobs, K. H., and Habermann, E. (1986). *Botulinum* C2 toxin ADP-ribosylates actin. *Nature (London)* **322,** 390–392.

Albanesi, J. P., Fujisaki, H., Hammer, J. A., III, Korn, E. D., Jones, R., and Sheetz, M. P. (1985). Monomeric *Acanthamoeba* myosins-I support movement in vitro. *J. Biol. Chem.* **260,** 8649–8652.

Albanesi, J. P., Hammer, J. A., III, and Korn, E. D. (1983). The interaction of F-actin with phosphorylated and unphosphorylated myosins IA and IB from *Acanthamoeba castellanii*. *J. Biol. Chem.* **258,** 10,176–10,181.

Allen, R. D. (1961). Ameboid movement. *In* "The Cell" (J. Brachet, and A. E. Mirsky, Eds.), Vol. 2, pp. 135–216, Academic Press, New York.

Allen, R. D., and Roslansky, J. D. (1959). The consistency of ameba cytoplasm and its bearing on the mechanism of ameboid movement. I. An analysis of endoplasmic velocity profiles of *Chaos chaos (L.)* *J. Biophys. Biochem. Cytol.* **6,** 437–446.

Allera, A., Beck, R., and Wohlfarth-Bottermann, K. E. (1971). Weitreichende fibrilläre Protoplasmadifferenzierungen und ihre Bedeutung für die Protoplasmaströmung. VIII. Identifizierung der plasmafilamente von *Physarum polycephalum* als F-actin durch Anlagerung von heavy meromyosin in situ. *Cytobiologie* **4,** 437–449.

Ampe, C., and Vandekerckhove, J. (1987). The F-actin capping proteins of *Physarum polycephalum:* cap42(a) is very similar, if not identical, to fragmin and is structurally and functionally very homologous to gelsolin; cap42(b) is *Physarum* actin. *EMBO J.* **6,** 4149–4157.

André, E., Lottspeich, F., Schleicher, M., and Noegel, A. (1988). Severin, gelsolin, and villin share a homologous sequence in regions presumed to contain F-actin severing domains. *J. Biol. Chem.* **263,** 722–727.

Arpin, M., Pringault, E., Finidori, J., Garcia, A., Jeltsch, J. M., Vandekerckhove, J., and Louvard, D. (1988). Sequence of human villin: A large duplicated domain homologous with other actin-severing proteins and a unique small carboxy-terminal domain related to villin specificity. *J. Cell Biol.* **107,** 1759–1766.

Ashino, N., Sobue, K., Seino, Y., and Yabuuchi, H. (1987). Purification of an 80 kDa Ca^{2+}-dependent actin modulating protein, which severs actin filaments, from bovine adrenal medulla. *J. Biochem.* **101,** 609–617.

Baelen, H. V., Bouillon, R., and De Moor, P. (1980). Vitamin D-binding protein (Gc-globulin) binds actin. *J. Biol. Chem.* **255,** 2270–2272.

Bamburg, J. R., Harris, H. E., and Weeds, A. G. (1980). Partial purification and characterization of an actin depolymerizing factor from brain. *FEBS Lett.* **121,** 178–182.

Banno, Y., Nakashima, T., Kumada, T., Ebisawa, K., Nonomura, Y., and Nozawa, Y. (1992). Effects of gelsolin on human platelet cytosolic phosphoinositide-phospholipase C isozymes. *J. Biol. Chem.* **267,** 6488–6494.

Bazari, W. L., Matsudaira, P., Wallek, M., Smeal, T., Jakes, R., and Ahmed, Y. (1988).

Villin sequence and peptide map identify six homologous domains. *Proc. Natl. Acad. Sci. USA* **85,** 4986–4990.

Begg, D. A., Rodewald, R., and Rebhun, L. I. (1978). The visualization of actin filament polarity in thin sections. *J. Cell Biol.* **79,** 846–852.

Bennett, H., and Condeelis, J. (1988). Isolation of an immunoreactive analogue of brain fodrin that is associated with the cell cortex of *Dictyostelium* amoebae. *Cell Motil. Cytoskel.* **11,** 303–317.

Blanchard, A., Ohanian, V., and Critchley, D. (1989). The structure and function of alpha-actinin. *J. Muscle Res. Cell Motil.* **10,** 280–289.

Bonder, E. M., Fishkind, D. J., and Mooseker, M. S. (1983). Direct measurement of critical concentrations and assembly rate constants at the two ends of an actin filament. *Cell* **34,** 491–501.

Bretscher, A. (1981). Fimbrin is a cytoskeletal protein that crosslinks F-actin in vitro. *Proc. Natl. Acad. Sci. USA* **78,** 6849–6853.

Bretscher, A., and Weber, K. (1980a). Villin is a major protein of the microvillus cytoskeleton which binds both G and F actin in a calcium-dependent manner. *Cell* **20,** 839–847.

Bretscher, A., and Weber, K. (1980b). Fimbrin, a new microfilament-associated protein present in microvilli and other cell surface structures. *J. Cell Biol.* **86,** 335–340.

Bretscher, A., and Weber, K. (1979). Villin: The major microfilament-associated protein of the intestinal microvillus. *Proc. Natl. Acad. Sci. USA* **76,** 2321–2325.

Brix, K., Kukulies, J., and Stockem, W. (1987). Studies on microplasmodia of *Physarum polycephalum*. V. Correlation of cell surface morphology, microfilament organization and motile activity. *Protoplasma* **137,** 156–167.

Brown, S. S., Yamamoto, K., and Spudich, J. A. (1982). A 40,000-dalton protein from *Dictyostelium discoideum* affects assembly properties of actin in a Ca^{2+}-dependent manner. *J. Cell Biol.* **93,** 205–210.

Bryan, J. (1988). Gelsolin has three actin-binding sites. *J. Cell Biol.* **106,** 1553–1562.

Bryan, J., and Kane, R. E. (1978). Separation and interaction of the major components of sea urchin actin gel. *J. Mol. Biol.* **125,** 207–224.

Bryan, J., and Kurth, M. C. (1984). Actin-gelsolin interactions. Evidence for two actin-binding sites. *J. Biol. Chem.* **259,** 7480–7487.

Bryan, J., and Hwo, S. (1986). Definition of an NH_2-terminal actin-binding domain and a COOH-terminal Ca^{2+} regulatory domain in human brevin. *J. Cell Biol.* **102,** 1439–1446.

Brzeska, H., Lynch, T. J., and Korn, E. D. (1988). Localization of the actin-binding sites of *Acanthamoeba* myosin-IB and effect of limited proteolysis on its actin-activated Mg^{2+}-ATPase activity. *J. Biol. Chem.* **263,** 427–435.

Brzeska, H., Lynch, T. J., and Korn, E. D. (1989). The effect of actin and phosphorylation on the tryptic cleavage pattern of *Acanthamoeba* myosin IA. *J. Biol. Chem.* **264,** 10,243–10,250.

Buckley, I. K. (1975). Three dimensional fine structure of cultured cells: Possible implication for subcellular motility. *Tissue Cell* **7,** 51–72.

Buckley, I. K. (1981). Fine-structural and related aspects of nonmuscle-cell motility. *In* "Cell and Muscle Motility, Vol. I" (R. M. Dowben and J. W. Shay, Eds.), pp. 135–203. Plenum, New York.

Buckley, I. K., and Porter, K. R. (1967). Cytoplasmic fibrils in living cultured cells. A light and electron microscope study. *Protoplasma* **64,** 349–380.

Burridge, K. (1981). Are stress fibers contractile? *Nature* (*London*) **294,** 691–692.

Burridge, K., and Feramisco, J. R. (1981). Non-muscle alpha actinins are calcium-sensitive actin-binding proteins. *Nature* (*London*) **294,** 565–567.

Burridge, K., Molony, L., and Kelly, T. (1987). Adhesion plaques: sites of transmembrane interaction between the extracellular matrix and the actin cytoskeleton. *J. Cell Sci. Suppl.* **8,** 211–229.

Byers, H. R., White, G. E., and Fujiwara, K. (1984). Organization and function of stress fibers in cells in vitro and situ. *In* "Cell and Muscle Motility, Vol. 5" (J. W. Shay, Ed.), pp. 83–137. Plenum, New York.

Carlsson, L., Markey, F., Blikstad, I., Persson, T., and Lindberg, U. (1979). Reorganization of actin in platelets stimulated by thrombin as measured by the DNase I inhibition assay. *Proc. Natl. Acad. Sci. USA* **76,** 6376–6380.

Carlsson, L., Nyström, L. -E., Sundkvist, I., Markey, F., and Lindberg, U. (1977). Actin polymerizability is influenced by profilin, a low molecular weight protein in non-muscle cells. *J. Mol. Biol.* **115,** 465–483.

Carron, C. P., Hwo, S., Dingus, J., Benson, D. M., Meza, I., and Bryan, J. (1986). A re-evaluation of cytoplasmic gelsolin localization. *J. Cell Biol.* **102,** 237–245.

Chaponnier, C., Janmey, P. A., and Yin, H. L. (1986). The actin filament-severing domain of plasma gelsolin. *J. Cell Biol.* **103,** 1473–1481.

Chaponnier, C., Yin, H. L., and Stossel, T. P. (1987). Reversibility of gelsolin/actin interaction in macrophages. Evidence of Ca^{2+}-dependent and Ca^{2+}-independent pathways. *J. Exp. Med.* **165,** 97–106.

Clarke, M., and Spudich, J. A. (1974). Biochemical and structural studies of actomyosin-like proteins from non-muscle cells. Isolation and characterization of myosin from amoebae of *Dictyostelium discoideum. J. Mol. Biol.* **86,** 209–222.

Coleman, T. R., Fishkind, D. J., Mooseker, M. S., and Morrow, J. S. (1989). Functional diversity among spectrin isomers. *Cell Motil. Cytoskel.* **12,** 225–247.

Collins, J. H., and Korn, E. D. (1980). Actin activation of Ca^{2+}-sensitive Mg^{2+}-ATPase activity of *Acanthamoeba* myosin II is enhanced by dephosphorylation of its heavy chains. *J. Biol. Chem.* **255,** 8011–8014.

Collins, J. H., and Korn, E. D. (1981). Purification and characterization of actin-activatable, Ca^{2+}-sensitive myosin II from *Acanthamoeba. J. Biol. Chem.* **256,** 2586–2595.

Collins, J. H., Kuznicki, J., Bowers, B., and Korn, E. D. (1982). Comparison of the actin binding and filament formation properties of phosphorylated and dephosphorylated *Acanthamoeba* myosin II. *Biochemistry* **21,** 6910–6915.

Condeelis, J. S., and Taylor, D. L. (1977). The contractile basis of amoeboid movement. V. The control of gelation, solation, and contraction in extracts from *Dictyostelium discoideum. J. Cell Biol.* **74,** 901–927.

Cooper, J. A., Bryan, J., Schwab, B., III, Frieden, C., Loftus, D. J., and Elson, E. L. (1987). Microinjection of gelsolin into living cells. *J. Cell Biol.* **104,** 491–501.

Cooper, J. A., Loftus, D. J., Frieden, C., Bryan, J., and Elson, E. L. (1988). Localization and mobility of gelsolin in cells. *J. Cell Biol.* **106,** 1229–1240.

Côté, G. P., Collins, J. H., and Korn, E. D. (1981). Identification of three phosphorylation sites on each heavy chain of *Acanthamoeba* myosin II. *J. Biol. Chem.* **256,** 12,811–12,816.

Craig, S. W., and Pollard, T. D. (1982). Actin-binding proteins. *Trends Biochem. Sci.* **7,** 88–92.

De Lozanne, A., and Spudich, J. A. (1987). Disruption of the *Dictyostelium* myosin heavy chain gene by homologous recombination. *Science* **236,** 1086–1091.

D'Haese, J., and Hinssen, H. (1979). Aggregation properties of non-muscle myosins. *In* "Cell Motility: Molecules and Organization" (S. Hatano, H. Ishikawa, and H. Sato, Eds.), pp. 105–118. Univ. of Tokyo Press, Tokyo.

Doberstein, S. K., and Pollard, T. D. (1989). Recombinant myosin-I fusion proteins bind isolated membranes. *J. Cell Biol.* **109,** 86a.

Doi, Y., Higashida, M., and Kido, S. (1987). Plasma-gelsolin-binding sites on the actin sequence. *Eur. J. Biochem.* **164,** 89–94.

Drenckhahn, D., Steffens, R., and Groschel-Stewart, U. (1980). Immunocytochemical localization of myosin in the brush border region of the intestinal epithelium. *Cell Tissue Res.* **205,** 163–166.

Ebashi, S., Ebashi, F., and Maruyama, K. (1964). A new protein factor promoting contraction of actomyosin. *Nature (London)* **203,** 645–646.

Fleischer, M., and Wohlfarth-Bottermann, K. E. (1975). Correlation between tension force generation, fibrillogenesis and ultrastructure of cytoplasmic actomyosin during isometric and isotonic contractions of protoplasmic strands. *Cytobiologie* **10,** 339–365.

Frieden, C. (1985). Actin and tubulin polymerization: The use of kinetic methods to determine mechanism. *Ann. Rev. Biophys. Biophys. Chem.* **14,** 189–210.

Friederich, E., Huet, C., Arpin, M., and Louvard, D. (1989). Villin induces microvilli growth and actin redistribution in transfected fibroblasts. *Cell* **59,** 461–475.

Früst, D. O., Osborn, M., Nave, R., and Weber, K. (1988). The organization of titin filaments in the half-sarcomere revealed by monoclonal antibodies in immunoelectron microscopy: A map of ten nonrepetitive epitopes starting at the Z line extends close to the M line. *J. Cell Biol.* **106,** 1563–1572.

Füchtbauer, A., Jockusch, B. M., Maruta, H., Kilimann, M. W., and Isenberg, G. (1983). Disruption of microfilament organization after injection of F-actin capping proteins into living tissue cells. *Nature (London)* **304,** 361–364.

Fujime, S., and Hatano, S. (1972). Plasmodium actin polymers studied by quasielastic scattering of laser light. *J. Mechanochem. Cell Motil.* **1,** 81–90.

Fujisaki, H., Albanesi, J. P., and Korn, E. D. (1985). Experimental evidences for the contractile activities of *Acanthamoeba* myosin-IA and myosin-IB. *J. Biol. Chem.* **260,** 1183–1189.

Fujiwara, K., and Pollard, T. D. (1976). Fluorescent antibody localization of myosin in the cytoplasm, cleavage furrow, and mitotic spindle of human cells. *J. Cell Biol.* **71,** 848–875.

Fukui, Y., Lynch, T. J., Brzeska, H., and Korn, E. D. (1989). Myosin I is located at the leading edges of locomoting *Dictyostelium* amoebae. *Nature (London)* **341,** 328–331.

Furuhashi, K., and Hatano, S. (1989). A fragmin-like protein from plasmodium of *Physarum polycephalum* that severs F-actin and caps the barbed end of F-actin in a Ca^{2+}-sensitive way. *J. Biochem.* **106,** 311–318.

Furuhashi, K., and Hatano, S. (1990). Control of actin filament length by phosphorylation of fragmin-actin complex. *J. Cell Biol.* **111,** 1081–1087.

Furuhashi, K., and Hatano, S. (1992). Actin kinase: a protein kinase that phosphorylates actin of fragmin-actin complex. *J. Biochem.* **111,** 366–370.

Furuhashi, K., Hatano, S., Ando, S., Nishizawa, K., and Inagaki, M. (1992). Phosphorylation by actin kinase of pointed end-domain on the actin molecule. *J. Biol. Chem.* **267,** 9326–9330.

Gassner, D., Shraideh, Z., and Wohlfarth-Bottermann, K. E. (1985). A giant titin-like protein in *Physarum polycephalum:* Evidence for its candidacy as a major component of an elastic cytoskeletal superthin filament lattice. *Eur. J. Cell Biol.* **37,** 44–62.

Geiger, B., Volk, T., Volberg, T., and Bendori, R. (1987). Molecular interactions in adherens-type contacts. *J. Cell Sci. Suppl.* **8,** 251–272.

Gettemans, J., De Ville, Y., Vandekerckhove, J., and Waelkens, E. (1992). *Physarum* actin is phosphorylated as the actin-fragmin complex at residues Thr203 and Thr202 by a specific 80 kDa kinase. *EMBO J.* **11,** 3185–3191.

Gettemans, J., De Ville, Y., Vandekerckhove, J., and Waelkens, E. (1993). Purification and partial amino acid sequence of the actin-fragmin kinase from *Physarum polycephalum*. *Eur. J. Biochem.* **214,** 111–119.

Giuliano, K. A., Khatib, F. A., Hayden, S. M., Daoud, E. W. R., Adams, M. E., Amorese,

D. A., Bernstein, B. W., and Bamburg, J. R. (1988). Properties of purified actin depolymerizing factor from chick brain. *Biochemistry* **27,** 8931–8938.

Glenney, Jr., J. R., Geisler, N., Kaulfus, P., and Weber, K. (1981). Demonstration of at least two different actin-binding sites in villin, a calcium-regulated modulator of F-actin organization. *J. Biol. Chem.* **256,** 8156–8161.

Glenney, Jr., J. R., Glenney, P., Osborn, M., and Weber, K. (1982). An F-actin- and calmodulin-binding protein from isolated intestinal brush borders has a morphology related to spectrin. *Cell* **28,** 843–854.

Glenney, Jr., J. R., and Weber, K. (1981). Calcium control of microfilaments: Uncoupling of the F-actin-severing and -bundling activity of villin by limited proteolysis in vitro. *Proc. Natl. Acad. Sci. USA* **78,** 2810–2814.

Goldman, R. D., Lazarides, E., Pollack, R., and Weber, K. (1975). The distribution of actin in non-muscle cells. The use of actin antibody in the localization of actin within the microfilament bundles of mouse 3T3 cells. *Exp. Cell Res.* **90,** 333–344.

Goldschmidt-Clermont, P. J., Machesky, L. M., Baldassare, J. J., and Pollard, T. D. (1990). The actin-binding protein profilin binds to PIP_2 and inhibits its hydrolysis by phospholipase C. *Science* **247,** 1575–1578.

Goodman, E. M. (1982). Myxoamoebae: Structure and physiology. *In* "Cell Biology of *Physarum* and *Didymium*" (H. C. Aldrich, and J. W. Daniel, Eds.), Vol. 2, pp. 101–128. Academic Press, New York.

Grazi, E., and Magri, E. (1979). Phosphorylation of actin and removal of its inhibitory activity on pancreatic DNAse I by liver plasma membranes. *FEBS Lett.* **104,** 284–286.

Griffith, L. M., Downs, S. M., and Spudich, J. A. (1987). Myosin light chain kinase and myosin light chain phosphatase from *Dictyostelium:* Effects of reversible phosphorylation on myosin structure and function. *J. Cell Biol.* **104,** 1309–1323.

Haarer, B. K., and Brown, S. S. (1990). Structure and function of profilin. *Cell Motil. Cytoskel.* **17,** 71–74.

Hammer, J. A., III, Albanesi, J. P., and Korn, E. D. (1983). Purification and characterization of a myosin I heavy chain kinase from *Acanthamoeba castellanii. J. Biol. Chem.* **258,** 10,168–10,175.

Hammer, J. A., III, Bowers, B., Paterson, B. M., and Korn, E. D. (1987). Complete nucleotide sequence and deduced polypeptide sequence of a nonmuscle myosin heavy chain gene from *Acanthamoeba:* Evidence of a hinge in the rod-like tail. *J. Cell Biol.* **105,** 913–925.

Hanson, J., and Lowy, J. (1963). The structure of F-actin and of actin filaments isolated from muscle. *J. Mol. Biol.* **6,** 46–60.

Harris, D. A., and Schwartz, J. H. (1981). Characterization of brevin, a serum protein that shortens actin filaments. *Proc. Natl. Acad. Sci. USA* **78,** 6798–6802.

Harris, H. E., and Weeds, A. G. (1983). Plasma actin depolymerizing factor has both calcium-dependent and calcium-independent effects on actin. *Biochemistry* **22,** 2728–2741.

Hartwig, J. H., and Kwiatkowski, D. J. (1991). Actin-binding proteins. *Curr. Opin. Cell Biol.* **3,** 87–97.

Hartwig, J. H., and Stossel, T. P. (1975). Isolation and properties of actin, myosin and a new actin-binding protein in rabbit alveolar macrophages. *J. Biol. Chem.* **250,** 5696–5705.

Hartwig, J. H., and Stossel, T. P. (1981). The structure of macrophage actin-binding protein molecules in solution and interacting with actin filaments. *J. Mol. Biol.* **145,** 563–581.

Hartwig, J. H., Chambers, K. A., and Stossel, T. P. (1989). Association of gelsolin with actin filaments and cell membranes of macrophages and platelets. *J. Cell Biol.* **108,** 467–479.

Hasegawa, T., Takahashi, S., Hayashi, H., and Hatano, S. (1980). Fragmin: A calcium ion sensitive regulatory factor on the formation of actin filaments. *Biochemistry* **19,** 2677–2683.

Hatano, S. (1970). Specific effect of Ca^{2+} on movement of plasmodial fragment obtained by caffeine treatment. *Exp. Cell Res.* **61,** 199–203.

Hatano, S. (1972). Conformational changes of plasmodium actin polymers formed in the presence of Mg^{++}. *J. Mechanochem. Cell Motil.* **1,** 75–80.

Hatano, S., and Ohnuma, J. (1970). Purification and characterization of myosin A from the myxomycete plasmodium. *Biochim. Biophys. Acta* **205,** 110–120.

Hatano, S., and Oosawa, F. (1966). Isolation and characterization of plasmodium actin. *Biochim. Biophys. Acta* **127,** 488–498.

Hatano, S., and Owaribe, K. (1976). Actin and actinin from myxomycete plasmodia. *In* "Cold Spring Harbor Conferences on Cell Proliferation. Vol. 3, Cell Motility" (R. Goldman, T. Pollard, and J. Rosenbaum, Eds.), pp. 499–511. Cold Spring Harbor Laboratory Press, Cold Spring Harbor, NY.

Hatano, S., and Owaribe, K. (1977). A simple method for the isolation of actin from myxomycete plasmodia. *J. Biochem.* **82,** 201–205.

Hatano, S., Sugino, H., and Ozaki, K. (1983). Regulatory proteins of actin polymerization from *Physarum* plasmodium and classification of actin-associated proteins. *In* "Actin: Structure and Function in Muscle and Non-Muscle Cells" (Dos Remedios, Ed.), pp. 277–284. Academic Press, Australia.

Hatano, S., and Tazawa, M. (1968). Isolation, purification and characterization of myosin B from myxomycete plasmodium. *Biochim. Biophys. Acta* **154,** 507–519.

Hatano, S., and Takahashi, K. (1971). Structure of myosin A from the myxomycete plasmodium and its aggregation at low salt concentrations. *J. Mechanochem. Cell Motil.* **1,** 7–14.

Hatano, S., Totsuka, T., and Oosawa, F. (1967). Polymerization of plasmodium actin. *Biochim. Biophys. Acta* **140,** 109–122.

Hatano, S., and Totsuka, T. (1972). The polymerization of plasmodium actin in the presence of divalent cations. *J. Mechanochem. Cell Motil.* **1,** 67–74.

Hellewell, S. B., and Taylor, D. L. (1979). The contractile basis of amoeboid movement. VI. The solation-contraction coupling hypothesis. *J. Cell Biol.* **83,** 633–648.

Herman, I. M., and Pollard, T. D. (1981). Electron microscopic localization of cytoplasmic myosin with ferritin-labeled antibodies. *J. Cell Biol.* **88,** 346–351.

Herman, I. M., Crisona, N. J., and Pollard, T. D. (1981). Relation between cell activity and the distribution of cytoplasmic actin and myosin. *J. Cell Biol.* **90,** 84–91.

Higashi-Fujime, S. (1991). Reconstitution of active movement in vitro based on the actin-myosin interaction. *Int. Rev. Cytol.* **125,** 95–138.

Hinssen, H. (1981a). An actin-modulating protein from *Physarum polycephalum*. I. Isolation and purification. *Eur. J. Cell Biol.* **23,** 225–233.

Hinssen, H. (1981b). An actin-modulating protein from *Physarum polycephalum*. II. Ca^{++}-dependence and other properties. *Eur. J. Cell Biol.* **23,** 234–240.

Hinssen, H., D'Haese, J., Small, J. V., and Sobieszek, A. (1978). Mode of filament assembly of myosins from muscle and nonmuscle cells. *J. Ultrastruct. Res.* **64,** 282–302.

Hinssen, H., Small, J. V., and Sobieszek, A. (1984). A Ca^{2+}-dependent actin modulator from vertebrate smooth muscle. *FEBS Lett.* **166,** 90–95.

Hirokawa, N., Tilney, L. G., Fujiwara, K., and Heuser, J. E. (1982). The organization of actin, myosin, and intermediate filaments in the brush border of intestinal epithelial cells. *J. Cell Biol.* **94,** 425–443.

Hitchcock, S. E., Carlsson, L., and Lindberg, U. (1976). Depolymerization of F-actin by deoxyribonuclease I. *Cell* **7,** 531–542.

Hoffmann, H.-U., Meyer, R., Gawlitta, W., and Wolf, K. V. (1981). Studies on microplasmodia of *Physarum polycephalum*. IV. Three-dimensional reconstruction of a dumb-bell-shaped microplasmodium. *Microscopica Acta* **85,** 59–67.

Holmes, K. C., Popp, D., Gebhard, W., and Kabsch, W. (1990). Atomic model of the actin filament. *Nature (London)* **347,** 44–49.

Howe, C. L., Mooseker, M. S., and Graves, T. A. (1980). Brush-border calmodulin. A major component of the isolated microvillus core. *J. Cell Biol.* **85,** 916–923.

Huxley, H. E. (1963). Electron microscope studies on the structure of natural and synthetic protein filaments form striated muscle. *J. Mol. Biol.* **7,** 281–308.

Huxley, H., and Hanson, J. (1954). Changes in the cross-striations of muscle during contraction and stretch and their structural interpretation. *Nature (London)* **173,** 973–976.

Huxley, A. F., and Niedergerke, R. (1954). Structural changes in muscle during contraction. Interference microscopy of living muscle fibres. *Nature (London)* **173,** 971–973.

Isenberg, G., and Wohlfarth-Bottermann, K. E. (1976). Transformation of cytoplasmic actin: Importance for the organization of the contractile gel reticulum and the contraction-relaxation cycle of cytoplasmic actomyosin. *Cell Tissue Res.* **173,** 495–528.

Isenberg, G., Rathke, P. C., Hülsmann, N., Franke, W. W., and Wohlfarth-Bottermann, K. E. (1976). Cytoplasmic actomyosin fibrils in tissue culture cells: Direct proof of contractility by visualization of ATP-induced contraction in fibrils isolated by laser micro-beam dissection. *Cell Tissue Res.* **166,** 427–443.

Ishigami, M. (1986). Dynamic aspects of the contractile system in *Physarum* plasmodium. I. Changes in spatial organization of the cytoplasmic fibrils according to the contraction-relaxation cycle. *Cell Motil. Cytoskel.* **6,** 439–447.

Ishigami, M., Kuroda, K., and Hatano, S. (1987). Dynamic aspects of the contractile system in *Physarum* plasmodium. III. Cyclic contraction-relaxation of the plasmodial fragment in accordance with the generation-degeneration of cytoplasmic actomyosin fibrils. *J. Cell Biol.* **105,** 381–386.

Ishigami, M., Nagai, R., and Kuroda, K. (1981). A polarized light and electron microscopic study of the birefringent fibrils in *Physarum* plasmodia. *Protoplasma* **109,** 91–102.

Ishigami, M., and Hatano, S. (1986). Dynamic aspects of the contractile system in *Physarum* plasmodium. II. Contractility of Triton cell models in accordance with the contraction and relaxation phases of plasmodia. *Cell Motil. Cytoskel.* **6,** 448–457.

Ishikawa, H., Bishoff, R., and Holtzer, H. (1969). Formation of arrowhead complexes with heavy meromyosin in a variety of cell types. *J. Cell Biol.* **43,** 312–328.

Itano, N., and Hatano, S. (1991). F-actin bundling protein from *Physarum polycephalum:* Purification and its capacity for co-bundling of actin filaments and microtubules. *Cell Motil. Cytoskel.* **19,** 244–254.

Janmey, P. A., Iida, K., Yin, H. L., and Stossel, T. P. (1987). Polyphoosphoinositide micelles and polyphosphoinositide-containing vesicles dissociate endogenous gelsolin-actin complexes and promote actin assembly from the fast-growing end of actin filaments blocked by gelsolin. *J. Biol. Chem.* **262,** 12,228–12,236.

Janmey, P. A., and Matsudaira, P. (1988). Functional comparison of villin and gelsolin: Effects of Ca^{2+}, KCl and polyphosphoinositides. *J. Biol. Chem.* **263,** 16,738–16,743.

Janmey, P. A., and Stossel, T. P. (1987). Modulation of gelsolin function by phosphatidylinositol 4,5-bisphosphate. *Nature (London)* **325,** 362–364.

Jung, G., Korn, E. D., and Hammer, J. A., III (1987). The heavy chain of *Acanthamoeba* myosin IB is a fusion of myosin-like and non-myosin-like sequences. *Proc. Natl. Acad. Sci. USA* **84,** 6720–6724.

Jung, G., Schmidt, C. J., and Hammer, J. A., III (1989). Myosin-I heavy-chain genes of *Acanthamoeba castellanii:* Cloning of a second gene and evidence for the existence of the third isoform. *Gene* **82,** 269–280.

Kabsch, W., Mannherz, H. G., Suck, G., Pai, E. F., and Holmes, K. C. (1990). Atomic structure of the actin: DNase I complex. *Nature (London)* **347,** 37–44.

Kamiya, N. (1959). Protoplasmic streaming. *Protoplasmatologia* **8,** 3a: 1–199.

Kamiya, N. (1973). Contractile characteristics of the myxomycete plasmodium. *Proc. 4th Int. Biophysics Congr. Moscow*, pp. 447–465.

Kamiya, N. (1979). Dynamic aspects of movement in the myxomycete plasmodium. *In* "Cell Motility: Molecules and Organization" (S. Hatano, H. Ishikawa, and H. Sato, Eds.), pp. 399–414. Univ. of Tokyo Press, Tokyo.

Kamiya, N., Allen, R. D., and Zeh, R. (1972). Contractile properties of the slime mold strand. *Acta Protozool.* **11,** 113–124.

Kamiya, N., and Kuroda, K. (1958). Studies on the velocity distribution of the protoplasmic streaming in the myxomycete plasmodium. *Protoplasma* **49,** 1–4.

Kamiya, N., and Kuroda, K. (1965). Rotational protoplasmic streaming in *Nitella* and some physical properties of the endoplasm. *Proc. 4th Int. Congr. Rheol.*, pp. 157–171.

Kamiya, N., and Yoshimoto, Y. (1972). Dynamic characteristics of the cytoplasm. A study on the plasmodial strand of a myxomycete. *In "Aspects of Cellular and Molecular Physiology"* (K. Hamaguchi, Ed.), pp. 167–189, Univ. Tokyo Press, Tokyo.

Kamiya, R., Maruyama, K., Kuroda, M., Kawamura, M., and Kikuchi, M. (1972). Mg-polymer of actin formed under the influence of beta-actinin. *Biochim. Biophys. Acta* **256,** 120–131.

Kane, R. E. (1975). Preparation and purification of polymerized actin from sea urchin egg extracts. *J. Cell Biol.* **66,** 305–315.

Kane, R. E. (1976). Actin polymerization and interaction with other proteins in temperature-induced gelation of sea urchin egg extracts. *J. Cell Biol.* **71,** 704–714.

Kasai, M. (1969). Thermodynamical aspect of G-F transformations of actin. *Biochim. Biophys. Acta* **180,** 399–409.

Kasai, M., Asakura, S., and Oosawa, F. (1962). The G-F equilibrium in actin solutions under various conditions. *Biochim. Biophys. Acta* **57,** 13–21.

Kersey, Y. M., Hepler, P. K., Palevitz, B. A., and Wessells, N. K. (1976). Polarity of actin filaments in *characean* algae. *Proc. Natl. Acad. Sci. USA* **73,** 165–167.

Kessler, D. (1982). Plasmodial structure and motility. *In* "Cell Biology of *Physarum* and *Didymium*" (H. C. Aldrich and J. W. Daniel, Eds.), Vol. 1, pp. 145–208. Academic Press, New York.

Kiehart, D. P. (1990). Molecular genetic dissection of myosin heavy chain function. *Cell* **60,** 347–350.

Kiehart, D. P., Mabuchi, I., and Inoue, S. (1982). Evidence that myosin does not contribute to force production in chromosome movement. *J. Cell Biol.* **94,** 165–178.

Kimura, S., Maruyama, K., and Huang, Y. P. (1984). Interactions of muscle beta-connectin with myosin, actin, and actomyosin at low ionic strengths. *J. Biochem.* **96,** 499–506.

Knecht, D. A., and Loomis, W. F. (1987). Antisense RNA inactivation of myosin heavy chain gene expression in *Dictyostelium discoideum*. *Science* **236,** 1081–1086.

Kobayashi, R., Bradley, W. A., Bryan, J., and Field, J. B. (1983). Identification and purification of calcium ion dependent modulators of actin polymerization from bovine thyroid. *Biochemistry* **22,** 2463–2469.

Kobayashi, T., Takagi, T., Konishi, K., Hamada, Y., Kawaguchi, M., and Kohama, K. (1988). Amino acid sequence of the calcium-binding light chain of myosin from the lower eukaryote, *Physarum polycephalum*. *J. Biol. Chem.* **263,** 305–313.

Kohama, K. (1985). The Ca^{2+} inhibition of actin-myosin-ATP interaction of *Physarum polycephalum*. *In* "Cell Motility: Mechanism and Regulation" (H. Ishikawa, S. Hatano, and H. Sato, Eds.) pp. 65–76. Univ. of Tokyo Press, Tokyo.

Kohama, K. (1987). Ca-inhibitory myosins: their structure and function. *Adv. Biophys.* **23,** 149–182.

Kohama, H., and Kendrick-Jones, J. (1986). The inhibitory Ca^{2+}-regulation of the actin-activated Mg-ATPase activity of myosin from *Physarum polycephalum*. *J. Biochem.* **99,** 1433–1446.

Kohama, K., Ishikawa, R., and Okagaki, T. (1992). Calcium inhibition of *Physarum* actomyosin system: Myosin-linked and actin-linked natures. *In* "Calcium Inhibition. A New Mode for Ca^{2+} Regulation" (K. Kohama, Ed.), pp. 91–107. Japan Sci. Soc. Press, Tokyo and CRC Press, Tokyo.

Kohama, K., Kobayashi, K., and Mitani, S. (1980). Effects of Ca ion and ADP on superprecipitation of myosin B from slime mold, *Physarum polycephalum*. *Proc. Jpn. Acad.* **56B,** 591–596.

Kohama, K., Kohama, T., and Kendrick-Jones, J. (1987). Effect of N-ethylmaleimide on Ca-inhibition of *Physarum* myosin. *J. Biochem.* **102,** 17–23.

Kohama, K., Tanokura, M., and Yamada, K. (1984). ^{31}P-nuclear magnetic resonance studies of intact plasmodia of *Physarum polycephalum*. *FEBS Lett.* **76,** 161–165.

Komnick, H., Stockem, W., and Wohlfarth-Bottermann, K. E. (1973). Cell motility: Mechanisms in protoplasmic streaming and amoeboid movement. *Int. Rev. Cytol.* **34,** 169–249.

Kondo, H., and Ishiwata, S. (1976). Uni-directional growth of F-actin. *J. Biochem.* **79,** 159–171.

Korn, E. D. (1982). Actin polymerization and its regulation by proteins from nonmuscle cells. *Physiol. Rev.* **62,** 672–737.

Korn, E. D., Carlier, M.-F., and Pantaloni, D. (1987). Actin polymerization and ATP hydrolysis. *Science* **238,** 638–644.

Korn, E. D., and Hammer, J. A., III (1988). Myosins of nonmuscle cells. *Ann. Rev. Biophys. Biophys. Chem.* **17,** 23–45.

Kreis, T. E., and Birchmeier, W. (1980). Stress fiber sarcomeres of fibroblasts are contractile. *Cell* **22,** 555–561.

Kuczmarski, E. R., Tafuri, S. R., and Parysek, L. M. (1987). Effect of heavy chain phosphorylation on the polymerization and structure of *Dictyostelium* myosin filaments. *J. Cell Biol.* **105,** 2989–2997.

Kuczmarski, E. R., and Spudich, J. A. (1980). Regulation of myosin self-assembly: Phosphorylation of *Dictyostelium* heavy chain inhibits formation of thick filaments. *Proc. Natl. Acad. Sci. USA* **77,** 7292–7296.

Kukulies, J., and Stockem, W. (1985). Function of the microfilament system in living cell fragments of *Physarum polycephalum* as revealed by microinjection of fluorescent analogs. *Cell Tissue Res.* **242,** 323–332.

Kukulies, J., Stockem, W., and Achenbach, F. (1984). Distribution and dynamics of fluorochromed actin in living stages of *Physarum polycephalum*. *Eur. J. Cell Biol.* **35,** 235–245.

Kuramochi, K., Mabuchi, I., and Owaribe, K. (1986). Spectrin from sea urchin eggs. *Biomed. Res.* **7,** 65–68.

Kuroda, K. (1979). Movement of cytoplasm in a membrane-free system. *In* "Cell Motility: Molecules and Organization" (S. Hatano, H. Ishikawa, and H. Sato, Eds.), pp. 347–361. Univ. of Tokyo Press, Tokyo.

Kuroda, K. (1990). Cytoplasmic streaming in plant cells. *Int. Rev. Cytol.* **121,** 267–307.

Kuroda, K., and Sonobe, S. (1981). Reactivation of a glycerinated model of amoeba. *Protoplasma* **109,** 127–142.

Kuroda, R., Hatano, S., Hiramoto, Y., and Kuroda, H. (1988). Change of cytosolic Ca-ion cocentration in the contraction and relaxation cycle of *Physarum* microplasmodia. *Protoplasma, Suppl.* **1,** 72–80.

Kurth, M. C., and Bryan, J. (1984). Platelet activation induces the formation of a stable gelsolin-actin complex from monomeric gelsolin. *J. Biol. Chem.* **259,** 7473–7479.

Kuznicki, J., Albanesi, J. P., Côté, G. P., and Korn, E. D. (1983). Supramolecular regulation of the actin-activated ATPase activity of filaments of *Acanthamoeba* myosin II. *J. Biol. Chem.* **258,** 6011–6014.

Kwiatkowski, D. J., Mehl, R., and Yin, H. L. (1988). Genomic organization and biosynthesis of secreted and cytoplasmic forms of gelsolin. *J. Cell Biol.* **106,** 375–384.

Kwiatkowski, D. J., Stossel, T. P., Orkin, S. H., Mole, J. E., Colten, H. R., and Yin, H. L. (1986). Plasma and cytoplasmic gelsolins are encoded by a single gene and contain a duplicated actin-binding domain. *Nature (London)* **323,** 455–458.

Labeit, S., Gautel, M., Lakey, A., and Trinick, J. (1992). Towards a molecular understanding of titin. *EMBO J.* **11,** 1711–1716.

Langanger, G., Moeremans, M., Daneels, G., Sobieszek, A., De Brabander, M., and De Mey, J. (1986). The molecular organization of myosin in stress fibers of cultured cells. *J. Cell Biol.* **102,** 200–209.

Lassing, I., and Lindberg, U. (1985). Specific interaction between phosphatidylinositol 4,5-bisphosphate and profilactin. *Nature (London)* **314,** 472–474.

Lassing, I., and Lindberg, U. (1988). Evidence that the phosphatidylinositol cycle is linked to cell motility. *Exp. Cell Res.* **174,** 1–15.

Lazarides, E. (1975). Tropomyosin antibody. The specific localization of tropomyosin in nonmuscle cells. *J. Cell Biol.* **65,** 549–561.

Lazarides, E., and Burridge, K. (1975). Alpha-actinin: Immunofluorescent localization of a muscle structural protein in nonmuscle cells. *Cell* **6,** 289–298.

Lazarides, E., and Weber, K. (1974). Actin antibody: the specific visualization of actin filaments in non-muscle cells. *Proc. Natl. Acad. Sci. USA* **71,** 2268–2272.

Levine, J., and Willard, M. (1981). Fodrin: Axonally transported polypeptides associated with the internal periphery of many cells. *J. Cell Biol.* **90,** 631–643.

Lind, S. E., Janmey, P. A., Chaponnier, C., Herbert, T.-J., and Stossel, T. P. (1987). Reversible binding of actin to gelsolin and profilin in human platelet extracts. *J. Cell Biol.* **105,** 833–842.

Lymn, R. W., and Taylor, E. W. (1971). Mechanism of adenosine triphosphate hydrolysis by actomyosin. *Biochemistry* **10,** 4617–4624.

Lynch, T. J., Albanesi, J. P., Korn, E. D., Robinson, E. A., Bowers, B., and Fujisaki, H. (1986). ATPase activities and actin-binding properties of subfragments of *Acanthamoeba* myosin IA. *J. Biol. Chem.* **261,** 17,156–17,162.

Lynch, T. J., Brzeska, H., Miyata, H., and Korn, E. D. (1989). Purification and characterization of a third isoform of myosin I from *Acanthamoeba castellanii*. *J. Biol. Chem.* **264,** 19,333–19,339.

Mabuchi, I. (1981). Purification from starfish eggs of a protein that depolymerizes actin. *J. Biochem.* **89,** 1341–1344.

Mabuchi, I. (1983). An actin-depolymerizing protein (depactin) from starfish oocytes: Properties and interaction with actin. *J. Cell Biol.* **97,** 1612–1621.

Mabuchi, I. (1986). Biochemical aspects of cytokinesis. *Int. Rev. Cytol.* **101,** 175–213.

MacLean-Fletcher, S. and Pollard, T. D. (1980). Viscometric analysis of the gelation of *Acanthamoeba* extracts and purification of two gelation factors. *J. Cell Biol.* **85,** 414–428.

Mannherz, H. G., Leigh, J. B., Leberman, R., and Pfrang, H. (1975). A specific 1:1 G-actin: DNAase I complex formed by the action of DNAase I on F-actin. *FEBS Lett.* **60,** 34–38.

Markey, F., Persson, T., and Lindberg, U. (1981). Characterization of platelet extracts before and after stimulation with respect to the possible role of profilactin as microfilament precursor. *Cell* **23,** 145–153.

Maruta, H., and Korn, E. D. (1977a). *Acanthamoeba* cofactor protein is a heavy chain kinase required for actin activation of the Mg^{2+}-ATPase activity of *Acanthamoeba* myosin I. *J. Biol. Chem.* **252,** 8329–8332.

Maruta, H., and Korn, E. D. (1977b). *Acanthamoeba* myosin II. *J. Biol. Chem.* **252,** 6501–6509.

Maruta, H., Gadasi, H., Collins, J. H., and Korn, E. D. (1979). Multiple forms of *Acanthamoeba* myosin I. *J. Biol. Chem.* **254,** 3624–3630.

Maruta, H., Isenberg, G., Schreckenbach, T., Hallmann, H., Risse, G., Shibayama, T., and Hesse, J. (1983). Ca^{2+}-dependent actin-binding phosphoprotein in *Physarum polycephalum*. I. Ca^{2+}/actin-dependent inhibition of its phosphorylation. *J. Biol. Chem.* **258,** 10,144–10,150.

Maruyama, K. (1973). Instability of F-actin under the influence of beta-actinin. *Biochim. Biophys. Acta* **305,** 679–683.

Maruyama, K., and Ebashi, S. (1965). Alpha-actinin, a new structural protein from striated muscle. II. Action on actin. *J. Biochem.* **58,** 13–19.

Maruyama, K., Kimura, S., Yoshidomi, H., Sawada, H., and Kikuchi, M. (1984). Molecular size and shape of beta-connectin, an elastic protein of striated muscle. *J. Biochem.* **95,** 1423–1433.

Maruyama, K., Mabuchi, I., Matsubara, S., and Ohashi, K. (1976b). An elastic protein from the cortical layer of the sea-urchin egg. *Biochim. Biophys. Acta* **446,** 321–324.

Maruyama, K., Murakami, F., and Ohashi, K. (1977). Connectin, an elastic protein of muscle. Comparative biochemistry. *J. Biochem.* **82,** 339–345.

Maruyama, K., Natori, R., and Nonomura, Y. (1976b). New elastic protein from muscle. *Nature (London)* **262,** 58–59.

Maruyama, K., Yoshioka, T., Higuchi, H., Ohashi, K., Kimura, S., and Natori, R. (1985). Connectin filaments link thick filaments and Z lines in frog skeletal muscle as revealed by immunoelectron microscopy. *J. Cell Biol.* **101,** 2167–2172.

Masuda, H., Owaribe, K., and Hatano, S. (1983). Contraction of Triton-treated culture cells. A calcium-sensitive contractile model. *Exp. Cell Res.* **143,** 79–90.

Masuda, H., Owaribe, K., Hayashi, H., and Hatano, S. (1984). Ca^{2+}-dependent contraction of human lung fibroblasts treated with Triton X-100: A role of Ca^{2+}-calmodulin-dependent phosphorylation of myosin 20,000-dalton light chain. *Cell Motil.* **4,** 315–331.

Matsudaira, P. T., and Burgess, D. R. (1979). Identification and organization of the components in the isolated microvillus cytoskeleton. *J. Cell Biol.* **83,** 667–673.

Matsumura, F., and Hatano, S. (1978). Reversible superprecipitation and bundle formation of plasmodium actomyosin. *Biochim. Biophys. Acta* **533,** 511–523.

Matsumura, F., Yoshimoto, Y., and Kamiya, N. (1980). Tension generation by actomyosin thread from a non-muscle system. *Nature (London)* **285,** 169–171.

Mimura, N., and Asano, A. (1979). Ca^{2+}-sensitive gelation of actin filaments by a new protein factor. *Nature (London)* **282,** 44–48.

Mimura, N., and Asano, A. (1986). Isolation and characterization of a conserved actin-binding domain from rat hepatic actinogelin, rat skeletal muscle, and chicken gizzard alpha-actinins. *J. Biol. Chem.* **261,** 10,680–10,687.

Mimura, N., and Asano, A. (1987). Further characterization of a conserved actin-binding 27-kDa fragment of actinogelin and alpha-actinins and mapping of their binding sites on the actin molecule by chemical cross-linking. *J. Biol. Chem.* **262,** 4717–4723.

Miyata, H., Bowers, B., and Korn, E. D. (1989). Plasma membrane association of *Acanthamoeba* myosin-I. *J. Cell Biol.* **109,** 1519–1528.

Mooseker, M. S. (1985). Organization, chemistry, and assembly of the cytoskeletal apparatus of the intestinal brush border. *Ann. Rev. Cell Biol.* **1,** 209–241.

Mooseker, M. S. (1976). Brush horder motility. Microvillar contraction in Triton-treated brush borders isolated from intestinal epithelium. *J. Cell Biol.* **71,** 417–433.

Mooseker, M. S., and Tilney, L. G. (1975). The organization of an actin filament-membrane

complex: Filament polarity and membrane attachment in the microvilli of intestinal epitherial cells. *J. Cell Biol.* **67,** 725–743.
Muneyuki, E., Nishida, E., Sutoh, K., and Sakai, H. (1985). Purification of cofilin, a 21,000 molecular weight actin-binding protein, from porcine kidney and identification of the cofilin-binding site in the actin sequence. *J. Biochem.* **97,** 563–568.
Nachmias, V. T. (1964). Fibrillar structures in the cytoplasm of *Chaos chaos*. *J. Cell Biol.* **23,** 183–188.
Nachmias, V. T. (1968). Further electron microscope studies on fibrillar organization of the ground cytoplasm of *Chaos chaos*. *J. Cell Biol.* **38,** 40–50.
Nachmias, V. T. (1972). Filament formation by purified *Physarum* myosin. *Proc. Natl. Acad. Sci. USA* **69,** 2011–2014.
Nachmias, V. T. (1979). The contractile proteins of *Physarum polycephalum* and actin polymerization in plasmodial extracts. *In* "Cell Motility: Molecules and Organization" (S. Hatano, H. Ishikawa, and H. Sato, Eds.), pp. 33–57. Univ. of Tokyo Press, Tokyo.
Nachmias, V. T., and Meyers, C. H. (1980). Cytoplasmic droplets produced by the effect of adenine on *Physarum* plasmodia. Comparison with caffeine droplets and effect of calcium. *Exp. Cell Res.* **128,** 121–126.
Nagai, R. (1993). Regulation of intracellular movements in plant cells by environmental stimuli. *Int. Rev. Cytol.* **145,** 251–310.
Nagai, R., Yoshimoto, Y., and Kamiya, N. (1978). Cyclic production of tension force in the plasmodial strand of *Physarum polycephalum* and its relation to microfilament morphology. *J. Cell Sci.* **33,** 205–225.
Nagai, R., and Kato, T. (1975). Cytoplasmic filaments and their assembly into bundles in *Physarum* plasmodium. *Protoplasma* **86,** 141–158.
Naib-Majani, W., Achenbach, F., Weber, K., Wohlfarth-Bottermann, K. E., and Stockem, W. (1984). Immunocytochemistry of the acelluLar slime mold *Physarum polycephalum*. IV. Differentiation and dynamics of the polygonal actomyosin system. *Differentiation* **26,** 11–22.
Naib-Majani, W., Stockem, W., Wohlfarth-Bottermann, K. E., Osborn, M., and Weber, K. (1982). Immunocytochemistry of the acellular slime mold *Physarum polycephalum*. II. Spatial organization of cytoplasmic actin. *Eur. J. Cell Biol.* **28,** 103–114.
Nakajima, H., and Allen, R. D. (1965). The changing pattern of birefringence in plasmodia of the slime mold, *Physarum polycephalum*. *J. Cell Biol.* **25,** 361–374.
Neuhaus, J.-M., Wanger, M., Keiser, T., and Wegner, A. (1983). Treadmilling of actin. *J. Muscle Res. Cell Motil.* **4,** 507–527.
Nishida, E., Maekawa, S., and Sakai, H. (1984). Cofilin, a protein in porcine brain that binds to actin filaments and inhibits their interactions with myosin and tropomyosin. *Biochemistry* **23,** 5307–5313.
Norberg, R., Thorstenssen, R., Utter, G., and Fagraeus, A. (1979). F-actin depolymerizing activity of human serum. *Eur. J. Biochem.* **100,** 575–583.
Ogihara, S., Ikebe, M., Takahashi, K., and Tonomura, Y. (1983). Requirement of phosphorylation of *Physarum* myosin heavy chain for thick filament formation, actin activation of Mg^{2+}-ATPase activity, and Ca^{2+}-inhibitory superprecipitation. *J. Biochem.* **93,** 205–223.
Ogihara, S., and Kuroda, K. (1979). Identification of a birefringent structure which appears and disappears in accordance with the shuttle streaming in *Physarum* plasmodia. *Protoplasma* **100,** 167–177.
Ohta, K., Toriyama, M., Miyazaki, M., Murofushi, H., Hosoda, S., Endo, S., and Sakai, H. (1990). The mitotic apparatus-associated 51-kDa protein from sea urchin eggs is a GTP-binding protein and is immunologically related to yeast polypeptide elongation factor 1alpha. *J. Biol. Chem.* **256,** 3240–3247.
Ohta, Y., Akiyama, T., Nishida, E., and Sakai, H. (1987). Protein kinase C and cAMP-

dependent protein kinase induce opposite effects on actin polymerizability. *FEBS Lett.* **222,** 305–310.

Ohtaki, T., Tsukita, S., Mimura, N., Tsukita, S., and Asano, A. (1985). Interaction of actinogelin with actin. No nucleation but high gelation activity. *Eur. J. Biochem.* **153,** 609–620.

Oosawa, F., and Asakura, S. (1975). "Thermodynamics of the Polymerization of Protein." Academic Press, New York/London.

Oosawa, F., and Kasai, M. (1962). A theory of linear and helical aggregations of macromolecules. *J. Mol. Biol.* **4,** 10–21.

Osborn, M., Weber, K., Naib-Majani, W., Hinssen, H., Stockem, W., and Wohlfarth-Bottermann, K. E. (1983). Immunocytochemistry of the acellular slime mold *Physarum polycephalum.* III. Distribution of myosin and the actin-modulating protein (fragmin) in sandwiched plasmodia. *Eur. J. Cell Biol.* **29,** 179–186.

Otto, J. J., and Bryan, J. (1981). The incorporation of actin and fascin into the cytoskeleton of filopodial sea urchin coelomocytes. *Cell Motil.* **1,** 179–192.

Otto, J. J., Kane, R. E., and Bryan, J. (1979). Formation of filopodia in coelomocytes: Localization of fascin, a 58,000 dalton actin cross-linking protein. *Cell* **17,** 285–293.

Otto, J. J., Kane, R. E., and Bryan, J. (1980). Redistribution of actin and fascin in sea urchin eggs after fertilization. *Cell Motil.* **1,** 31–40.

Owaribe, K., and Hatano, S. (1975). Induction of antibody against actin from myxomycete plasmodium and its properties. *Biochemistry* **14,** 3024–3029.

Ozaki, K., and Maruyama, K. (1980). Connectin, an elastic protein of muscle. A connectin-like protein from the plasmodium *Physarum polycephalum. J. Biochem.* **88,** 883–888.

Ozaki, K., Sugino, H., Hasegawa, T., Takahashi, S., and Hatano, S. (1983). Isolation and characterization of *Physarum* profilin. *J. Biochem.* **93,** 295–298.

Pagh, K., and Adelman, M. R. (1982). Identification of a microfilament-enriched, motile domain in amoeboflagellates of *Physarum polycephalum. J. Cell Sci.* **54,** 1–21.

Pies, N. J., and Wohlfarth-Bottermann, K. E. (1984). Reactivation of NBD-phallacidin-labeled actomyosin fibrils in cryosections of *Physarum polycephalum:* a new cell-free model. *Cell Biol. Int. Rep.* **8,** 1065–1068.

Pollard, T. D., Stafford, III, W. F., and Porter, M. E. (1978). Characterization of a second myosin from *Acanthamoeba castellanii. J. Biol. Chem.* **253,** 4798–4808.

Pollard, T. D. (1981). Cytoplasmic contractile proteins. *J. Cell Biol.* **91,** 156s–165s.

Pollard, T. D. (1982). Structure and polymerization of *Acanthamoeba* myosin-II filaments. *J. Cell Biol.* **95,** 816–825.

Pollard, T. D. (1984). Purification of a high molecular weight actin filament gelation protein from *Acanthamoeba* that shares antigenic determinants with vertebrate spectrins. *J. Cell Biol.* **99,** 1970–1980.

Pollard, T. D. (1986). Rate constants for the reactions of ATP- and ADP-actin with the ends of actin filaments. *J. Cell Biol.* **103,** 2747–2754.

Pollard, T. D. (1987). The myosin crossbridge problem. *Cell* **48,** 909–910.

Pollard, T. D., and Cooper, J. A. (1984). Quantitative analysis of the effect of *Acanthamoeba* profilin on actin filament nucleation and elongation. *Biochemistry* **23,** 6631–6641.

Pollard, T. D., and Cooper, J. A. (1986). Actin and actin-binding proteins. A critical evaluation of mechanisms and functions. *Ann. Rev. Biochem.* **55,** 987–1035.

Pollard, T. D., Doberstein, S. K., and Zot, H. G. (1991). Myosin-I. *Ann. Rev. Physiol.* **53,** 653–681.

Pollard, T. D., and Ito, S. (1970). Cytoplasmic filaments of *Amoeba proteus.* I. The role of filaments in consistency changes and movements. *J. Cell Biol.* **46,** 267–289.

Pollard, T. D., and Korn, E. D. (1971). Filaments of *Amoeba proteus.* II. Binding of heavy meromyosin by thin filaments in motile cytoplasmic extracts. *J. Cell Biol.* **48,** 216–219.

Pollard, T. D., and Korn, E. D. (1973a). *Acanthamoeba* myosin. I. Isolation from *Acanthamoeba castellanii* of an enzyme similar to muscle myosin. *J. Biol. Chem.* **248,** 4682–4690.

Pollard, T. D., and Korn, E. D. (1973b). *Acanthamoeba* myosin. II. Interaction with actin and with a new cofactor protein required for actin activation of Mg^{2+} adenosine triphosphatase activity. *J. Biol. Chem.* **248,** 4691–4697.

Pollard, T. D., and Mooseker, M. S. (1981). Direct measurement of actin polymerization rate constants by electron microscopy of actin filaments nucleated by isolated microvillus cores. *J. Cell Biol.* **88,** 654–659.

Pope, B., Way, M., and Weeds, A. G. (1991). Two of the three actin-binding domains of gelsolin bind to the same subdomain of actin. Implications for capping and severing mechanisms. *FEBS Lett.* **280,** 70–74.

Reines, D., and Clarke, M. (1985). Immunochemical analysis of the supramolecular structure of myosin in contractile cytoskeleton of *Dictyostelium* amoebae. *J. Biol. Chem.* **260,** 14,248–14,254.

Rodewald, R., Newman, S. B., and Karnovskyu, M. J. (1976). Contraction of isolated brush borders from the intestinal epithelium. *J. Cell Biol.* **70,** 541–554.

Safer, D., Elzinga, M., and Nachmias, V. T. (1991). Thymosin β_4 and Fx, an actin-sequestering peptide, are indistinguishable. *J. Biol. Chem.* **266,** 4029–4032.

Safer, D., Golla, R., and Nachmias, V. T. (1990). Isolation of a 5-kilodalton actin-sequestering peptide from human blood platelets. *Proc. Natl. Acad. Sci. USA* **87,** 2536–2540.

Salles-Passador, I., Moisand, A., Planques, V., and Wright, M. (1991). *Physarum* plasmodia do contain cytoplasmic microtubules! *J. Cell Sci.* **100,** 509–520.

Sanger, J. M., and Sanger, J. W. (1980). Banding and polarity of actin filaments in interphase and cleaving cells. *J. Cell Biol.* **86,** 568–575.

Sato, H., Hatano, S., and Sato, Y. (1981). Contractility and protoplasmic streaming preserved in artificially induced plasmodial fragments, the "caffeine drops." *Protoplasma* **109,** 187–208.

Schliwa, M. (1981). Proteins associated with cytoplasmic actin. *Cell* **25,** 587–590.

Schweiger, A., Mihalache, O., Ecke, M., and Gerisch, G. (1992). Stage-specific tyrosine phosphorylation of actin in *Dictyostelium discoideum* cells. *J. Cell Sci.* **102,** 601–609.

Sekiya, K., Takeuchi, K., and Tonomura, Y. (1967). Binding of H-meromyosin with F-actin at low ionic strength. *J. Biochem.* **61,** 567–579.

Sheetz, M. P., and Spudich, J. A. (1983). Movement of myosin-coated fluorescent beads on actin cables in vitro. *Nature (London)* **303,** 31–35.

Shimaoka, S., Oosawa, M., and Maruyama, K. (1991). Magnesium polymer of actin if formed by beta-actinin but not by gelsolin-actin complex. *Zoological Sci.* **8,** 499–504.

Shimmen, T., and Yano, M. (1984). Active sliding movement of latex beads coated with skeletal muscle myosin on *Chara* actin bundles. *Protoplasma* **121,** 132–137.

Sinard, J. H., and Pollard, T. D. (1989). The effect of heavy chain phosphorylation and solution conditions on the assembly of *Acanthamoeba* myosin-II. *J. Cell Biol.* **109,** 1525–1535.

Sinard, J. H., and Pollard, T. D. (1990). *Acanthamoeba* myosin-II minifilaments assemble on a millisecond time scale with rate constants greater than those expected for a diffusion limited. *J. Biol. Chem.* **265,** 3654–3660.

Sinard, J. H., Stafford, W. F., and Pollard, T. D. (1989). The mechanism of assembly of *Acanthamoeba* myosin-II minifilaments: Minifilaments assemble by three successive dimerization steps. *J. Cell Biol.* **109,** 1537–1547.

Sobieszek, A. (1972). Cross-bridges on self-assembled smooth muscle myosin filaments. *J. Mol. Biol.* **70,** 741–744.

Sonobe, S., Hatano, S., and Kuroda, K. (1985). Cytoplasmic movement in a glycerinated

Model of *Amoeba proteus*. *In* "Cell Motility: Mechanism and Regulation" (H. Ishikawa, S. Hatano, and H. Sato, Eds.), pp. 271–282. Univ. of Tokyo Press, Tokyo.

Sonobe, S., Takahashi, S., Hatano, S., and Kuroda, K. (1986). Phosphorylation of *Amoeba* G-actin and its effect on actin polymerization. *J. Biol. Chem.* **261,** 14,837–14,843.

Spooner, B. S., Yamada, K. M., and Wessells, N. K. (1971). Microfilaments and cell locomotion. *J. Cell Biol.* **49,** 595–613.

Stockem, W., and Brix, K. (1994). Analysis of microfilament organization and contractile activities in *Physarum*. *Int. Rev. Cytol.* **149,** 145–215.

Stossel, T. P., Chaponnier, C., Ezzell, R. M., Hartwig, J. H., Janmey, P. A., Kwiatkowski, D. J., Lind, S. E., Smith, D. B., Southwick, F. S., Yin, H. L., and Zaner, K. S. (1985). Nonmuscle actin-binding proteins. *Ann. Rev. Cell Biol.* **1,** 353–402.

Stossel, T. P., and Hartwig, J. H. (1976). Interactions of actin, myosin, and a new actin-binding protein of rabbit pulmonary macrophages. II. Role in cytoplasmic movement and phagocytosis. *J. Cell Biol.* **68,** 602–619.

Sugino, H., and Hatano, S. (1982). Effect of fragmin on actin polymerization: Evidence for enhancement of nucleation and capping of the barbed end. *Cell Motil.* **2,** 457–470.

Sugino, H., and Matsumura, F. (1983). Fragmin induces tension reduction of actomyosin threads in the presence of micromolar levels of Ca^{2+}. *J. Cell Biol.* **96,** 199–203.

Sutoh, K. (1982). Identification of myosin-binding sites on the actin sequence. *Biochem.* **21,** 3654–3661.

Sutoh, K., and Hatano, S. (1986). Actin-fragmin interactions as revealed by chemical cross-linking. *Biochemistry* **25,** 435–440.

Sutoh, K., Iwane, M., Matsuzaki, F., Kikuchi, M., and Ikai, A. (1984). Isolation and characterization of a high molecular weight actin-binding protein from *Physarum polycephalum* plasmodia. *J. Cell Biol.* **98,** 1611–1618.

Sutoh, K., and Mabuchi, I. (1984). N-terminal and C-terminal segments of actin participate in binding depactin, an actin-depolymerizing protein from starfish oocytes. *Biochemistry* **23,** 6757–6761.

Sutoh, K., and Yin, H. L. (1989). End-label fingerprintings show that the N- and C-terminal of actin are in the contact site with gelsolin. *Biochemistry* **28,** 5269–5275.

Suzuki, A., Goll, D. E., Singh, I., Allen, R. E., Robson, R. M., and Stromer, M. H. (1976). Some properties of purified skeletal muscle alpha-actinin. *J. Biol. Chem.* **251,** 6860–6870.

Tada, M., and Tonomura, Y. (1967). Superprecipitation of myosin B induced by ATP as a nucleated growth process. *J. Biochem.* **61,** 123–135.

Takagi, T., Mabuchi, I., Hosoya, H., Furuhashi, K., and Hatano, S. (1990). Primary structure of profilins from two species of *Echinoidea* and *Physarum polycephalum*. *Eur. J. Biochem.* **192,** 777–781.

Takiguchi, K. (1991). Heavy meromyosin induces sliding movements between antiparallel actin filaments. *J. Biochem.* **109,** 520–527.

Tawada, K., and Sekimoto, K. (1991). A physical model of ATP-induced actin-myosin movement in vitro. *Biophys. J.* **59,** 343–356.

Taylor, D. L., and Condeelis, J. S. (1979). Cytoplasmic structure and contractility in amoeboid cells. *Int. Rev. Cytol.* **56,** 57–144.

Taylor, D. L., Condeelis, J. S., Moore, P. L., and Allen, R. D. (1973). The contractile basis of amoeboid movement. I. The chemical control of motility in isolated cytoplasm. *J. Cell Biol.* **59,** 378–394.

Tellam, R. L., Morton, D. J., and Clarke, F. M. (1989). A common theme in the amino acid sequence of actin and many actin-binding proteins? *Trends Biochem. Sci.* **14,** 130–133.

Thompson, C. M., and Wolpert, L. (1963). The isolation of motile cytoplasm from *Amoeba proteus*. *Exp. Cell Res.* **32,** 156–160.

Tilney, L. G., and Mooseker, M. (1971). Actin in the brush-border of epithelial cells of the chicken intestine. *Proc. Natl. Acad. Sci. USA* **68,** 2611–2615.

Tobacman, L. S., Brenner, S. L., and Korn, E. D. (1983). Effect of *Acanthamoeba* profilin on the pre-steady state kinetics of actin polymerization and on the concentration of F-actin at steady state. *J. Biol. Chem.* **258,** 8806–8812.

Tobacman, L. S., and Korn, E. D. (1982). The regulation of actin polymerization and the inhibition of monomeric actin ATPase activity by *Acanthamoeba* profilin. *J. Biol. Chem.* **257,** 4166–4170.

Tonomura, Y. (1972). "Muscle Proteins, Muscle Contraction and Cation Transport." Univ. of Tokyo Press, Tokyo.

Tonomura, Y. (1986). "Energy-Transducing ATPase—Structure and Kinetics." Cambridge Univ. Press, London/New York.

Toriyama, M., Ohta, K., Endo, S., and Sakai, H. (1988). 51-kD protein, a component of microtubule-organizing granules in the mitotic apparatus involved in aster formation in vitro. *Cell Motil. Cytoskel.* **9,** 117–128.

Totsuka, T., and Hatano, S. (1970). ATPase activity of plasmodium actin polymer formed in the presence of Mg^{2+}. *Biochim. Biophys. Acta* **223,** 189–197.

Trinick, J., Knight, P., and Whiting, A. (1984). Purification and properties of native titin. *J. Mol. Biol.* **180,** 331–356.

Tseng, P. C.-H., and Pollard, T. D. (1982). Mechanism of action of *Acanthamoeba* profilin: Demonstration of actin species specificity and regulation by micromolar concentrations of $MgCl_2$. *J. Cell Biol.* **94,** 213–218.

Tyler, J. M., Anderson, J. M., and Branton, D. (1980). Structural comparison of several actin-binding macromolecules. *J. Cell Biol.* **85,** 489–495.

Uyeda, T. Q. P., Hatano, S., Kohama, K., and Furuya, M. (1988a). Purification of myxamoebal fragmin, and switching of myxamoebal fragmin to plasmodial fragmin during differentiation of *Physarum polycephalum*. *J. Muscle Res. Cell Motil.* **9,** 233–240.

Uyeda, T. Q. P., Hatano, S., and Furuya, M. (1988b). Involvement of myxamoebal fragmin in the Ca^{2+}-induced reorganization of the microfilamentous cytoskeleton in flagellates of *Physarum polycephalum*. *Cell Motil. Cytoskel.* **10,** 410–419.

Uyeda, T. Q. P., and Furuya, M. (1986). Effects of low temperature and calcium on microfilament structure in flagellates of *Physarum polycephalum*. *Exp. Cell Res.* **165,** 461–472.

Uyeda, T. Q. P., and Furuya, M. (1985). Cytoskeletal changes visualized by fluorescence microscopy during amoeba-to-flagellate and flagellate-to-amoeba transformations in *Physarum polycephalum*. *Protoplasma* **126,** 221–232.

Vandekerckhove, J. S., Kaiser, D. A., and Pollard, T. D. (1989). *Acanthamoeba* actin and profilin can be cross-linked between glutamic acid 364 of actin and lysine 115 of profilin. *J. Cell Biol.* **109,** 619–626.

Vandekerckhove, J., and Weber, K. (1978a). Actin amino-acid sequences. Comparison of actins from calf thymus, bovine brain, and SV40-transformed mouse 3T3 cells with rabbit skeletal muscle actin. *Eur. J. Biochem.* **90,** 451–462.

Vandekerckhove, J., and Weber, K. (1978b). Mammalian cytoplasmic actins are the products of at least two genes and differ in primary structure in at least 25 identified positions from skeletal muscle actins. *Proc. Natl. Acad. Sci. USA* **75,** 1106–1110.

Waites, G. T., Graham, I. R., Jackson, P., Millake, D. B., Patel, B., Blanchard, A. D., Weller, P. A., Eperon, I. C., and Critchley, D. R. (1992). Mutually exclusive splicing of calcium-binding domain exons in chick alpha-actinin. *J. Biol. Chem.* **267,** 6263–6271.

Wang, K. (1977). Filamin, a new high-molecular-weight protein found in smooth muscle and non muscle cells. Purification and properties of chicken gizzard filamin. *Biochemistry* **16,** 1857–1865.

Wang, K., Ash, J. F., and Singer, S. J. (1975). Filamin, a new high-molecular-weight protein found in smooth muscle and non-muscle cells. *Proc. Natl. Acad. Sci. USA* **72,** 4483–4486.

Wang, K., McClure, J., and Tu, A. (1979). Titin: Major myofibrillar components of striated muscle. *Proc. Natl. Acad. Sci. USA* **76,** 3698–3702.

Wang, L.-L., and Spudich, J. A. (1984). A 45,000-mol-wt protein from unfertilized sea urchin eggs severs actin filaments in a calcium-dependent manner and increases the steady-state concentration of nonfilamentous actin. *J. Cell Biol.* **99,** 844–851.

Warrick, H. M., Lozanne, A., Leinwand, L. A., and Spudich, J. A. (1986). Conserved protein domains in a myosin heavy chain gene from *Dictyostelium discoideum. Proc. Natl. Acad. Sci. USA* **83,** 9433–9437.

Warrick, H. M., and Spudich, J. A. (1987). Myosin structure and function in cell motility. *Ann. Rev. Cell Biol.* **3,** 379–421.

Way, M., Gooch, J., Pope, B., and Weeds, A. G. (1989). Expression of human plasma gelsolin in *Escherichia coli* and dissection of actin binding sites by segmental deletion mutagenesis. *J. Cell Biol.* **109,** 593–605.

Way, M., and Weeds, A. G. (1988). Nucleotide sequence of pig plasma gelsolin. Comparison of protein sequence with human gelsolin and other actin-severing proteins shows strong homologies and evidence for large internal repeats. *J. Mol. Biol.* **203,** 1127–1133.

Weber, K., and Gröschel-Stewart, U. (1974). Antibody to myosin: the specific visualization of myosin-containing filaments in non muscle cells. *Proc. Natl. Acad. Sci. USA* **71,** 4561–4564.

Weeds, A. (1982). Actin-binding proteins—regulators of cell architecture and motility. *Nature (London)* **296,** 811–816.

Wegner, A., and Isenberg, G. (1983). 12-fold difference between the critical monomer concentrations of the two ends of actin filaments in physiological salt concentrations. *Proc. Natl. Acad. Sci. USA* **80,** 4922–4925.

Whiting, A., Wardale, J., and Trinick, J. (1989). Does titin regulate the length of muscle thick filaments? *J. Mol. Biol.* **205,** 263–268.

Wohlfarth-Bottermann, K. E. (1962). Weitreichende, fibrilläre Protoplasmadifferenzierungen und ihre Bedeutung für die Protoplasmaströmung. I. Elektronmikroskopischer Nachweis and Feinstruktur. *Protoplasma* **54,** 514–539.

Wohlfarth-Bottermann, K. E. (1963). Weitreichende, fibrilläre Protoplasmadifferenzierungen und ihre Bedeutung für die Protoplasmaströmung. II. Lichtmikrokopische Darstellung. *Protoplasma* **57,** 747–761.

Wohlfarth-Bottermann, K. E. (1964). Cell structures and their significance for amoeboid movement. *Int. Rev. Cytol.* **16,** 61–131.

Wohlfarth-Bottermann, K. E. (1965). Weitreichende, fibrilläre Protoplasmadifferenzierungen und ihre Bedeutung für die Protoplasmaströmung. III. Entstehung und experimentell induzierbare Musterbildungen. *Roux' Archiv* **156,** 371–403.

Wohlfarth-Bottermann, K. E. (1974). Plasmalemma invaginations as characteristic constituents of plasmodia of *Physarum polycephalum. J. Cell Sci.* **16,** 23–37.

Wohlfarth-Bottermann, K. E. (1975). Tensiometric demonstration of endogenous, oscillating contractions in plasmodia of *Physarum polycephalum. Z. Pflanzenphysiol.* **76,** 14–27.

Wohlfarth-Bottermann, K. E., and Fleischer, M. (1976). Cycling aggregation patterns of cytoplasmic F-actin coordinated with oscillating tension force generation. *Cell Tissue Res.* **165,** 327–344.

Wohlfarth-Bottermann, K. E. (1977). Oscillating contractions in protoplasmic strands of *Physarum:* Simultaneous tensiometry of longitudinal and radial rhythms, periodicity analysis and temperature dependence. *J. Exp. Biol.* **67,** 49–59.

Wolosewick, J. J., and Porter, K. R. (1976). Stereo high-voltage electron microscopy of whole cells of the human diploid line WI-38. *Am. J. Anat.* **147,** 303–323.

Woodrum, D. T., Rich, S. A., and Pollard, T. D. (1975). Evidence for biased bidirectional polymerization of actin filaments using heavy meromyosin prepared by an improved method. *J. Cell Biol.* **67,** 231–237.

Yamamoto, K., Pardee, J. D., Reidler, J., Stryer, L., and Spudich, J. A. (1982). Mechanism of interaction of *Dictyostelium* severin with actin filaments. *J. Cell Biol.* **95,** 711–719.

Yamashiro-Matsumura, S., and Matsumura, F. (1985). Purification and characterization of an F-actin-bundling 55-kilodalton protein from HeLa cells. *J. Biol. Chem.* **260,** 5087–5097.

Yang, F., Demma, M., Warren, V., Dharmawardhane, S., and Condeelis, J. (1990). Identification of an actin-binding protein from *Dictyostelium* as elongation factor 1α. *Nature (London)* **347,** 494–496.

Yin, H. L., Albrecht, J. H., and Fattoum, A. (1981). Identification of gelsolin, a Ca^{2+}-dependent regulatory protein of actin gel-sol transformation, and its intracellular distribution in a variety of cells and tissues. *J. Cell Biol.* **91,** 901–906.

Yin, H. L., and Stossel, T. P. (1979). Control of cytoplasmic actin gel-sol transformation by gelsolin, a calcium-dependent regulatory protein. *Nature (London)* **281,** 583–586.

Yin, H. L., and Stossel, T. P. (1980). Purification and structural properties of gelsolin, a Ca^{2+}-activated regulatory protein of macrophages. *J. Biol. Chem.* **255,** 9490–9493.

Yin, H. L., Iida, K., and Janmey, P. A. (1988). Identification of a polyphosphoinositide-modulated domain in gelsolin which binds to the sides of actin filaments. *J. Cell Biol.* **106,** 805–812.

Yin, H. L., Janmey, P. A., and Schleicher, M. (1990). Severin is a gelsolin prototype. *FEBS Lett.* **264,** 78–80.

Yin, H. L., Kwiatkowski, D. J., Mole, J. E., and Cole, F. S. (1984). Structure and biosynthesis of cytoplasmic and secreted variants of gelsolin. *J. Biol. Chem.* **259,** 5271–5276.

Yin, H. L., Zaner, K. S., and Stossel, T. P. (1980). Ca^{2+} control of actin gelation. Interaction of gelsolin with actin filaments and regulation of actin gelation. *J. Biol. Chem.* **255,** 9494–9500.

Yonezawa, N., Nishida, E., and Sakai, H. (1985). pH control of actin polymerization by cofilin. *J. Biol. Chem.* **260,** 14,410–14,412.

Yoshimoto, Y., and Kamiya, N. (1978a). Studies on contraction rhythm of the plasmodial strand. I. Synchronization of local rhythms. *Protoplasma* **95,** 89–99.

Yoshimoto, Y., and Kamiya, N. (1978b). Studies on contraction rhythm of the plasmodial strand. IV. Site of active oscillation in an advancing plasmodium. *Protoplasma* **95,** 123–133.

Yumura, S., and Fukui, Y. (1985). Reversible cyclic AMP-dependent change in distribution of myosin thick filaments in *Dictyostelium. Nature (London)* **314,** 194–196.

The Cell Biology of Nematocysts

Glen M. Watson and Patricia Mire-Thibodeaux[1]
Department of Biology, University of Southwestern Louisiana, Lafayette, Louisiana 70504-2451

I. Introduction

Nematocysts, the most widely used term to describe the "stinging capsules" characteristic of the phylum Cnidaria, constitute the best-studied and most diverse group of cnidae, the secretory products of cnidocytes (Watson, 1988). All cnidae consist of a collagenous capsule containing an eversible tubule (Fig. 1) (Mariscal, 1974, 1984). Approximately 30 different types of cnida have been identified on the basis of morphology, with a single animal typically possessing three or more distinct cnida types in its tissues. The true diversity of cnida types may be higher due to the presence of distinct size classes of a single nematocyst type in different tissues of the animal. Since almost nothing is known for certain about the functions of the specific cnida types (or of size classes), an analysis based on function might reveal a greater diversity than currently is appreciated. All cnidae discharge by everting the tubules to contact the target. Tubule eversion can be extremely rapid (<3 msec; Holstein and Tardent, 1984) or fairly slow (seconds), depending on the cnida type. The tubules may be specialized for penetrating the target (Tardent and Holstein, 1982) to permit injection of potent toxins into the target tissue (Kem, 1988), or for adhering to the surface of the target (Mariscal, 1974).

II. Chemistry of Nematocysts

A. Structural Components

Classic studies revealed that nematocysts consist of a collagen-like protein containing a large percentage of hydroxyproline, proline, and glycine residues (Lenhoff, et al., 1957) cross-linked by disulfide bonds (Blanquet and

[1] Present Address: Pennington Biomedical Research Center, Louisiana State University, Neurosciences Division, 6400 Perkins Road, Baton Rouge, Louisiana 70808-4124.

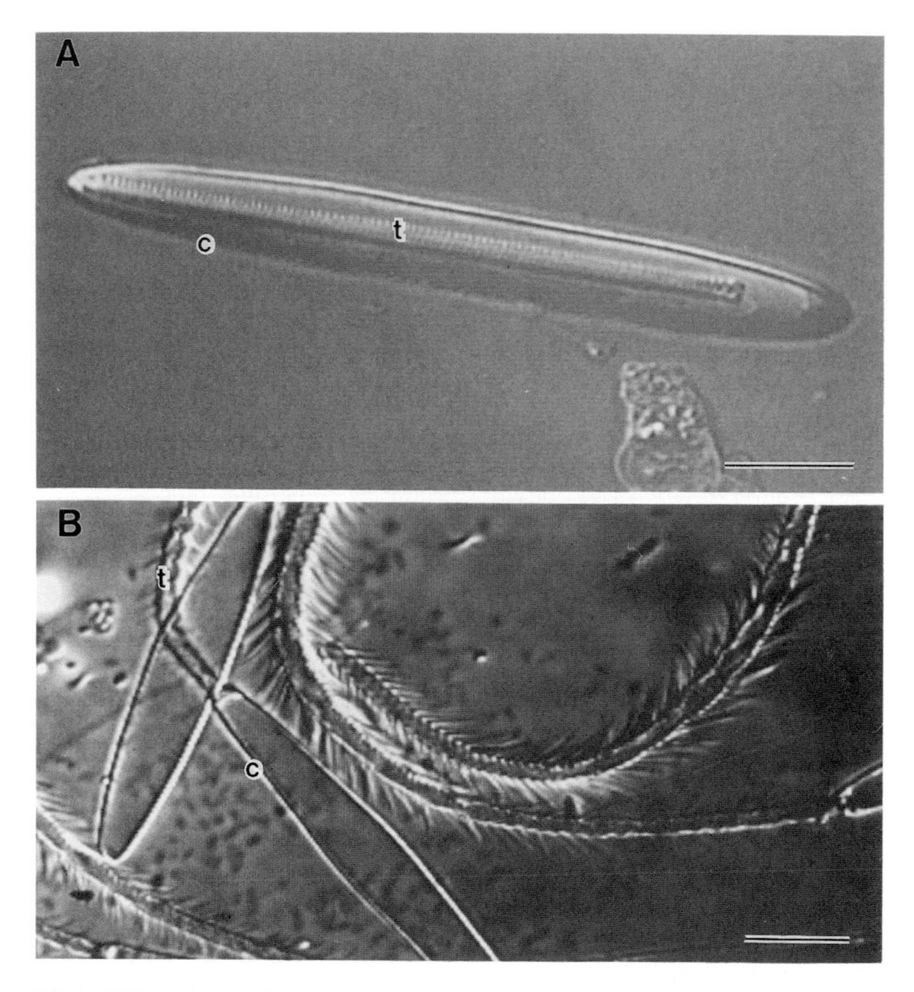

FIG. 1 Differential interference contrast photomicrographs of microbasic p-mastigophore nematocysts isolated from the sea anemone, *Haliplanella luciae*. (A) An undischarged nematocyst capsule (c) containing a tubule (t). (B) Several discharged nematocysts featuring everted tubules (t). Bars = 5 μm.

Lenhoff, 1966; Fishman and Levy, 1967; Mariscal, 1971). The relative sulfur content of the tubule wall is less than that of the capsule wall (Mariscal, 1980, 1984; Phelan and Blanquet, 1985). Cytochemical evidence suggests that at least a considerable proportion of the tubule sulfates are not in the form of disulfide bonds (Watson and Mariscal, 1984b). Such apparent differences in the relative incidence of disulfide bonds in the capsule and tubule walls are consistent with the predicted physical properties of these regions of the nematocyst.

The capsule is likely to be more rigid than the tubule; the tubule must be flexible in order to evert after being held in a highly folded state prior to discharge. During cnidogenesis, the tubule forms as an extension of the forming capsule. Once formed, the tubule folds as it inverts into the capsule (Skaer, 1973; Holstein, 1981; Watson and Mariscal, 1984a,b). Some morphological and cytochemical evidence suggests that formation of disulfide bonds continues after the deposition of wall material onto the capsule primordium (Watson and Mariscal, 1984b). Proline hydroxylase activity was reported for the contents of mature nematocysts, which is consistent with the possibility of post-translational modifications to nematocyst collagens within the nematocyst primordium (Blanquet, 1988). No evidence is yet available concerning the possibility of structurally distinct domains within the tubule wall although distinct domains may exist, for example, to enable the correct positioning of spines forming along the tubule wall (see later discussion). Typically, spines are helically and regularly arranged along the length of the tubule (Skaer and Picken, 1965).

Sodium dodecylsulfate–polyacrylamide gel electrophoresis (SDS–PAGE) of washed anemone nematocysts performed in the presence of disulfide reducing agents revealed a major 32-kDa polypeptide along with several minor polypeptides (Phelan and Blanquet, 1985). Four minicollagens have been characterized from hydra nematocysts (Kurz *et al.*, 1991). Sequencing of cDNA probes showed that the minicollagens possess Gly-X-Y triplets found in other collagens that are related to forming a triple-helix structure and that they also contain numerous proline residues that occur both in polyproline stretches and in proline-enriched stretches.

For two of four minicollagens investigated, cysteine residues are repeated every 5 amino acids in the carboxy- and amino-terminal domains of the peptides. For the other two minicollagens, cysteine residues are abundant in the flanking regions of the peptides, but are not regularly arranged. *In situ* hybridization of the cDNA probes localized to nematoblasts producing all types of nematocysts, but not to other types of cells. Mature nematocytes are unstained.

SDS-PAGE of isolated nematocysts solubilized in dithiothreitol produced a complex mixture of proteins ranging in size from 12 to 200 kDa. Pronase digestion of isolated nematocysts removed the electron-dense outer wall while apparently leaving intact the inner, electron-lucent wall of the nematocysts. SDS-PAGE of pronase-treated nematocysts yielded lower molecular-weight proteins ranging in size from 12 to 40 kDa. Tritiated proline was incorporated into 12–16 kDa and 40-kDa proteins, suggesting that minicollagens are constituents of these proteins, and that these proteins reside in the electron-lucent capsule wall.

B. Contents

Nematocysts contain complex mixtures of ions in high concentrations, polyamino acids, and proteins. Some of the proteins are potent toxins. The protein toxins fall into two broad categories: those that modulate voltage-gated sodium channels (Narahashi and Herman, 1992) and those that perturb membrane integrity (Kem, 1988). In the early 1980s, Lubbock and colleagues discovered that certain nematocysts contain high concentrations of Ca^{2+} estimated to be approximately 0.5 M (Lubbock and Amos, 1981; Lubbock *et al.* 1981; Gupta and Hall, 1984). This extraordinary finding was interpreted as forming the basis for a mechanism by which the intracapsular osmotic pressure could be rapidly increased to trigger nematocyst discharge. Many types of cnida, but certainly not all, contain high concentrations of one or several of the following elements: calcium, magnesium, phosphorus, and potassium (Lubbock and Amos, 1981; Lubbock *et al.,* 1981; Mariscal, 1984, 1988; Gerke *et al.,* 1991). The cations are believed to be complexed to venoms (Mariscal, 1988) and/or to nonvenomous intracapsular proteins (Greenwood, 1992), or to low-molecular-weight polyamino acids (Weber, 1990).

Proteins in nematocysts have high proportions of aspartic and glutamic acid residues (Lane and Dodge, 1958; Blanquet, 1968; Phelan and Blanquet, 1985), similar to those of known calcium-binding proteins such as calsequestrin and calbindins (Gross and Kumar, 1990). Several calcium-binding proteins have been partially characterized from the fluid contents of nematocysts isolated from the anemone, *Metridium senile* (Greenwood, 1992). At least two proteins ranging in molecular weight from 70 to 85 kDa bind Ca^{2+} and exhibit different electrophoretic mobilities according to whether Ca^{2+} is bound. On the other hand, homopolymers of L-glutamic acid isolated from hydra nematocysts vary in length but fall into two size classes, the larger class having a mean molecular mass of 11–14 kDa and the smaller class having a mean molecular mass of from 2.1 to 2.9 kDa (Weber, 1990). Some evidence suggests that the two classes of polyglutamate are nematocyst type-specific, with the smaller class occurring in desmonemes and isorhizas, and the larger class of polyglutamates occurring in stenoteles (Weber, 1990).

III. Mechanism of Nematocyst Discharge

Four different mechanisms for nematocyst discharge have been proposed: the contraction hypothesis, the tension hypothesis, the osmotic hypothesis, and the stopper hypothesis.

The contraction hypothesis proposes that intrinsic contraction of the capsule wall generates high pressure that causes tubule eversion. Recent proponents of this hypothesis (Cormier and Hessinger, 1980) have evoked the involvement of extracapsular contractile elements to generate the necessary pressure. Generally speaking, the contraction hypothesis has few proponents because there is no evidence for any contractile activity by isolated nematocysts, which in the absence of cellular contractile elements retain the capability of discharging in a manner that is viewed as essentially normal.

The tension hypothesis proposes that the nematocyst capsule and/or tubule stores energy that is released during discharge. Evidence in support of this hypothesis comes from microscopic observations of segments of inverted nematocyst tubules that spill out of damaged, isolated nematocysts. Exposure to the extracapsular medium induces the tubule segments to spontaneously and rapidly unwind (Carre', 1980), indicating that inverted tubules store potential energy. Upon analyzing nematocyst discharge in hydra through ultrarapid microcinematography (40,000 frames/sec), Holstein and Tardent (1984) concluded that at least some parts of discharge most likely occur as the result of a release of stored energy. More recently, Tardent (1988) raised the interesting question as to "how this mechanical force is built up during cnidogenesis and how the tubule is kept in this state of permanent tension" prior to discharge. As the tubule inverts during cnidogenesis, it folds into a triskelian that tightens further after the tubule arrives in the capsule (Skaer, 1973; Carre', 1980; Holstein, 1981; Watson and Mariscal, 1984a,b). Portions of inverted tubules spilled from isolated nematocysts move in the presence of 10 m*M* ATP (Fig. 2A), suggesting that the intrinsic motility of the tubule may be attributable to ATPases located on the tubule wall. It is possible that such ATPases normally establish tension in the tubule wall to be released at discharge. In this context, it is interesting to note that the segments of inverted tubules spilled from undischarged nematocysts decrease in diameter as they move in the presence of ATP (Fig. 2B), probably as a consequence of reestablishing maximum tightening of the folding tubule. Everted tubules of discharged nematocysts neither demonstrate motility nor change diameter in the presence of ATP (Fig. 2B).

Thus, the nematocyst tubule ATPase, if it exists, may be a substrate-dependent ATPase similar to other ATPases involved in cell motility such as myosin and kinesin. The substrate for such ATPases would most likely be a polypeptide constituent of the tubule wall that is more accessible on the inverted tubule than on the everted tubule, perhaps due to folding of the tubule. ATPase activity is demonstrable by cytochemical methods for the capsule and tubule walls of immature nematocysts (G. M. Watson,

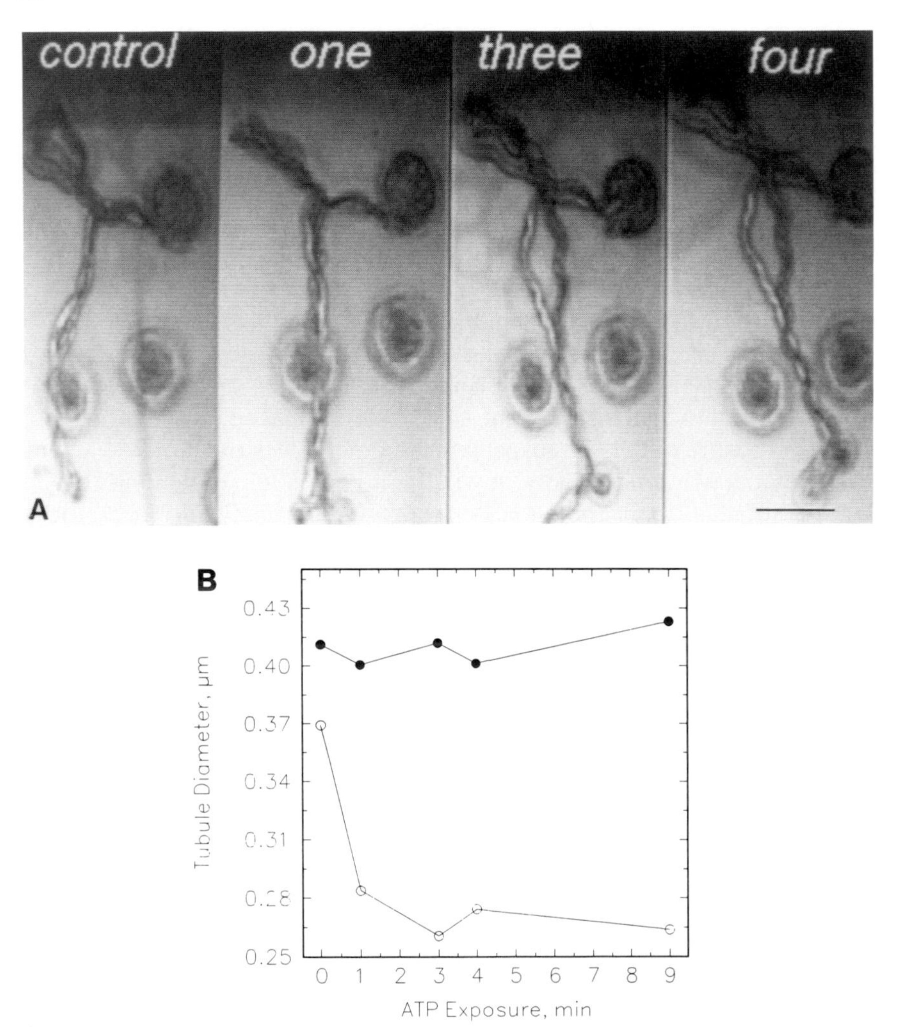

FIG. 2 Apparent motility of a segment of tubule spilled from a ruptured holotrichous isorhiza nematocyst isolated from the Portuguese man-of-war, *Physalia*. (A) The tubule moves several microns after 4 min (right panel) in 10 mM ATP. Bar = 5 μm. (B) Measurements of tubule diameter before and after exposure to ATP for an uneverted tubule (open circles) such as that shown in Fig. 2A, and for an everted tubule (closed circles) from a discharged nematocyst.

unpublished observations). However, given the complexity of nematocysts, such ATPase activity could be unrelated to tubule motility.

Some isolated nematocysts contain measurable phosphorus (Mariscal, 1984). Isolated, mature nematocysts deplete ATP from solutions (Greenwood *et al.*, 1989). Undischarged nematocysts deplete ATP more rapidly than discharged nematocysts. One of the two suspensions of discharged nematocysts tested did not appear to deplete ATP from solutions. This apparent difference in ATPase activity may be due to intrinsic differences between the nematocyst types originating from two different species of anemone. On the other hand, the observed difference in ATP depletion may be attributable to a 10-fold concentration difference between the nematocyst suspensions (i.e., the less concentrated nematocyst suspension was the one that failed to deplete ATP) (Greenwood *et al.*, 1989).

The observation that at least one of the suspensions of discharged nematocysts depleted ATP appears to rule out the possibility that ATPase activity is restricted to enzymes contained in the soluble contents expelled at discharge. It is possible that intertubule contacts were too rare in the less concentrated suspension of nematocysts to result in significant substrate-dependent ATPase activity. In the mature nematocyst, the inverted tubule may be maintained in a folded, high-energy configuration by putative ATPases on the tubule wall that exist in a state similar to that of rigor for the myosin of sarcomeres, stabilizing the structure of the folded tubule in the near absence of ATP and presence of Ca^{2+}. During discharge, a local dissociation of Ca^{2+} from the everting tubule tip may release the "rigor" of these putative tubule wall ATPases so that the tubule can unfold. Intracapsular osmotic forces (see later discussion) may induce the tubule to evert as it unfolds rather than allowing it simply to unfold (i.e., to backfire). We emphasize that no study has yet directly confirmed the existence of ATPases on the tubule wall of nematocysts.

The osmotic hypothesis requires that the osmotic pressure of the nematocyst increase prior to discharge. The increased osmotic pressure is believed to overwhelm the resistance of the nematocyst to discharge. Abundant evidence supports the existence of an osmotic component to discharge. For example, several reports have indicated that undischarged nematocysts have a high internal pressure measuring approximately 140 atm (Picken and Skaer, 1966; Lubbock and Amos, 1981; Weber, 1990). Furthermore, the volume of the nematocyst capsule increases just prior to discharge by approximately 15–20% (Holstein and Tardent, 1984) or more (Robson, 1973). Such swelling of the capsule can be mimicked by manipulating extracapsular cations (Hidaka and Afuso, 1993; Weber, 1989).

Soon after discharge begins, the capsule volume decreases to below predischarge values, indicating that the capsule wall is somewhat elastic (Holstein and Tardent, 1984; Weber, 1989). The decrease in capsule volume is partially, if not fully, compensated for by an increase in tubule volume during its eversion. Thus, the osmotic pressure appears to increase prior to discharge and then to decrease during discharge. It is possible that some osmotic pressure is dissipated during discharge by the passage of molecules through pores in the tubule wall.

Based on hemolytic activity detected in the vicinity of partially everted nematocyst tubules, it was suggested that toxins (and presumably other osmotically active compounds) pass through pores in the everting tubule wall from the lumen of the tubule (Tardent *et al.*, 1985; Klug *et al.*, 1988). Of several nematocyst types examined, only one exhibited hemolytic activity restricted to the end of the fully everted tip, suggesting that a porous tubule wall may be a feature common to many, but not all, types of nematocyst (Tardent *et al.*, 1985; Klug *et al.*, 1988).

On the other hand, the existence of such pores has not yet been demonstrated by other means. An alternative explanation is that the hemolytic toxins may be packaged within the lumen of the inverted tubule so that they are released at all stages of tubule eversion (Tardent *et al.*, 1985; Klug *et al.*, 1988). Nevertheless, the possibility of large pores in the tubule wall is intriguing, especially since they would constitute a second, special structural domain of the tubule wall, with the attachment point for spines being the other possible structural domain.

Salleo and colleageus (1988a,b) demonstrated that nematocysts release most of their calcium prior to discharge. These findings are consistent with predictions of the osmotic hypothesis as revised by Lubbock and colleagues (Lubbock and Amos, 1981; Lubbock *et al.*, 1981). According to this version of the osmotic hypothesis, Ca^{2+} complexes macromolecules into large aggregates that exhibit low osmotic pressure. Prior to discharge, most of the Ca^{2+} dissociates from these macromolecules to increase the osmotic pressure. However, this increase in osmotic pressure alone may be insufficient to overwhelm the resistance to discharge from structures that may seal the operculum (or tripartite apical flaps in anthozoan nematocysts) (Salleo *et al.*, 1988a,b) or that may hold the tubule in its folded configuration (Watson and Mariscal, 1985). Antimonate cytochemistry indicated Ca^{2+}-binding particles at the apical flaps and tubule walls of nematocysts (Fig. 3) (Watson and Mariscal, 1985). Salleo and colleagues (1988a,b) proposed that Ca^{2+} release leading to discharge must come from these sealing structures at the operculum. Although it may prove difficult to test whether Ca^{2+} plays a functional role in sealing the capsule operculum (or apical flaps), as was proposed (Salleo *et al.*, 1988a,b), the tubule may be more accommodating to experimentation since it can be isolated from nematocysts. Certain hydra nematocysts appear to contain potassium as the chief cation inside the capsule matrix. Potassium can be experimentally substituted for with divalent cations, in which case the osmotic pressure decreases (Gerke *et al.*, 1991).

Perhaps the most satisfactory current hypotheses for nematocyst discharge are those that combine osmotic mechanisms for driving nematocyst

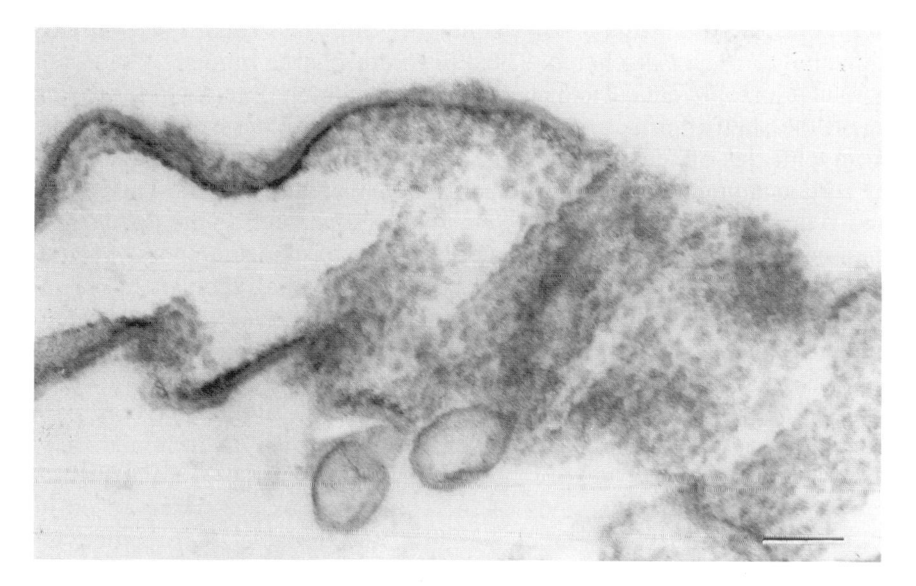

FIG. 3 Electron micrograph of an everted nematocyst tubule from a holotrichous isorhiza nematocyst from a catch tentacle of the sea anemone, *Haliplanella luciae*, stained with potassium pyroantimonate to demonstrate calcium binding domains on the nematocyst tubule. Antimonate-stained domains appear electron dense. See Watson and Mariscal (1985) for methods. Bar = 100 nm.

discharge with the release of stored tension (Holstein and Tardent, 1984) or with the opening of stoppers (Salleo *et al.*, 1988a,b). However, as discussed earlier, none of these hypotheses has yet adequately emphasized the dynamic role of the tubule in discharge. The tubule must first unfold from a triskelian configuration to a cylinder and then evert. Unfolding of the tubule must occur only at the everting tip to minimize friction (Tardent, 1988). The molecules on the tubule wall that establish and maintain the tubule in its folded configuration prior to discharge and then release the tubule at the everting tip during discharge must play a key role in discharge.

IV. Regulation of Discharge

A. Contact-Dependent Nematocyst Discharge

Classic studies revealed that many cnidarians regulate nematocyst discharge. Very few nematocysts discharge when clean glass rods are touched to the tentacles. However, in the presence of extracts of their normal prey, massive discharge follows contact by the glass rods. Prey extracts alone fail to trigger discharge. Thus, *in situ* discharge requires the proper combination of chemical and mechanical stimulation (Pantin, 1942). Using an off-the-shelf approach, it was determined that discharge of nematocysts from tentacles of sea anemones is sensitized by *N*-acetylated sugars and by certain amino compounds, including the imino acid proline (Thorington and Hessinger, 1988a,b). Unlike some cnidarians, such as the Portuguese man-of-war, *Physalia,* sea anemones discharge baseline numbers of nematocysts in the absence of chemical stimulation. Adding agonists to seawater increases levels of discharge to a maximum that is two- to threefold higher than seawater controls. Further increases in agonist concentrations decrease discharge to seawater control levels (Fig. 4).

Colloidal gold conjugates of bovine submaxillary mucin, a glycoprotein bearing many terminal sugars, including *N*-acetylneuraminic acid (NANA), were used to localize the chemoreceptors. After 1 hr of exposure to mucin-gold, more than 99% of the gold particles are observed on or within supporting cells, epithelial cells of the tentacle, with cnidocytes binding 0.25% or less of the mucin-gold. The binding of mucin-gold is blocked in the presence of excess free mucin (2 mg/ml) (Watson and

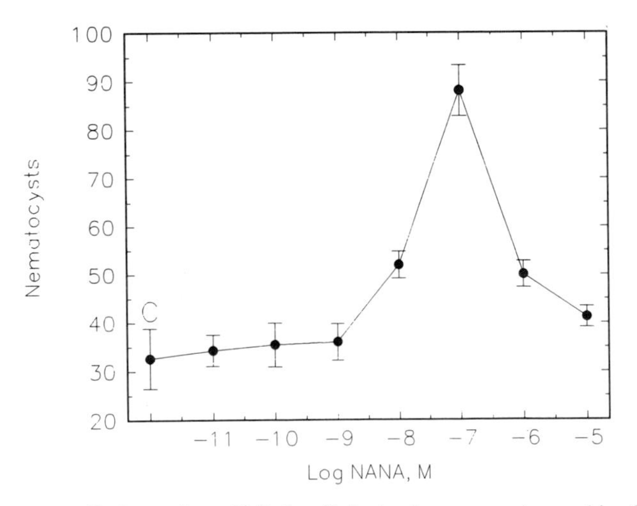

FIG. 4 Nematocyst discharge from *Haliplanella luciae* in response to combined mechanical and chemical stimulation using *N*-acetylneuraminic acid. Data points depict the mean number of nematocysts (± SEM) discharged into gelatin-coated test probes touched to tentacles after 10 min of exposure to seawater containing NANA at the concentration indicated. Eight test probes were touched to the tentacles for each experimental condition. The probes were fixed in glutaraldehyde and then prepared as wet mounts. Nematocysts were counted on the probes for a single microscopic field of view at 400× magnification. See Watson and Hessinger (1991) for methods.

Hessinger, 1988). Mucin-gold sensitizes discharge of nematocysts into test probes, indicating that it is a biologically active cytochemical marker. With time, mucin-gold particles decrease in abundance at the surface of supporting cells because of receptor-mediated endocytosis (Watson and Hessinger, 1987). During an hour of continuous exposure to mucin-gold, large decreases in surface labeling are observed at 25 and at 45 min. Interestingly, in the continuous presence of mucin-gold, the numbers of nematocysts discharged into test probes touched to tentacles of intact sea anemones vary along a similar time course, with large decreases detected at 25 and at 45 min (Watson and Hessinger, 1989b). The time courses are significantly correlated, suggesting that binding events at surface receptors of supporting cells affect the discharge of nematocysts from cnidocytes. Colloidal gold conjugates of polyproline were used to localize the proline chemoreceptors. The polyproline-gold particles also are observed almost exclusively on or within supporting cells. Binding of polyproline-gold is blocked by excess free proline or by free polyproline (G. M. Watson and J. Roberts, unpublished observations).

Chemical stimulation alone is insufficient to trigger discharge. Whereas the existence of a contact-sensitive mechanoreceptor is inferred by the absolute requirement for contact to trigger discharge, its composition and cellular distribution are at present unknown. The newly described stretch-gated channels are one interesting possibility, but none of these has yet been identified in a mechanosensory system. Furthermore, none is yet known to be modulated by second messengers or by G-proteins (Morris, 1990; French, 1992). However, too little is yet known about these channels to rule them out.

Another possibility is that the anemone "contact sensitive mechanoreceptors" comprise two different types of channels working in concert: stretch-gated channels and voltage-gated channels. Whereas the stretch-gated channels may be insensitive to chemoreceptor activation, the voltage-gated channels may be activated (i.e., become more sensitive to voltage fluctuations) following chemoreceptor activation. In this scenario, voltage-gated channels may amplify a depolarizing signal generated by stretch-gated channels located elsewhere in the membrane.

In many cell systems, voltage-gated ion channels are modulated by second messengers and/or by G-proteins (Hosey and Lazdunski, 1988; Kaupp and Altenhofen, 1992; Schultz *et al.*, 1990). Stretch-gated channels are inhibited by gadolinium (Yang and Sachs, 1989), which blocks discharge of nematocysts triggered by touching gelatin-coated probes to isolated tentacles of *Pelagia noctiluca* (Salleo *et al.*, 1994). Thus, stretch-gated channels appear to play a role in *in situ* nematocyst discharge. It is not yet known whether these channels are modulated following chemoreceptor activation.

B. Vibration-Dependent Nematocyst Discharge

In seawater alone, the greatest numbers of nematocysts are discharged by anemones into test probes vibrating at 50–55, 65, and 75 Hz. Activating chemoreceptors for *N*-acetylated sugars using 10^{-7} *M* mucin or 10^{-7} *M* NANA shifts maximal discharge to the lower frequencies of approximately 5, 15, 30, and 40 Hz, frequencies corresponding to those produced by swimming planktonic crustaceans (i.e., prey) (Watson and Hessinger, 1989a, 1991).

Activation of proline receptors does not shift the frequency response of discharge compared with seawater controls (i.e., maximal discharge occurs at 50–55, 65, and 75 Hz in the presence or absence of 10^{-8} *M* proline). However, if proline receptors are activated (using 10^{-8} *M* proline) after those for *N*-acetylated sugars, maximal discharge shifts by 60 Hz to the higher frequencies of 65, 75, 90, and 100 Hz. This effect of proline on the frequency response of nematocyst discharge is dose dependent and progressive. The response is extremely sensitive, with frequency tuning beginning at 10^{-18} *M* proline.

The chemosensory system of sea anemones can adapt to background levels of proline as high as 10^{-8} *M* proline and still respond to increases of 10^{-18} *M* proline by frequency tuning nematocyst discharge (Watson and Hessinger, 1994a). However, the chemosensory system must first adapt to the background proline before such performance is achieved. For a background concentration of 10^{-8} *M* proline, approximately 70 min are required. As the background proline concentration is reduced, the time required for adaptation decreases (Watson and Hudson, 1994).

In addition to modulating frequency responsiveness, proline shifts the amplitude responsiveness of nematocyst discharge. In seawater alone, the greatest numbers of nematocysts are discharged over a broad range of amplitudes—spanning ± 300 μm at 50–55, 65, and 75 Hz. Upon activating chemoreceptors for *N*-acetylated sugars, maximal discharge narrows to a small range of small amplitudes ($\pm$50–75 μm) at 5, 15, 30, and 40 Hz. However, activating chemoreceptors for proline after those for *N*-acetylated sugars increases the optimum amplitude for stimulating nematocyst discharge to ± 200 μm at low frequencies (e.g., 5 and 15 Hz) (Watson and Hudson, 1994).

These observations were incorporated into the following model in an attempt to explain the interactions between prey and anemone chemosensory and mechanosensory systems that regulate nematocyst discharge. *N*-Acetylated sugars diffuse from the surface of the prey so that local levels are higher in the vicinity of the prey. As the prey approaches the tentacle, the *N*-acetylated sugars activate the chemoreceptors that tune the hair bundles regulating nematocyst discharge to the low frequencies and small amplitudes corresponding to movements produced by relatively small, calmly swimming prey. If the prey swims into contact with a tentacle, it is stung by many nematocysts, including penetrant ones. Proline is released from the hemolymph of the wounded prey (Gilles, 1979). These proline molecules activate chemoreceptors that tune the hair bundles regulating nematocyst discharge to higher frequencies at small amplitudes and to larger amplitudes at low frequencies; movements presumed to correspond to those produced by wounded prey struggling to escape (Watson and Hudson, 1994).

After this point, two mechanisms may conserve nematocysts. First, as the prey succumbs to the injected toxins, it most likely produces small-amplitude movements ($<\pm 200\ \mu m$) that fail to elicit significant nematocyst discharge (Watson and Hudson, 1994). Second, as local proline levels continue to increase because of massive wounds to the prey, nematocyst discharge becomes tuned to higher frequencies that are biologically irrelevant (Watson and Hessinger, 1994a). Thus chemoreceptors on the anemone regulate nematocyst discharge so that discharge is maximal into fresh prey and into wounded, struggling prey (Watson and Hudson, 1994).

1. Anemone Hair Bundles as Vibration-Sensitive Mechanoreceptors

The tentacles of sea anemones and other cnidarians are covered with conical arrays of stereocilia morphologically similar to vertebrate hair bundles (Mariscal *et al.*, 1978; Hausmann and Holstein, 1985; Peteya, 1975). Treating specimens with $2 \times 10^{-5}\ M$ cytochalasin B disrupts hair bundles on tentacles and abolishes the frequency selectivity of nematocyst discharge. After cytochalasin treatment, no peaks of discharge are detected at 50–55, 65, and 75 Hz in seawater alone. Similarly, in $10^{-7}\ M$ NANA, no peaks of discharge are detected at 5, 15, 30, and 40 Hz (Watson and Hessinger, 1991). Interestingly, nematocyst discharge into nonvibrating targets appears to be unaffected by cytochalasin treatment. Dose-response curves to NANA of nematocyst discharge into nonvibrating test probes are statistically indistinguishable from untreated controls (Watson and Hessinger, 1991). Thus, disrupting hair bundles on the tentacles selectively interferes with the ability to distinguish vibration frequencies without appearing to affect chemoreceptor-sensitized nematocyst discharge into nonvibrating targets.

2. Morphodynamic Anemone Hair Bundles

Activating chemoreceptors for *N*-acetylated sugars induces hair bundles on the tentacle to elongate by approximately 1–2 μm while shifting nematocyst discharge to lower frequencies. Subsequently activating chemoreceptors for proline induces hair bundles to shorten to control lengths or shorter while shifting nematocyst discharge to higher frequencies (Watson and Hessinger, 1989a, 1991, 1994a).

Two types of hair bundles are present in the epithelium of anemone tentacles: CSCC hair bundles arising from cnidocyte–supporting cell complexes (Mariscal *et al.,* 1978) and SNSC hair bundles arising from sensory cell–supporting cell complexes (Peteya, 1975). In each case, supporting cells, epithelial cells bearing surface chemoreceptors for *N*-acetylated sugars and for proline, contribute stereocilia to the periphery of the hair bundle. In the CSCC bundle, the cnidocyte contributes a kinocilium to the center of the bundle. In the SNSC bundle, the sensory cell contributes a kinocilium surrounded by 5–10 large-diameter stereocilia. Sensory cells are purported to be surface neurons that interconnect to the underlying nerve net of the tentacle (Peteya, 1975; van Marle, 1990). Until recently, CSCC bundles were presumed to be the tunable mechanoreceptors (Watson and Hessinger, 1989a). Video-enhanced differential interference contrast (DIC) microscopy enabled anemone hair bundles to be distinguished according to type on living, isolated tentacles and, thus permitted the determination that activating chemoreceptors for *N*-acetylated sugars causes SNSC bundles to elongate while not affecting CSCC hair bundles (Mire-Thibodeaux and Watson, 1994).

3. Relationship of Bundle Morphodynamics to Function

In certain vertebrate hair bundles, longer bundles respond to lower frequencies of vibration, suggesting that these bundles select frequency by a mechanical means (Frishkopf and DeRosier, 1983; Holton and Hudspeth, 1983). In this case, the hair bundle is viewed as a rigid rod attached at its base to the substratum that resonates about its base in response to stimulation. Changing the length of the bundle changes its resonant frequency and, consequently, the frequency to which the bundle responds (Roberts *et al.,* 1988).

In certain vertebrates, frequency selectivity also can be regulated by an electrical means. For these hair bundles, depolarizing stimuli produce an electrical resonance of the hair cell plasma membrane near the characteristic frequency (the frequency to which the hair cell responds) (Hudspeth and Lewis, 1988). The electrical resonance results from interplay between Ca^{2+} channels and Ca^{2+}-dependent K^{+} channels. The time dependence on gating of these ion channels limits the electrical mechanisms for specifying frequency responsiveness to 800 Hz or below (Ashmore, 1991).

Based on the morphodynamic nature of SNSC hair bundles and the effects of low doses of cytochalasin D on the frequency responsiveness of nematocyst discharge (described in detail later; Mire-Thibodeaux and Watson, 1994), it appears that frequency selectivity in anemones is determined in large part by a mechanical mechanism. It is not yet known whether frequency selectivity of SNSC hair bundles also is regulated, at least in part, by an electrical mechanism.

C. Second Messengers

Santoro and Salleo (1991) concluded that discharge of nematocysts from isolated acontia requires an influx of calcium ions based upon observations that NaSCN-induced discharge requires 10 m*M* extracellular calcium and is blocked by La^{3+}, Co^{2+}, and Cd^{2+}. Such Ca^{2+}-dependent discharge is insensitive to inhibitors of L-type calcium channels (Santoro and Salleo, 1991).

In the presence of *N*-acetylated sugars, discharge of nematocysts from tentacles into nonvibrating targets exhibits a biphasic dose response (Fig. 4). As the levels of *N*-acetylated sugars are increased, discharge increases to a maximum and then decreases to seawater control levels. In the presence of optimal levels of *N*-acetylated sugars, reducing seawater Ca^{2+} to 1 m*M* decreases discharge of nematocysts from anemone tentacles to 10% of seawater control values (Watson and Hessinger, 1994b), suggesting that *in situ* discharge of nematocysts from tentacles also requires extracellular calcium ions. However, it is difficult to determine whether Ca^{2+} is involved in effector processes (i.e., in *triggering* discharge), in receptor processes (i.e., as a second messenger for chemoreceptors and/or for mechanoreceptors), or in both effector and receptor processes. Evidence for an involvement of Ca^{2+} as a second messenger comes from calcium imaging of intact, isolated sea anemone tentacles. Activation of chemoreceptors for *N*-acetylated sugars induces a two- to threefold increase in the abundance of the relatively rare cells exhibiting the highest levels of intracellular Ca^{2+} levels, an increase comparable in magnitude to the increase in nematocyst discharge into nonvibrating test probes stimulated by NANA (Mire-Thibodeaux and Watson, 1993).

The phenylalkylamine and dihydropyridine Ca^{2+} channel inhibitors, verapamil and nifedipine, and the divalent cations, cadmium and nickel, block the enhancement in discharge into nonvibrating test probes normally induced by activating chemoreceptors for *N*-acetylated sugars, but do not decrease discharge below that of untreated controls (over the range of concentrations tested) (Watson and Hessinger, 1994b). The dihydropyridine Ca^{2+} channel activator, Bay K-8644, mimics chemoreceptor activation both by enhancing discharge at relatively low concentrations and then by decreasing discharge to seawater control levels at higher concentrations.

These findings raise the interesting possibility that the "desensitizing" portion of the dose-response curve to *N*-acetylated sugars is caused by a further activation of calcium channels that seem to be present in this system. To test this possibility, Ca^{2+} channel inhibitors were employed in combination with the IC_{100} concentration of NANA (i.e., the lowest concentration of NANA that decreases nematocyst discharge to seawater control levels). Under these conditions, discharge could be rescued at certain concentrations of Ca^{2+} channel inhibitors so that it is comparable to that normally detected at the EC_{100} concentration of NANA (i.e., the lowest concentration of NANA that maximally enhances discharge) (Watson and Hessinger, 1994b).

Although these channel modulators have been most extensively studied for their effects on voltage-gated channels, they also affect certain ligand-gated channels, including cAMP-gated Ca^{2+} channels in vertebrate olfaction (Firestein, 1992). The possibility of a direct interaction between the Ca^{2+} channel inhibitors and NANA receptors appears to be unlikely since the tuning of discharge to low frequencies by NANA is unaffected by Ca^{2+} channel inhibitors.

Vibration-dependent nematocyst discharge is selectively blocked by the aminoglycoside antibiotics, gentamicin and streptomycin (Watson and Hessinger, 1994b). Discharge into nonvibrating test probes is unaffected by the aminoglycosides. In vertebrates, these antibiotics are ototoxic in part because they occlude the transduction channels located at the tips of the stereocilia of the hair bundle (Kroese *et al.*, 1989). Thus SNSC hair bundles of anemones appear to be functionally and pharmacologically similar to hair bundles of vertebrates. In hydra, the situation may be different. Aminoglycosides are known to inhibit capture of prey (Wilby, 1976). Based on sensitivity to aminoglycosides of nematocyst discharge into test targets, Gitter and colleagues (1993) concluded that mechanoreceptors in hydra are similar to vertebrate hair bundles. This conclusion may be somewhat premature since the test targets employed apparently were nonvibrating (Gitter *et al.*, 1993) and vibration-dependent discharge of nematocysts in hydra has not yet been demonstrated.

In addition to calcium, cyclic-AMP (cAMP) has been implicated as a second messenger in nematocyst discharge. Increasing intracellular cAMP with dibutyryl cAMP or with photoactivation of caged cAMP mimics the activation of receptors for *N*-acetylated sugars by sensitizing nematocyst discharge into test probes and by inducing anemone hair bundles to elongate (Watson and Hessinger, 1992; Thibodeaux and Watson, 1992). Similar results are obtained upon activating adenylyl cyclase with forskolin, or upon activating stimulatory G-proteins with cholera toxin. Transmission electron microscopy (TEM) enzyme cytochemistry localizes adenylyl cyclase activity to the cytoplasmic leaflet of the plasma membrane of supporting cells but not cnidocytes (Watson and Hessinger, 1992). The site of action for cAMP is unclear although cAMP-gated calcium channels are known to occur in vertebrate olfaction (Firestein, 1992) and cAMP-gated cation channels occur in vertebrate hair bundles (Kolesnikov *et al.*, 1991). The interaction between cAMP and other intracellular molecules in relation to the elongation of anemone hair bundles has been examined in detail and will be the topic of a forthcoming report (Mire-Thibodeaux and Watson, submitted).

With respect to vibration-dependent nematocyst discharge, activated NANA receptors appear to act antagonistically to proline receptors in that NANA tunes nematocyst discharge to lower frequencies whereas proline tunes nematocyst discharge to higher frequencies. Furthermore, NANA induces SNSC hair bundles to elongate whereas proline (after NANA treatment) induces the bundles to shorten (Watson and Hessinger, 1994a). Some pharmacological evidence suggests that the antagonistic relationship between sugar and proline receptors occurs at the level of second messengers. The response to *N*-acetylated sugars can be mimicked by activating G_s proteins, leading to the production of cAMP. It stands to reason that proline receptors may employ G_i proteins to inhibit cAMP production. This may in fact be the case since pertussis toxin, which inhibits G_i proteins, blocks the ability of proline to tune nematocyst discharge to higher frequencies (Watson and Hessinger, 1994a).

D. Integration

One of the most challenging problems for the cnidarian most likely involves determining exactly when to discharge cnidae and how many of a particular type should be discharged. The classic explanation for the regulation of nematocyst discharge is the independent effector (IE) hypothesis posed by Parker (1916). According to this hypothesis, responsibility is delegated to the cnidocyte so that discharge of the nematocyst is under the direct control of the cnidocyte without requiring nervous intervention. Thus, the cnidocyte functions both as a receptor cell and an effector cell.

Later interpretations of the IE hypothesis followed experiments in which it was shown that the physiological state of the cnidarian could profoundly affect discharge (Conklin and Mariscal, 1976). In revising the IE hypothesis, the nervous system was envisioned as establishing the overall responsiveness of the animal and mainly exerting a negative effect on discharge. The cnidocyte was still portrayed as an independent effector cell that could be activated or inactivated by the nervous system.

The characterization of specific chemoreceptors involved in regulating nematocyst discharge in anemones and their eventual localization to supporting cells surrounding cnidocytes, and not to cnidocytes, forced another revision of the IE hypothesis (Thorington and Hessinger, 1988a,b; Watson and Hessinger, 1987, 1988, 1989b). According to this latest revision, supporting cells and cnidocytes form receptor–effector complexes that regulate nematocyst discharge. The role of the nervous system is unchanged from the previous revision (i.e., the nervous system may set the overall sensitivity, with cnidocyte–supporting cell complexes regulating nematocyst discharge in response to local stimulation).

New findings from two studies seriously challenge the validity of the IE hypothesis. The first study established that activation of chemoreceptors for *N*-acetylated sugars increases the abundance of relatively rare epidermal cells that exhibit high intracellular Ca^{2+} (Mire-Thibodeaux and Watson, 1993). The abundance of these cells increases to two to three times that of controls, an increase comparable in magnitude to the increase in number of nematocysts discharged into test probes. Furthermore, time course experiments show that the abundance of these cells changes in parallel with the abundance of activated receptors at the surface of supporting cells, and with the numbers of nematocysts discharged into test probes. Thus, the chemoreceptor-dependent enhancement in nematocyst discharge most likely involves Ca^{2+} signaling by relatively rare cells estimated to be approximately 1/7 as abundant as the responding cnidocytes and 1/21 as abundant as the supporting cells associated with these c-nidocytes. The rare cells are only about half as abundant as sensory cells, neurons that also form complexes with supporting cells (P. Mire-Thibodeaux, unpublished observations). Although the exact identity of the rare cells has not yet been determined, sensory cells are considered to be the most likely candidate among these possibilities: sensory cells and/or supporting cells of SNSCs, cnidocytes, and/or supporting cells of CSCCs (Mire-Thibodeaux and Watson, 1993).

The second study employed video-enhanced DIC microscopy to distinguish hair bundles arising from CSCCs from thoese arising from SNSCs. It was determined that activating chemoreceptors for *N*-acetylated sugars induces SNSC bundles to elongate by approximately 2 μm while CSCC bundles remain unchanged in length (Mire-Thibodeaux and Watson, 1994). In the presence of submicromolar cytochalasin D, existing actin filaments are undisturbed while actin polymerization is blocked (Cooper, 1987). Submicromolar cytochalasin D blocks the elongation of SNSC hair bundles and the downward frequency shift in nematocyst discharge normally induced by NANA. Increasing the concentration of cytochalasin D to 1 μM induces SNSC hair bundles to shorten to lengths below those of untreated controls whether or not NANA is present, while shifting nematocyst discharge to higher frequencies.

Thus, SNSC hair bundles (and not the CSCC hair bundles as was originally thought; Watson and Hessinger, 1989a, 1991, 1992) are the tunable mechanoreceptors involved in vibration-dependent nematocyst discharge (Mire-Thibodeaux and Watson, 1994). Taken together, the results of these studies strongly suggest that sensory cell neurons participate in *in situ* regulation of nematocyst discharge. At this point it is not clear how the nervous system communicates with the cnidocytes/and or supporting cells although lipophilic fluorescent dyes that label the plasma membranes of neurons show apparent interconnections between sensory cells and cnidocytes. These dyes also show interconnections between sensory cells (Fig. 5) (Mire-Thibodeaux and Watson, 1994). Reports of chemical and electrical synapses between neurons of anthozoans have appeared (Westfall, 1970, 1973; van Marle, 1990).

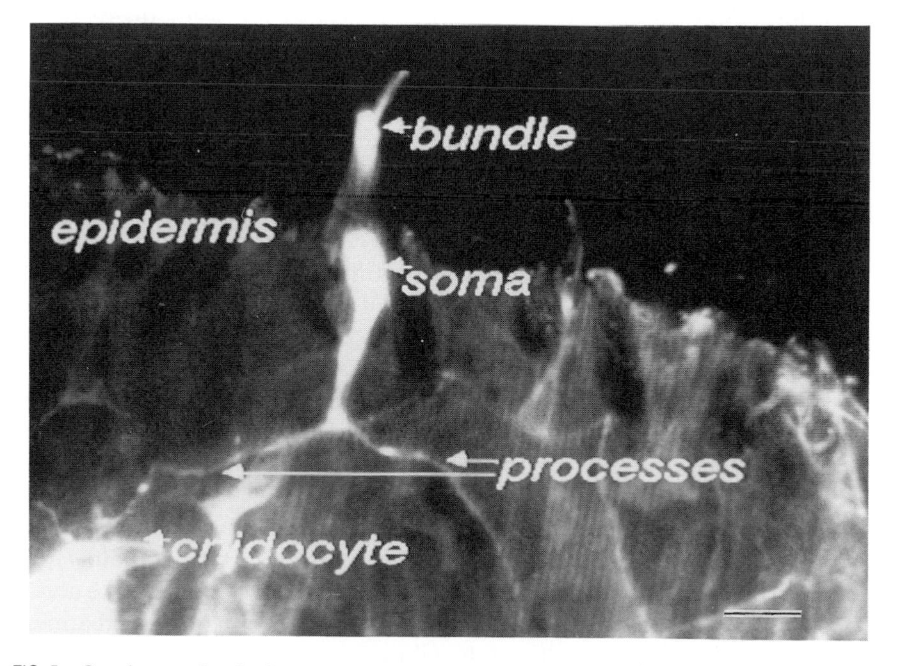

FIG. 5 Overlay confocal micrograph of the tentacle epidermis of *Haliplanella luciae* labeled with DiI. The hair bundle projects into the seawater from the epidermal surface of the tentacle. The soma and basal processes of the sensory cell are visible. Some processes extend toward a cnidocyte. See Mire-Thibodeaux and Watson (1994) for methods. Bar = 5 μm.

The next question concerns how sensory input is integrated to regulate nematocyst discharge. Activating chemoreceptors for *N*-acetylated sugars increases by two- to threefold the numbers of nematocysts discharged into nonvibrating test probes, and shifts discharge into vibrating test probes to the lower frequencies of 5, 15, 30, and 40 Hz. Test probes vibrating at frequencies other than these key frequencies (to a resolution of 5 Hz) elicit few nematocysts to discharge despite the presence of *N*-acetylated sugars. Thus, three forms of mechanical stimulation apparently are perceived by the anemone:

1. Objects vibrating at stimulatory frequencies
2. Objects vibrating at inhibitory frequencies
3. Objects not vibrating at all

The available evidence suggests that either of the vibrating stimuli take precedence over nonvibrating stimuli. For example, the stimulatory effects of NANA on discharge into nonvibrating targets are masked simply by vibrating the test probe at a frequency other than 5, 15, 30, or 40 Hz. At such inhibitory frequencies (i.e., other than at 5, 15, 30, or 40 Hz in NANA), levels of discharge are comparable to those obtained for nonvibrating test probes in seawater alone (baseline) (Watson and Hessinger, 1989a). In this context, the term ''inhibitory frequency'' seems appropriate since, at these frequencies, the presence of NANA is insufficient to elicit high levels of nematocyst discharge.

At stimulatory frequencies (i.e., 50–55, 65, and 75 Hz in seawater alone and 5, 15, 30, and 40 Hz in NANA), levels of discharge are high and comparable to those obtained for nonvibrating test probes at optimal levels of NANA. Notice that NANA is not required in order to elicit such high levels of discharge, but only to shift discharge to lower frequencies. Thus, the term, ''stimulatory frequency'' seems appropriate here.

Exposure to 2×10^{-5} *M* cytochalasin B disrupts anemone hair bundles and reduces discharge at stimulatory frequencies to baseline control levels (i.e., comparable to discharge obtained by touching tentacles with nonvibrating test probes in seawater alone; Watson and Hessinger, 1991). Thus, the enhancement in discharge normally produced by vibrating a test probe at a stimulatory frequency is abolished by cytochalasin treatment. However, since discharge over the range of 5–100 Hz does not increase above baseline control levels, even in the presence of NANA, such treatment does not also abolish sensitivity to vibrations at inhibitory frequencies. At 0 Hz, however, discharge to NANA is unaffected by cytochalasin treatment (Watson and Hessinger, 1991).

Thus, SNSC hair bundles constitute frequency-selective stimulatory mechanoreceptors. The inhibitory vibration-sensitive mechanoreceptors are not yet identified but appear to be cytochalasin insensitive and frequency nonselective. As described earlier, the contact-sensitive mechanoreceptors that seem to be modulated by chemoreceptors are not yet identified, but also are cytochalasin insensitive. Finally, the cytochalasin experiments suggest a hierarchy with respect to mechanical sensory input:

$$\text{Stimulatory freq.} > \text{inhibitory freq.} > \text{contact alone}$$

Morphological evidence for possible neuronal involvement in regulating discharge of nematocysts in hydra was reviewed by Westfall (1988). The role of the nervous system in regulating nematocyst discharge in hydra was tested using a very promising experimental system based on animals that possess cnidocytes but apparently lack neurons (Aerne *et al.*, 1991). It was determined that these animals are capable of at least limited nematocyst discharge in response to contact with prey.

Aerne and colleagues (1991) found that electrical stimulation, which bypasses cellular regulation of discharge, results in normal levels of discharge, indicating that the experimental tissue contains normal levels of functional nematocysts. It would be interesting to determine whether contact with prey stimulates comparable numbers of nematocysts to discharge in experimental animals as in untreated control animals, or whether discharge can be modulated by such chemical stimuli as *N*-acetylated sugars, or vibrating mechanical stimuli. In hydra, the role of chemoreceptors and mechanoreceptors in regulating nematocyst discharge is largely unknown.

V. Concluding Remarks

Although our understanding of the cell biology of nematocysts has improved in recent years, it still is primitive. Many fundamental questions concerning the regulation and mechanisms of nematocyst discharge remain unanswered. Nevertheless, it is clear that certain cells of the cnidarian exert a complex and sophisticated control over the utilization of nematocysts. In the end, the success of cnidarians as preditors can be attributed largely to the ingenuity of nematocysts and to the many forms of cellular control over their use.

Acknowledgments

This work was supported by the National Science Foundation, grant DCB-9105058.

References

Aerne, B. L., Stidwell, R. P., and Tardent, P. (1991). Nematocyst discharge in *Hydra* does not require the presence of nerve cells. *J. Exp. Zool.* **258,** 137–141.

Ashmore, J. F. (1991). The electrophysiology of hair cells. *Annu. Rev. Physiol.* **53,** 465–476.

Blanquet, R. S. (1968). Properties and composition of the nematocyst toxin of the sea anemone, *Aiptasia pallida. Comp. Biochem. Physiol.* **25,** 893–902.

Blanquet, R. S. (1988). The chemistry of cnidae. *In* "The Biology of Nematocysts" (D. A. Hessinger and H. M. Lenhoff, Eds.), pp. 407–425. Academic Press, New York.

Blanquet, R. S., and Lenhoff, H. M. (1966). Disulfide-linked collagenous protein of nematocyst capsules. *Science* **154,** 152–153.

Carre', D. (1980). Hypothesis on the mechanism of cnidocyst discharge. *Eur. J. Cell Biol.* **20,** 265–271.

Conklin, E. J., and Mariscal, R. N. (1976). Increase in nematocyst and spirocyst discharge in a sea anemone in response to mechanical stimulation. *In* "Coelenterate Ecology and Behavior" (G. O. Mackie, Ed.), pp. 549–588. Plenum, New York.

Cooper, J. A. (1987). Effects of cytochalasin and phalloidin on actin. *J. Cell Biol.* **105,** 1473–1478.

Cormier, S. M., and Hessinger, D. A. (1980). Cnidocil apparatus: Sensory receptor of *Physalia* nematocytes. *J. Ultrastruct. Res.* **72,** 13–19.

Firestein, S. (1992). Electrical signals in olfactory transduction. *Curr. Opinion Neurobiol.* **2,** 444–448.

Fishman, L., and Levy, M. (1967). Studies on nematocyst capsule protein from the sea anemone *Metridium marginatum. Biol. Bull.* **133,** 464–465.

French, A. (1992). Mechanotransduction. *Annu. Rev. Physiol.* **54,** 135–152.

Frishkopf, L. S., and DeRosier, D. J. (1983). Mechanical tuning of free-standing stereociliary bundles and frequency analysis in the alligator lizard cochlea. *Hearing Res.* **12,** 393–404.

Gerke, I., Zeirold, K., Weber, J., and Tardent, P. (1991). The spatial distribution of cations in nematocytes of *Hydra vulgaris. Hydrobiologia* **261/217,** 661–669.

Gilles, R. (1979). Intracellular organic osmotic effectors. *In* "Mechanisms of Osmoregulation in Animals: Maintenance of Cell Volume" (R. Gilles, Ed.), pp. 111–154. Wiley, New York.

Gitter, A. H., Oliver, D., and Thurm, U. (1993). Streptomycin inhibits nematocyte discharge in *Hydra vulgaris* by blockage of mechanosensitivity. *Naturwissenschaften* **80,** 273–276.

Greenwood, P. G., and Ellis, R. A. (1992). Calcium binding proteins in acontial nematocysts of the sea anemone *Metridium senile. Mol. Biol. Cell* **3,** 341a.

Greenwood, P. G., Johnson, L. A., and Mariscal, R. N. (1989). Depletion of ATP in suspensions of isolated cnidae: A possible role of ATP in the maturation and maintenance of anthozoan cnidae. *Comp. Biochem. Physiol. A* **93,** 761–765.

Gross, M., and Kumar, R. (1990). Physiology and biochemistry of vitamin D-dependent calcium binding proteins. *Am. J. Physiol.* **259,** F195–F209.

Gupta, B. L., and Hall, T. A. (1984). Role of high concentrations of Ca, Cu and Zn in the maturation and discharge *in situ* of sea anemone nematocysts as shown by X-ray microanalysis. *In* "Toxins, Drugs and Pollutants in Marine Animals" (L. Bolis, J. Zadunaisky, and R. Gilles, Eds.), pp. 77–95. Springer-Verlag, New York/Berlin.

Hausmann, K., and Holstein, T. (1985). Bilateral symmetry in the cnidocil-nematocyst complex of the freshwater medusa *Craspedacusta sowerbii* Lankester (Hydrozoa, Limnomedusae). *J. Ultrastruct. Res.* **90,** 89–104.

Hidaka, M., and Afuso, K. (1993). Effects of cations on the volume and elemental composition of nematocysts isolated from acontia of the sea anemone *Calliactis polypus. Biol. Bull.* **184,** 97–104.

Holstein, T. (1981). The morphogenesis of nematocysts in *Hydra* and *Forskalia:* An ultrastructural study. *J. Ultrastruct. Res.* **75,** 276–290.

Holstein, T., and Tardent, P. (1984). An ultrahigh-speed analysis of exocytosis: nematocyst discharge. *Science* **223,** 830–833.

Holton, T., and Hudspeth, A. J. (1983). A micromechanical contribution to cochlear tuning and tonotopic organization. *Science* **222,** 508–510.

Hosey, M. M., and Lazdunski, M. (1988). Calcium channels: Molecular pharmacology, structure and regulation. *J. Membr. Biol.* **104,** 81–105.

Hudspeth, A. J., and Lewis, R. S. (1988). A model for electrical resonance and frequency tuning in saccular hair cells of the bull-frog *Rana catesbeiana. J. Physiol.* **400,** 275–297.

Kaupp, U. B., and Altenhofen, W. (1992). Cyclic nucleotide-gated channels of vertebrate photoreceptor cells and olfactory epithelium. *In* "Sensory Transduction" (D. P. Corey and S. D. Roper, Eds.), pp. 133–150. Rockefeller Univ. Press, New York.

Kem, W. R. (1988). Sea anemone toxins: Structure and action. *In* "The Biology of Nematocysts" (D. A. Hessinger and H. M. Lenhoff, Eds.), pp. 375–405. Academic Press, New York.

Klug, M., Weber, J., and Tardent, P. (1988). Direct observation of hemolytic activity associated with single nematocysts. *In* "The Biology of Nematocysts" (D. A. Hessinger and H. M. Lenhoff, Eds.), pp. 543–550. Academic Press, New York.

Kolesnikov, S. S., Rebrik, T. I., Zhainazarov, A. B., Tavartkiladze, G. A., and Kalamkarov, G. R. (1991). A cyclic-AMP-gated conductance in hair cells. *FEBS Lett.* **290,** 167–170.

Kurz, E., Holstein, T. W., Petri, B. M., Engel, J., and David, C. N. (1991). Mini-collagens in hydra nematocysts. *J. Cell Biol.* **115,** 1159–1169.

Lane, C. E., and Dodge, E. (1958). Toxicity of *Physalia* nematocysts. *Biol. Bull.* **115,** 219–225.

Lenhoff, H. M., Kline, E. S., and Hurly, R. (1957). A hydroxyproline-rich, intracellular, collagen-like protein of *Hydra* nematocysts. *Biochem. Biophys. Acta* **26,** 204–205.

Lubbock, R., and Amos, W. B. (1981). Removal of bound calcium from nematocyst contents causes discharge. *Nature (London)* **290,** 500–501.

Lubbock, R., Gupta, B. L., and Hall, T. A. (1981). Novel role of calcium in exocytosis: Mechanism of nematocyst discharge as shown by X-ray microanalysis. *Proc. Natl. Acad. Sci. USA* **78,** 3624–3628.

Mariscal, R. N. (1971). Effect of a disulfide reducing agent on the nematocyst capsules from some coelenterates with an illustrated key to nematocyst classification. *In* "Experimental Coelenterate Biology" (H. M. Lenhoff, L. Muscatine, and L. V. Davis, eds.), pp. 157–168. Univ. of Hawaii Press, Honolulu.

Mariscal, R. N. (1974). Nematocysts. *In* "Coelenterate Biology: Reviews and New Perspectives" (L. Muscatine and H. M. Lenhoff, eds), pp. 129–178. Academic Press, New York.

Mariscal, R. N. (1980). The elemental composition of nematocysts as determined by X-ray microanalysis. *In* "Developmental and Cellular Biology of Coelenterates" (P. Tardent and R. Tardent, Eds.), pp. 337–342. Elsevier, Amsterdam/New York.

Mariscal, R. N. (1984). Cnidaria: Cnidae. *In* "Biology of the Integument" vol. I, Invertebrates (J. Bereiter-Hahn, A. G. Maltoltsy, and K. S. Richards, eds.), pp. 57–67. Springer-Verlag, Berlin/New York.

Mariscal, R. N. (1988). X-ray microanalysis and perspectives on the role of calcium and other elements in cnidae. *In* "The Biology of Nematocysts" (D. A. Hessinger and H. M. Lenhoff, Eds.), pp. 95–113. Academic Press, Orlando.

Mariscal, R. N., Conklin, E. J., and Bigger, C. H. (1978). The putative sensory receptors associated with the cnidae of cnidarians. *Scanning Electron Microsc.* **2;** 959–966.

Mire-Thibodeaux, P., and Watson, G. M. (1993). Direct monitoring of intracellular calcium ions in sea anemone tentacles suggests regulation of nematocyst discharge by remote rare epidermal cells. *Biol. Bull.* **185,** 335–345.

Mire-Thibodeaux, P., and Watson, G. M. (1994). Morphodynamic hair bundles arising from sensory cell/supporting cell complexes frequency-tune nematocyst discharge in sea anemones. *J. Exp. Zool.* **268,** 282–292.

Morris, C. E. (1990). Mechanosensitive ion channels. *J. Membr. Biol.* **113,** 93–107.

Narahashi, T., and Herman, M. D. (1992). Overview of toxins and drugs as tools to study excitable membrane ion channels: I. Voltage-activated channels. *Methods Enzymol.* **207,** 620–643.

Pantin, C. F. A. (1942). The excitation of nematocysts. *J. Exp. Biol.* **19,** 294–310.

Parker, G. H. (1916). The effector system of actinians. *J. Exp. Zool.* **21,** 461–484.

Peteya, D. J. (1975). The ciliary-cone sensory cell of anemones and cerianthids. *Tissue Cell* **7,** 243–252.

Phelan, M. A., and Blanquet, R. S. (1985). Characterization of nematocyst proteins from the sea anemones *Aiptasia pallida* and *Pachycerianthus torreyi* (Cnidaria: Anthozoa). *Comp. Biochem. Physiol. B* **81,** 661–666.

Picken, L. E. R., and Skaer, R. J. (1966). A review on researches on nematocysts. *Symp. Zool. Soc. London* **16,** 19–50.

Roberts, W. M., Howard, J., and Hudspeth, A. J. (1988). Hair cells: Transduction, tuning, and transmission in the inner ear. *Annu. Rev. Cell Biol.* **4,** 63–92.

Robson, E. A. (1973). The discharge of nematocysts in relation to properties of the capsule. *Publ. Seto Mar. Lab.* **20,** 653–673.

Salleo, A., LaSpada, G., and Barbera, R. (1994). Gadolinium is a powerful blocker of the activation of nematocytes of *Pelagia noctiluca*. *J. Exp. Biol.* **187,** 201–206.

Salleo, A., LaSpada, G., and Denaro, M. G. (1988a). Release of free Ca^{2+} from the nematocysts of *Aiptasia mutabilis* during the discharge. *Physiol. Zool.* **61,** 272–279.

Salleo, A., LaSpada, G., Denaro, M. G., and Falzea, G. (1988b). Dynamics of release of free calcium during the discharge of holotrichous isorhiza nematocysts of *Pelagia noctiluca*. *In* "The Biology of Nematocysts" (D. A. Hessinger and H. M. Lenhoff, Eds.), pp. 551–565. Academic Press, New York.

Santoro, G., and Salleo A. (1991). The discharge of *in situ* nematocysts of the acontia of *Aiptasia mutabilis* in a Ca^{2+}-induced response. *J. Exp. Biol.* **156,** 173–185.

Schultz, G., Rosenthal, W., and Hescheler, J. (1990) Role of G proteins in calcium channel modulation. *Annu. Rev. Physiol.* **52,** 275–292.

Skaer, R. J. (1973). The secretion and development of nematocysts in a siphonophore. *J. Cell Sci.* **13,** 371–393.

Skaer, R. J., and Picken, L. E. R. (1965). The structure of the nematocyst thread and the geometry of discharge in *Corynactis viridis* (Allman). *Philos. Trans. R. Soc. London, Ser. B* **250,** 131–164.

Tardent, P. (1988). History and current state of knowledge concerning discharge of cnidae. *In* "The Biology of Nematocysts" (D. A. Hessinger and H. M. Lenhoff, Eds.), pp. 309–332. Academic Press, New York.

Tardent, P., and Holstein, T. (1982). Morphology and morphodynamics of the stenotele nematocyst of *Hydra attenuata* Pall. (Hydrozoa, Cnidaria). *Cell Tissue Res.* **224,** 269–290.

Tardent, P., Holstein, T., Weber, J., and Klug, M. (1985). The morphodynamics and actions of stenotele nematocytes in *Hydra*. *Arch. Sci. Geneve* **38,** 401–418.

Thibodeaux, P. M., and Watson, G. M. (1992). Tuning of hair bundles on sea anemone tentacles by photoactivation of caged cyclic AMP. *Mol. Biol. Cell* **3,** 361a.

Thorington, G. U., and Hessinger, D. A. (1988a). Control of cnida discharge: I. Evidence for two classes of chemoreceptor. *Biol. Bull.* **174,** 163–171.

Thorington, G. U., and Hessinger, D. A. (1988b). Control of discharge: Factors affecting discharge of cnidae. *In* "The Biology of Nematocysts" (D. A. Hessinger and H. M. Lenhoff, Eds.), pp. 233–253. Academic Press, New York.

van Marle, J. (1990). Catecholamines, related compounds and the nervous system in the tentacles of some anthozoans. *In* "Evolution of the First Nervous Systems" (P. A. V. Anderson, Ed.), pp. 129–140. Plenum, New York.

Watson, G. M. (1988). Ultrastructure and cytochemistry of developing nematocysts. *In* "The Biology of Nematocysts" (D. A. Hessinger and H. M. Lenhoff, Eds.), pp. 143–164. Academic Press, New York.

Watson, G. M., and Hessinger, D. A. (1987). Receptor-mediated endocytosis of a chemoreceptor involved in triggering the discharge of cnidae in a sea anemone tentacle. *Tissue Cell* **19,** 747–755.

Watson, G. M., and Hessinger, D. A. (1988). Localization of a purported chemoreceptor involved in triggering cnida discharge in sea anemones. *In* "The Biology of Nematocysts" (D. A. Hessinger and H. M. Lenhoff, Eds.), pp. 255–272. Academic Press, New York.

Watson, G. M., and Hessinger, D. A. (1989a). Cnidocyte mechanoreceptors are tuned to the movements of swimming prey by chemoreceptors. *Science* **243,** 1589–1591.

Watson, G. M., and Hessinger, D. A. (1989b). Cnidocytes and adjacent supporting cells form receptor-effector complexes in anemone tentacles. *Tissue Cell* **21,** 17–24.

Watson, G. M., and Hessinger, D. A. (1991). Chemoreceptor mediated elongation of stereocilium bundles tunes vibration-sensitive mechanoreceptors on cnidocyte-supporting cell complexes to lower frequencies. *J. Cell. Sci.* **99,** 307–316.

Watson, G. M., and Hessinger, D. A. (1992). Receptors for N-acetylated sugars may stimulate adenylate cyclase to sensitize and tune mechanoreceptors involved in triggering nematocyst discharge. *Exp. Cell Res.* **198,** 8–16.

Watson, G. M., and Hessinger, D. A. (1994a). Evidence for calcium channels involved in regulating nematocyst discharge. *Comp. Biochem. Physiol. A* **107,** 473–481.

Watson, G. M., and Hessinger, D. A. (1994b). Antagonistic frequency tuning of hair bundles by different chemoreceptors regulates nematocyst discharge. *J. Exp. Biol.* **187,** 57–73.

Watson, G. M., and Hudson, R. R. (1994). Frequency and amplitude tuning of nematocyst discharge by proline. *J. Exp. Zool.* **268,** 177–185.

Watson, G. M., and Mariscal, R. N. (1984a). Calcium cytochemistry of nematocyst development in catch tentacles of the sea anemone *Haliplanella luciae* (Cnidaria: Anthozoa) and the molecular basis for tube inversion into the capsule. *J. Ultrastruct. Res.* **86,** 202–214.

Watson, G. M., and Mariscal, R. N. (1984b). Ultrastructure and sulfur cytochemistry of nematocyst development in catch tentacles of the sea anemone *Haliplanella luciae* (Cnidaria: Anthozoa). *J. Ultrastruct. Res.* **87,** 159–171.

Watson, G. M., and Mariscal, R. N. (1985). Ultrastructure of nematocyst discharge in catch tentacles of the sea anemone *Haliplanella luciae* (Cnidaria: Anthozoa). *Tissue Cell* **17,** 199–213.

Weber, J. (1989). Nematocysts (stinging capsules of *Cnidaria*) as Donnon-potential-dominated osmotic systems. *Eur. J. Biochem.* **184,** 465–476.

Weber, J. (1990). Poly (γ-glutamic acid)s are the major constituents of nematocysts in hydra (*Hydrozoa, Cnidaria*). *J. Biol. Chem.* **265,** 9664–9669.

Westfall, J. A. (1970). Synapses in a sea anemone *Metridium* (Anthozoa). Electron Microscopy Proc. Int. Congr. 7th, Societe Francaise de Microscopie Electronique, Paris 3, 717–718.

Westfall, J. A. (1973). Ultrastructural evidence for neuromuscular systems in coelenterates. *Amer. Zool.* **13,** 237–246.

Westfall, J. A. (1988). Presumed neuronematocyte synapses and possible pathways controlling discharge of a battery of nematocysts in hydra. *In* "The Biology of Nematocysts" (D. A. Hessinger and H. M. Lenhoff, Eds.), pp. 41–51. Academic Press, New York.

Wilby, O. K. (1976). Nematocyst discharge and the effects of antibodies on feeding in hydra. *Nature* (*London*) **262,** 387–388.

Yang, X. C., and Sachs, F. (1989). Block of stretch-activated ion channels in *Xenopus* oocytes by gadolinium and calcium ions. *Science* **243,** 1068–1071.

Impact of Altered Gravity on Aspects of Cell Biology

Dale E. Claassen and Brian S. Spooner[1]
Kansas State University, Division of Biology, Ackert Hall, Manhattan, Kansas 66506-4901

I. Introduction

One environmental factor that has remained constant throughout evolution is the force of earth's gravitational field. Since all life evolved within this field, biological mechanisms have developed that allow life to exist within the pull of gravity. Some of these mechanisms are apparent. For example, in higher life forms, the vestibular system detects the gravity vector and directs the musculoskeletal system to counteract gravitational force in order to maintain posture and motility. The cardiovascular system counteracts gravity when it pumps blood to the upper body, but also uses the pull of gravity when it distributes fluid to the lower extremities.

These well-known adaptations to gravity become clearly evident during manned orbital spaceflight, where the inertial acceleration caused by gravitational force is virtually canceled and the gravity vector is no longer detectable. In this microgravity environment, where mass is nearly weightless, astronauts experience space motion sickness, muscle atrophy and bone demineralization, as well as cardiovascular deconditioning and the redistribution and pooling of body fluids in the upper body.

Given that organ systems have envolved mechanisms that counteract or use gravity, does it follow that cellular and subcellular systems developed analogous mechanisms? For example, does the cytoskeleton and extracellular matrix counteract the pull of gravity? Do cellular systems use gravity-driven convections to transfer signals and metabolites? When gravitational force is reduced or increased, does normal cell physiology change, and if so, which cellular systems and components are directly affected by the alterations in gravity?

This chapter summarizes the progress that has been made in determining

[1] To whom all correspondence should be addressed.

and understanding the effect of gravity on cell function. It describes the results from three general areas of spaceflight research, and within each area, focuses on specific cell systems for which sufficient data have been accumulated. Included are (1) cell physiology, with a focus on immune cell activation; (2) cell development, with a focus on plant cell differentiation; and (3) the physiology of unicellular organisms. In addition, we describe the underlying principles that govern spaceflight research on the cell, and examine experimental approaches used to investigate the effects of altered gravity on cell function.

In an attempt to identify the cellular targets of gravitational force, we analyze experimental results from cultured immune cells in both spaceflight and ground-based research. Similarities between the behavior of immune cells that are activated in microgravity and those that are activated on earth in the presence of phorbol esters are identified. These similarities support a hypothesis recently proposed independently by several investigators—that a serine/threonine kinase, protein kinase C (PKC), is a cellular effector of alterations in gravity. In order to substantiate or invalidate this hypothesis, we compare spaceflight effects with the known effects of PKC antagonists and agonists in ground-based studies. Where sufficient data exist, we present evidence to demonstrate that cells exhibiting drug-enhanced PKC activity on earth mimic the behavior of cells in microgravity, an indication that PKC activity is modulated in cellular systems that respond to alterations in gravity.

II. Altered Gravity Research on the Cell

Experiments analyzing the reaction of living systems to altered gravity have been performed since early in the ninteenth century, although for many years the only way effects could be measured were to increase the gravitational force by centrifugation, and to alter gravity's directional pull by repositioning living systems with respect to the gravitational vector. Not until the advent of the space age and the development of free falling spaceflight vehicles did it become possible to test the effects of sustained and substantial reductions in the force of gravity.

Over the past 30 years, a number of experiments have been performed on cellular systems in the reduced gravity environment of orbital spaceflight. These experiments have been published in meeting proceedings, as technical reports in space technology journals, or in space agency publications. From these sources, comprehensive listings of test organisms, test parameters, and results have been compiled by Cogoli and Tschopp (1982), Dickson (1991), and Taylor (1977). Rather than duplicate these lists, we have selected and presented experiments that investigated the effects of spaceflight on related systems. Published reviews on gravity and the cell include Cogoli and Gmünder (1991), Halstead and Dutcher (1987), and Lewis and Hughes-Fulford (1994).

The responses of living systems to reductions in the force of gravity are often not obvious. Many of the early spaceflight experiments in the 1960s, which were conducted more to determine the survivability of organisms in space than to specifically investigate cellular effects, reported that spaceflight had no impact on cellular or embryonic systems. Thus, the first decade of flight experiments yielded the overall conclusion that growth, development, and mutagenesis are not significantly affected by spaceflight (Cogoli and Tschopp, 1982). In the following two decades, however, the effects of spaceflight on living systems became more evident as the length of flights increased, and as the analysis of experimental results improved. Results from these more sophisticated experiments eventually led to the conclusion that spaceflight has specific effects on a wide range of living systems, even at the level of the single cell.

Clearly, conclusions drawn from recent spaceflight experiments contradict the early results. In some instances, this may be due to differences in the length of time that systems were exposed to the spaceflight environment. Many of the early flights in the 1960s lasted from 1 to 5 days, whereas nearly all spaceflights carrying biological experiments since 1970 have lasted for at least a week, with more than 50% orbiting for more than 10 days. A comparison of the biological responses from short-term and long-term missions suggests a possible correlation between the appearance of cellular and molecular effects and the duration of the flight. For example, Dickson (1991) lists six flights lasting 1 day or less in which *Escherichia coli* phage production was found to be unchanged or slightly increased, compared with ground controls. However, on seven of seven flights that lasted more than 3 days, significant changes in phage levels were observed; the number of particles increased by nearly twofold on spaceflights of 3 days, and by three- to fourfold on flights lasting 5 or more days.

Some results from early spaceflight experiments were based on general observations that overlooked cellular changes in biological systems. For instance, the development of plant seedlings during spaceflight appeared unaltered when the analysis was based on differences in gross anatomy, but when seedlings were analyzed ultrastructurally, developmental anomalies were found. Furthermore, a summary based on preliminary observations of results from the 1967 Biosatellite II flight stated that orbital spaceflight did not influence small biological systems; however, that conclusion

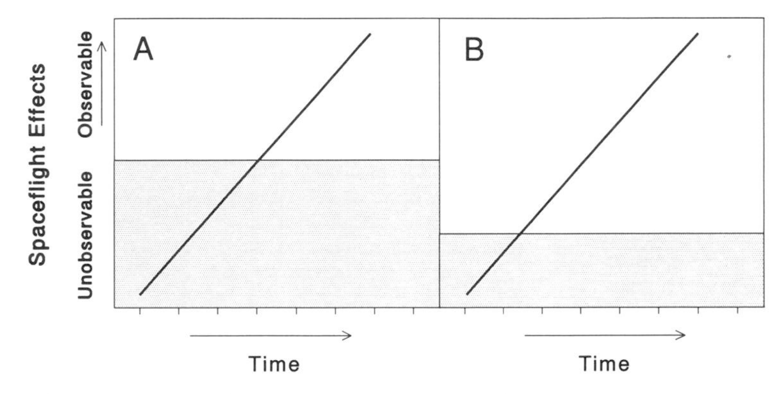

FIG. 1 Theoretical explanation for recent observations of the effects of spaceflight on living systems. (A) Observations of effects have become more numerous as flight times have increased. (B) In addition, previously unnoticed effects have been observed as assays have become more sensitive and as experimental analyses have improved.

was not supported in the completed analyses, which provided some of the first evidence that biological responses may be altered during spaceflight (Brown, 1991).

Still, not all contradictory results from 30 years of spaceflight can be attributed to flight length or experimental analysis. In some cases, results may differ because of variabilities in the spacecraft environment. Successful reproduction of experimental data can prove difficult in an environment where the levels of light, temperature, radiation, vibration, and accelerative g-force fluctuate during flight and between missions. For example, various levels of accelerative g-forces are exerted on all free falling vehicles, and since the sensitivity of biological systems to g-forces is not known, it is possible that a system may experience accelerative forces on some flights and not on others (Brown, 1991). While there are experimental variabilities in earth-based research, the inaccessibility of free falling vehicles greatly hinders the accumulation of data, and increases the impact of spaceflight variabilities on conclusions and extrapolations.

Yet, even when spaceflight effects are clearly identified and reproducible, as in the increased production of phage particles by *E. coli,* the question remains of which spaceflight factor(s) is responsible for the change in cellular function. For example, is the increase in phage levels due solely to a reduced gravitational force on the bacterial cell, or is space radiation involved in inducing the expression of phage particles? This question was addressed in high-altitude balloon flights, where bacteriophage were exposed to space radiation while at unit gravity. In this case, balloon flights revealed that cosmic rays had a detrimental, not stimulatory, effect on phage viability (Hotchin *et al.,* 1965). In addition, spaceflight studies found that phage particles were even more sensitive to cosmic rays in microgravity (Spizizen *et al.,* 1975; Taylor *et al.,*1974).

Although these studies suggest that radiation does not increase the cellular production of phage particles, additional investigations using radioprotectant drugs during spaceflight generated results that varied with the length of the flight. For example, on flights of 3 and 5 days, radioprotectant drugs were effective in lowering the level of phage particles; but, on a 22-day flight, increased expression was not blocked by such drugs (Zhukov-Verezhnikov *et al.,* 1965, 1968), an indication that the effects of microgravity become more observable during longer spaceflights. This example shows the difficulties in determining the role of reduced gravity in the effects of spaceflight. Additional examples in which attempts were made to identify the factors responsible for changes in cellular functions are discussed in the following sections.

A. Effects of Altered Gravity on Cell Physiology

The first cells from a multicellular organism to be studied in the altered gravity of spaceflight were of the HeLa human cell line. As chronicled by Dickson (1991), HeLa cells were often carried on spaceflights in the 1960s, while amnion cells, fibroblasts, leukocytes, and cells from human conjunctival and synovial linings were carried less frequently. In the past 20 years, established lines from hamster cells, human embryonic cells, and mouse cells have been cultivated during spaceflight, and the effects of microgravity have also been investigated using primary cultures of plant cells and plant protoplasts from the carrot, *Daucus carota, Hapolpappus,* and the anise plant, *Pimpinella.* More recently, the activation of cellular responses during spaceflight has been studied in human red blood cells, and in human and mouse immune cultures containing lymphocytes and macrophages. It is interesting that cells that undergo activation during spaceflight appear to be more affected by the weightless environment.

1. Activation of Immune Cells in Microgravity

A primary cell culture that has been studied under conditions of altered gravity and that continues to be of much interest are the cells that generate an immune response. Spaceflight experiments on immune cultures have differed from most other cell experiments in that immune cells usually have been analyzed during their activation response to immune stimulation. Elements of the immune response that have been investigated during spaceflight include the proliferation of T lymphocytes, and the release of cytokines from T lymphocytes and accessory cells (monocytes and macrophages).

The results from spaceflight experiments reveal that immune cells in microgravity exhibit responses that differ from cells cultured at $1 \times g$. In ground-based cultures, T-cell mitogens such as concanavalin A (Con A) bind and stimulate the T-cell antigen receptor (TCR) and, in the presence of substrate-adherent immune accessory cells, activate immune responses. Within 24–48 hr, T lymphocytes and accessory cells produce cytokines, express cytokine receptors, and induce T-cell proliferation. In contrast, identical immune cell cultures in microgravity showed little response to mitogens, indicating that the immune response was suppressed during spaceflight.

Results from human peripheral blood immune cells cultured on Spacelab 1, Spacelab D-1, and SLS-1 demonstrate that T-cell proliferation, as measured by the incorporation of radioactive thymidine into DNA, decreased by 55–97% compared with control cultures on the ground or in the flight centrifuge (Bechler *et al.*, 1992; Cogoli *et al.*, 1984; Cogoli *et al.*, 1987a). Interestingly, the failure of T cells to proliferate appeared to be caused by a 97% reduction in the receptor for the autocrine growth factor, interleukin-2 (IL-2), and not by reductions in the growth factor itself (Cogoli *et al.*, 1993). This is unusual since IL-2 secretion normally follows expression of its receptor, and secretion normally results in T-cell proliferation. In fact, the appearance of IL-2 often is used as an indicator of T-cell activation and proliferation.

The inhibition of T-cell proliferation in immune cultures during spaceflight does not appear to result from space radiation, since T cells underwent normal proliferation on the in-flight centrifuge, and since proliferation increased rather than decreased when immune cells were exposed to cosmic radiation during a stratospheric balloon flight (Cogoli *et al.*, 1992). Furthermore, ground-based clinorotation experiments, which alter the gravity vector, mimic the spaceflight results by inducing a 50% reduction in T-cell proliferation; in addition, centrifugation experiments showed an increase of 30% in lymphocyte proliferation in hypergravity conditions at $3\text{–}10 \times g$ (Cogoli *et al.*, 1979; Tschoop and Cogoli, 1983).

One explanation for the inhibitory effect of spaceflight on lymphocyte proliferation is that essential interactions between T cells and monocytes could not take place in microgravity conditions (Meehan, 1987). The activation of T cells depends on direct contact with substrate-adherent monocyte accessory cells; these are themselves stimulated upon attaching to the culture dish (Horgan *et al.*, 1990) but undergo full activation upon encountering stimulated T cells (Davis and Lipsky, 1985; Landis *et al.*, 1991). Activated accessory cells cross-link and immobilize the Con A-engaged T-cell antigen receptor, and transiently bind to the ensuing high-affinity T-cell adhesion receptors, along with other transmembrane cosignaling molecules such as CD28 (Geppert *et al.,* 1990; Horgan *et al.,*1990; Shahinian *et al.,* 1993; Weiss *et al.,* 1986).

Stimulation of these coreceptors, and of receptors to membrane-bound monocyte cytokines, activates lymphokine production and T-cell proliferation (Davis and Lipsky, 1993; Kawakami *et al.,* 1989; Landis *et al.,* 1991; Shimizu *et al.,* 1990). Thus, if mitogen-stimulated T cells are inhibited from interacting with accessory cells, the result is a reduction in immune cell activation and T cell proliferation (Dransfield *et al.,* 1990; Hogg *et al.,* 1993; Horgan *et al.,* 1990) similar to that seen during spaceflight.

At the cell concentrations used in spaceflight immune response assays (10^6 cells/ml), T cell–accessory cell interactions are dependent on gravitational force. Ground-based experiments have demonstrated that sedimentation increases the density of T cells in the immediate vicinity of the bottom-adherent accessory cells, and permits the cell–cell interactions that are necessary for immune cell activation and T cell proliferation (Meehan, 1987). Indeed, binding assays use low-speed centrifugation or a time period for sedimentation to allow suspended T cells to encounter adherent accessory cells (Van Seventer *et al.,* 1990). Thus, since gravity increases the possibility of immune cell interactions, the absence of gravity during spaceflight may inhibit immune responses *in vitro* by simply not assisting suspended T cells, which are nonadherent before activation, to encounter surface-adherent monocytes, which attach prior to launch (see Fig. 2A). In addition, this dependence of T-cell activation on accessory

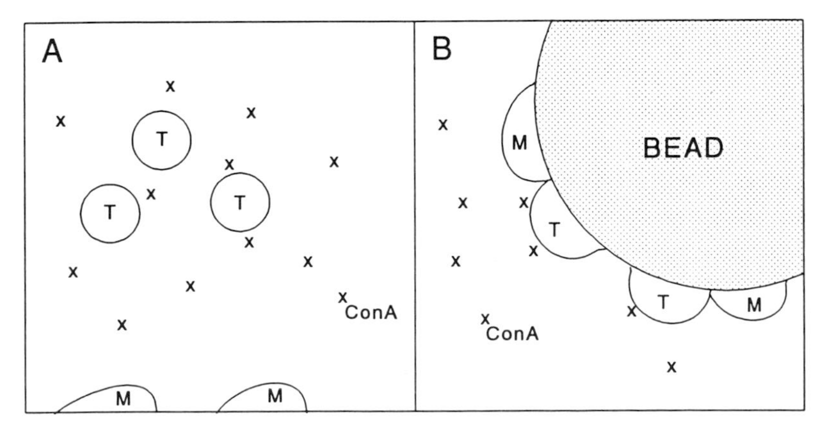

FIG. 2 Activation of immune cells in spaceflight cultures. (A) In microgravity, Con A-stimulated T cells (T) remain suspended and unable to interact with the monocyte accessory cells (M), which spontaneously attach to the culture flask prior to launch. (B) When T cells and monocytes are attached to microcarrier beads prior to launch, the stimulated T cells are able to interact with monocytes and undergo proliferation in microgravity.

cell contact may be one explanation for the altered proliferation responses observed during centrifugation and clinorotation. Whereas stimulatory interactions between immune cells would be assisted by centrifugation, forces generated by the rotating fluid during clinorotation most likely would decrease T-cell interactions with the adherent accessory cells.

Since accessory cells are stimulated to produce interleukin-1 by physically interacting with activated T cells (Landis *et al.*, 1991), the reduced levels of IL-1 from spaceflight (SLS-1) also suggest that immune cell interactions were inhibited during flight. Compared with Con A-stimulated ground controls, IL-1 secretion in microgravity decreased by 50–80%, and was similar to ground cultures that received no Con A stimulation (Cogoli *et al.*, 1993). Although this large reduction in monocyte IL-1 secretion has led some researchers to suggest that microgravity directly inhibits monocyte function, the similarity between IL-1 production in unstimulated ground controls and in the SLS-1 immune cultures may also indicate that monocytes received no stimulation from T cells during spaceflight. The low levels of IL-1 that are present in unstimulated control cultures and in spaceflight cultures most likely were triggered by monocyte attachment to the culture flask (Fuhlbrigge *et al.*, 1987; Landis *et al.*, 1991), which occurred before launch and following cell loading. Thus, spaceflown monocytes did not malfunction during spaceflight, but were unable to form stimulatory contacts with mitogen-bound T cells in the weightless environment (see Fig. 2A).

Results from SLS-1 immune cultures also reveal a 97% reduction in the surface expression of the IL-2 receptor (IL-2R) on T lymphocytes (Cogoli *et al.*, 1993). This absence of IL-2R explains the failure of T cells to proliferate during spaceflight even though the levels of the autocrine growth factor, IL-2, were nearly normal. Furthermore, IL-2R expression and IL-2 responsiveness in peripheral blood T-cells are dependent on the cross-linking and immobilization of the TCR, and/or on costimulation through T-cell signaling molecules; both functions are performed through physical contact with accessory cells (Davis *et al.*, 1989; Granelli-Piperno *et al.*, 1989; Kawakami *et al.*, 1989; Kern *et al.*, 1986; Van Seventer *et al.*, 1990; Wakasugi *et al.*, 1985). Thus, the absence of IL-2R in spaceflown T cells also indicates that interactions between T cells and monocytes were inhibited during spaceflight.

The strongest evidence that microgravity inhibits immune cell interactions comes from an additional spaceflight experiment in which immune cultures were attached to Cytodex 1 microcarrier beads (Cogoli *et al.*, 1993). Cytodex 1 beads contain a positive charge and will bind resting, nonadherent lymphocytes as well as adherent accessory cells, thereby positioning immune cells in closer proximity on a surface that supports cell–cell interactions. It is assumed that the addition of beads should allow interactions between T cells and monocytes to be established in microgravity, overcoming the inhibitory effects of spaceflight (Fig. 2B). Indeed, results from SLS-1 show that Con A-induced T-cell proliferation and monocyte IL-1 production were restored in spaceflight cultures that contained microcarrier beads (Cogoli *et al.*, 1993).

Surprisingly, however, the proliferative response of T cells during spaceflight was not only restored when cells were cultured on microcarrier beads, it also was enhanced by nearly twofold over $1 \times g$ controls that also contained beads (Cogoli *et al.*, 1993). In addition, spaceflown T cells on microcarrier beads showed a 2.5-fold increased secretion of IL-2 after 46 hr of Con A stimulation, and exhibited a twofold increase in interferon (IFN-γ), compared with identically cultured $1 \times g$ control cells (Cogoli *et al.*, 1993).

Similar increases in lymphokine production have been reported in other spaceflight studies. On spaceflight STS-43, human peripheral blood immune cells that were stimulated with Con A for 24 hr secreted threefold more IFN-γ than ground controls (Chapes *et al.*, 1992). In addition, mouse lymph node immune cells on the same flight released three times more IFN-γ in response to Con A stimulation than did identical ground control cultures (Chapes *et al.*, 1992). However, since these increases in lymphokine levels on STS-43 were obtained from immune cells cultured in the absence of microcarrier beads, the results appear to contradict those from SLS-1, where IFN-γ levels increased only when cells were cultured on beads. In fact, on SLS-1, Cogoli *et al.* (1993) found that in the absence of beads, T cells secreted much less IFN-γ (42 IU/ml) than $1 \times g$ controls (1000 IU/ml), most likely due to the lack of costimulatory interactions. Likewise, on STS-37, Chapes *et al.* (1992) found that in the absence of beads, T cells released low levels of IFN-γ (35 IU/ml); however, in this case the ground control cultures contained even less IFN-γ (11 IU/ml).

This low level of IFN-γ secretion in STS-37 ground controls most likely was due to the requirement that culture containers be coated with silicon, which created a barrier to immune cell–substratum adhesion even at $1 \times g$. Such inhibition of cell adhesion by the culture container has been shown in ground-based experiments in which Con A-induced T-cell proliferation and IFN-γ production was severely reduced, even when cell–cell interactions were possible (Gmünder *et al.*, 1992). Indeed, immune cell adhesion molecules that are involved in substratum adherence have been shown to costimulate the immune response (Larson and Springer, 1990; Patarroyo *et al.*, 1990). Thus, these spaceflight studies demonstrate that microgravity directly enhances the production of IFN-γ if costimulatory interactions are identical in spaceflight and control cultures. Furthermore, these results indicate that both cell–substrate and cell–cell interactions are required for the optimum immune response.

Spaceflight studies have found that the release of other cytokines also is enhanced in microgravity. On Salyut 6, human immune cells exposed to various T-cell activators (e.g., poly I:C, inactivated virus) for 4 and 6 days produced a level of antiviral interferon activity that exceeded ground control levels by four-to eightfold (Bátkaj *et al.,* 1988). Similarly, on spaceflight STS-37, antiviral interferon (IFN-α) levels, released by murine splenic immune cells that were stimulated with poly I:C, increased by 2.8-fold above ground controls after 14 hr (Chapes *et al.,* 1992).

In studies on STS-37 and STS-43, levels of IL-1 were found to increase by two-to threefold when an anchorage-dependent macrophage cell line (B6MP102), cultured on microcarrier beads, was activated with lipopolysaccharide (LPS) for 24 hr (Chapes *et al.,* 1992). In contrast, IL-1 levels from T-cell-stimulated monocytes in SLS-1 cultures were not greater than controls, possibly due to high uptake rates by T-cells undergoing enhanced activation (Cogoli *et al.,* 1993), or to differences in the response of monocytes to LPS and T cells (see the following discussion).

Additional studies on STS-37 and STS-43 demonstrated that levels of tumor necrosis factor-α (TNF-α) increased by two- to threefold when B6MP102 cells were treated with LPS (Chapes *et al.,* 1992). Likewise, in SLS-1 immune cultures containing beads, the level of secreted TNF increased by twofold over $1 \times g$ controls during 46 hr of a Con A-induced immune response (Cogoli *et al.,* 1993). Although levels of TNF increased during spaceflight, a subsequent experiment on STS-50 revealed that the cytotoxicity of TNF on a TNF-sensitive cell line (LM929) was completely suppressed during spaceflight. While ground control cultures exhibited a 22–59% decrease in cell proliferation after 12 hr of exposure to TNF, spaceflight cultures showed no change between cells cultured with or without TNF; in fact, even in the presence of TNF, flight cultures proliferated at a greater rate than did ground controls that were not exposed to TNF (Chapes *et al.,* 1994; Woods and Chapes, 1994).

Unquestionably, immune cell responses to mitogens, antigens, and cytokines can be influenced by the weightless environment of spaceflight. While some immune responses are suppressed, often by impaired cell–cell or cell–substrate contact in the absence of sedimentation, other immune cell responses are enhanced, and these appear to be more directly influenced by weightlessness. For example, when T cell–monocyte interactions were inhibited in the absence of microcarrier beads, as indicated by their failure to express IL-2R and IL-1, the T cells were able to secrete IL-2 nevertheless (Cogoli *et al.,* 1993). This is an unexpected result since the production of IL-2 in immune cultures normally is dependent on accessory cell costimulation above that required for IL-2R expression (Granelli-Piperno *et al.,* 1989; Kawakami *et al.,* 1989).

In an attempt understand how spaceflight might enhance certain T-cell responses,we have identified situations from ground-based immunological studies in which immune cells responded as if in weightlessness. Interestingly, these ground-based situations, which mimic immune cell responses during spaceflight, all appear to involve the increased activation of the serine/threonine kinase, protein kinase C (PKC). Recently, several investigators have proposed that PKC activity during spaceflight is either enhanced (Chapes *et al.*, 1992, 1993) or inhibited (de Groot *et al.*, 1990; Limouse *et al.*, 1991; Wilfinger and Hymer, 1992).

a. Role of Protein Kinase C (PKC) The treatment of ground-based immune cultures with PKC-activating compounds, such as diacylglycerol (DAG) analogs or phorbol esters, enhances the reponse of T cells to stimulation of the antigen receptor. This enhanced response also is seen in the absence of accessory cells, when the T-cell receptor is engaged by mitogenic lectin or antibody. Although accessory cells normally are required to costimulate IL-2R expression, IL-2 production, and T-cell proliferation, the substitution of diacylglycerol analogs (100 μM) or phorbol esters at low concentrations (0.5–5 ng/ml), while not sufficient for T-cell proliferation, has been shown to replace some accessory cell functions (Berry *et al.*, 1989; Davis and Lipsky, 1985). For example, when T cells are blocked from interacting with accessory cells in ground-based studies, TCR stimulation triggers little IL-2 secretion and T-cell proliferation compared with control cultures that contain accessory cells (Davis and Lipsky, 1993; Dougherty and Hogg, 1987; Farrar *et al.*, 1980); however, upon addition of low concentrations of phorbol esters to the isolated TCR-stimulated T cells, IL-2 secretion is stimulated to control levels, although proliferation remains about 50–60% below controls (Davis and Lipsky, 1993; Farrar *et al.*, 1980; Kawakami *et al.*, 1989). This is similar to results from SLS-1 in which stimulation of the TCR during spaceflight, in the absence of microcarrier beads, triggered IL-2 secretion at control levels and T-cell proliferation at about 50–80% below controls (Cogoli *et al.*, 1993).

In addition, ground-based studies have shown that when accessory cell costimulation is present, PKC activators enhance the response of T cells to TCR stimulation. For example, addition of phorbol esters to costimulated T cells was found to increase IL-2 levels by 5- to 20-fold and increased proliferation by 80–90% (Cerbrián *et al.*, 1988; Davis and Lipsky, 1993; Granelli-Piperno *et al.*, 1989). Again, this is similar to the results from spaceflight in which Con A-stimulated T cells, in the presence of accessory cells, exhibited a 2.3-fold increase in IL-2 secretion and an 85% increase in T cell proliferation over ground controls (Cogoli *et al.*, 1993).

We have identified additional similarities between immune cell responses during spaceflight and cell responses in ground-based studies using PKC activators. For example, during spaceflight on Spacelab D-1, lymphocytes formed stable aggregates in Con A-stimulated immune cultures that lacked microcarrier beads (Bechler *et al.*, 1986). Normally, peripheral blood lymphocytes do not form stable aggregates at 1 x *g*, although ground-based studies have shown that the addition of phorbol esters to Con A-stimulated T cells increases the avidity of cell–cell adhesion molecules and induces the formation of stable aggregates of lymphocytes, even in the absence of accessory cell costimulation (Dustin and Springer, 1989; van Kooyk *et al.*, 1989; Patarroyo *et al.*, 1990). In addition, the increased response of a macrophage cell line to lipopolysaccharides also suggests enhanced activation of PKC in microgravity. On spaceflights STS-37 and STS-43, the anchorage-dependent B6MP102 macrophage line, cultured on microcarrier beads, exhibited a two- to three-fold increase in IL-1 secretion over ground controls (Chapes *et al.*, 1992).

Although PKC is not required for LPS stimulation of monocytes and macrophages, it is activated by LPS binding and induces signaling factors that function in the LPS stimulatory pathway (Müller *et al.*, 1993). Furthermore, ground-based studies have shown that the secretion of IL-1 from LPS-stimulated monocytes is enhanced by the addition of T-cell cytokines (Landis *et al.*, 1991), and one cytokine responsible for that enhancement, IFN-γ, increases the activity of PKC (Becton *et al.*, 1985; Trinchieri *et al.*, 1987). The activation of PKC by phorbol ester prolongs the transient expression of message for IL-1 in LPS-stimulated monocytes and macrophages, possibly by inactivating a repressor protein (Fenton *et al.*, 1988).

Spaceflight studies also have reported that immune cells secrete increased levels of the cytokines IFN-γ and TNF when stimulated in microgravity (Chapes *et al.*, 1992; Cogoli *et al.*, 1993). Likewise, ground-based studies have shown that phorbol esters enhance the release of IFN-γ from activated T cells and the secretion of TNF from activated monocytes and macrophages (Johnson and Torres,1982; Pennica *et al.*, 1984; Trinchieri *et al.*, 1987). In addition, a TNF-sensitive cell line exhibited increased resistance to the cytotoxic effects of TNF during spaceflight (Chapes *et al.*, 1994; Woods and Chapes, 1994). This also appears to involve the enhanced activation of PKC during spaceflight since additional studies on STS-54 and STS-57 demonstrated that, in the presence of a PKC inhibitor, cells were no longer resistant and TNF cytotoxicity was restored to control levels (Woods and Chapes, 1994). Similar results were observed in ground-based studies in which phorbol ester-induced PKC activity increased the resistance of a TNF-sensitive cell line to TNF by triggering the rapid downregulation of TNF receptors, an effect that was reversed in the presence of PKC inhibitors (Galeotti *et al.*, 1993).

Microgravity also has been shown to enhance the action of phorbol ester on the secretion of superoxide from *P. acnes*-induced murine peritoneal inflammatory cells. During the 10–20 sec of free fall on parabolic flight aircraft, inflammatory cells treated with phorbol ester released superoxide at four times the level observed in phorbol ester-treated 1 x *g* controls (Fleming *et al.*, 1991). In ground-based studies, phorbol esters have been shown to costimulate the rapid release of superoxide from antigen-stimulated neutrophils by activating PKC (Muzykantov *et al.*, 1993; Pontremoli *et al.*, 1986; Wilson *et al.*, 1986). Thus, the increase in superoxide secretion during parabolic flight suggests that the activation of the PKC pathway by phorbol esters is enhanced in microgravity.

While some immune cell responses are known to be unaltered by spaceflight, these responses are not directly influenced by phorbol esters in ground-based experiments. On spaceflight SLS-1, for example, Con A-stimulated T cells on microcarrier beads with costimulating monocytes expressed the IL-2 receptor at levels similar to ground controls (Cogoli *et al.*, 1993). The failure of microgravity to alter IL-2R expression is consistent with ground-based studies that show IL-2R levels in naive peripheral blood T cells to be marginally affected by low concentrations of phorbol esters (Cebrián *et al.*, 1988; Isakov and Altman, 1985) or diacylglycerol analogs (Berry *et al.*, 1989).

Several spaceflight studies have reported cell responses in which PKC activity appeared to be inhibited by weightlessness. For example, on Cosmos 2044, a human monocytic leukemia cell line, THP-1, secreted IL-1 at levels 85% less than ground controls when it was stimulated with a phorbol ester (Limouse *et al.*, 1991). However, this decrease in IL-1 may have resulted from the inability of phorbol ester-activated THP-1 cells to sediment and attach to the substratum during spaceflight. Phorbol esters alone trigger only low levels of IL-1 (Fenton *et al.*, 1988), but they are known to induce monocytic phenotype expression and substrate attachment in monocytic cell lines (Trinchieri *et al.*, 1987), and substrate attachment is a known inducer of IL-1 secretion in monocytes (Landis *et al.*, 1991). Indeed, in the absence of phorbol esters, the nonadherent THP-1 cell can be stimulated to express IL-1 solely by the antibody-induced clustering of its low avidity adhesion molecules (Lin *et al.*, 1993). Thus, the reduced levels of IL-1 in THP-1 spaceflight cultures may have been due to the absence of sedimentation, which caused the phorbol ester-differentiated THP-1 cells to remain suspended and unstimulated by adhesion molecules.

On the same Cosmos flight, THP-1 monocytes secreted normal levels of IL-1 when they were stimulated with activated Jurkat T cells (Limouse *et al.*, 1991). Since both THP-1 and Jurkat cells grow in suspension, and the Jurkat T-cell activator (anti-CD3 antibody) was added with mixing, physical interactions between T cells and monocytes were able to form and costimulate cytokine release. The failure of spaceflight to enhance the release of IL-1 from Jurkat-stimulated THP-1 cells is similar to results from SLS-1, in which peripheral blood monocytes, on microcarrier beads with T-cells, produced normal levels of IL-1 (Cogoli *et al.,* 1993).

This differs from the increased secretion of IL-1 that was seen when macrophages were stimulated with LPS on STS-37 and STS-43 (Chapes *et al.,* 1992). Possibly this disparity is due to differences in the stimulation by LPS compared with T cells. For example, it may not be possible to further enhance the secretion of IL-1 in response to T cell stimulation owing to the presence of costimulating T-cell cytokines, some of which are potent PKC activators known to stimulate IL-1 (Becton *et al.,* 1985; Landis *et al.,* 1991). Since T-cell cytokines are not present when monocytes and macrophages are stimulated with LPS, PKC may not be maximally activated and may be responsive to further stimulation in microgravity. This is consistent with ground-based experiments in which macrophages stimulated with LPS exhibited only one fifth of the PKC activity induced by a T-cell cytokine, IFN-γ (Hamilton *et al.,* 1985). Furthermore, when T-cell cytokines were added to LPS-stimulated monocytes, which presumedly increased PKC activity, the secretion of IL-1 increased by 10-fold over LPS stimulation alone (Landis *et al.,* 1991).

Experiments on Cosmos 2044 also revealed that the production of IL-2 was completely blocked in the Jurkat T-cell line when it was stimulated by phorbol ester and calcium ionophore (Limouse *et al.,*1991). In ground experiments, phorbol ester-induced PKC activation and ionophore-mediated calcium influx are sufficient to trigger the release of IL-2 from Jurkat cells with no additional requirement for costimulation from accessory cells.

Thus, the failure of Jurkat cells to release IL-2 in microgravity appears to suggest that PKC was inhibited and to contradict results from peripheral blood T-cells in which secretion of IL-2 was enhanced during spaceflight. However, the mature T cell of peripheral blood differs from the immature, transformed T cell from the Jurkat line (Khan *et al.,* 1993; Lucas *et al.,* 1990; Ward and Cantrell, 1990), and these differences may alter cell responses to spaceflight. For instance, in Jurkat T cells, PKC activation inhibits the signal transduction pathway for IL-2 by exerting negative feedback control on phospholipase Cγ (Fig. 3). Phospholipase Cγ induces T-cell activation by hydrolyzing phosphatidyl inositol 4,5-bisphosphate (PIP_2) into second messengers, inositol trisphosphate (IP_3) and diacylglycerol, which mobilize intracellular stores of calcium and activate PKC (Berridge, 1987; Desai *et al.,* 1990). Strong feedback inhibition by PKC on phospholipase Cγ, and on IL-2 synthesis, is seen when Jurkat cells are preincubated with phorbol ester for as little as 15 min; in contrast, peripheral blood T cells show no such inhibition even after 40 hr of phorbol ester treatment (Park *et al.,* 1992; Ward and Cantrell, 1990).

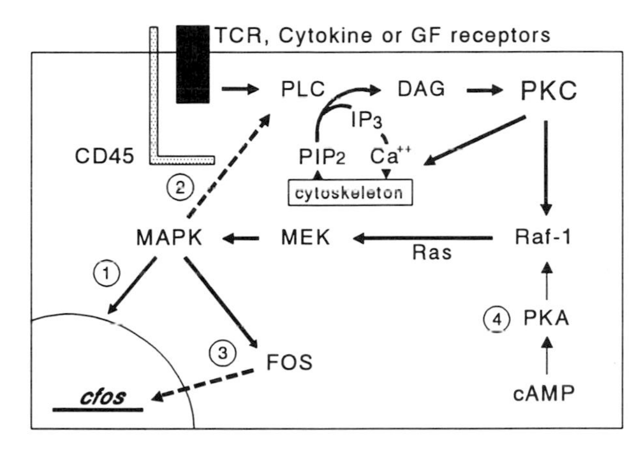

FIG. 3 Schematic representation of signal transduction pathways that have been activated in spaceflight experiments. Depending on cell type, receptor-mediated pathways that stimulate PKC induction of MAPK may trigger (1) cell proliferation; (2) feedback inhibition of receptor kinase of PLC activity, which can be blocked by CD45; or (3) inhibition of immediate early genes such as *c-fos*. PKC-induced activation of these pathways was enhanced in microgravity; however, activation of analogous pathways by PKA (4) was not affected.

Although calcium ionophore and phorbol ester have been thought to bypass early steps (e.g., phospholipase Cγ) in the IL-2 pathway, the influx of extracellular calcium is known to activate phospholipase Cγ (Lucas *et al.*, 1990; Ward and Cantrell., 1990), and even in the presence of calcium ionophore and phorbol ester, induction of IL-2 may require the activation of phospholipase Cγ for the hydrolysis of PIP_2. PIP_2 is a regulator of the actin binding proteins, including α-actinin, vinculin, severin, protein 4.1, gCap39, gelsolin, and profilin; and the hydrolysis of PIP_2 influences cytoskeletal reorganization (Fukami *et al.*, 1994; Isenberg, 1991).

Profilin, for example, is inactive when bound by PIP_2, but when PIP_2 is removed by phospholipase Cγ, profilin binds to G-actin to promote and stabilize actin filament formation by catalyzing actin nucleotide exchange (Aderem, 1992; Goldschmidt-Clermont and Janmey, 1991). Interestingly, profilin has been shown to be involved in intracellular transport mechanisms (Cooley *et al.*, 1992; Smith and Portnoy, 1993), which also may be required for signal transduction. Indeed, inhibition of phospholipase Cγ by dibutyryl cAMP suppressed the stimulation of spleen T cells by phorbol ester and calcium ionophore (Isakov and Altman, 1985), an indication that the role of phospholipase Cγ in signal transduction goes beyond PKC activation and calcium mobilization. Thus, the failure of Jurkat T cells to produce IL-2 during spaceflight may have resulted from enhanced PKC inhibition of phospholipase Cγ.

Additional experiments on Cosmos 2044 found that this inhibition of Jurkat cells in producing IL-2 during spaceflight could be overcome by accessory cell costimulation. When Jurkat T cells were mixed with THP-1 accessory cells and the TCR–CD3 complex was stimulated with an antibody to CD3, IL-2 was secreted at levels similar to ground controls (Limouse *et al.,* 1991). Since both Jurkat and THP-1 cells do not adhere to the substrate prior to stimulation, the addition of an anti-CD3 antibody with mixing allowed the suspended cells to make contact and form costimulatory interactions that apparently overcame the PKC-induced inhibition of phospholipase Cγ.

Recent ground-based studies suggest possible pathways in which PKC inhibits IL-2, and in which accessory cells may overcome that inhibition (Fig. 3). In the Jurkat T-cell line, activation of PKC triggers a tight association with the plasma membrane (Manger *et al.,* 1987; Weiss *et al.,* 1987), and initiates a signal transmission pathway that includes Raf-1, *ras,* and mitogen-activated/microtubule-associated protein kinases (e.g., MEK, MAPK) that phosphorylate serine or threonine (Amaral *et al.,* 1993; Wotton *et al.,* 1993). Since membrane-associated phospholipase Cγ is inhibited by serine phosphorylation (Park *et al.,* 1992), one of the MAP kinases, MAPK, may be the serine kinase that mediates PKC inhibition of PIP_2 hydrolysis. Indeed, in nonimmune cell lines, activated MAPK has been shown not only to translocate to the nucleus (Lenormand *et al.,* 1993) but also to localize at the plasma membrane (Gonzalez *et al.,* 1993) and to phosphorylate and regulate the activity of membrane proteins such as phospholipase A_2 (Nemenoff *et al.,* 1993) and the EGF receptor (Takishima *et al.,* 1991).

The suppression of MAPK-induced phosphorylation of phospholipase Cγ may be a mechanism by which THP-1 accessory cells overcome Jurkat T cell inhibition during spaceflight (Fig. 3). This suppression of MAPK may be mediated by CD45, a transmembrane tyrosine phosphatase that associates with the TCR–CD3 complex during receptor cross-linking by accessory cells (Abraham *et al.,* 1992; Tonks and Charbonneau, 1989). Jurkat T-cells are known to require CD45 for activation,and the absence of CD45 results in the failure of TCR stimulation to activate phospholipase Cγ (Koretzky *et al.,* 1992). Not only is CD45 involved in activating the TCR-linked tryosine kinase that activates phospholipase Cγ (Abraham *et al.,* 1992), CD45 also is capable of dephosphorylating MAPK directly and blocking the inhibition of phospholipase Cγ (Amaral *et al.,* 1993; Anderson *et al.,* 1990). Thus, accessory cells may overcome spaceflight inhibition by cross-linking CD45 with the TCR complex, thereby blocking PKC-induced feedback inhibition and allowing phospholipase C activation and IL-2 production.

2. Growth Factor Activation in Microgravity and the Role of PKC

Signal transduction pathways that mediate the actions of growth factors also have been shown to involve activation of PKC (Berridge, 1993; Sadowski *et al.*, 1993; Silvennoinen *et al.*, 1993). It may be expected, then, that growth factor-induced pathways would be affected by changes in gravitational force. Indeed, when A431 epidermoid carcinoma cells were stimulated with epidermal growth factor (EGF) for 6 min during suborbital rocket flights, the cells exhibited a 50% reduction in the level of transcripts produced by the immediate early genes, *c-fos* and *c-jun*; moreover, the stimulation of A431 cells with phorbol ester (instead of EGF) during microgravity also resulted in a 25–50% reduction in *c-fos* and *c-jun* RNA compared with ground controls (de Groot *et al.*, 1990, 1991b). Similar results were reported in the simulated low gravity of clinorotation, where levels of *c-fos* transcripts in response to EGF or phorbol ester fell by 25–30% compared with stationary cultures (de Groot *et al.*, 1991a).

These reductions in the levels of immediate early gene transcripts during rocket flight have been suggested by some workers to be the result of a microgravity-induced inhibition of PKC (de Groot *et al.*, 1991b). However, ground-based research on A431 cells and fibroblast cell lines suggests that reductions in *c-fos* and *c-jun* transcription are the result of enhanced PKC activation (Fig. 3). When PKC is activated by EGF stimulation, PKC induces negative regulation of cell responses to EGF (Davis, 1988; Miyata *et al.*, 1989; Welsh *et al.*, 1991), including the initiation of cell proliferation, a response that is dependent on the transcription of *c-fos* and *c-jun* (Lechner and Kaighn, 1980; Kerr *et al.*, 1992). Moreover, negative regulation by PKC is consistent with ground-based studies that show *c-fos* transcription in response to EGF or phorbol ester to be highly transient, with transcripts increasing 15- to 20-fold and returning to initial levels within 30–60 min (Greenberg and Ziff, 1984; Kruijer *et al.*, 1984; Müller *et al.*, 1984). In addition, this negative regulation of *c-fos* expression most likely also would lead to a decrease in *c-jun* transcription, since reductions in the *c-fos* protein (c-Fos) limit formation of the c-Fos:c-Jun heterodimers that optimally stimulate *c-jun* promoter activity (Lin *et al.*, 1993b; Lucibello *et al.*, 1989).

One known site of regulation in the EGF signaling pathway is at the EGF receptor, where a PKC-activated negative feedback loop, which may involve MAPK (Takishima *et al.*, 1991), phosphorylates the EGF receptor and decreases receptor affinity, receptor-linked tyrosine kinase activity, and cell responses to EGF stimulation (Davis, 1988; Friedman *et al.*, 1984; Hunter *et al.*, 1984; Lin *et al.*, 1986; Miyata *et al.*, 1989; Welsh *et al.*,

1991; Whiteley and Glaser, 1986). In addition, ground-based studies recently have revealed that PKC negatively regulates *c-fos* expression by inducing transrepression of the *c-fos* promoter (Fig. 3). Stimulation of PKC activity results in phosphorylation of the *c-fos* protein c-Fos, which feeds back to inhibit promoter activity of its gene (Lin *et al.*, 1993; Ofir *et al.*, 1990; Sassone-Corsi *et al.*, 1988). Indeed, when PKC-induced phosphorylation is blocked by amino acid substitutions in c-Fos, PKC loses regulatory control over *c-fos* transcription, and the activity of the *c-fos* promoter increases fivefold (Ofir *et al.*, 1990).

Additional studies conducted in the simulated microgravity of clinorotation suggest that the reductions in *c-fos* transcripts during rocket flight were the result of decreases in the activity of the serum response element (SRE) in the *c-fos* promoter (de Groot *et al.*, 1990). Interestingly, ground-based studies have shown that the SRE in the *c-fos* promoter is a major target of PKC-induced transrepression by c-Fos (Lucibello *et al.*, 1989; Sassone-Corsi *et al.*, 1988; Treisman, 1992). This supports the explanation that the level of *c-fos* transcripts decreased during rocket flight as a result of enhanced PKC activation and increased negative regulation of *c-fos* promoter activity.

Importantly, results from rocket flight experiments also showed that not all activation pathways for *c-fos* and *c-jun* are affected by microgravity (Fig. 3). The response of both immediate early genes to activation by a calcium ionophore, and of *c-fos* to activation by forskolin, was normal in A431 cells during rocket flight (de Groot *et al.*, 1991b). Similar results were reported in the simulated low gravity of clinorotation, where A431 cells expressed normal levels of *c-fos* following activation with calcium ionophore or forskolin (de Groot *et al.*, 1990). Since forskolin and calcium ionophores are known to activate a protein kinase A (PKA) pathway that also triggers *c-fos* transrepression by c-Fos phosphorylation (Sheng *et al.*, 1988; Tratner *et al.*, 1992), these microgravity results demonstrate that the activity of a PKC pathway is enhanced by weightlessness, while an analogous PKA pathway is unaffected (Fig. 3).

3. Nonactivated Cells in Altered Gravity

The very first cell line to be cultured in orbit showed little response to initial flights, but, according to Dickson (1991), this human uterine cancer line, HeLa, exhibited increased proliferation and increased viability when reflown on subsequent flights. Additional Soviet spaceflights flew cell lines derived from hamster tissues (BHK-21, FAF 28, and V-79) and mouse tissue (L strain), and Dickson (1991) lists these lines as exhibiting increases in mitotic index and nuclear size, with no increases in mutation rates, chromosomal aberrations, karyotype, or ploidy. Furthermore, after the

20-day flight of Cosmos 782, a cell count of FAF 28 and L cells indicated enhanced proliferation since, although the number of living cells was similar to ground controls, the number of dead cells was twice that of controls (Planel *et al.*, 1982).

More evident increases in cell proliferation were exhibited in the LM929 mouse connective tissue cell line, which showed a 63% increase over ground controls in the incorporation of [^{3}H]thymidine during flight STS-50 (Chapes *et al.*, 1994). In contrast, an AM2 mouse hybridoma cell line cultured during the Spacelab D-1 flight showed no significant difference in population size compared with ground controls, although increased numbers of mitochondria were observed (d'Augères *et al.*, 1987). Similarly, a human embryonic lung fibroblast cell line, WI-38, demonstrated no changes in mitotic index, cell cycle, growth curve, or cellular ultrastructure, although a 37% decrease in the consumption of glucose by the flight cells was detected (Montgomery *et al.*, 1978). In addition, a trend indicating slower replication rates in the WI-38 cells during spaceflight was noted in an analysis of the cell doubling time (Hughes-Fulford, 1991).

Alterations in cell numbers and in [^{3}H]thymidine incorporation have been detected in cell lines cultured under hypergravity conditions. A 74% decrease in cell count was reported when the AM2 hybridoma line was cultured at 1.4 x *g* for 6 days (d'Augères *et al.*, 1987). In contrast, HeLa, V-79 lung, and JTC-12 kidney cell lines exhibited a 20–80% increase in proliferation when cultured for 3–4 days at 18, 35, or 70 x *g*, with no observed changes in cell morphology (Kumei *et al.*, 1987). Similarly, after 2–6 days in culture at 10 x *g*, HeLa, SGS-3, and FBU-3b cells showed as much as a 20–30% increase in either cell numbers or in [^{3}H]thymidine incorporation, with no changes in cell morphology (Tschopp and Cogoli, 1983). These hypergravity studies also found that, at least in HeLa cells, cell migration decreased, glucose consumption per cell dropped by 25% (Tschopp and Cogoli, 1983), and the increase in proliferation rates correlated with a reduction in the G1 phase of the cell cycle, and with enhanced phosphoinositide turnover and *c-myc* expression (Kumei *et al.*, 1989a,b).

In addition to established cell lines, primary cells from both plants and animals have also been exposed to altered gravity conditions. Cultures containing crown gall tumor tissue from the carrot plant, *Daucus carota,* flew on Soviet Cosmos flights in the 1970s. According to Dickson (1991), gall cells exhibited depressed respiratory activity, increased flow of electrolytes out of the cells, and lower dry weights, compared with controls in the flight centrifuge. More recently on Cosmos 2044, cultures of protoplasts from *Daucus carota* exhibited an 18% decrease in intracellular calcium levels and an increase in the number of membrane-bound inclusions (Klimchuk et al., 1992). In addition, the cell walls that regenerated during spaceflight contained lower amounts of the structural components cellulose (71% less) and hemicellulose (33% less), and showed a reduction of 47% in the activity of peroxidases, which are involved in the synthesis of cell wall polymers (Iversen *et al.*, 1992). The growth rates of *Daucus carota* protoplasts averaged lower in flight cultures, but were not significantly different than ground controls.

Protoplasts isolated from the rapeseed, *Brassica napus*, also were flown on Cosmos 2044. During spaceflight, rapeseed cells underwent changes similar to carrot cells, including alterations in cell wall composition, peroxidase activity, and the frequency of microbodies; however, in this case, the proliferation of rapeseed protoplasts was strongly depressed and a 44% reduction in biomass production was reported (Iversen *et al.*, 1992). In addition, analysis of flight samples revealed that the thickness of regenerated cell walls decreased by 60%, the level of calcium associated with the cell wall decreased by 24% and intracellular calcium dropped by 46%, and mitochondria decreased in size and exhibited a less electron-dense matrix (Klimchuk *et al.*, 1992). Moreover, rapeseed cultures grown during spaceflight showed a 71% decrease in protein per unit biomass (Iversen *et al.*, 1992).

One of the first primary cell cultures to be isolated from animal tissue and cultivated during orbital spaceflight (STS-8) was from human embryonic kidney. These cells were not analyzed during spaceflight, and in the postflight analysis, no modifications in either urokinase production or cellular morphology were noted (Todd *et al.*, 1985; Tschopp *et al.*, 1984); however, mouse pituitary cells that also were cultured on STS-8 showed a 95% reduction in growth hormone release at 1 x *g* following spaceflight (Hymer and Grindeland, 1991).

Spaceflight also resulted in the abnormal aggregation of red blood cells taken from normal donors and from patients with ischemic heart disease. Whereas pathologic red blood cells aggregate on earth in dense compact clumps that are detrimental to blood flow, in weightlessness on STS-51C, these cells were found to form the normal rouleaux-type aggregates of nonpathologic blood; conversely, during spaceflight cells from nonpathologic blood failed to form the normal rouleaux aggregates seen in ground controls (Dintenfass, 1986). To our knowledge, this interesting observation has not been further addressed in spaceflight experiments.

B. Effects of Altered Gravity on Cell Development

The development of multicellular organisms in reduced gravity seldom results in major alterations in gross morphology, but may lead to subtle changes in growth dynamics and ultrastructural morphology. For example, many of the experiments that studied the development of plants during spaceflight found no changes in the appearance of seedlings, but did identify discrete developmental anomalies. Unfortunately, these anomalies varied and although a relatively large number of plants underwent germination and/or seedling development during spaceflight, no clear consensus regarding the effects of reduced gravity on plant development has materialized.

1. Plant Development in Microgravity

The first developmental plant study to orbit in space investigated the growth of wheat seedlings during the 2-day flight of the unmanned Biosatellite II in 1968. In that pioneering study, and in others that followed, seeds were wetted and germination was initiated prior to flight, but timed so that the seedlings emerged after reaching orbit. More recently, in U.S. shuttle and Soviet space station experiments, seeds imbibed water and initiated germination during spaceflight, thereby allowing the seed embryo to complete development in microgravity. Several spaceflights have also included plant somatic embryos of different ages. These were generated from single dedifferentiated totipotent plant cells and are capable of fully developing into normal plants.

The results from these developmental studies show that plant shoots and roots may grow better than, lesser than, or the same as the control plants at unit gravity. This variability was demonstrated in lettuce seedlings (*Lactuca sativa*) that were flown on several Soviet spaceflights. After 4–5 days of development, the stems (hypocotyls) of the lettuce seedlings were found to be about 55% longer in three experiments on Salyut 6, and 15% longer than in-flight centrifuge controls on Salyut 7 (Merkys *et al.*, 1983; Merkys and Laurinavicius, 1990). In addition, the lettuce stems grew somewhat taller (10%) than ground controls during 7 days on Cosmos 1667 (Merkys, 1990). However, on two Soyus flights, lettuce seedlings showed either no significant difference in stem length (Merkys and Laurinavicius, 1990), or a reduction of nearly 12% (Merkys, 1990). Furthermore, conflicting results have been reported on the development of oat seedlings (*Avena sativa*), which when grown for 8 days on IML-1, were found to be noticeably taller than ground controls (Brown *et al.*, 1992), but when grown for 8 days on STS-3, displayed a 10–30% decrease in stem height compared with ground controls (Cowles *et al.*, 1982).

An increase in shoot growth over in-flight 1 x *g* controls was found in seedlings of garden cress (*Lepidium sativum*) and mustard cress (*Arabidopsis thaliana*) on Salyut 6 (Merkys and Laurinavicius, 1990), and in *Arabidopsis* seedlings on Cosmos 1667, which exhibited a 30% increase in shoot length over ground controls after 7 days (Merkys, 1990). But after 8 days on STS-3, mung bean seedlings (*Vigna mungo*) and pine seedlings (*Pinus elliotti*) showed a 14% and 22% decrease in stem length compared with ground controls (Cowles *et al.*, 1984), and stems of pine seedlings that developed for 20 days on Cosmos 690 exhibited reductions of 35% compared with controls (Merkys, 1990). In addition, plantlets derived from *Haplopappus gracilis* displayed a 26% decrease in primary shoot regrowth during 5 days on STS-29, but compensated with an increase in secondary growth so that the total shoot production was similar to ground controls (Levine and Krikorian, 1992).

Some experiments found no effect of spaceflight on the length of seedling shoots. Sunflower (*Helianthus annuus*) seedlings exhibited stem lengths similar to ground controls following 8 days on STS-3 (Brown and Chapman, 1982); corn (*Zea mays*) seedlings were no different than controls in shoot weight following 5 days of development on STS-34 (Schulze *et al.*, 1992); and wheat seedlings did not differ significantly from ground controls in shoot growth after 2 days on Biosatellite II (Lyon, 1968a). However, wheat seedlings were found to be noticeably taller than ground controls following 8 days on IML-1 (Brown *et al.*, 1992), and barley (*Hordeum vulgare*) shoots were shown to be 50% longer than in-flight controls after 3–4 days of development on Salyut 6 (Merkys, 1990). Nevertheless, a 42% decrease in shoot length was observed in plantlets of the daylily, *Hemerocallis,* after 5 days on STS-29 (Levine and Krikorian, 1992).

Clearly, there appears to be no correlation between changes in shoot length and spaceflight duration or plant type. However, there may be a correlation between the length of shoots and an exposure of the plant to gravitational force during development. For example, of the 22 spaceflight investigations listed above, 13 studied plants in which development was initiated hours before launch and/or was terminated hours after landing, thereby exposing the growing plant to unit gravity. Of those 13 studies, 11 reported that shoot lengths were reduced or unchanged after spaceflight, and only 2 (15%) reported an increase in growth (both on Cosmos 1667). On the other hand, of the 9 experiments in which seed germination and seedling fixation occurred in orbit, all 9 studies (100%) reported increases in seedling shoot lengths compared with the 1 x *g* controls. Apparently even a short-term exposure of the developing seed embryo or growing seedling to gravitational force can cancel the enhancing effects of weightlessness on seedling shoot length (Table I).

Root growth during spaceflight also has produced variable results. For example, the length of roots in lettuce (*Lactuca sativa*) seedlings was reported to increase by 25% during 4–5 days of growth on Salyut 6 (Merkys *et al.*, 1983), but decreased by 5% after 7 days on Cosmos 1667 and by 17% in 4–5 days on Salyut 7 (Merkys, 1990; Merkys and Laurinavicius, 1990). Moreover, lettuce roots were not significantly different than control seedlings in two experiments lasting 4–5 days on Salyut 6, as well as in

TABLE I
Plant Stem Growth in Spaceflight Experiments

Experimental conditions	Stem lengths in spaceflight seedlings relative to controls[a]		
	Decreased	Equal	Increased
Seedlings exposed to 1 x *g* pre- or postflight	*H. gracilis*	Sunflowers	*Arabidopsis*
	Lettuce	Lettuce	Lettuce
	Mung beans	Wheat	
	Daylilies	Corn	
	Pine		
	Pine		
	Oats		
Seedlings grown only in microgravity			Mustard cress
			Garden cress
			Lettuce
			Lettuce
			Lettuce
			Lettuce
			Barley
			Wheat
			Oats

[a] Controls were grown on the ground or in flight in a 1 x *g* centrifuge.

two experiments conducted on Soyuz 12 and 13 that lasted for 2 days (Merkys and Laurinavicius, 1990). Root lengths in seedlings of garden cress (*Lepidium sativum*) and mustard cress (*Arabidopsis thaliana*) were similar to the in-flight centrifuge controls when grown for 32 hr on Spacelab D-1 or for 4–5 days on Salyut 6 (Merkys and Laurinavicius, 1990; Volkmann *et al.*, 1986). However, root growth decreased by 15% when *Arabidopsis* seedlings were cultivated for 7 days on Cosmos 1667 (Merkys, 1990), and by 64% when callus cells of *Arabidopsis* developed during an 8-day spaceflight (Podlutsky, 1992b).

Plantlets from callus cells of *Haplopappus gracilis* showed a reduction in the regrowth of roots during 5 days on STS-29 (Levine *et al.*, 1990), while a review by Halstead and Dutcher (1987) lists somatic embryos and carrot plantlets (*Daucus carota*) as producing longer roots than in-flight controls after 20 days on Cosmos 782. Pea seedlings (*Pisum sativum*) exhibited increases in root length of 30% after 2 days on Soyuz 12 (Merkys, 1990), but root growth decreased slightly in mung bean (*Vigna mungo*) seedlings that developed for 8 days on STS-3 (Cowles *et al.*, 1982), and the lengths of lentil (*Lens culinaris*) roots grown for 1–1.5 days on Spacelab D-1 were similar to ground and in-flight controls (Perbal *et al.*, 1987). The developing roots of pine (*Pinus elliotti*) showed a 45% reduction in length following growth for 20 days on Cosmos 690 (Merkys, 1990), and pine roots were 39% shorter than controls after 8 days on STS-3 (Cowles *et al.*, 1984). Wheat (*Triticum aestivum*) seedlings displayed no significant difference in root length after 2 days on Biosatellite II (Lyon, 1968b), and the weight of corn (*Zea mays*) roots was no different than controls following 5 days on STS-34 (Schulze *et al.*, 1992). However, oat seedlings produced slightly shorter roots (6%) during 8 days on STS-3 (Cowles *et al.*, 1984), and barley seedlings exhibited a 13% reduction in root growth following 4–5 days of development on Salyut 6 (Merkys, 1990).

In contrast to the response of seedling shoots to spaceflight, root growth may be affected by the duration of exposure to weightlessness (Table II). Of the eight spaceflight studies that lasted for 7 or more days, seven (88%) reported that root lengths were shorter than in ground controls. On the other hand, of the seven shorter studies lasting 1–2 days, none (0%) reported decreased root lengths, while six found no significant difference. Although all of the longer experiments exposed developing roots to gravity before or after launch, it does not appear that root growth was influenced

TABLE II
Plant Root Growth in Spaceflight Experiments

	Root lengths in spaceflight seedlings relative to controls[a]		
Experiment length	Decreased	Equal	Increased
1–2 days		Mustard cress Garden cress Lettuce Lettuce Lentil Wheat	Peas
4–5 days	*H. gracilis* Lettuce Barley	Mustard cress Garden cress Lettuce Lettuce Corn	Lettuce
7–8 days and 20 days	*Arabidopsis* *Arabidopsis* Mung beans Lettuce Pine Pine Oats		Carrots

[a] Controls were grown on the ground or in flight in a 1 x *g* centrifuge.

negatively by alterations in gravity. In flights lasting 5 days or less, no difference was seen in the root lengths of plants exposed to gravity and plants which developed solely in weightlessness.

A widely reported observation of seedling development in weightlessness is the reduction of mitotic activity in the root meristem following spaceflight. For example, the number of root cells undergoing mitosis in sunflower seedlings on STS-2 was found to be reduced by 50% when analyzed during the first cycle of cell division after landing, and division in the root tips of oat seedlings decreased by nearly 98% following the return of STS-3 (Krikorian and O'Connor, 1982). On Salyut 6, the mitotic index in the apical meristem of barley roots decreased by 33% compared with 1 x *g* in-flight controls (Merkys and Laurinavicius, 1990), and the number of dividing cells in the apical meristem of corn roots fell by 30% following flight on Cosmos 1667 (Grif *et al.*, 1988). In contrast, a similar study on corn seedlings, which were carried on the same Cosmos flight, reported that the mitotic index of the root meristem increased by 72% (Darbelley *et al.*, 1986); however, important differences in experimental design may have existed since the mitotic activity in ground control seedlings from the two studies differed by more than eightfold.

Additional experiments have reported reductions in the mitotic index of developing roots during spaceflight. On STS-29, the regrowth of roots in plantlets of *Haplopappus* was associated with a 38% decrease in cell division (Levine and Krikorian, 1992). In addition, five experiments conducted on Soyuz 12 and 13, and Salyut 6 found that the mitotic index in the apical meristem of lettuce roots decreased by 15–51% (Merkys *et al.*, 1983; Merkys, 1990; Merkys and Laurinavicius, 1990). Similarly, meristematic cells in the root apices of pea seedlings exhibited reductions of 25% in two experiments that flew on Soyuz 12 and 13 (Merkys, 1990; Merkys and Laurinavicius, 1990). The number of dividing cells in the roots of mung beans decreased by 60% after flight on STS-3 (Krikorian and O'Connor, 1982).

Even though experiments on spaceflights of shorter duration (1–5 days) often reported reductions in the mitotic index of root meristems, the root lengths often appeared to be normal (Table II). For example, in five lettuce studies on Soyuz 12 and 13, and Salyut 6, in which both mitotic activity and root lengths were measured, the mitotic index of root meristems decreased by 15–51%, but root lengths remained similar to controls, and in one case a root was 25% longer than controls (Merkys *et al.*, 1983; Merkys, 1990; Merkys and Laurinavicius, 1990). Moreover, on Soyuz 12, the roots of pea plants increased in length by 30% even though mitotic activity dropped by 25% (Merkys, 1990).

One possible explanation for this apparent discrepancy comes from a study of lentil development conducted on Spacelab D-1 (Darbelley *et al.*,

1989). The lentil experiment, which was initiated and terminated in orbit, showed that after 25 and 32 hr of growth there was no decrease in the mitotic index of the apical meristem (0–0.4 mm from the rootcap junction). However, when the 25- or 32-hr seedlings were placed on the in-flight 1 x *g* centrifuge for 3 hr, the mitotic index fell by 35%. Furthermore, the number of dividing cells throughout the whole meristem (0–1.2 mm) was found to be enhanced by 22% prior to centrifugation, but was inhibited by 43% following exposure to 1 x *g*. This decrease in cell division does not appear to be associated with the curvature of the root in response to gravitational force, since the mitotic indexes in the lower and upper regions of the root did not differ significantly (Darbelley *et al.*, 1989).

Thus, these results suggest that meristematic cell division is sensitive to alterations in gravity, and that the reductions in mitosis observed in spaceflown seedlings may be the direct result of living roots being exposed to gravity during reentry and landing.

However, reductions in the mitotic activity of root meristems also have been observed in spaceflight seedlings not exposed to gravity. On Salyut 6, for example, the mitotic index of the apical meristem was reduced in lettuce roots, even though the seedlings were fixed during spaceflight (Merkys *et al.*, 1983). Further analysis of these roots revealed an additional explanation for reductions in mitotic activity, namely, that the number of dividing cells in the meristem decreased because of premature differentiation. Normally, meristematic cells divide several times before undergoing differentiation, but cells in the spaceflown lettuce roots prematurely stopped dividing and underwent elongation, the first step in forming basipetal root tissues. The consequence of this premature differentiation was a 21% decrease in the number of meristematic cells, but no decrease in root length because of the advanced elongation of root cells.

According to a review by Halstead and Dutcher (1987), similar associations between premature elongation and reduced mitotic activity were found in pea seedlings grown during a Soyuz flight, and in wheat seedlings grown on BioSatellite II, where roots of normal length were believed to have resulted from the combined effects of increased elongation and reduced cell division. Moreover, meristematic cells stopped dividing one division cycle early and prematurely differentiated into root cap columella cells in corn seedlings flown on Cosmos 1667 (Grif *et al.*, 1988). This advanced formation of root statocytes was associated with a 33% decrease in the size of the apical meristem, and a 30% decrease in mitotic activity.

Overall, the largest decrease in mitotic activity during spaceflight occurred in roots of oat and mung bean seedlings, which experienced a 60% and 98% decline in the number of dividing meristematic cells (Krikorian and O'Connor, 1982). Interestingly, these seedlings not only were exposed to gravity upon reentry and landing, but also were grown for 8 days on

STS-3, the longest spaceflight study we found in which the mitotic index of the root meristem was analyzed. At the other extreme, no reduction in mitotic activity was observed in the lentil study on Spacclab D-1, where roots were not exposed to gravity and development was terminated after only 25 and 32 hr of growth (Darbelley *et al.*, 1989).

While cells in the elongation zones did elongate faster than in-flight controls in this study, the point where cells started elongating was the same distance from the rootcap junction in both experimental and control seedlings, an indication that the meristem was not reduced in size or in cell number during this short growth period. Thus, the mitotic index of meristematic cells appears to decrease during spaceflight if (1) roots are exposed to 1 x *g* before fixation (e.g., landing, on-board centrifuge); or (2) roots are allowed to develop in microgravity for more than 2–4 days, resulting in premature cessation of cell division and advanced differentiation.

Furthermore, although the early elongation of meristematic cells appears to maintain normal root lengths in spaceflight experiments lasting less than 5 days, early elongation apparently cannot compensate for reductions in the meristem during longer periods in microgravity (>7 days), and shorter root lengths often result (Table II).

Longer periods of root growth during spaceflight may result not only in premature differentiation but also in the early development of mature root tissues. For example, the Salyut 6 experiments that reported premature elongation in lettuce root meristems also found that the elongating cells prematurely formed root hairs before completing the elongation stage, thereby shortening the distance between emerging root hairs and the root tip by 25% (Merkys *et al.*, 1983; Merkys, 1990). *Crepis capillaris* seedlings from Cosmos 936 also exhibited root hairs closer to the root tip (Parfyonov *et al.*, 1979a), and the presence of root hairs in photographs of *Zea mays* seedlings grown on STS-34 (Schulze *et al.*, 1992) indicates premature organ formation in corn roots during spaceflight. Furthermore, advanced lateral root growth was observed on Salyut 6, where the number of cress seedlings with lateral roots increased by 25% over flight controls (Merkys and Laurinavicius, 1990), and the number of roots per seedling reportedly increased by 8% in oat seedlings and by 19% in mung bean seedlings during 8 days of development on STS-3 (Cowles *et al.*, 1984).

Advanced development also occurred in somatic embryoids of anise, *Pimpinella anisurum,* where root primordia appeared 33% earlier on Spacelab D-1 than in ground controls (Theimer *et al.*, 1986). Roots emerged early from preexisting root primordia during the regrowth of roots on STS-29 in plantlets derived from *Haplopappus gracilis* (Levine and Krikorian, 1992).

In addition to plant roots, shoots and lateral organs also show advanced development during spaceflight. For example, the coleoptiles of both wheat and oats showed accelerated embryonic shoot development by emerging early from seeds that were germinated on IML-1 (Brown *et al.*, 1992). Somatic embryoids grown from totipotent cells of carrots exhibited advanced shoot development after 18 days on Cosmos 1129 (Krikorian *et al.*, 1981). Moreover, somatic embryoids of anise formed leaf primorida in 4 days on Spacelab D-1 instead of the usual 6 days required on earth, and developed green leaflets at twice the rate of ground cultures, even though the light intensity in Spacelab was one-half of the light on the ground (Theimer *et al.*, 1986). Spaceflight effects on later stages of plant development, such as the formation of flowering structures, are discussed in a review by Halstead and Dutcher (1987).

While spaceflight may advance differentiation of plant tissues, it seldom results in conspicuous developmental anomalies. In fact, in most spaceflight studies the gross morphology of recovered seedlings was nearly identical to ground controls. However, closer analysis at the ultrastructural level has identified differences between flight and control root cells. For example, in the root meristem of garden cress seedlings that developed for 26 hr on Spacelab D-1, lipid droplets were fewer in number but were much larger in size due to confluence, and the digestion of storage protein was increased, as indicated by a 60% reduction in protein crystals and a parallel increase in vacuoles (Volkmann *et al.*, 1988; Volkmann and Sievers, 1990).

Meristematic cells in the roots of pea seedlings grown on Salyut 6 displayed dictyosomes with altered cisternae at the secretory face, and mitochondria with atypical electron-dense matrices and well-developed, well-ordered cristae (Kordyum and Sytnik, 1983). Moreover, after 5 days of growth on STS-61C, calyptrogen cells in the root meristem of corn seedlings showed a 20-fold increase in the relative volume of lipid bodies, a 31% decrease in the volume of vacuoles, a reduction of 80% in dictyosomes, and a 31% decrease in the volume of mitochondria compared with ground controls (Moore *et al.*, 1987a). Chromosomal alterations and mitotic spindle disturbances in spaceflown root cells are discussed in a review by Halstead and Dutcher (1987).

The differentiation of root meristematic cells into root cap columella cells (statocytes) during spaceflight also resulted in the appearance of developmental abnormalities at the ultrastructural level. After 32 hr of development on Spacelab D-1, columella cells in garden cress roots displayed a 10-fold increase in the diameter of lipid droplets, and up to a fourfold increase in endoplasmic reticulum (ER), which was displaced from, and less parallel to, the plasmalemma at the bottom of the cell (Volkmann *et al.*, 1986; Volkmann and Sievers, 1990). After 5 days of growth on STS-61C, maize columella cells maintained a normal size, but exhibited a 4.6-fold increase in the volume of lipid bodies, a 93% decrease in dictyosomal volume, a 37% reduction in the volume of amyloplasts, and a clumping of the ER that often was not parallel with the cell membrane as it is in ground controls (Moore, 1988; Moore *et al.,* 1987a). Likewise, after developing for 5 days on STS-61C, columella cells of the mustard spinach, *Brassica perviridis,* showed an 11-fold increase in the volume of lipid bodies, an 80% decrease in dictyosomes, a reduction of 37% in the volume of mitochondria, a 50% decrease in the volume of amyloplasts, and displacement of ER throughout the cell (Moore, 1990). Interestingly, the disruption of the normal sheet-like distribution of ER at the bottom of columella cells, which occurred in seedlings that emerged from seed during spaceflight, was not observed in seedlings that emerged prior to launch (Moore *et al.,* 1987b).

One of the most commonly observed ultrastructural changes in columella cells during spaceflight is that the gravity-sensing, starch-filled amyloplasts, which normally sediment under gravity to the distal part of the cells (Fig. 4), fail to localize at the lowest surfaces in microgravity (Fig. 5) and are found throughout the cells (Gray and Edwards, 1968; Grif *et al.,* 1988; Kordyum and Sytnik, 1983; Moore *et al.,* 1987a,b; Perbal *et al.,* 1987; Volkmann *et al.,* 1988).

In addition, the number and volume of starch grains in the amyloplast of columella cells decreases during spaceflight, and that decrease appears to be dependent on the amount of time a growing seedling is exposed to weightlessness. For example, after 25 hr of seedling development on Spacelab D-1, no changes were seen in the starch content of columella amyloplasts from lentils (Perbal *et al.,* 1987). However, after 32 hr of growth on the same flight, cress seedling amyloplasts contained smaller grains of starch, resulting in an approximately 30% reduction in starch compared with ground controls (Volkmann *et al.,* 1988; Volkmann and Sievers, 1990). After 5 days of seedling growth on STS-24, the relative volume of starch in amyloplasts of mustard spinach also decreased (39%), while the number of starch grains remained constant (Moore, 1990). Likewise, after 5 days on STS-61C, starch volumes decreased by 38% in columella cells of maize seedlings, while the number of starch grains per amyloplast remained similar to controls (Moore *et al.,* 1986). However, after 7 days on a Soviet flight, both the volume and the number of starch grains decreased in statocytes from pea and *Arabidopsis* seedlings, and after 18 days on a Cosmos flight, starch grains were completely absent or were reduced to a single grain (Kordyum and Sytnik, 1983).

The further differentiation of root cap columella cells into peripheral secretory cells involves a functional transition from gravity sensing to mucilage production, secretion, and expulsion, and ends with the peripheral cells being sloughed from the root cap (Moore *et al.,* 1987a). A number

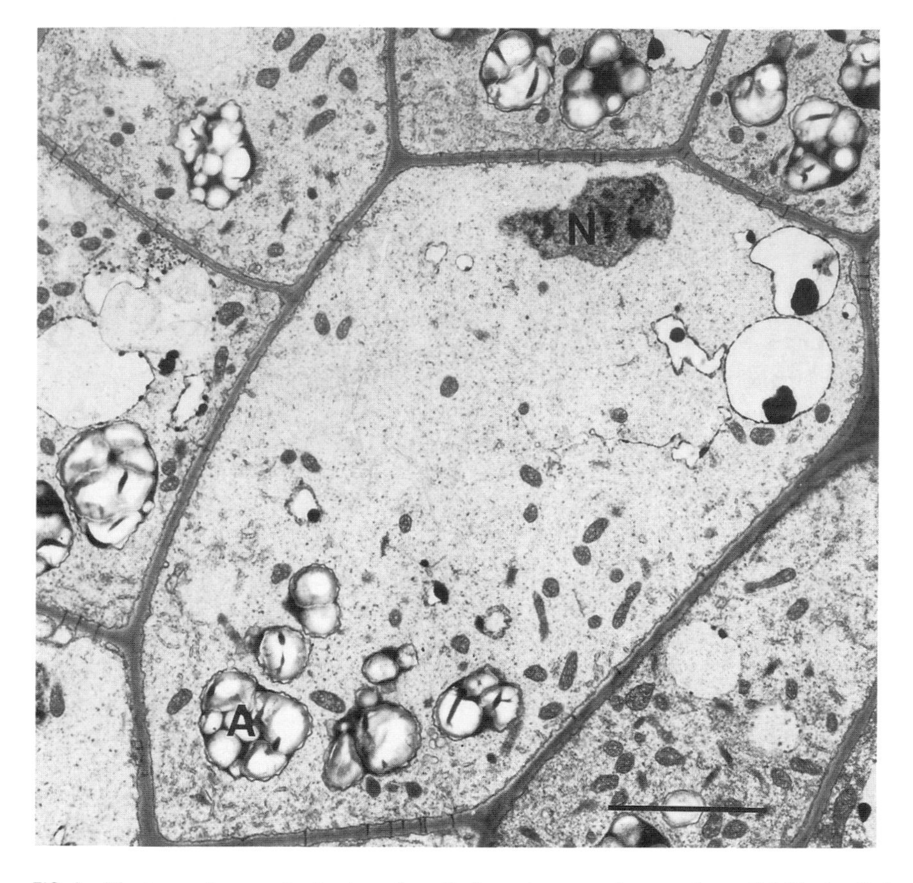

FIG. 4 Electron micrograph of columella cells from the root of sweet clover (*Melilotis alba*) grown at 1 x *g*. Amyloplasts (A) sediment under gravity to the distal part of the cell. The nucleus (N) is located at the proximal end. Bar = 5 μm. (Micrograph courtesy of E. Hilaire and J. A. Guikema.)

of studies have suggested that spaceflight alters not only the ultrastructure but also the performance of peripheral cells, resulting in abnormal root cap function. On STS-61C, the production and secretion of the polysaccharide mucilage in peripheral cells of maize seedlings decreased by 79%, and the volume of dictyosomes, which normally are present in high levels for mucilage production, was 58% less than ground controls (Moore *et al.*, 1987a). Furthermore, in garden cress seedlings on Spacelab D-1, the mucilage that was secreted between the plasmalemma and the cell wall failed to pass through the cell wall, resulting in root tips that lacked a covering of slime (Volkmann and Sievers, 1990).

Since the coating of mucilage is believed to prevent desiccation of the

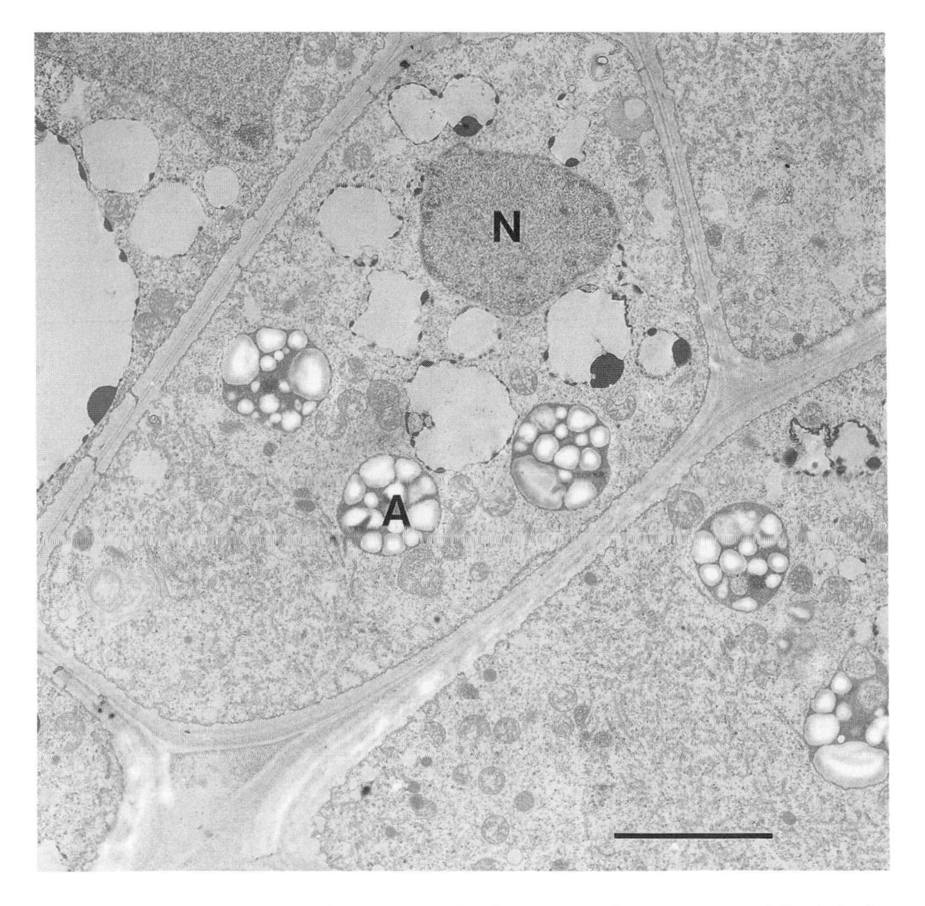

FIG. 5 Electron micrograph of columella cells from sweet clover grown and fixed during spaceflight. Amyloplasts (A) do not sediment in reduced gravity, and are found throughout the cell. The nucleus (N) remains in the proximal part of the cell. Bar = 5 μm. (Micrograph courtesy of E. Hilaire and J. A. Guikema.)

root tip and aid the exchange of nutrients (Esau, 1977), the absence of the coating may explain observations from STS-3, where root tips from oat seedlings appeared discolored and stunted, and where root cap peripheral cells from mung beans appeared degraded and compressed (Slocum and Galston, 1982; Slocum *et al.*, 1984; Cowles *et al.*, 1984). Furthermore, the roots of oat and mung bean seedlings on STS-3 lost their ability to penetrate the agar substrate (Cowles *et al.*, 1984). Since the root cap is required for penetration, a nonfunctioning cap may be indicated in these plants. In addition, according to Halstead and Dutcher (1987), an increase in the number of lateral roots, as reported in oat and mung bean seedlings on STS-3 and in oat seedlings on SLS-2, also may indicate damage to the primary root tip.

Interestingly, experiments that investigated the sensitivity of early embryonic root development to weightlessness found that *de novo* formation of the root cap during spaceflight was abnormal. Roots generated in cultures of dedifferentiated mustard cress callus cells appeared to be normal after 8 days of growth in orbit; however, although the root caps contained meristematic cells and peripheral secretory cells, the central zone of gravity-sensing columella cells was missing, an omission that did not occur in caps that formed on earth (Podlutsky, 1992a,b). Moreover, on STS 61-C, maize seedlings that were launched with decapped root tips were unable to regenerate root caps, a developmental process which occurred normally in ground controls (Moore *et al.,* 1987b). The maize roots, which did lengthen during flight, contained undifferentiated callus-like cells at the tips.

The effects of weightlessness on plant shoot development at the cellular level have not been as widely studied as root formation, but there is evidence that shoot development may be altered. For instance, meristematic tissue in shoots of wheat seedlings grown for 2 days on Biosatellite II exhibited a significant reduction in dry weight (Conrad, 1968). Since the shoot meristem is the tissue from which lateral organs develop, it is interesting that the further development of wheat plants over 16 days on Mir resulted in leaves that appeared to be normal, but the thickness of the leaf mesophyll and epidermal cuticle was reduced by 44% and 450%, respectively, and the density and size of chloroplasts were significantly decreased (Nedukha and Mashinskiy, 1992).

Furthermore, cell walls of plant shoots may be altered in weightlessness, since decreases have been reported in both the activity of cell wall enzymes and the level of structural polymers. After 8 days of flight on SLS-2, mung bean seedlings that developed from seeds in microgravity exhibited reductions in the activities of peroxidase (15%) and phenylalanine ammonia lyase (30%), which are enzymes in the biosynthetic pathway of lignin, a cell wall structural polymer (Cowles *et al.,* 1986). Similar reductions in the activities of peroxidase (19%) and phenylalanine ammonia lyase (22%) were reported in stems of pine plants that were launched as seedlings and grown for 8 days on STS-3 (Cowles *et al.,* 1984). However, the actual content of lignin in cell walls differed greatly in these two experiments. In mung bean seedlings, which developed from seed during spaceflights, lignin was reduced by 18% on STS-3 and by 25% on SLS-2. In pine seedlings, which developed into seedlings before launch, lignin content in the whole stem was not significantly altered. Nevertheless, lignin in the pine's upper stem tissue that developed during spaceflight showed a decrease of 24% (Cowles *et al.*, 1982, 1984).

In Soviet spaceflight experiments on moss, described in a review by Halstead and Dutcher (1987), the cell wall polymer cellulose displayed altered organization and cell walls were 33% thinner than ground controls; in addition, pea plants grown on Salyut 4 contained 54% less cellulose in their cell walls, and on Salyut 6, cell walls in pea root meristems appeared thinner than in ground controls. Furthermore, reductions in cellulose content of up to 20% were observed in mung bean, oat, and pine seedlings after 8 days of growth on SLS-2 (Cowles *et al.*, 1986).

These results have led to the hypothesis that cell wall synthesis is reduced in weightlessness (Cowles *et al.*, 1986; Halstead and Dutcher, 1987). Indeed, the failure of peripheral secretory cells to expulse mucilage through the cell wall in maize and garden cress root tips may indicate that the structure of root cell walls also is altered during spaceflight.

a. Role of a Putative PKC in Plants A plant enzyme analogous to the PKC enzyme in animals has not yet been identified; however, phosphoinositol pathways that include phospholipase C, PIP_2, PI, and diacylglycerol, are known to exist and to be involved in signal transduction (Coté and Crain, 1993; Gross and Boss, 1993). Furthermore, phorbol esters and DAG analogs, known activators of PKC, have been used recently to demonstrate that a PKC-related enzyme may play a role in plant development and growth. For example, phorbol ester treatment of primary roots of maize induces biphasic increases in primary root cell elongation rates by 60%–210%; this action of phorbol ester is duplicated by a synthetic DAG and is inhibited by staurosporine, an inhibitor of phospholipid/Ca^{2+}-dependent kinases (Kim *et al.*, 1991). Phorbol esters may activate a pathway stimulated by axin since phorbol ester-induced cell elongation is similar to that triggered by low concentrations of auxin (Kim *et al.*, 1992).

Low concentrations of phorbol ester also enhance root cell elongation by twofold in immature maize embryo cultures, and increase the number of embryogenic cells in carrot suspension cultures (Yang *et al.*, 1992). In addition, the induction of callus growth in tobacco leaf is inhibited by phorbol ester, while the regeneration of tobacco shoots is enhanced by low concentrations of phorbol ester and DAG (Yang *et al.*, 1993). These results from ground-based studies on phorbol ester activation of plant signaling pathways are consistent with the theory that PKC activation is enhanced in microgravity. As presented earlier, spaceflight increases seedling shoot lengths, enhances root cell elongation, and decreases the growth of callus tissue.

2. Animal Development in Altered Gravity

Compared with plants, the development of animals under microgravity conditions has not been as widely investigated, especially at the cellular level. Nevertheless, several spaceflight studies suggest that early stages of animal development may be sensitive to alterations in gravity. For example, fertilization of eggs from the African clawed toad, *Xenopus laevis,* during 6–7 min of suborbital rocket flight appeared to affect continued development upon their return to earth, since the majority of embryos died between gastrulation and neurulation, and the few that reached larval stage expressed abnormal morphology (Ubbels, 1992). In addition, when *Xenopus* eggs were fertilized on IML-1, development of the embryos continued only to gastrulation, and the size of the blastocoel cavity was greatly reduced (Ubbels *et al.,* 1992).

Similar results were reported with frog eggs on Spacelab J, where fertilization during spaceflight resulted in altered gastrulation in which the dorsal lip of the blastopore was nearer the vegetal pole and the blastocoel roof was thicker, although in this case defects were apparently corrected since neurulation and subsequent development in microgravity appeared to be nearly normal (Black *et al.,* 1993). As reviewed by Souza (1987), no such alterations were reported in later stages of amphibian development during seven U.S. and Soviet spaceflights in which eggs were fertilized on the ground before launch.

A sensitivity to gravity during early stages of amphibian development also has been suggested by results from clinorotation studies. For example, clinorotation during early *Xenopus* development relocates the third cleavage furrow and the blastocoel, changes the number of cell layers in the blastocoel roof, and alters the position and shape of the dorsal blastopore lip; nevertheless, further development must compensate for these anomalies since tadpoles appeared normal except for changes in head dimensions and eye structures (Yokota *et al.,* 1994). Other studies also have detected changes in early stages of amphibian development during clinorotation (Serova and Denisova, 1982; Nace and Tremor, 1981).

Recent spaceflight experiments on avian embryos suggest that earlier stages of development are more sensitive to altered gravity conditions. On STS-29, all of sixteen chicken eggs that had been incubated for 2 days prior to launch failed to survive during the 5-day flight; in fact, 81% stopped developing after just 48 h in microgravity (Hullinger, 1993). However, on the same flight, all of sixteen eggs that had been incubated for 9 days prior to launch survived the flight and hatched normal chicks upon their return to earth (Hullinger, 1993). Furthermore, the centrifugation of early-staged chicken eggs at 4 x *g* generates similar mortality rates (78%) after 4 days, with most deaths occurring by the second day (Smith and Abbott, 1981).

Comparable results have been reported for developing quail eggs on Cosmos 1129 in which eggs were launched after a 3–5-day incubation, and also on the Mir space station in which quail eggs were fertilized in microgravity. Of the 60 eggs on Cosmos 1129, 56% ceased development within the first 4 days of spaceflight, and of the 43 eggs fertilized on Mir, 51% did not develop beyond early developmental stages and 81% failed to survive past 17 days (Bodá *et al.*, 1991, 1992).

Eggs from the fish *Fundulus heteroclitus* also exhibited sensitivity to spaceflight and to clinorotation during early stages of development. For example, clinorotation at the midblastula stage resulted in a number of abnormalities such as altered cell movements, impaired epibolic coordination, and altered differentiation of the axial complex; nevertheless, these defects were largely overcome during continued development on the clinostat or during exposure of the clinorotated embryos to spaceflight, although 20% of spaceflown fish eggs exhibited frontal body asymmetry (Parfyonov *et al.*, 1979b). As with the results from chicken and quail, fish embryos exhibited no abnormalities when eggs underwent clinorotation during later stages (after 52 h) of development (Parfyonov *et al.*, 1979b).

Invertebrates also appear to show increased sensitivity to spaceflight during early stages of development. For example, on Spacelab D-1, *Drosophila melanogaster* eggs that underwent oogenesis in microgravity exhibited a 1-day delay in hatching, with only 10% of embryos emerging on schedule; these embryos also showed a tenfold increase in head and thoracic defects (Vernós *et al.*, 1987). In addition, the proliferation rate of the oocytes increased by 1.5-fold during spaceflight, and the eggs increased in size and showed more alterations in yolk granule deposition, compared with 1 x *g* controls on the in-flight centrifuge (Vernós *et al.*, 1989). These spaceflight effects on egg number, size, and hatching rates were observed only in embryos that had undergone oogenesis in weightlessness.

On the same flight, hatching rates also declined in eggs of the stick insect, *Carausius morosus,* which were exposed to microgravity during early stages of development (Bücker *et al.*, 1987). Since the incubation period for *Carausius* eggs is 75–105 days (compared with 1 day for *Drosophila* embryos), eggs at various stages of development were flown on Spacelab D-1. When eggs classified at launch as stage 1 (4 days of age) and stage 2 (16 days of age) were exposed to spaceflight on the 6-day mission, hatching rates decreased by 17% and 41%, respectively (Bücker *et al.*, 1987). These results are supported by studies on Cosmos 1887, where eggs launched at 24 days of age exhibited a reduction in hatching rate of 43% (Bücker *et al.*, 1989; Ushakov and Alpatov, 1992). Both studies found that hatching rates were not altered when embryos at later stages of development were exposed to spaceflight.

The Spacelab D-1 study also suggested that the effects of microgravity were synergetic with effects of cosmic radiation. For example, *Carausius* eggs in the 1 x *g* in-flight centrifuge showed no decrease in hatching rates and no increase in developmental anomalies when hit by high-charge Z nuclei and high-energy (HZE) particles, but stage 1 and stage 2 eggs that were hit by HZE particles while in microgravity showed a decrease in hatching rates of 39% and 50%, a greater effect than that due to microgravity alone (Bücker *et al.*, 1987). In addition, the incidence of abnormalities in larvae increased substantially when stage 2 eggs suffered HZE hits while in microgravity. Whereas few developmental anomalies were found when stage 2 eggs were exposed to either cosmic radiation or microgravity alone, more than 50% of newly hatched larvae showed deformation of abdominal segments or extremities when exposed to HZE particles in microgravity (Bücker *et al.*, 1987).

Results also have been obtained in spaceflight studies on another arthropod, the brine shrimp, *Artemia*. Brine shrimp are launched as encysted gastrula-stage embryos, and are rehydrated during spaceflight to reactivate embryonic development. The fact that early stages of development occur before encystment may explain results indicating normal hatching rates and normal patterns of larval development, morphology, and behavior (DeBell *et al.*, 1992, Spooner *et al.*, 1994a). Nevertheless, experiments on STS-54 and USML-1 have suggested that spaceflight may accelerate brine shrimp development (Fig. 6), since larvae in spaceflight cultures were developmentally advanced by 1 day (Spooner *et al.*, 1994c). For example, in cultures in which cysts were reactivated and then fixed 2.25 days later, 80% of spaceflight larvae were in the third instar, whereas 80% of ground controls were second instar larvae. Moreover, 70% of brine shrimp terminated after 5 days of spaceflight were between the fifth and sixth instar, whereas 67% of ground controls only had developed to the fourth instar. On USML-1, 60% of brine shrimp larvae had developed to the sixth instar, while 66% of ground controls had reached only the fifth instar (Spooner *et al.*, 1994c).

Recent results suggest that microgravity may also enhance the growth and development of cultured embryonic mammalian organs during spaceflight. On STS-54, cultures containing embryonic mouse premetatarsal mesenchyme were used to study the effects of microgravity on early stages of embryonic bone formation (Fig. 7). After 5 days in culture, the isolated mesenchyme had differentiated into chondrocytes and organized into cartilage rods that were morphologically normal, but that had grown at 2.6 times the rate of ground controls (Klement and Spooner, 1994). Similar results were obtained from STS-54 spaceflight organ cultures containing rudiments of developing mouse pancreas. Not only were flight rudiments larger after 3 days in microgravity, 67% of pancreatic acini in spaceflight cultures contained normal differentiated exocrine cells with zymogen gran-

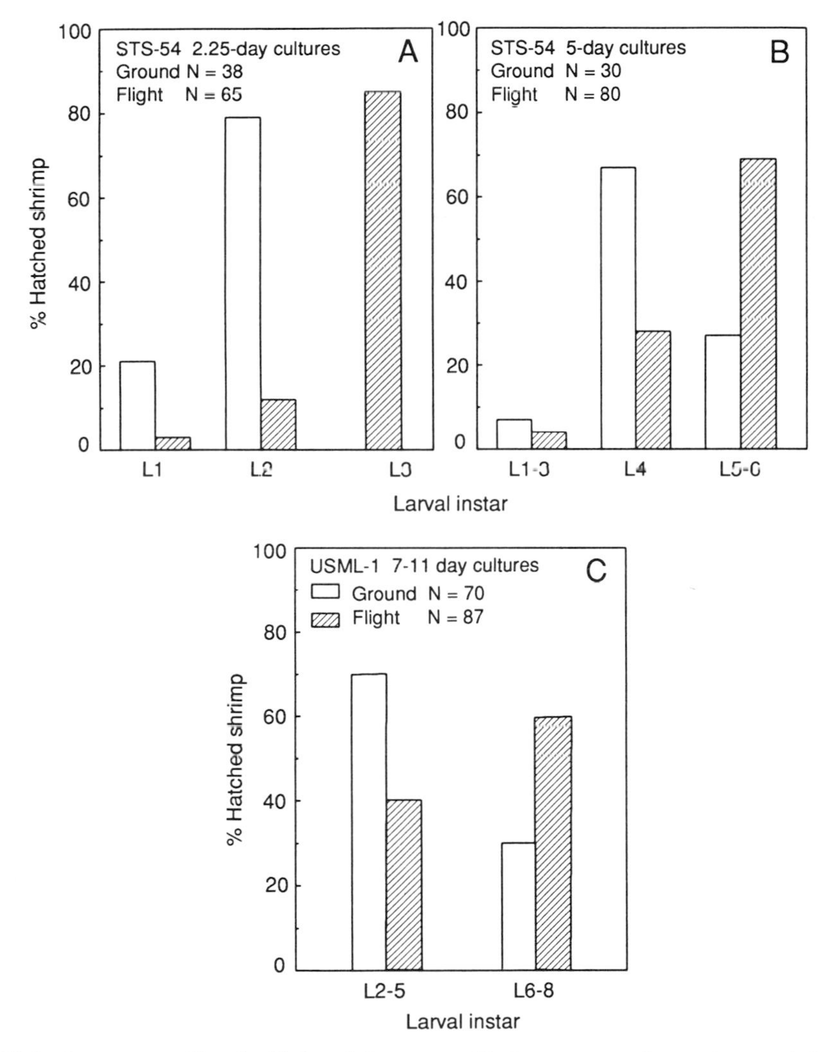

FIG. 6 Precentage of hatched brine shrimp at specific instar stages after various periods of development in microgravity. (A) Shrimp developed to the third instar during 2.25 days of spaceflight, whereas ground control shrimp reached only the second instar during the same period. (B) Advanced development also was observed in 5-day cultures, where 70% of flight shrimp and 27% of control shrimp had reached the fifth and sixth instar. (C) on USML, more than twice the percentage of flight shrimp developed to larval stages L6-8 than did ground control shrimp during the same period. (Data from Spooner *et al.*, 1994c.)

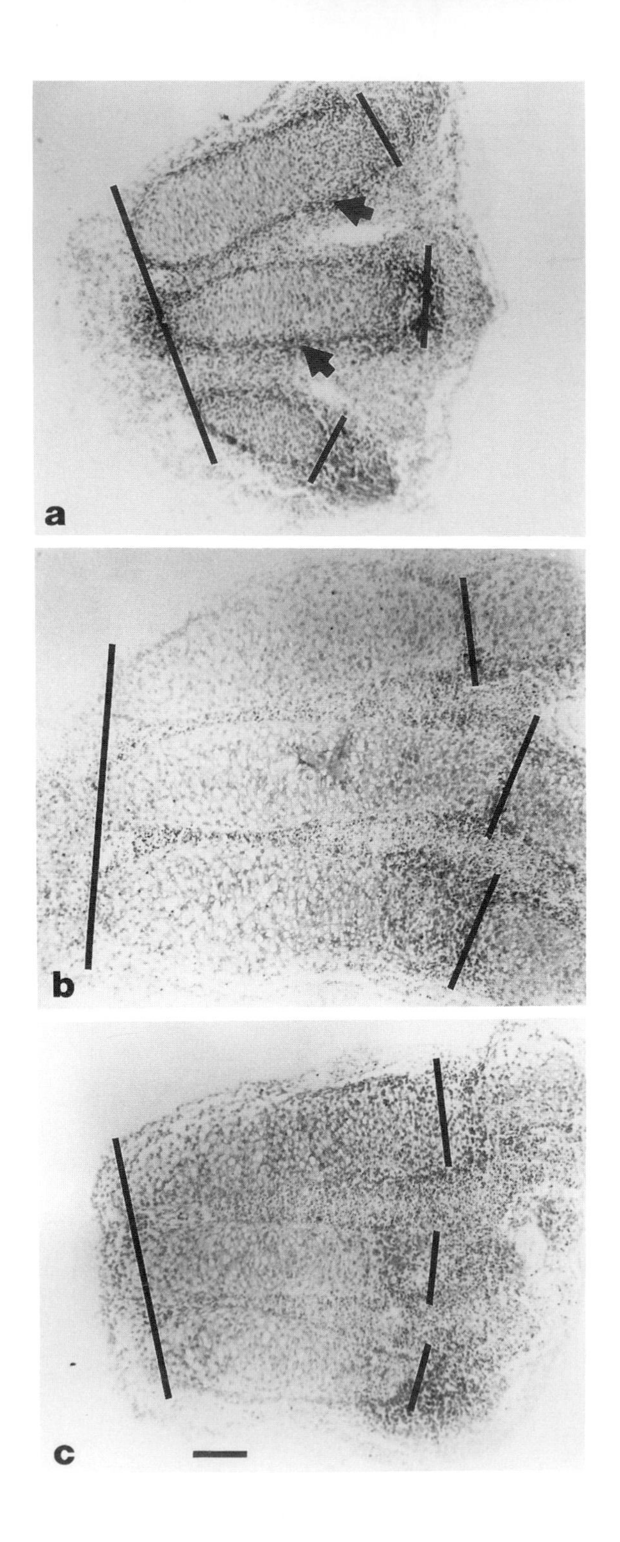
a
b
c

ules, compared with just 34% of ground controls (Spooner *et al.*, 1994b). These observations suggest that the embryonic growth and development of animals may be enhanced in microgravity. Clearly, additional ground-based and spaceflight studies are required to determine the underlying cause(s).

C. Effects of Spaceflight on Unicellular Organisms

Microorganisms were carried on many of the first Soviet and U.S. spaceflights in the 1960s. The Gram-negative prokaryotes *Escherichia coli* and *Salmonella* sp., and the Gram-positive *Bacillus* sp., and *Clostridium* sp., were frequently used in early spaceflight experiments. The algae *Chlorella* sp., the fungal mold *Neurospora crassa,* and the yeast *Saccharomyces* sp. were commonly flown eukaryotes. In the 1970s and 1980s, additional prokaryotes such as the Gram-negative *Proteus vulgaris* and the Gram-positive *Staphylococcus* sp. were investigated, and other unicellular eukaryotic organisms such as the yeast, *Rhodotorula rubra,* and the protozoans, *Paramecium* sp. and *Euglena gracilis,* were studied along with algae and molds. In general, the results from three decades of research establish that spaceflight can enhance proliferation of unicellular organisms, and can elicit specific and unique responses in many of the microorganisms tested (Lewis and Hughes-Fulford, 1994).

Of the prokaryotes, the bacterium *Bacillus subtilis* has been used as a test organism since the earliest spaceflights, mainly because of its ability to form spores and germinate upon cue. *Bacillus* spores continue to be used in radiobiology studies in which spores are germinated following their return to earth in order to measure the effect of space radiation on viability and mutation (Graul *et al.*, 1991). In addition, spore germination can be activated in flight, allowing cell growth studies to be conducted without requiring low temperatures during the launch phase. Results from *Bacillus* cell growth studies reveal that the production of biomass is in-

FIG. 7 Premetatarsal tissue isolated from mouse embryos at 13.7 days of gestation was cultured on STS-54 for 5 days. Zero time control explants (a) were fixed during the first movement, 16 hr after launch. Chondrocytes are present and have already formed short rods of cartilage. The perichondrium area is wide and very dense (arrowheads). After 5 days of culture during spaceflight (b), rod expansion is clearly evident in length (ends of rods are marked) and width. The cartilage rod morphology is also well preserved. The perichondrium has matured to form a distinct separation between the cartilage and mesenchyme tissues. Ground control explants (c) show an increase in length that is less than flight samples. The cartilage rod morphology is maintained, and the perichondrium has matured as in (b). Bar = 100 μm. (From Klement and Spooner, 1994, courtesy John Wiley & Sons, Inc.)

creased during spaceflight above that seen in ground controls (Cogoli and Tschopp, 1982; Mennigann and Lange, 1986). However, the increased biomass may not be due to an increase in the overall rate of growth, but instead to a more rapid onset of spore germination following inoculation of the growth medium (Bergter *et al.*, 1985; Mennigann and Lange, 1987). In ground control experiments, *Bacillus* spores convert into dividing vegetative cells after 13.5 hr, but in Spacelab D-1 experiments, biomass readings began to increase after just 30 min (Mennigann and Lange, 1987). Although it is possible that spore germination was influenced by cosmic radiation, a number of space radiobiology studies have reported that radiation has either no effect or a detrimental effect on the viability of *Bacillus* spores (Bücker *et al.*, 1975; Facius *et al.*, 1978; Horneck *et al.*, 1984, 1992).

Other bacteria, such as *Actinomyces, Methylbacterium* sp., *Staphlyococcus aureus, E. coli,* and *Salmonella,* have been listed by Dickson (1991) as exhibiting increases in cell growth during spaceflight experiments lasting 1–2 days. On Biosatellite II, the population densities of *E. coli* and *Salmonella* were significantly higher than in ground controls (Mattoni, 1968); and *E. coli* cultures on Salyut 6 exhibited a higher number of dividing cells than did ground controls (Tixador *et al.*, 1985b). Moreover, results from Spacelab D-1 experiments showed that after only 3 hr at 37°C, cultures of *E. coli* exhibited a 36% increase in cell numbers over in-flight control cultures on the 1 x *g* centrifuge (Ciferri *et al.*, 1986). Just how a reduction in gravity could cause an increase in the proliferation of bacteria has not been resolved, although one possibility is that when the force of gravity is removed, cellular energy reserves are required less for motility, and thus can be utilized more for cell reproduction (Lewis and Hughes-Fulford, 1994).

The increased proliferation of bacteria during spaceflight may be responsible for another flight effect, the enhanced resistance of bacteria to antibiotics (Tixador *et al.*, 1985b). A number of antibiotics with different mechanisms of action have been tested on bacteria during spaceflight, and all have shown reduced effectiveness. For example, experiments aboard Salyut 7 and Spacelab D-1 tested for the minimal antibiotic concentration that would inhibit *E. coli* growth, and discovered that two to four times the normal concentrations of kanamycin and colistin were required to be effective (Moatti *et al.*, 1986; Tixador *et al.*, 1985b). In the Spacelab experiment, the number of *E. coli* that survived in cultures containing a subinhibitory concentration of colistin increased by fourfold over identical cultures on the ground (Lapchine *et al.*, 1987). In addition, *E. coli* also exhibited enhanced resistance to streptomycin during flight on STS-42 (Tixador *et al.*, 1992). Likewise on Salyut 7, *Staphylococcus aureus* was found to be less sensitive to the antibiotics oxacillin, chloramphenicol, and erythromycin (Tixador *et al.*, 1985b).

Since all the antibiotics that were tested during spaceflight displayed a similar loss in effectiveness regardless of their mechanisms of action, and since bacteria increased in number during spaceflight, the reduction in antibiotic effectiveness may be due solely to the increase in cell proliferation (Tixador *et al.*, 1985b). However, although density may be a factor in cases where resistance is due to secretion of drug-altering enzymes, density is not the only factor in developing resistance to antibiotics (Fig. 8).

The rate-limiting factor in antibiotic delivery to the bacteria cell is membrane penetration, which can be altered not only by enzymatic modification of the antibiotic, but also by changes in the membrane itself or in membrane-associated electrolytes (Joklik *et al.*, 1992). For example, the ineffectiveness of antibiotics in microgravity may be due to alterations in the cellular envelope that result in decreased antibiotic absorption and/or in decreased antibiotic permeability through the envelope (Tixador *et al.*, 1981, 1985a). Indeed, alterations in the levels of extracellular and intracellular electrolytes, particularly calcium, were discovered in spaceflown *E. coli*, and a thickening of the membrane or cell wall has been reported in *E. coli*, *Staphylococcus aureus*, and *Pseudomonas aeruginosa*, all of which also exhibited increased resistance to antibiotics (Dickson, 1991; Lapchine *et al.*, 1987; Tixador *et al.*, 1985a).

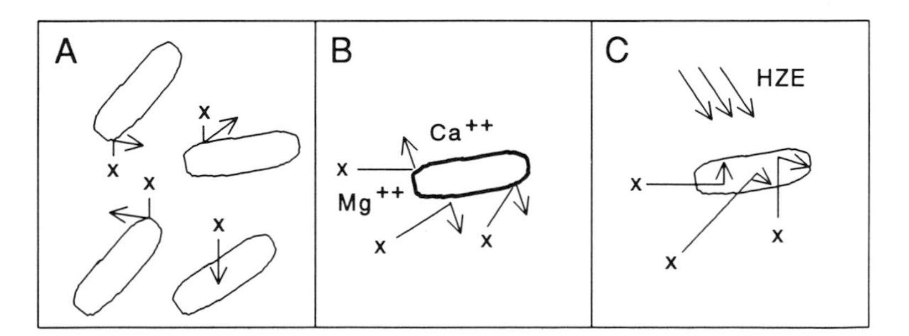

FIG. 8 A schematic representation showing possible mechanisms of antibiotic resistance in spaceflight bacterial cultures. (A) Increased proliferation of bacteria during spaceflight may increase the production of drug-altering enzymes that reduce the membrane penetration of certain antibiotics (X). (B) Increased levels of membrane-associated polyvalent ions stabilize the membrane and reduce antibiotic penetration. Increases in the thickness of the cell envelope also may interfere with antibiotic movement into the cell. (C) Cosmic radiation and/or microgravity may induce an increase in drug-resistant mutants, in which the drug target is altered in its susceptibility to the antibiotic.

Furthermore, additional flight experiments have reported changes in the cellular envelopes of other bacteria, although the consequences of those changes were not addressed. For example, spaceflight causes the cyanobacterium, *Anabaena azollae,* to show a decrease in the electron density of its cell wall (Popova, 1987). Spaceflight also induces *Proteus vulgaris* to consistently exhibit ultrastructural changes in the cell wall and membrane, with membrane gyrosity increasing under optimum aerobic growth conditions, and evaginations of the cell wall and cell membrane increasing under anaerobic growth conditions (Kordyum and Sytnik, 1983).

Not only are the cellular mechanisms of antibiotic resistance unclear but the spaceflight factors that elicit such mechanisms are unclear as well. On Spacelab D-1, *E. coli* cultures grown on colistin in the 1 x *g* in-flight centrifuge exhibited an increase in antibiotic resistance that was similar to flight cultures grown in orbit (Lapchine *et al.,* 1987). While this may be an indication that factors other than weightlessness increased resistance to antibiotics, one possibility is that controls were sufficiently exposed to microgravity during the 6 hr before Spacelab began operating and during the periods when the centrifuge was stopped for other experiments (Cogoli, 1993). It is unlikely that radiation-induced mutations were responsible since antibiotic resistance was present in bacteria only during flight, and was not found in spaceflown cultures back on the ground following flight (Lapchine *et al.,* 1987).

Eukaryotic unicellular organisms react to spaceflight with responses that are often similar to the prokaryotes. In the wild-type stain of the algae *Chlamydomonas reinhardii,* cell proliferation increased by 58% on Spacelab D-1, compared with ground controls (Mergenhagen and Mergenhagen, 1987). Moreover, the protozoan *Paramecium tetraurelia* increased its population size by two to threefold on Salyut 6 and Spacelab D-1 (Cogoli and Tschopp, 1982; Planel *et al.,* 1982; Richoilley *et al.,* 1986).

However, as with bacteria, the spaceflight factors involved in eliciting increased cell proliferation are not clearly evident. For example, low doses of gamma radiation have been shown to stimulate proliferation of paramecia in ground-based studies, as has the exposure to cosmic radiation in high-altitude balloon flights; however, then paramecia were grown in a 1 x *g* centrifuge during spaceflight, cell proliferation was nearly the same as ground controls, evidence that reduced gravity, not radiation, stimulates proliferation during spaceflight (Richoilley *et al.,* 1986). Since paramecia normally exhibit negative geotactism, the investigators suggest that reduced gravity decreases the cellular energy required for cell motility, and thereby increases the energy available for cell division (Richoilley *et al.,* 1992). Whether or not energy requirements decrease, it does appear that energy levels increase, since the cellular ATP levels increase by 23% in the simulated low gravity of clinorotation, where paramecia proliferation increased by 20%. Moreover, when cultures were exposed to hypergravity at 2 x *g*, paramecia ATP levels dropped by 23% and growth rates decreased by nearly 40% (Richoilley *et al.*, 1992).

In addition to increased proliferation rates, the spaceflight environment elicits other physiological responses in paramecia. On the Salyut 6 space station, the contents of *Paramecium* culture media exhibited altered electrolytes; levels of magnesium, calcium, potassium, and phosphate were 20–60% higher than in ground control cultures (Planel *et al.*, 1982, 1990). Furthermore, measurements of intracellular electrolytes revealed that magnesium, calcium, and potassium levels were 15–56% lower than in ground controls (Tixador *et al.*, 1981). Paramecia flown on Spacelab D-1 also showed lower magnesium levels (10–13%); lower calcium levels were indicated, but were not significantly different than in ground controls (Richoilley *et al.*, 1987; Planel *et al.*, 1990). Additional studies on paramecia from Salyut 6 demonstrated that cell volumes increased by as much as 40%, while cell protein levels decreased by 38% and the dry weight of cells decreased by 34% (Planel *et al.*, 1982, 1990). Electrolytes and protein levels were not altered following exposure of paramecia to cosmic rays in high-altitude balloon flights, an indication that microgravity influenced these changes (Richoilley *et al.*, 1986).

Other unicellular eukaryotes have also exhibited responses to spaceflight. The protozoan *Tetrahymena* showed as much as a three- to fourfold increase in cell numbers in 48 hr on three Cosmos flights (Alpatov *et al.*, 1991). On Spacelab D-1, the slime mold, z*Physarum polycephalum,* exhibited a 110% increase in protoplasmic streaming velocity, and distinct changes in the rhythmic contractions of the ectoplasmic envelope that generates ameboid migration (Briegleb *et al.*, 1986). Altered contractions of the envelope did not affect *Physarum* migration on the 7-day Spacelab flight, but, according to Dickson (1991), migration was inhibited on the 18-day Cosmos 1129 flight. Cellular responses that appear to intensify as spaceflight duration increases have been also noted in the alga *Chlorella.* Dickson (1991) lists 20 flights that lasted less than 2 weeks in which *Chlorella* showed little or no response to spaceflight, although a tendency for reductions in plastid volume and in the number of starch grains was observed in some flights (Kordyum and Sytnik, 1983). However, on two flights lasting 28 and 30 days, cultures exhibited enhanced biomass gain, larger mitochondria and cristae, numerous vesicles in the plastid stroma, outgrowth of plastids to the plasmalemma and cell wall, and more lipid drops in the cytoplasm, indicating an increase in the rate of aging (Kordyum and Sytnik, 1983; Sytnik *et al.*, 1992).

III. The Search for Underlying Principles

Even though there is growing evidence that changes in gravitational force can influence biological systems at the cellular and subcellular levels, and that protein kinase C can act as an effector of that influence, little is known of the specific mechanism(s) by which gravity affects the cell. To reveal the basis for gravitational effects on the cell, a number of physical phenomena that are directly induced by gravity must be considered. The most evident gravity-induced phenomenon occurs when the attractive force of gravity acts on a body's mass to generate weight. Moreover, when a body (e.g., the cell or cell organelle) is surrounded by a fluid environment, the pull of gravity on the fluid generates hydrostatic pressure and buoyancy forces. Hydrostatic pressure is caused by the weight of the fluid on a body, and buoyancy is caused by the weight of the fluid that a body displaces. The buoyancy force, which opposes the pull of gravity on the body, is equal to the weight of the displaced fluid. If that weight is less than the weight of the body, gravitational force overcomes the buoyancy force, and the body sediments.

In a microgravity environment, the cell, its components, and the surrounding fluids are nearly weightless, and the gravity-induced forces that normally act on cellular structures, that is, buoyancy, sedimentation, and hydrostatic pressure, are reduced. The decrease in these forces may possibly be detected by the cell in several ways. For example, the cell may be sensitive to a change in tension of the cytoskeleton and a change in stress of the cytoplasmic membrane that, in microgravity, could be caused by the unloading of attached cellular components or the absence of fluid weight (hydrostatic pressure). Such a sensitivity to alterations in gravity-induced forces could cause the cell to alter normal activities.

There has been discussion, however, as to whether a single cell can detect and be influenced by alterations in gravitational force (Albrecht-Buehler, 1990, 1991; Gordon, 1964; Kondepudi, 1991; Kondo, 1968; Pollard, 1965; Todd, 1989). In the mid-1960s, the U.S. National Aeronautics and Space Administration's (NASA) Office of Space Science sought to define the size threshold below which gravitational force would have no effect, and the absence of gravity would not be perceived. Pollard (1965) and other theoretical physicists (Gordon, 1964; Kondo, 1968) calculated that, given the densities of the cell environment, cells of less than 10 μm in diameter and all subcellular components were physically incapable of sedimentation, and therefore would not be affected by a reduction in gravity. Although their conclusion that the weightless environment would have no effect on the single cell has been shown to be incorrect, the theory that gravity cannot influence the cell via sedimentation forces has not been disproved. One possibility is that for sedimentation to affect the cell, organelles (e.g., mitochondria, nucleus, ribosomes) would have to be sufficiently displaced so that their movement could be transferred through the cytoskeleton to the membrane, where deformation of the membrane would initiate a cellular response.

Recalculations by Pollard (1971) suggested that a 1-μm-diameter organelle could exhibit a change in distribution in altered gravity, but only if the movement-restrictive properties of the cytoskeleton were ignored and freedom of movement within the cell was assumed. Others have argued that cellular organelles associated with the cytoskeleton would have to be of greater mass before gravitational pull could displace them sufficiently that the cytoskeleton would be transformed and the membrane would experience significant deformation. Small displacements of cellular structures may not be detected by the cell according to Albrecht-Buehler (1991), who calculates that cells up to 10 μm in diameter contain too little mass to be influenced by sedimentation, since the cellular environment includes contractile, electrical, chemical, and polymerization forces that are many orders of magnitude larger than the sedimentary forces acting on the cell. However, a theoretical study by Kondepudi (1991) recalculated the lower size limit for gravitational effect to be 0.5 μm, twenty times smaller than that previously proposed, by assuming that the cell is not in thermodynamic equilibrium. A dynamic nonequilibrium mechanism would be more sensitive to a weak gravitational signal because of its ability to detect a small systematic force (gravity) that is embedded in the much larger, random, and fluctuating cellular forces. Thus, even though sedimentation forces may be completely overpowered by other forces in the cell, it is possible that the constant force of gravity may still exert an influence through the cellular noise, much in the way gravity influences the path of a butterfly in hurricane.

Given the difficulties of demonstrating an effect of gravity-induced sedimentation, buoyancy, or hydrostatic forces on cell function, attention has been given to fluid phenomena that result when gravity acts on droplets or fluid zones of different densities. Fluid zones that differ in density, due to concentration, temperature, solute or fluid differences, react to gravitational pull as would a body with mass. That is, zones of higher densities sediment, while those of lesser densities rise due to buoyancy. The movement of these zones results in a number of fluid phenomena, including convection currents, streaming potential, and particle streaming, all of which cause mixing of the fluid and of solutes within the fluid (Todd, 1989).

For the cell in a low-gravity environment, the absence of fluid movements due to differences in density might result in the abnormal distribution of cellular nutrients and metabolites. For example, cell signaling molecules might accumulate at the site of production, being able to spread to their site of action only by the slower, gravity-independent process of diffusion. Moreover, since heat convection currents would not occur, thermodynamic processes in the cell may be altered due by a decrease in the dissipation of heat. In addition, the absence of gravity-dependent flow has been suggested as a possible cause for the detrimental effects of spaceflight on cells in culture, as well as on bone decalcification and muscle loss (Albrecht-Buehler, 1991). Without the movement of the immediate fluid environment that surrounds cells, nutrients could be depleted and wastes could accumulate, ultimately resulting in diminished cell growth, responsiveness, and vitality. However, this theory is not supported by spaceflight results on bacterial cells, plant cells, or animal cells, either in culture or in the organism.

If the effects of reduced gravity cannot be attributed solely to the absence of any one gravity-induced phenomenon, and if sedimentation cannot be disregarded, then what are possible cellular targets of gravity-induced phenomena that would induce changes in cell function? One proposition (Spooner, 1992) is that the intracellular cytoskeleton, the extracellular matrix (ECM), and the plasma membrane represent compartments of sufficient macromolecular organization to be sensitive to gravity-induced phenomena (Fig. 9). The cytoskeleton and ECM are indispensable

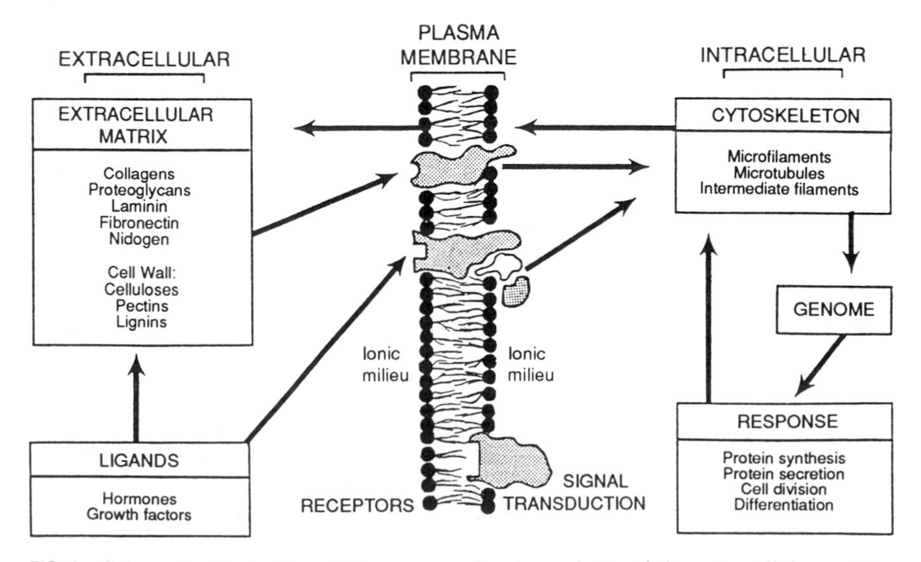

FIG. 9 Schematic illustration of the macromolecular makeup of the extracellular matrix, the plasma membrane, and the intracellular cytoskeleton. Directional relationships between the extracellular matrix and the cytoskeleton are mediated by the plasma membrane. These cellular compartments may be of sufficient mass to be gravity sensitive.

for cellular and developmental processes (Spooner, 1973, 1975; Spooner and Thompson-Pletscher, 1986; Spooner *et al.*, 1986), and are in essential communication through the plasma membrane. Incoming information (e.g., a growth factor) binds to the cell via specific receptors and initiates a signal transducing cascade that involves the cytoskeleton. The incoming signal also may be the ECM, which can interact with specific receptors called integrins (Juliano and Haskill, 1993). Furthermore, it is now becoming clear that even soluble signals, like growth factors, interact with the ECM as a necessary prelude to eliciting a cellular response (Ingber and Folkman, 1989). The response often includes synthesis and secretion of ECM molecules, a process which requires the cytoskeleton and which alters the composition of the ECM, initiating another cellular response. Given the central role of the ECM, the plasma membrane signal transduction system, and the intracellular cytoskeleton, any perturbations in one or another compartment could well alter normal cell function.

This possibility applies to both plant and animal cellular and developmental systems, from single-celled to multicellular organisms. The cytoskeletal systems of plant and animal cells are essential for life functions, such as mitotic cell division, where microtubules assemble as the spindle apparatus, moving chromosomes in both plant and animal cells. In animal cells, microtubules also comprise the astral rays that stimulate the cell cortex and initiate cytokinesis or cell cleavage (Devore *et al.*, 1989). While asters are absent in plant cells, microtubules are components of the phragmoplast (Gunning and Wick, 1985), which establishes the cell plate and ultimately completes cytokinesis. Microfilaments, composed of F-actin and various actin-binding proteins, form the contractile ring of animal cells (Rappaport, 1986), which mechanically constricts the cell into two, dividing it from the outside in. In plant cells, there is no contractile ring, but microfilaments are involved in the action of the phragmoplast in performing cytokinesis, which occurs from the inside out (Schmit and Lambert, 1990). It is of interest that the Golgi apparatus is structurally tied to the microtubule system in animal cells (Kreis, 1990), since in plant cells the phragmoplast is composed of Golgi complex, microtubules, and microfilaments. The cytoskeletal compartment has crucial cellular and developmental functions, readily illustrated by plant and animal cell division. Other functions include intracellular transport, endo- and exocytosis, and single cell and tissue motility; these are all actvities that are potentially gravity sensitive through the cytoskeleton (Cipriano, 1993). Indeed, flight results suggest that the polymerization of actin and tubulin into microfilaments and microtubules, respectively, may be altered in microgravity (Moos *et al.*, 1989; Rijken *et al.*, 1992).

Plant and animal systems also require highly organized, yet dynamically remodeled, extracellular matrices. In animal systems, the ECM occurs as the basement membrane of epithelial cells and as the interstitial matrix of the mesenchymal cells. Both compartments contain classes of macromolecules in supramolecular self-assembling arrays. They differ in the collagen types present, the kinds of proteoglycan present, and in the respective presence of laminin of fibronectin. The ECM of plant cells is the cell wall. Cell wall ECM molecules include celluloses, pectins, lignins, and other molecules, deposited and organized in precise macromolecular arrays that differ with developmental age of the cell wall. In both plant and animal cells, the ECM mediates cell–cell interactions, regulates morphogenesis, and processes incoming information that alters gene expression. Evidence from spaceflight and clinorotation experiments indicates that altered gravity influences both the ECM in skeletal tissue, where abnormal collagen deposition and mineralization are observed (Morey-Holton and Cone, 1990), and the ECM in plant tissue, where changes in the composition of the cell wall are observed (Halstead and Dutcher, 1987; Iversen *et al.*, 1992).

The plasma membrane is the intermediary between the ECM and the cytoskeleton, through which they are directly and/or indirectly linked. Resident molecules of the plasma membrane include receptors, cytoskeletal anchoring proteins, and other proteins involved in signal transduction. The central role of the plasma membrane and signal transducing systems in cell function requires evaluation of the effects of gravity at this level. In addition to effects related to the ECM and cytoskeleton, as in mitogenic stimulation, the membranes themselves are potentially gravity sensitive (Schatz *et al.*, 1994), and, in fact, there is some evidence of changes in membrane lipid composition in altered gravity environments (Polyulyakh, 1988).

Since protein kinase C appears to be an effector of gravity's influence on certain cells, is there an association between PKC and the ECM, cytoskeleton, or cell membrane? Indeed, recent ground-based studies have revealed that, in addition to receptor-mediated activation, PKC also can be activated upon cell interactions with ECM molecules (Juliano and Haskill, 1993). Furthermore, PKC associates with the cytoskeleton upon activation (Gregorio *et al.*, 1992; Miyata *et al.*, 1989) and is translocated by the cytoskeleton to locations that include the cell membrane (Manger *et al.*, 1987; Murray *et al.*, 1993), where the activity of PKC can be influenced by membrane composition (Nishizuka *et al.*, 1983; 1992). At the cell membrane, PKC alters the function of membrane-associated molecules such as receptors, enzymes, and integrin molecules (Liscovitch *et al.*, 1992; Miyata *et al.*, 1989; Nishizuka, 1992; Tonks and Charbonneau, 1989).

IV. Approaches to Manipulation of Gravity

To study gravity's influence on the cell, it is necessary to analyze cell function in environments where the gravitational "dose" is increased or decreased. It is possible to artificially increase the pull of gravity on specially designed low-speed centrifuges that allow cell and tissue cultures and whole organisms to function at hypergravity, and such experiments have been performed since the early 1800s (Cogoli and Tschopp, 1982; Montgomery *et al.*, 1978). Decreasing the pull of gravity, however, is more difficult. Gravity is generated by the earth's mass, and although the force decreases as one moves away, a true microgravity environment (10^{-6} x g) can only be achieved at least 400,000 miles from earth (163,000 miles past the moon). Since traveling that distance is impractical, several approaches have been used to artificially generate a microgravity environment by canceling out the gravity vector. The most effective and successful approach is to free fall within earth's gravitational field, that is, to accelerate at 9.8 m—sec^2 in the gravity vector, toward the center of the earth. A mass in free fall has an accelerative force that equals and cancels the gravitational force. Therefore, the apparent gravitational pull on the falling mass approaches zero, and the mass appears to be weightless.

A number of vehicles have been designed to free fall under the force of gravity and generate a weightless environment. The simplest of these is a platform that falls from a drop tower or from the top of a mine shaft for a few seconds of weightlessness. Drops from high-altitude balloons can provide 30–60 sec of free fall in the upper atmosphere before reaching terminal velocity. Other more useful free-fall vehicles (see Fig. 10), which are discussed in the following sections, are parabolic flight airplanes that repeatedly generate 20–30 sec of near weightlessness, and unmanned sounding rockets that fly in a suborbital parabolic arc at an altitude of approximately 300 miles to deliver 7–14 min of free fall. The space shuttle and unmanned research platforms also attain an altitude of 200–300 miles, but their capacity to free fall around the earth (i.e., orbit) can provide from a week to several years of weightlessness. Finally, the clinostat is a ground-based instrument that attempts to artificially reproduce the weightless environment by averaging the gravity vector or by simulating free fall in a rotating fluid.

A. Parabolic Flight

A number of different airplanes, from small prop-driven aerobatic planes to large KC-135 turbojet transports, have been used to generate a reduced gravity environment by flying in a parabolic trajectory. Small propeller

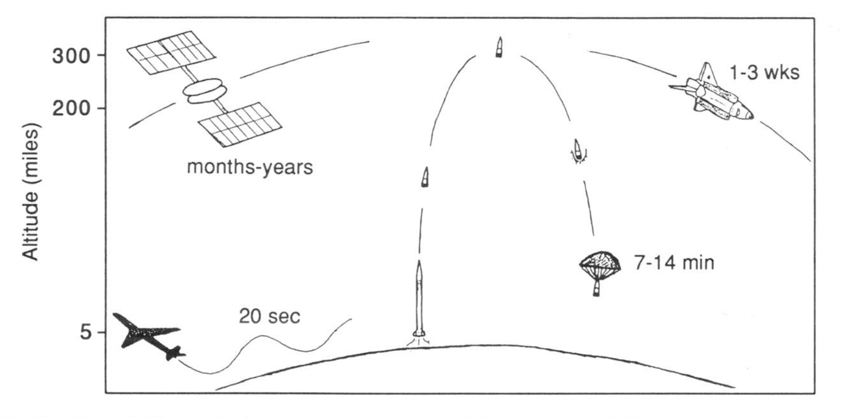

FIG. 10 Free falling vehicles generate near weightlessness by falling under the force of gravity. Free falls can be maintained for 20 sec in parabolic aircraft, 7–14 min in suborbital rocket flights, 1–3 weeks in orbital spaceflight, and for years in orbiting space stations.

aircraft can produce 5–8 sec of free fall, and small nonmilitary jets such as the Lear jet and the Nagoya MU-300, can yield up to 20 sec of 5×10^{-2} x g (Mori *et al.*, 1987). Specially modified parabolic flight airplanes, operated by the European Space Agency (ESA), the Canadian Space Agency, and the U.S. National Aeronautics and Space Administration, are able to produce 20–30 sec of $1–3 \times 10^{-2}$ x g.

NASA operates both a Boeing KC-135 transport with 23 seats and 60 feet of padded test space, and a small two-person F-104 supersonic jet fighter, with 2 ft^3 of test space. The F-104 flies a parabolic trajectory from 25,000 feet to 65,000 feet that produces about 30 sec of 3×10^{-2} x g or 60 sec of less than 1×10^{-1} x g. The larger KC-135 aircraft (NASA-930) flies in a similar but smaller parabolic trajectory between the altitudes of 26,000 and 34,000 feet. A NASA-930 parabola is initiated by high-speed entry into a climb that generates 2 g-forces. As the climb approaches 34,000 feet, power is reduced until thrust equals drag and the plane pushes over the top of the arc and falls into a steep dive to 26,000 feet, where it pulls out and begins the climb for the next parabola. Low-gravity conditions are experienced during the "push over" stage near the apex of the parabolic arc. Other flight profiles can be selected that provide lunar gravity for 30 sec, and martian gravity for 40 sec (Walker *et al.*, 1992). The parabolas are usually flown roller-coaster style so that, during a single flight, approximately 40 free-fall episodes are produced in 2 hr, with a 5–10 min turnaround period separating each set of 10 parabolas. A NASA-930 flight mission includes three flights flown over 3 days, and provides a total of 120 parabolas. Although NASA's primary use for the plane is training astronauts and testing spaceflight hardware, five to six missions a year have been available for scientific research.

Experimentation on parabolic flight airplanes has several advantages over other freefall vehicles. First, the handling of biological materials is less of a problem since fresh material can be collected immediately prior to the flight. In addition, 1 x *g* control experiments can be performed on board under flight conditions during the 5–10-min turnaround period. Also, since a flight mission occurs over several days, problems encountered in early tests can be addressed, and experimental procedures and hardware can be modified for flights on subsequent days. Finally, and most important, the parabolic airplane is the only free-fall vehicle in which investigators can personally perform their experiments and conduct real-time analysis of the results.

Parabolic flight also has its disadvantages. The continuous cycling from low to high g-forces can create difficulties for both instrumentation and investigator. To ensure that periods of hypo- and hypergravity, and inadvertent periods of "negative" gravity, do not impair instrument operation, flight hardware should be tested on the ground in a number of positions, including 180 degrees from normal. Since all free-fall vehicles require that experimental fluids and biologicals be confined within closed systems, operating the hardware at various positions will also reveal the possibility of fluid leaks. Pressure variations should be tested on the closed systems containing air, since fluctuations in cabin pressure occur during the rapid changes in altitude associated with parabolic flight. In addition, research equipment may need to be strengthened and modified for flight. On the NASA-930, flight hardware is mounted on the floor of the aircraft, and the hardware and mountings must be rated to withstand 9–12 x *g*.

More important, however, is the well-being of the investigator. Since parabolic maneuvers hinder normal movements, experiments should be as automated as possible and not require the operator to perform complex activities. In addition, the setup for repeated experimental runs should be simplified, since those preparations usually are conducted during the physically stressful 2 x *g* pullout and climb. All fliers on the NASA parabolic aircraft are required to complete an Air Force Flight Class III physical examination and the 2-day Advanced Physiological Training Program of the Air Force Space Command (Williams, 1987). Approximately half of all fliers become ill, and more than half of those are physically impaired and unable to perform the simplest operations. In general, executing an experimental protocol during parabolic flight is not unlike preparing and performing an experiment during the ultimate roller coaster ride.

Cell biology experiments that are amenable to parabolic flight focus on cellular and subcellular activities that can be activated to measurable levels within the 20 sec of freefall. Such studies may measure the cellular responses to stimuli, and the intermediate steps that couple stimuli and responses, for example, receptor–ligand binding, signal transduction, and enzyme–substrate and protein-protein interactions.

B. Sounding Rocket Flight (Suborbital)

Sounding rockets fly a suborbital trajectory, often landing less than 100 miles from the launch site, but they are capable of reaching heights that yield 7–14 min of free fall and produce a low gravity environment of 1×10^{-5} x *g* (Kassemi and Ostrach, 1992). Cellular experiments conducted on sounding rockets are similar to experiments flown on parabolic airplanes, but the increased period of higher quality weightlessness broadens the experimental opportunities and improves the detection of cellular responses.

The first sounding rocket flights to contain biological experiments were launched from the White Sands Proving Ground aboard German V-2 rockets in 1946–47. Carrying fungus spores and fruit flies, the rockets reached an altitude of over 105 miles before falling back to earth and ejecting the payload for parachute recovery. In the early 1950s, the U.S. Air Force launched an Aerobee rocket that contained two mice, and photographic records provided the first documentation of mammals experiencing up to several minutes of a low-gravity environment. The mice were seen floating within their containment from the moment the rocket expended its fuel and deaccelerated under gravitational pull. Near weightlessness continued as the rocket progressed under its own momentum to the apex of the parabolic trajectory, through the arc, and into free fall toward earth, until atmospheric drag reduced acceleration. These early sounding rocket flights produced a microgravity environment that lasted for nearly 4 min, and by the mid-1950s, U.S. and Russian sounding rockets were placing monkeys and dogs in microgravity for up to 10 min (Taylor and Munson, 1972).

For several decades after the advent of orbital spaceflight in 1957, few suborbital rocket flights carried biological payloads. Since 1985, however, ESA and NASA have supported a number of sounding rocket flights for biological research. The European Maser and Texas and the U.S. Starfire are 50-foot rockets that reach an altitude of 200 miles and provide 6–7 min of free fall with proven reliability. Rockets developed for higher altitude flights (330 miles), which generate 14 min of free fall, have as yet to prove successful (Gurkin, 1992).

From launch to the parachute landing 60–200 miles downrange, modern sounding rocket flights last a total of 15–25 min. During rocket launch and rapid ascent, a period lasting nearly 60 sec, accelerative forces on the payload gradually increase from less than 5 x *g* to as much as 15 x *g*, and upon reentry into the atmosphere, deacceleration produces g-forces that may peak as high as 40 x *g* for a few seconds. The factor most deleterious to biological experiments, however, is the temperature of the payload during the reentry and recovery phases. Even with heat shields and insulation, air friction heats payloads to temperatures that may exceed 45°C during reentry, and further temperature extremes may occur post-landing, while the payload is being located and recovered. This problem is compounded by the location of rocket launch sites: the Maser and Texas are launched in the cold of northern Sweden, and the Starfire is launched in the desert of New Mexico.

The payloads carried by sounding rockets are limited in their size and weight. For example, the power supply and hardware for a dozen or more different experiments may weigh only 300 pounds and be packed into an area 40 inches in diameter by 40 inches in height. Experiments must be fully automated, with activation triggered upon payload entry into the reduced gravity phase of the flight, and termination occurring shortly before atmospheric drag slows free fall during reentry. In order to study cellular activities independent of reentry and landing effects, termination of the experiment often involves chemical fixation of the samples, with the analysis occurring postflight. Real-time observations are less common owing to hardware and electrical power limitations.

Although sounding rockets have generally flown fluid physics and material processing experiments, their use for biological research has increased in the past decade. Payload modules are being equipped with 1 x *g* centrifuges for flight controls, and with improved downlink telemetry and uplink ground telecommand capabilities. Laboratory preparation space, which initially was nonexistent, is now available at most launch sites. In addition, the availability of late access to the payload has greatly improved. Previously, investigators turned experiments over for payload loading 8–18 hr before scheduled launch, whereas now biological experiments may be loaded as late as 20–60 min before launch.

C. Orbital Spaceflight

Spacecraft that orbit the earth are not beyond the earth's gravitational field; they are pulled toward the earth by nearly 90% of nominal gravity. Instead of falling straight down, however, the forward velocity of the spacecraft causes it to fall in a curved path around the earth. As long as the forward velocity is maintained, the spacecraft will continue to free fall in an orbital path. If forces such as friction from the upper atmosphere or thruster firings by the spacecraft counter the forward velocity and reduce the altitude of the orbit, then "reentry" of the spacecraft into the denser, lower atmosphere will eventually occur.

On manned spacecraft in orbit, the quality of the reduced gravity environment largely depends on the level of vibration produced by astronaut movement and on-board equipment (Kassemi and Ostrach, 1992). Accelerative forces may be relatively large, ranging widely from 5×10^{-5} to 1×10^{-3} x g, which is as much as 100 times greater than the forces produced during sounding rocket flight. On unmanned orbiting platforms, however, g-levels may be as much as 200 times lower than on the suborbital rockets, ranging from 5×10^{-8} to 1×10^{-5} x g. Thus, the unmanned orbiting spacecraft is the only free-fall vehicle capable of achieving a sustained "microgravity" environment; but even on these spacecraft, equipment vibration, spacecraft rotation, gravity gradients, and solar pressure create accelerative forces that prevent the attainment of a true "zero-gravity" environment (Kassemi and Ostrach, 1992).

The first orbital spaceflights occurred in 1957–1958, when the Soviets, and then the United States, placed artificial satellites into near-earth orbits. During this period the Soviet Sputnik 1 successfully launched and recovered a dog, in what was the first animal experiment to test for effects of orbital spaceflight. In the 1960s, with the achievement of manned spaceflight, a number of simple experiments that studied the effects of short duration (5–96 hr) spaceflight on microorganisms, cultured cells, and amphibian eggs were carried on both U.S. and Soviet manned spacecraft. Since little or no astronaut time was available for these early spaceflight experiments, and since experimental hardware consisted of bags, petri dishes, or sealed ampoules, these studies were limited to general observations, usually made far from the ocean landing site, of cell or organism viability, growth, and morphology (Cogoli and Tschopp, 1982). Nevertheless, this was the beginning of biological experimentation in the extended weightlessness of orbital spaceflight.

By the 1970s, NASA's interest in gravitational biology appeared to diminish, owing to a lack of data showing gross changes in cell physiology during spaceflight, and the conclusion by several physicists that most cells were too small to be affected by changes in gravity (Gordon, 1964; Kondo, 1968; Pollard, 1965). In the first 7 years of orbital spaceflight, eleven flights had carried cell biology experiments; whereas in the next 14 years, only four flights, all Skylab missions, carried such experiments. Interest in gravitational biology was renewed, however, when the shuttle program began in the early 1980s. Nearly every Shuttle mission has carried basic biological experiments as secondary payloads, and several Spacelab missions have been dedicated to life science research. Furthermore, astronaut crews now include payload specialists who undergo months of intensive training in conducting flight experiments.

In addition to experiments on manned space vehicles, several unmanned research platforms have been placed into orbit in the last 25 years. Most recently, the European Retrievable Carrier (EURECA), an automated research satellite, was released in July 1992, and was recently recovered by the Shuttle. Unmanned orbiting platforms achieve the highest quality low gravity of all free-fall vehicles, but since research platforms orbit for months to years, their biological investigations are generally limited to developmentaly arrested organisms, such as seeds, cysts, and spores, and their studies are concerned with the effects of the microgravity environment and space radiation.

1. Space Radiation

Space radiation, and its influence on biological systems, is an important problem which has long hampered the unambiguous interpretation of spaceflight results. The primary source of space radiation, cosmic rays, are blocked from the surface of the earth by the earth's magnetic field, but radiation intensity greatly increases with altitude. Cosmic rays are high-energy nuclear particles that are emitted from the sun (solar particle radiation) and from outside our solar system (galactic cosmic rays). Upon reaching the earth's upper atmosphere, the high-energy rays produce cascades of radioactive particles (secondary radiation) that can become trapped in the magnetic field, creating regions of intense radiation, especially at higher altitudes (McCormack, 1988).

Except for rare periods of high solar flare activity, galactic cosmic rays are the largest contributor to space radiation levels, but the level and the impact of galactic cosmic particles on biological systems in oribiting spacecraft are not fully known. Measurements taken inside the Shuttle show that the average dose rate from low-energy space radiation ranges from 0.005 to 0.01 rad/day (Ikenaga *et al.*, 1987), and the total dose rate from both low- and high-energy rays is calculated to be less than 0.5 rad/day (Petrov, 1992). This is equivalent to an absorbed dose of 18 rems/year, which is only 3.5 times the limit for workers in the nuclear power industry. Moreover, in tests where doses of low-energy radiation equal to a 40-day spaceflight (2 rads) were used in a developmental bioassay measuring somatic mutations in *Drosophila* wings, only small increases in genetic modifications were observed (Ikenaga *et al.*, 1987). Still, even though spaceflight radiation levels do not appear to be damaging in ground-based tests, the biological effect of cosmic radiation may pose a greater risk than an equivalent dose of X-rays or gamma rays since cosmic radiation contains HZE particles, which possess very high energy transfer and high biological effectiveness that may cause deleterious effects, particularly during the early stages of embryonic development (Bonting, 1983).

Although cosmic radiation may be potentially deleterious to biological systems, little radiation damage has been reported from spaceflight experiments that date back to the earliest rocket flights (Cogoli and Tschopp, 1982). More recently, in long-term studies conducted on an orbiting research platform for nearly 6 years, development and mutation rates were mixed when dormant seeds, spores, and cysts were activated following their return to earth (Graul *et al.,* 1991; Horneck *et al.,* 1991; Kranz *et al.,* 1991; Reitz and Bücker, 1992). In Shuttle and BioCosmos studies on actively developing *C. elegans, Drosophila,* and *Artemia,* physical abnormalities were not significantly increased above controls, suggesting that the radiation levels on board orbiting spacecraft do not strongly affect developing systems (Marco *et al.,* 1992; Nelson *et al.,* 1994; Spooner *et al.,* 1994a,c). It is possible, however, that low levels of cosmic radiation induce biological modifications that are not detrimental, and do not result in cell death or somatic mutations (Cogoli, 1991). Such effects of radiation may then be attributed to the weightless condition. For example, the increased growth rate exhibited by *Paramecium* during spaceflight was believed to be a response to the reduced gravity environment; however, results from high-altitude balloon flights showed that cosmic radiation could induce a similar increase in *Paramecium* proliferation (Richoilley *et al.,* 1986).

To aid in distinguishing radiation effects from effects due to weightlessness, the in-flight centrifuge was developed in an attempt to separate out spaceflight conditions. In the centrifuge, which is available on some Spacelab Shuttle missions, biological samples are exposed to the spaceflight environment while spinning at 1 x *g*. Several Spacelab studies, including one on *Paramecium* proliferation, have shown that alterations in growth and development during spaceflight do not occur when the sample is placed in the artificial 1 x *g* environment of the centrifuge, indicating that the observed spaceflight effects were caused by the reduced gravity environment, and not be radiation (Bücker *et al.,* 1986; Marco *et al.,* 1986; Richoilley *et al.,* 1986).

While these studies suggest that the centrifuge is useful for isolating spaceflight conditions, the distinction between radiation and reduced gravity effects may not be that clear. In an experiment that tracked heavy ion hits on developing eggs of the stick insect, cosmic particles were found to exert a greater detrimental impact on eggs developing in weightlessness than on eggs in the 1 x *g* centrifuge (Bücker *et al.,* 1986). Other studies have also proposed that cells are more sensitive to radiation damage, or less able to repair such damage, in a microgravity environment (McCormack, 1988). These results imply that ground-based radiation studies may not be able to fully duplicate the biological damage caused by radiation during spaceflight. Moreover, the possibility of a synergistic effect of reduced gravity and radiation suggests that the flight centrifuge alone may not be sufficient to control for biological effects of radiation in a low-gravity environment.

D. Clinorotation

At this time, the use of free-fall flight vehicles for basic biological research is limited, and it is not uncommon for a reflight of Shuttle experiments to take nearly a decade. Flight numbers may increase for those experiments that can conform to the limitations of parabolic aircraft flight or suborbital rocket flight, or to the restrictions associated with the generic bioprocessing hardware that flies on the Shuttle as a secondary payload. However, conforming to these flight limitations may mean that experiments are performed within specific time constraints, at less than optimum conditions (e.g., temperature, light, air, etc.), on equipment that was designed for other purposes. Even if these limitations can be complied with, only one to two free-fall flights per year would be likely, and this lack of flight opportunities imposes extraordinary challenges for the researcher.

Due to the difficulties of conducting experiments on free-fall vehicles, the possibility of simulating weightlessness in the laboratory has been investigated. Clinorotation, although still controversial, is thought to elicit at least some responses from biological systems that mimic responses from the reduced gravity environment. Clinorotation involves the placement of cells, tissues, organs, or organisms in a vertically rotating vessel that turns around a horizontal axis. Depending on the placement of the sample, clinorotation has been used to simulate weightlessness by two different methods. In the classic method, the sample is held in the center of the vessel, either by attachment or containment, and the vessel is rotated in an attempt to confuse the directional force of gravity on the sample, or to average out the pull of gravity by applying it equally to all sides of the sample. Since the sample is near the axis of rotation, centrifugal force is minimal. Such devices have been used to study the effects of altered gravity on plant and animal development, as well as on cultured cell physiology (Cogoli, 1992; Gruener and Hoeger, 1988, 1991; Lorenzi and Perbal, 1990; Sievers and Hejnowicz, 1992; Yokota *et al.*, 1994).

In a newly developed clinorotation method, the sample is suspended in a fluid environment, and the vessel is rotated at a speed at which the sample falls continuously through the fluid in which it is carried. In this simulation of free fall, the sample, which usually is cells or tissues in suspension or attached to beads, moves with the rotating fluid stream but falls in a circular path with respect to the fluid stream line, moving away from the stream line on the downward motion of rotation and moving back toward the stream line on the upward motion (Wolf and Schwartz, 1991, 1992). With this technique, the rotation rate of the vessel is of primary importance, and is dependent on the sample's sedimentation rate. As the rotation rate decreases, deviation of the sample from the fluid stream increases, eventually resulting in the sample sedimenting to the vessel wall. As the rotation rate increases, sample deviation from the fluid stream line decreases; however, at too high a rate of rotation, centrifugal force drives the sample toward the vessel wall. At the correct rotation rate, the sample essentially falls continuously within the fluid in which it is carried (Wolf and Schwartz, 1991). This type of clinorotation has stimulated plant tissues to respond as if they were in reduced gravity (Cogoli and Tschopp, 1982; Moore, 1990).

V. Concluding Remarks

After 30 years of spaceflight research, answers concerning the impact of gravitational force on cell biology remain controversial. Results from spaceflight investigations often have been considered observational anecdotes, rather than experimental data, owing to the complexities of conducting research in a weightless environment, the lack of opportunities for reproducing spaceflight data, and, because of the few spaceflight opportunities, a reluctance of investigators to design experiments that test a specific hypothesis. The goals of this chapter were to organize and analyze spaceflight and ground-based data in order to reveal similarities and consensus, and ultimately, to assist in the formation of experimental hypotheses for future flights.

To summarize, the data indicate that (1) the activity of PKC is enhanced in microgravity; (2) spaceflight effects are more evident in cells undergoing activation in response to growth factors, cytokines, and TCR stimulation, most likely due to the enhanced activation of PKC; (3) in nonstimulated cells, flights of a longer duration may be required in order to reveal the effects of spaceflight; however, these effects may also involve additional factors, such as high-energy cosmic radiation; (4) windows of gravitational sensitivity exist in plants and animals during embryogenesis and juvenile development, with the possibility of detrimental or enhancing effects; and (5) although changes in gravity may induce developmental anomalities, the organism often is resilient in overcoming the effects of altered gravity.

Acknowledgments

Funding for original research from our laboratory cited in this chapter, and for support during its preparation, came from the Kansas Health Foundation Cancer Scholar Program and the NSCORT in Gravitational Biology (NASA grant NAGW-2328).

References

Abraham, R. T., Karnitz, L. M., Secrist, J. P., and Leibson, P. J. (1992). Signal transduction through the T-cell antigen receptor. *Trends Biochem. Sci.* **17,** 434–438.

Aderem, A. (1992). Signal transduction and the actin cytoskeleton: The roles of MARCKS and profilin. *Trends Biochem. Sci.* **17,** 438–443.

Albrecht-Buehler, G. (1990). In defense of "non-molecular" cell biology. *Int. Rev. Cytol.* **120,** 191–241.

Albrecht-Buehler, G. (1991). Possible mechanisms of indirect gravity sensing by cells. *Am. Soc. Gravitational Space Biol. Bull.* **4**(2), 25–34.

Alpatov, A. M., Tairbekov, M. G., and Ushakov, I. A. (1991). Gravitational biology experiments aboard the Biosatellites "Cosmos" #1887 and #2044. *Physiologist* **34**(1), S78–S79.

Amaral, M. C., Casillas, A. M., and Nel, A. E. (1993). Contrasting effects of two tumour promoters, phorbol myristate acetate and okadaic acid, on T-cell responses and activation of p43 MAP-kinase/ERK-2. *Immunology* **79,** 24–31.

Anderson, N. G., Maller, J. L., Tonks, N. K., and Sturgill, T. W. (1990). Requirement for integration of signals from two distinct phosphorylation pathways for activation of MAP kinase. *Nature* **343,** 651–653.

Bátkai, L., Tálas, Stöger, Nagy, K., Hiros, L., Konstantinova, I., Rykova, M., Mozgovaya, I., Guseva, O., and Kozharinov, V. (1988). *In vitro* interferon production by human lymphocytes during spaceflight. *Physiologist* **31,** S50–S51.

Bechler, B., and Cogoli, A. (1986). Lymphozyten sind schwerkraftempfindlich. *Naturwissenschaften* **73,** 400–403.

Bechler, B., Cogoli, A., Cogoli-Grueter, M., Müller, O., Hunzinger, E., and Criswell, S. B. (1992). Activation of microcarrier-attached lymphocytes in microgravity. *Biotechnol. Bioeng.* **40,** 991–996.

Becton, D. L., Adams, D. O., and Hamilton, T. A. (1985). Characterization of protein kinase C activity in interferon gamma treated murine peritoneal macrophages. *J. Cell. Physiol.* **125,** 485–491.

Bergter, F., Harz, D., Müller, P. J., Mund, K., Gunther, U., Hesse, T., Hartmann, R., Wanke, G., Tairbedov, G. G., Parfyonov, G. P., and Paknomov, A. I. (1985). Determination of increment of *Bacillus subtilis* biomass in weightlessness. *Space Biol. Aerospace Med.* **19,** 92–95.

Berridge, M. J. (1987). Inositol triphosphate and diacylglycerol: Two interacting second messengers. *Annu. Rev. Biochem.* **56,** 159–193.

Berridge, M. J. (1993). Inositol trisphosphate and calcium signalling. *Nature* **361,** 315–325.

Berry, N., Ase, K., Kikawa, U., Kishimoto, A., and Nishizuka, Y. 1989. Human T cell activation by phorbol esters and diacylglycerol analogues. *J. Immunol.* **143,** 1407–1413.

Black, S., Larkin, K., and Souza, K. (1993). Results from the frog embryology experiment: Development of swimming tadpoles occurs at microgravity. *Am. Soc. Gravitational Space Biol. Bull.* **7**(1), 32a.

Boda, K., Meleško, G. I., Sabo, V., Šepelev, E. J.a., Gurjevak T. S., Juràni,M., and Koštàl,

L. (1991). Embryonic development of Japanese quail under microgravity conditions. *Physiologist* **34**(1), S59–S61.

Boda, K., Sabo, V., Juráni, M., Guryeva, T. S., Koštàl, L., Lauková, A., and Dadasheva, O. A. (1992). Embryonic development and behaviour of Japanese quail exposed to microgravity. *Acta Vet. Brno* **61,** 99–107.

Bonting S. (1983). Space biology research. *Trends Biochem. Sci.* **8,** 265–267.

Briegleb, W., Block, I., Sobick, V., and Wohlfarth-Bottermann, K. E. (1986). Steady compensation of gravity effects in *Physarum polycephalum. Naturwissenschaften* **73**(7), 422–424.

Brown, A. H. (1991). From gravity and the organism to gravity and the cell. *Am. Soc. Gravitational Space Biol.* **4**(2), 7–18.

Brown, A. H., and Chapman, D. K. (1982). The first plants to fly on shuttle. *Physiologist* **25**(6), S5–S8.

Brown, A. H., Chapman, D. K., and Heathcote, D. G. (1992). Characterization of precocious seedling development observed during IML-1 mission. *Am. Soc. Gravitational Space Biol.* **6**(1), 58a.

Bücker, H., Facius, R., Hildebrand, D., and Horneck, G. (1975). Results of the *Bacillus subtilus* of the Biostack II experiment: Physical characteristics and biological effects of individual HZE particles. *Life Sci. Space Res.* **13,** 161–166.

Bücker, H., Horneck, G., Reitz, G., Graul, E. H., Berger, H., Höffken, H., Rüther, W., Heinrich, W., and Beaujean, R. (1986). Embryogenesis and organogenesis of *Carausius morosus* under spaceflight conditions. *Naturwissenschaften* **73**(7), 433–434.

Bücker, H., Facius, R., Horneck, G., and Reitz, G. (1987). Embryogenesis and organogenesis of *Carausius morosus* under spaceflight conditions. *In* "Biorack on Spacelab D-1" (N. Longdon and V. David, eds.), pp. 135–145. European Space Agency, Paris.

Bücker, H., Reitz, G., Facius, R., Horneck, G., Graul, E. H., Berger, H., Rüther, W., Heinrich, W., Beaujean, R., Enge, W., Alpatov, A. M., Ushakov, I. A., Zachvatkin, Y. A., and Mesland, D. A. M. (1989). Influence of cosmic radiation and/or microgravity on development of *Carausius morosus. Am. Soc. Gravitational Space Biol.* **3**(1), 103a.

Cebrián, M., Yagüe, E., Rincón, M., López-Botet, M., de Landàzuri, M. O., and Sànchez-Madrid, F. (1988). Triggering of T cell proliferation through AIM, an activation inducer molecule expressed on activated human lymphocytes. *J. Exp. Med.* **168,** 1621–1637.

Chapes, S. K., Morrison, D. R., Guikema, J. A., Lewis, M. A., and Spooner, B. S. (1992). Cytokine secretion by immune cells in space. *J. Leukocyte Biol.* **52,** 104–110.

Chapes, S. K., Woods, K. M., and Armstrong, J. W. (1993). Ground-based experiments complement microgravity flight opportunities in the investigation of the effects of spaceflight on the immune response: Is protein kinase C gravity sensitive? *Trans. Kansas Acad. Sci.* **96**(1-2), 74–79.

Chapes, S. K., Morrison, D. R., Guikema, J. A., Lewis, M. L., and Spooner, B. S. (1994). Production and action of cytokines in space. *Adv. Space Res.* **14**(8), (8)5–(8)9.

Ciferri, O., Tiboni, O., Di Pasquale, G., Orlandoni, A. M., and Marchesi, M. L. (1986). Effects of microgravity on genetic recombination in *Escherichia coli. Naturwissenschaften* **73**(7), 418–421.

Cipriano, L. F. (1993). An overlooked gravity sensing mechanism. *Physiologist* **34**(1), S72–S75.

Cogoli, A. (1991). Changes observed in lymphocyte behavior during gravitational unloading. *Am. Soc. Gravitational Space Biol.* **4**(2), 107–115.

Cogoli, A. (1993). The effect of hypogravity and hypergravity on cells of the immune system. *J. Leukocyte Biol.* **54,** 250–268.

Cogoli A., and Gmünder, F. K. (1991). Gravity effects on single cells: Techniques, findings, and theory. *Adv. Space Biol. Med.* **1,** 183–248.

Cogoli, A., and Tschopp, A. (1982). Biotechnology in space laboratories. *In* "Advances in Biochemical Engineering" (A. Feichter, ed.), vol 22, pp. 1–50. Springer-Verlag, Berlin.

Cogoli, A., Valluchi, M., Reck, J., Müller, M., Briegleb, W., Cordt, I., and Michel, Ch. (1979). Human lymphocyte activation is depressed at low-g and enhanced at high-g. *Physiologist* **22**(6), S29–S30.

Cogoli, A., Tschopp, A., and Fuchs-Bislin, P. (1984). Cell sensitivity to gravity. *Science* **225,** 228–230.

Cogoli, A., Bechler, B., Lorenzi, G., Gmuender, F. K., and Cogoli, M. (1987a). Cell cultures in space: From basic research to biotechnology. *In* "Biological Sciences in Space 1986" (S. Watanabe, G. Mitarai, and S. Mori, eds.), pp. 225–232. Internat. Symp. on Biol. Sci. in Space, MYU Research, Tokyo.

Cogoli, A., Bechler, B., Müller, O., and Hunzinger, E. (1987b). Effect of microgravity on lymphocyte activation. *In* "Biorack on Spacelab D-1" (N. Longdon and V. David eds.), pp. 89–100, European Space Agency, Paris.

Cogoli, A., Bechler, B., Cogoli-Greuter, M., Criswell, S. B., Joller, H., Jollber, P., Hunzinger, E., and Müller, O. (1993). Mitogenic singal transduction in T lymphocytes in microgravity. *J. Leukocyte Biol.* **53,** 569–575.

Cogoli, M. (1992). The fast rotating clinostat: A history of its use in gravitational biology and a comparison of ground-based and flight experiment results. *Am. Soc. Gravitational Space Biol.* **5**(2), 59–67.

Cogoli, M., Bechler, B., Cogoli, A., Arena, N., Barni, S., Pippia, P., Sechi, G., Valora, N., and Monti, R. (1992). Lymphocytes on sounding rockets. *Adv. Space Res.* **12**(1), (1)141–(1)144.

Conrad, H. M. (1968). Biochemical changes in the developing wheat seedling in the weightless state. *BioScience* **18,** 645–651.

Cooley, L., Verheyen, E., and Ayers, K. (1992). *chickadee* encodes a profilin required for intercellular cytoplasm transport during *Drosophila* oogenesis. *Cell* **69,** 173–184.

Coté, G. G., and Crain, R. C. (1993). Biochemistry of phosphoinositides. *Annu. Rev. Physiol. Plant Mol. Biol.* **44,** 333–356.

Cowles, J. R., Scheld, H. W., Peterson, C., and LeMay, R. (1982). Lignification in young plants exposed to the near-zero gravity of space flight. *Physiologist* **25**(6), S129–S130.

Crowles, J. R., Scheld, H. W., LeMay, R., Peterson, C. (1984). Growth and lignification in seedlings exposed to eight days of microgravity. *Ann. Bot.* **54**(3), 33–48.

Cowles, J., Jahns, G., LeMay, R., Omran, R. (1986). Cell wall related synthesis in plant seedlings grown in the microgravity environment of the space shuttle. *Plant Physiol.* **80,** 9a.

d'Augères, C. B., Arnoult, J., Bureau, J., Duie, P., Dupuy-Coin, A. M., Gèraud, G., Laquerrière, F., Masson, C., Pestmal, M., and Bouteille, M. (1987). The effect of microgravity on mammalian cell polarisation at the ultrastructural level. *In* "Biorack on Spacelab D-1" (N. Longdon and V. David eds.), pp. 101–105, European Space Agency, Paris.

Darbelley, N., Driss-Ecole, D., and Perbal, G. (1986). Differenciation et proliferation cellulaires dans des racines de mais cultive en microgravite (Biocosmos 1985) *Adv. Space Res.* **6**(12), (6)157–(6)160.

Darbelley, N., Driss-Ecole, D., and Perbal, G. (1989). Elongation and mitotic activity of cortical cells in lentil roots grown in microgravity. *Plant Physiol. Biochem.* **27**(3), 341–347.

Davis, L. S., and Lipsky, P. E (1985). Signals involved in T cell activation. I. Phorbol esters enhance responsiveness but cannot replace intact accessory cells in the induction of mitogen-stimulated T-cell proliferation. *J. Immunol.* **135**(5), 2946–2952.

Davis, L. S., and Lipsky, P. E. (1993). Co-stimulation of T-lymphocyte activation by adhesion molecules. *In* "Structure, Function, and Regulation of Molecules Involved in Leukocyte Adhesion" (P. E. Lipsky, R. Rothlein, T. K. Kishimoto, R. B. Faanes, and C. W. Smith, eds.), pp. 256–272. Springer-Verlag, New York.

Davis, L. S., Wacholtz, M. C., and Lipsky, P. E. (1989). The induction of T cell unresponsiveness by rapidly modulating CD3. *J. Immunol.* **142,** 1084–1094.

Davis, R. J. (1988). Independent mechanisms account for the regulation by protein kinase C of the epidermal growth factor receptor affinity and tyrosine-protein kinase activity. *J. Biol. Chem.* **263**(19), 9462–9469.

de Groot, R. P., Rijken, P. J., Den Hertog, J., Boonstra, Verkleij, A. J., de Laat, S. W., and Kruijer, W. (1990). Microgravity decreases c-*fos* induction and serum response element activity. *J. Cell Sci.* **97,** 33–38.

de Groot, R. P., Rijken, P. J., Boonstra, J., Verkleij, A. J., de Laat, S. W., and Kruijer, W. (1991a). Epidermal growth factor-induced expresssion of c-*fos* is unfluenced by altered gravity conditions. *Aviation, Space Environ. Med.* **62,** 37–40.

de Groot, R. P., Rijken, P. J., den Hertog, J., Boonstra, J., Verkleij, A. J., de Laat, S. W., and Kruijer, W. (1991b). Nuclear responses to protein kinase C signal transduction are sensitive to gravity changes. *Exp. Cell Res.* **197,** 87–90.

DeBell, L., Paulsen, A., and Spooner, B. (1992). Scanning electron microscope observations of brine shrimp larvae from space shuttle experiments. *Scanning Micros.* **6**(4), 1129–1135.

Desai, D. M., Newton, M. E., Kadlecek, T., and Weiss, A. (1990). Stimulation of the phosphatidylinositol pathway can induce T-cell activation. *Nature* **348,** 66–69.

Devore, J. J., Conrad, G. W., and Mesland, D. (1989). A model for astral stimulation of cytokinesis in animal cells. *J. Cell Biol.* **109,** 2225–2232.

Dickson, K. J.. (1991). Summary of biological spaceflight experiments with cells. *Am. Society for Gravitational and Space Biol. Bull.* **4,** 151–260.

Dintenfass, L. (1986). Execution of "ARC" experiment on space shuttle "Discovery" STS 51-C: Some results on aggregation of red blood cells under zero gravity. *Biorheology* **23,** 331–347.

Dougherty, G. J., and Hogg, N. (1987). The role of monocyte lymphocyte function-associated antigen 1 (LFA-1) in accessory cell function. *Eur. J. Immunol.* **17,** 943–947.

Dransfield, I., Buckle, A.-M., and Hogg, N. (1990). Early events of the immune response mediated by leukocyte integrins. *Immunol. Rev.* **114,** 29–44.

Dustin, M. L., and Springer, T. A. (1989). T-cell receptor cross-linking transiently stimulates adhesiveness through LFA-1. *Nature* **341,** 619–624.

Esau, K. (1977). "Anatomy of Seed Plants." 2nd ed. pp. 550 Wiley, New York.

Facius, R., Bücker, H., Horneck, G., Reitz, G., and Schafer, M. (1978). Dosimetric and biological results from the *Bacillus subtilis* biostack experiment with the Apollo-Soyuz test project. *Life Sci. Space Res.* **17,** 123–128.

Farrar, J. J., Mizel, S. B., Fuller-Farrar, J., Farrar, W. L., and Hilfiker, M. L. (1980). Macrophage-independent activation of helper T-cells. I. Production of Interleukin 2. *J. Immunol.* **125**(2), 793–798.

Fenton, M. J., Vermeulen, M. W., Clark, B. D., Webb, A. C., and Auron, P. E. (1988). Human pro-Il-1β gene expression in monocytic cells is regulated by two distinct pathways. *J. Immunol.* **140,** 2267–2273.

Fleming, S., Edelman, L., and Chapes, S. K. (1991). Effects of corticosterone and microgravity on inflammatory cell production of superoxide. *J. Leukocyte Biol.* **50,** 69–76.

Freidman, B. A., Frackelton, A. R., Jr., Ross, A. H., Connors, J. M., Fujiki, H., Sugimura, T., and Rosner, M. R. (1984). Tumor promoters block tyrosine-specific phosphorylation of the epidermal growth factor receptor. *Proc. Nat. Acad. Sci. U.S.A.* **81,** 3034–3038.

Fuhlbrigge, R. C., Chaplin, D. D., Kiely, J. M., and Unanue, E. R. (1987). Regulation of Interleukin-1 gene expression by adherence and lipopolysaccharide. *J. Immunol.* **138,** 3799–3802.

Fukami, K., Endo, T., Imamura, M., Takenawa, T. (1994). α- Actinin and vinculin are PIP_2-binding proteins involved in signaling by tyrosine kinase. *J. Biol. Chem.* **269,** (2), 1518–1522.

Galeotti, T., Boscoboinik, D., and Azzi, A. (1993). Regulation of the TNF-α receptor in human osteosarcoma cells: Role of microtubules and of protein kinase C. *Arch. Biochem. Biophys.* **300**(1), 287–292.

Geppert, T. D., Davis, L. S., Gur, H., Wacholtz, M. C., and Lipsky, P. E. (1990). Accessory cell signals involved in T-cell activation. *Immunol. Rev.* **117,** 5–66.

Gmünder, F. K., Kiess, M., Sonnenfeld, G., Lee, J., and Cogoli, A. (1992). Reduced lymphocyte activation in space: Role of cell-substratum interactions. *Adv. Space Res.* **12**(1), (1)55–(1)61.

Goldschmidt-Clermont, P. J., and Janmey, P. A. (1991). Profilin, a weak CAP for actin and RAS. *Cell* **66,** 419–421.

Gonzalez, F. A., Seth, A., Raden, D. L., Bowman, D. S., Fay, F. S., and Davis, R. J. (1993). Serum-induced translocation of mitogen-activated protein kinase to the cell surface ruffling membrane and the nucleus. *J. Cell Biol.* **122**(5), 1089–1101.

Gordon, S. A. (1964). Gravity and plant development: Bases for experiment. *In* "Space Biology" (F. A. Gilfillan, ed.), pp. 75–105. Oregon State University Press, Corvallis, OR.

Granelli-Piperno, A., Keane, M., and Steinman, R. M. (1989). Growth factor production and requirements during the proliferation response of human T lymphocytes to anti-CD3 monoclonal antibody. *J. Immunol.* **142,** 4138–4143.

Graul, E. H., Rüther, W., and Hiendl, C. O. (1991). Preliminary results of the *Artemia salina* experiments in Biostack on LDEF. *In* "LDEF—69 Months in Space" (A. Levine, ed.) pp. 1661–1665. NASA Langley Research Center, Langley, VA.

Gray, S. W., and Edwards, B. F. (1968). The effect of weightlessness on wheat seedling morphogenesis and histochemistry. *BioScience* **18**(6), 638–644.

Greenberg, M. E., and Ziff, E. B. (1984). Stimulation of 3T3 cells induces transcription of the c-*fos* proto-oncogene. *Nature* **311,** 433–438.

Gregorio, C. C., Kubo, R. T., Bankert, R. B., and Repasky, E. A. (1992). Translocation of spectrin and protein kinase C to a cytoplasmic aggregate upon lymphocyte activation. *Cell Biol.* **89,** 4947–4951.

Grif, V. G., Tairbekov, M. G., and Barmicheva, E. M. (1988). Cell morphology and ultrastructure of maize root meristem in microgravity. *Physiologist* **31**(1), S88–S91.

Gross, W., and Boss, W. F. (1993). Inositol phospholipids and signal transduction. *In* "Control of Plant Gene Expression," pp. 17–50. CRC Press, Inc., Boca Raton, FL.

Gruener, R., and Hoeger, G. (1988). Does vector-free gravity stimulate microgravity? Functional and morphological attributes of clinorotated nerve and muscle grown in cell culture. *Physiologist* **31,** S48–S49.

Gruener, R., and Hoeger, G. (1991). Vector-averaged gravity alters myocyte and neuron properties in cell culture. *Aviation, Space, Environ. Med.* **62**(12), 1159–1165.

Gunning, B. E. W., and Wick, S. M. (1985). Preprophase bands, phragmoplasts, and spatial control of cytokinesis. *J. Cell Sci.* **2,** 157–179.

Gurkin, L. W. (1992). The NASA sounding rocket program and space sciences. Am. Soc. Gravitational Space Biol. Bull. **6,** 113–120.

Halstead, T. W., and Dutcher, F. R. (1987). Plants in space. *Annu. Rev. Plant Physiol.* **38,** 317–345.

Hamilton, T. A., Becton, D. L., Somers, S. D., Gray, P. W., and Adams, D. O. (1985). Interferon-γ modulates protein kinase C activity in murine peritoneal macrophages. *J. Biol. Chem.* **260**(3), 1378–1381.

Hogg, N., Cabañas, C., and Dransfield, I. (1993). Leukocyte integrin activation. *In* "Structure, Function, and Regulation of Molecules Involved in Leukocyte Adhesion" (P. E.

Lipsky, R. Rothlein, T. K. Kishimoto, R. B. Faanes, and C. W. Smith, eds.), pp. 3–13. Springer-Verlag, New York.

Horgan, K. J., Van Seventer, G. A., Shimizu, Y., and Shaw, S. (1990). Hyporesponsiveness of "naive" (CD45RA+) human T cells to multiple receptor-mediated stimuli but augmentation of responses by co-stimuli. *Eur. J. Immunol.* **20,** 1111–1118.

Horneck, G., Bücker, H., Reitz, G., Requardt, H., Dose, K., Martens, K. D., Mennigmann, H. D., and Weber, P. (1984). Microorganisms in the space environment. *Science* **225,** 226–228.

Horneck, G., Bücker, H., and Reitz, G. (1991). Long-term exposure of bacterial spores to space. *In* "LDEF—69 Months in Space" (A. Levine, ed.), pp. 1667–1669. NASA Langley Research Center, Langley, VA.

Horneck, G., Drasavin, E., and Kozubek, S. (1992). Mutagenic effects of heavy ions in bacteria. *In* "The World Space Congress" (IAF and COSPAR, eds), pp. 548a. 29th Plenary Meeting of the Committee on Space Research, Washington, DC.

Hotchin, J., Lorenz, P., and Hemenway, C. (1965). Survival of micro-organisms in space. *Nature* **206,** 442–445.

Hughes-Fulford, M. (1991). Altered cell function in microgravity. *Exp. Gerontol.* **26,** 247–256.

Hullinger, R. L. (1993). The avian embryo responding to microgravity of space flight. *Physiologist* **36**(1), S42–S45.

Hunter, T., Ling, N., and Cooper, J. A. (1984). Protein kinase C phosphorylation of the EGF receptor at a threonine residue close to the cytoplasmic face of the plasma membrane. *Nature* **311,** 480–483.

Hymer, W. C., and Grindeland, R. E. (1991). The pituitary: Aging and spaceflown rats. *Exp. Gerontol.* **26,** 257–265.

Ikenaga, M., Yoshikawa, I., Ayaki, T., and Ryo, H. (1987). Biological effects of radiation during space flight. *In* "Biological Sciences in Space 1986" (S. Watanabe, G. Mitarai, and S. Mori, eds.), pp. 47–52. Internat. Symp. on Biol. Sci. in Space, MYU Research, Tokyo.

Ingber, D. E., and Folkman, J. (1989). Mechanochemical switching between growth and differentiation during fibroblast growth factor-stimulated angiogenesis *in vitro:* Role of extracellular matrix. *J. Cell Biol.* **109,** 317–330.

Isakov, N., and Altman, A. (1985). Tumor promoters in conjunction with calcium ionophores mimic antigenic stimulation by reactivation of alloantigen-primed murine T lymphocytes. *J. Immunol.* **135**(6), 3674–3680.

Isenberg, G. (1991). Actin binding proteins—lipid interactions. *J. Muscle Res. Cell Motil.* **12,** 136–144.

Iversen, T.-H., Rasmussen, O., Gmünder, F., Baggerud, C., Kordyum, E. L., Lozovaya, V. V., and M. Tairbekov. (1992). The effect of microgravity on the development of plant protoplasts flown on Biokosmos 9. *Adv. Space Res.* **12**(1), (1)123–(1)131.

Johnson, H. M., and Torres, B. A. (1982). Phorbol ester replacement of helper cell and interleukin 2 requirements in gamma interferon production. *Inf. Imm.* **36,** 911–914.

Joklik, W. K., Willett, H. P., Amos, D. B., and Wilfert, C. M. (1992). "Zinsser Microbiology." 20th ed. 1294 pp. Appleton and Lange, Norwalk, CT.

Juliano, R. L., and Haskill, S. (1993). Signal transduction from the extracellular matrix. *J. Cell Biol.* **120**(3), 577–585.

Kassemi, S. A., and Ostrach, S. (1992). Nature of buoyancy-driven flows in a reduced-gravity environment. *Am. Inst. Aeronaut. Astronaut. J.* **7,** 1815–1818.

Kawakami, K., Yamamoto, Y., Kakimoto, K., and Onoue, K. (1989). Requirement for delivery of signals by physical interaction and soluble factors from accessory cells in the induction of receptor-mediated T cell proliferation. *J. Immunol.* **142,** 1818–1825.

Kern, J. A., Reed, J. C., Daniele, R. P., and Nowell, P. C. (1986). The role of the accessory cell in mitogen-stimulated human T cell gene expression. *J. Immunol.* **137**(3), 764–769.

Kerr, L. D., Inoue, J.-I., and Verma, I. M. (1992). Signal transduction: The nuclear target. *Curr. Opin. Biol.* **4,** 496–501.

Khan, M. A., Jeremy, J. Y., Hallinan, T., Tateson, J. E., Hoffbrand, A. V., and Wickremasinghe, R. G. (1993). Antioxidants impair the coupling of cell-surface ligand receptors to the inositol lipid signaling pathway in human T lymphocytes but not in Jurkat T lymphoblastic leukemia cells. Evidence that leukotrienes are not involved in the coupling mechanism. *Biochim. Biophys. Acta.* **1178,** 215–220.

Kim, S. Y., Lee, J. S., and Mulkey, T. J. (1991). Effect of phorbol derivatives and staurosporine on growth and calcium responses in primary root of maize. *Am. Soc. Gravitational Space Biol. Bull.* **5**(1), 59a.

Kim, S. Y., Lee, J. S., and Mulkey, T. J. (1992). Effect of TPA and IAA on elongation and protein phosphorylation in maize roots. *Am. Soc. Gravitational Space Biol. Bull.* **6**(1), 39a.

Klement, B. J., and Spooner, B. S. (1994). Pre-metatarsal skeletal development in tissue culture at unit- and microgravity. *J. Exp. Zool.* **269,** 230–241.

Klimchuk, D. A., Kordyum, E. L., Danevich, L. A., Tarnavskaya, E. B., Tairbekov, M. G., Iversen, T.-H., Baggerud, C., and Rasmussen, O. (1992). Structural and functional organisation of regenerated plant protoplasts exposed to microgravity on Biokosmos 9. *Adv. Space Res.* **12**(1), (1)133–(1)140.

Kondepudi, D. D. (1991). Detection of gravity through nonequilibrium mechanisms. *Am. Soc. Gravitational Space Biol. Bull.* **4,** 119–124.

Kondo, S. (1968). Possibility and impossibility for genetic effects of weightlessness. *Jpn. J. Genet.* **43,** 467–488.

Kordyun, E. L., and Sytnik, K. M. (1983). Biological effects of weightlessness at cellular and subcellular levels. *Physiologist* **26**(6), S141–142.

Koretzky, G. A., Kohmetscher, M. A., Kadleck, T., and Weiss, A. (1992). Restoration of T cell receptor-mediated signal transduction by transfection of CD45 cDNA into a CD45-deficient variant of the Jurkat T cell line. *J. Immunol.* **149**(4), 1138–1142.

Kranz, A. R., Zimmermann, M. W., Stadler, R., Gartenbach, K. E., and Pickert, M. (1991). Total dose effects (TDE) of heavy ionizing radiation in fungus spores and plant seeds: Preliminary investigations. *In* "LDEF—69 Months in Space," (A. Levine, ed.), pp. 1651–1660. NASA Langley Research Center, Langley, VA.

Kreis, T. E. (1990). Role of microtubules in the organization of the Golgi apparatus. *Cell Motil. Cytoskeleton* **15,** 67–70.

Krikorian, A. D., Dutcher, F. R., Quinn, C. E., and Steward, F. C. (1981). Growth and development of cultured carrot cells and embryos under spaceflight conditions. *Adv. Space Res.* **1,** 117–127.

Krikorian, A. D., and O'Conner, S. A. (1982). Some karyological observations on plants grown in space. *Physiologist* **25**(6), S125–S132.

Kruijer, W., Cooper, J. A., Hunter, T., and Verma, I. M. (1984). Platelet-derived growth factor induces rapid but transient expression of the c-*fos* gene and protein. *Nature* **312,** 711–716.

Kumei, Y., Sato, A., Ozawa, K., Nakajima, T., and Yamashita, N. (1987). Effects of hypergravity on cultured mammalian cells. *In* "Biological Sciences in Space 1986" (S. Wantanabe, G. Mitarai, and S. Mori, eds.), pp. 291–296. MYU Research, Tokyo.

Kumei, Y., Nakajima, T., Sato, A., Kamata, N., and Enomoto, S. (1989a). Reduction of G_1 phase duration and enhancement of *c-myc* gene expression in HeLa cells at hypergravity. *J. Cell Sci.* **93,** 221–226.

Kumei, Y., Whitson, P. A., and Cintron, N. M. (1989b). Rapid increase in inositol triphosphate in HeLa cells after hypergravity exposure. *Am. Soc. Gravitational Space Biol.* **3**(1), 39a.

Landis, R. C., Friedman, M. L., Fisher, R. I., and Ellis, T. M. (1991). Induction of human monocyte IL-1 mRNA and secretion during anti-CD3 mitogenesis requires two distinct T cell-derived signals. *J. Immunol.* **146,** 128–135.

Lapchine, L., Moatti, N., Richoilley, G., Templier, J., Gasset G., and Tixador, R. (1987). The antibio experiment. *In* "Biorack on Spacelab D1" (N. Longdon and V. David eds.), pp. 45–51, European Space Agency, Paris.

Larson, R. S., and Springer, T. A. (1990). Structure and function of leukocyte integrins. *Immunol. Rev.* **114,** 181–217.

Lechner, J. F., and Kaighn, M. E. (1980). EGF growth promoting activity is neutralized by phorbol esters. *Cell Biol. Int. Rep.* **4**(1), 23–28.

Lenormand, R., Pagès, G., Sardet, C., L'Allemain, G., Meloche, S. and Pouyssègur. (1993). MAP kinases: Activation, subcellular localization and role in the control of cell proliferation. *In* "Advances in Second Messenger and Phosphoprotein Research" (B. L. Brown and P. R. M. Dobson, eds.), vol. 28, pp. 237–244. Raven Press, New York.

Levine, H. G., and Krikorian, A. D. (1992). Shoot growth in aseptically cultivated daylily and *Haplopappus* plantlets after a 5-day spaceflight. *Physiol. Plant.* **86,** 349–359.

Levine, H. G., Kann, R. P., and Krikorian, A. D. (1990). Plant development in space: Observations on root formation and growth. *In* "Proceedings of the Fourth European Symposium on Life Sciences Research in Space," pp. 503–508. European Space Agency, Paris.

Lewis, M. L., and Hughes-Fulford, M. (1994). Cellular responses to microgravity. *In* "Fundamentals of Space Life Science" (S. Churchill, ed.), Kriejer Publishing Co., Orlando, FL (in press).

Limouse, M., Manié, S., Konstantinova, I., Ferrua, B., and Schaffer, L. (1991). Inhibition of phorbol ester-induced cell activation in microgravity. *Exp. Cell Res.* **197,** 82–86.

Lin, T. H., Haskill, H., and Juliano, R. L. (1993a). Integrin clustering induces tyrosine phosphorylation and monocyte immediate-early gene expression in human monocytic THP-1 cells. *Mol. Biol. Cell* **4,** 5363a.

Lin, A., Smeal, T., Binetruy, B., Deng, T., Chambard, J.-C., and Karin, M. (1993b). Control of AP-1 activity by signal transduction cascades. *In* "Advances in Second Messenger and Phosphoprotein Research" (B. L. Brown, and P. R. M. Dobson, eds.), vol. 28, pp. 255–260. Raven Press, New York.

Lin, C. R., Chen, W. S., Lazar, C. S., Carpenter, C. D., Gill, G. N., Evans, R. M., and Rosenfeld, M. G. (1986b). Protein kinase C phosphorylation at Thr 654 of the unoccupied EGF receptor and EGF binding regulate functional receptor loss by independent mechanisms. *Cell* **44,** 839–848.

Liscovitch, M. (1992). Crosstalk among multiple signal-activated phospholipases. *Trends Biochem. Sci.* **17,** 393–399.

Lorenzi, G., and Perbal, G. (1990). Root growth and statocyte polarity in lentil seedling roots grown in microgravity or on a slowly rotating clinostat. *Physiol. Plant.* **78,** 532–537.

Lucas, S., Marais, R., Graves, J. D., Alexander, D., Parker, P., and Cantrell, D. A. (1990). Heterogeneity of protein kinase C expression and regulation in T lymphocytes. *FEBS Lett.* **260**(1), 53–56.

Lucibello, F. C., Lowag, C., Neuberg, M., and Müller, R. (1989). Trans-repression of the mouse c-*fos* promoter: A novel mechanism of *fos*-mediated *trans*-regulation. *Cell* **59,** 999–1007.

Lyon, C. J. (1968a). Growth physiology of the wheat seedling in space. *BioScience* **18**(6), 633–637.

Lyon, C. J. (1968b). Wheat seedling growth in the absence of gravitational force. *Plant Physiol.* **43,** 1002–1007.

Manger, B., Weiss, A., Imboden, J., Laing, T., and Stobo, J. D. (1987). The role of protein kinase C in transmembrane signaling by the T cell antigen receptor complex. *J. Immunol.* **139,** 2755–2760.

Marco, R., Vernós, I., Gonzalez, J., and Calleja, M. (1986). Embryogenesis and aging of *Drosophila melanogaster* flown in the space shuttle. *Naturwissenschaften* **73,** 431–432.

Marco, R., González-Jurado, J., Calleja, M., Garesse, R., Maroto, M., Ramírez, E., Holgado, M. C., de Juan, E., and Miquel, J. (1992). Microgravity effects on *Drosophila melanogaster* development and aging: Comparative analysis of the results of the fly experiment in the Biokosmos 9 Biosatellite flight. *Adv. Space Res.* **12**(1), (1)157–(1)166.

Mattoni, R. H. T. (1968). Space-flight effects and gamma radiation interaction on growth and induction of lysogenic bacteria, a preliminary report. *BioScience* **18**(6), 602–608.

McCormack, P. (1988). *In* "Terrestrial Space Radiation and its Biological Effects" (P. McCormack, C. Swenberg, and H. Bücker, eds.), pp. 1–71. NATO ASI Series, vol. 154, Plenum Publishing, New York.

Meehan, R. (1987). Human mononuclear cell *in vitro* activation in microgravity and post-spaceflight. *In* "Advances in Experimental Medicine and Biology" (M. Z. Atassi, ed.), vol. 225, pp. 273–286. Plenum Publishing, New York.

Mennigmann H. D., and Lange, M. (1986). The circadian rhythm in *Chlamydomonas reinhardii* in a zeitgeber-free environment. *Naturwissenschaften* **73**(7), 410–412.

Mennigmann H. D., and Lange, M. (1987). Growth and differentiation of *Bacillus subtilis* under microgravity conditions. *In* "Biorack on Spacelab D 1" (N. Longdon and V. David eds.), pp. 37–44. European Space Agency, Paris.

Mergenhagen D., and Mergenhagen, E. (1987). The biological clock of *Chlamydomonas reinhardii* in space. *In* "Biorack on Spacelab D-1" (N. Longdon and V. David eds.), pp. 75–86, European Space Agency, Paris.

Merkys, A. (1990). Plant growth under microgravity conditions: Experiments and problems. *In* "Proceedings of the Fourth European Symposium on Life Sciences Research in Space" pp. 509–515, European Space Agency, Paris.

Merkys, A. J., Laurinavichius, R. S., Rupainene, O. J., Savichene, E. K., Jaroshius, A. V., Shvegzhdene, D. V., and Bendoraityte, D. P. (1983). The state of gravity sensors and peculiarities of plant growth during different gravitational loads. *Adv. Space Res.* **3**(9), (9)211–(9)219.

Merkys, A. J., and Laurinavicius, R. S. (1990). Plant growth in space. *In* "Fundamentals of Space Biology" (M. Asashima and G. M. Malacinski eds.), pp. 69–83. Springer-Verlag, Berlin.

Miyata, Y., Nishida, E., Koyasu, S., Yahara, I., and Sakai, H. (1989). Protein kinase C-dependent and -independent pathways in the growth factor-induced cytoskeletal reorganization. *J. Biol. Chem.* **264**(26), 15565–15568.

Moatti, N., Lapchine, L., Gasset, G., Richoilley, G., Templier, J., Tixador, R. (1986). Preliminary results of the "antibio" experiment. *Naturwissenschaften* **73**(7), 413–415.

Montgomery, P.O'B., Jr., Cook, J. E., Reynolds, R. C., Paul, J. S., Hayflich, L., Stock, D., Schulz, W. W., Kinsey, S., Thirolf, R. G., Rogers, T., and Campbell, D. (1978). The response of single human cells to zero gravity. *In Vitro* **14**(2), 165–173.

Moore, R. (1988). How gravity affects plant growth and development. *Biol. Digest* **14,** 11–15.

Moore, R. (1990). Comparative effectiveness of a clinostat and a slow-turning lateral vessel at mimicking the ultrastructural effects of microgravity in plant cells. *Ann. Bot.* **66,** 541–549.

Moore, R., Fondren, W. M., Koon, E. C., and Wang, C.-L. (1986). The influence of gravity on the formation of amyloplasts in columella cells of *Zea mays* L. *Plant Physiol.* **82,** 867–868.

Moore, R., Fondren, W. M., McClelen, E., and Wang, C.-L. (1987a). Influence of microgravity on cellular differentiation in root caps of *Zea mays. Am. J. Bot.* **74**(7), 1006–1012.

Moore, R., McClelen, C. E., Fondren, W. M., and C.-L., Wang. (1987b). Influence of microgravity on root-cap regeneration and the structure of columella cells in *Zea mays. Am. J. Bot.* **74**(2), 218–223.

Moos, P. J., Graff, K., Edwards, M., Stodieck, L. S., Einhorn, R., and Luttges, M. W.

(1989). Gravity-induced changes in microtubule formation. *Am. Soc. Gravitational Space Biol.* **2,** 48a.

Morey-Holton, E. R., and Cone, C. M. (1990). Bone as a model system for organ/tissue responses to microgravity. *In* "Fundamentals of Space Biology" (M. Asashima and G. M. Malacinski, eds.), pp. 113–122. Japan Sci. Soc. Press, Tokyo.

Mori, S., Watanabe, S., Takabayashi, A., Sakakibara, M., Koga, K., Takagi, S., and Usui, S. (1987). Behavior and brain activity of carp during parabolic-flight low gravity. *In* "Biological Sciences in Space 1986" (S. Watanabe, G. Mitarai, and S. Mori, eds.), pp. 155–162. Internat. Symp. on Biol. Sci. in Space, MYU Research, Tokyo.

Müller, J. M., Ziegler-Heitbrock, H. W., and Baeuerle, P. A. (1993). Nuclear factor kappa B, a mediator of lipopolysaccharide effects. *Immunobiology* **187,** 233–256.

Müller, R., Bravo, R., Burckhardt, J., and Curran, T. (1984). Induction of c-*fos* gene and protein by growth factors precedes activation of c-*myc*. *Nature* **312,** 716–720.

Murray, N. R., Baumgardner, G. P., Burns, D. J., and Fields, A. P. (1993). Protein kinase C isotypes in human erythroleukemia (K562) cell proliferation and differentiation. *J. Biol. Chem.* **268**(21), 15847–15853.

Muzykantov, V. R., Kushnareva, T. A., Smirnov, M. D., and Ruuge, E. K. (1993). Avidin attachment to biotinylated human neutrophils induces generation of superoxide anion. *Biochim. Biophys. Acta* **1177,** 229–235.

Nace, G. W., and Tremor, J. W. (1981). Clinostat exposure and symmetrization of frog eggs. *Physiologist* **24**(6), S77–S78.

Nedukha, E., and Mashinskiy, A. L. (1992). Sixteen days influence microgravity on the ultrastructure Triticum durum L. leaves cells. *In* "The World Space Congress" (IAF and COSPAR, eds.), p. 528a. 29th Plenary Meeting of the Committee on Space Research, Washington, DC.

Nelson, G. A., Schubert, W. W., Kazarians, G. A., and Richards, G. F. (1994). Development and chromosome mechanics in nematodes: Results from IML-1. *Adv. Space Res.* **14**(8), (8)209–(8)214.

Nemenoff, R. A., Winitz, S., Qian, N.-X., Van Putten, V., Johnson, G. L., and Heasley, L. E. (1993). Phosphorylation and activation of a high molecular weight form of phospholipase A2 by p42 microtubule-associated protein 2 kinase and protein kinase C. *J. Biol. Chem.* **268,** 1960–1964.

Nishizuka, Y. (1983). Phospholipid degradation and signal translation for protein phosphorylation. *Trends Biochem. Sci.* **17,** 713–716.

Nishizuka, Y. (1992). Intracellular signaling by hydrolysis of phospholipids and activation of protein kinase C. *Science* **258,** 607–613.

Ofir, R., Dwarki, V. J., Rashid, D., and Verma, I. M. (1990). Phosphorylation of the C terminus of Fos protein is required for transcriptional transrepression of the c-*fos* promoter. *Nature* **348,** 80–82.

Parfyonov, G. P., Platonova, R. N., Tairbekov, M. G., Zhvalikovskaya, V. P., Mozgovaya, I. E., Rostopshina, A. V., and Rozov, A. N. (1979a). Biological experiments carried out aboard the biological satellite Cosmos-936. *Life Sci. Space Res.* **17,** 297–299.

Parfyonov, G. P., Platonova, R. N., Tairbekov, M. G., Belenev, Y. N., Oklhovenko, V. P., Rostopshina, A. V., and Oigenblick, E. A. (1979b). Biological investigations aboard biosatellite Cosmos-782. *Acta Astronaut.* **6,** 1235–1238.

Park, D. J., Min, H. K., and Rhee, S. G. (1992). Inhibition of CD3-linked phospholipase C by phorbol ester and by cAMP is associated with decreased phosphotyrosine and increased phosphoserine contents of PLC-γ1. *J. Biol. Chem.* **267**(3), 1496–1501.

Patarroyo, M., Prieto, J., Rincon, J., Timonen, T., Lundberg, C., Lindom, L., Åsjö, B., and Gahmberg, C. G. (1990). Leukocyte-cell adhesion: A molecular process fundamental in leukocyte physiology. *Immunol. Rev.* **114,** 67–108.

Pennica, D., Nedwin, G. E., Hayflick, J. S., Seeburg, P. H., Derynck, R., Palladino, M. A., Kohr, W. J., Aggarwal, B. B., and Goeddel, D. V. (1984). Human tumour necrosis factor: Precursor structure, expression and homology to lymphotoxin. *Nature* **312,** 724–729.

Perbal, G., Driss-Ecole, E., and Salle, G. (1987). Graviperception of lentil seedling roots grown in space. *In* "Biorack on Spacelab D-1" (N. Longdon and V. David eds.), pp. 109–117, European Space Agency, Paris.

Petrov, V. (1992). Overview on experience to date on human exposure to space radiations. *In* "The World Space Congress" (IAF and COSPAR, eds.), p. 553a. 29th Plenary Meeting of the Committee on Space Research, Washington, DC.

Planel, H., Tixador, R., Nefodov, Y., Gretchko, G., Richoilley, G. (1982). Effects of space flight factors at the cellular level: Results of the cytos experiment. *Aviation, Space Environ. Med.* **53,** 370–374.

Planel, H., Tixador, R., Richoilley, G., and Gasset, G. (1990). Effects of space flight on a single-cell organism: *Paramecium tetraurelia*. *In* "Fundamentals of Space Biology" (M. Asashima and G. M. Malacinski eds.), pp. 85–96. Springer-Verlag, Berlin.

Podlutsky, A. G. (1992a). Roots formed from callus cultures as a perspective model for gravibiological studies. *In* "The World Space Congress" (IAF and COSPAR, eds.), p. 526a. 29th Plenary Meeting of the Committee on Space Research, Washington, DC.

Podlutsky, A. G. (1992b). Ultrastructural analysis of organization of roots obtained from cell cultures at clinostating and under microgravity. *Adv. Space Res.* **12**(1), (1)93–(1)102.

Pollard, E. C. (1965). Theoretical considerations on living systems in the absence of mechanical stress. *J. Theor. Biol.* **8,** 113–123.

Pollard, E. C. (1971). Physical determinants of receptor mechanisms. *In* "Gravity and the Organism" (S. A. Gordon and M. J. Cohen, eds.), pp. 25–34. University of Chicago Press, Chicago.

Polulyakh, Y. A. (1988). Phospholipid and fatty acid composition of the pea root cell plasma membrane under clinostating. *Dokl. Akad. Nauk, UKR SSR* **10,** 67–69.

Pontremoli, S., Melloni, E., Salamino, F., Sparatore, B., Michetti, M., Sacco, O., and Horecker, B. L. (1986). Activation of NADPH oxidase and phosphorylation of membrane proteins in human neutrophils: Coordinate inhibition by a surface antigen-directed monoclonal antibody. *Biochem. Biophys. Comm.* **140**(3), 1121–1126.

Popova, A. F. (1987). Submicroscopic organization of *Anabaena azollae* Strasb. exposed space flight. *Space Biology and Biotechnology* **1987,** 23–28. NASA TT-20033.

Rappaport, R. (1986). Establishment of the mechanism of cytokinesis in animal cells. *Int. Rev. Cytol.* **105,** 245–281.

Reitz, G., and Bücker, H. (1992). The radiobiological experiments on LDEF. *In* "The World Space Congress" (IAF and COSPAR, eds.), p. 568a. 29th Plenary Meeting of the Committee on Space Research, Washington, DC.

Richoilley, G., Tixador, R., Gasset, G., Templier, J., and Planel, H. (1986). Preliminary results of the "Paramecium" experiment. *Naturwissenschaften* **73**(7), 404–406.

Richoilley, G., Tixador, R., Templier, J., Bes, J. C., Gasset, G., and Planel, H. (1987). The *Paramecium* experiment. *In* "Biorack on Spacelab D-1" (N. Longdon and V. David eds.), pp. 69–73, European Space Agency, Paris.

Richoilley, G., Pouech, Ph., Gasset, G., and Planel, H. (1992). Effects of gravity on multiplication, motility and ATP content in *Paramecium tetraurelia*. *In* "The World Space Congress" (IAF and COSPAR, eds.), p. 534a. 29th Plenary Meeting of the Committee on Space Research, Washington, DC.

Rijken, P. J., de Groot, R. P., Bruijer, W., de Laat, S. W., Verkleij, A. J., and Boonstra, J. (1992). Identification of specific gravity sensitive signal transduction pathways in human A431 carcinoma cells. *Adv. Space Res.* **12**(1), (1)145–(1)152.

Sadowski, H. B., Shuai, K., Darnell, J. E., Jr., and Gilman, M. Z. (1993). A common nuclear signal transduction pathway activated by growth factor and cytokine receptors. *Science* **261,** 1739–1744.

Sassone-Corsi, P., Sisson, J. C., and Verma, I. M. (1988). Transcriptional autoregulation of the proto-oncogene *fos*. *Nature* **334,** 314–319.

Schatz, A., Reitstetter, R., Linke-Hommes, A., Briegleb, W., Slenzka, K., and Rahmann, H. (1994). Gravity effects on membrane processes. *Adv. Space Res.* **14**(8), (8)35–(8)43.

Schmit, A. C., and Lambert, A. M. (1990). Microinjected fluorescent phalloidin *in vivo* reveals the F-actin dynamics and assembly in higher plant mitotic cells. *Plant Cell* **2,** 129–138.

Schulze, A., Jensen, P. J., Desrosiers, M., Buta, J. G., and Bandurski, R. S. (1992). Studies on the growth and indole-3-acetic acid and abscisic acid content of *Zea mays* seedlings grown in microgravity. *Plant Physiol.* **100,** 692–698.

Serova, L. V., and Denisova, L. A. (1982). The effect of weightlessness on the reproductive function of mammals. *Physiologist* **25**(6), S9–S12.

Shahinian, A., Pfeffer, K., Lee, K. P., Kündig, T. M., Kishihara, K., Wakeham, A., Kawai, K., Ohashi, P. S., Thompson, C. B., and Mak, T. W. (1993). Differential T cell constimulatory requirements in CD28-deficient mice. *Science* **261,** 609–612.

Sheng, M., Dougan, S. T., McFadden, G., and Greenberg, M. E. (1988). Calcium and growth factor pathways of c-*fos* transcriptional activation require distinct upstream regulatory sequences. *Mol. Cell Biol.* **8,** 2787–2796.

Shimizu, Y., van Seventer, G. A., Horgan, K. J., and Shaw, S. (1990). Roles of adhesion molecules in T-cell recognition: Fundamental similarities between four integrins on resting human T cells (LFA-1, VLA-4, VLA-5, VLA-6) in expression, binding and costimulation. *Immunol.* **114,** 109–144.

Sievers, A., and Hejnowicz, Z. (1992). How well does the clinostat mimic the effect of microgravity on plant cells and organs? *Am. Soc. Gravitational Space Biol.* **5**(2), 69–75.

Silvennoinen, O., Schindler, C., Schlessinger, J., and Levy, D. E. (1993). Ras-independent growth factor signaling by transcription factor tyrosine phosphorylation. *Science* **261,** 1736–1738.

Slocum, R. D., and Galston, A. W. (1982). A comparative study of monocot and dicot root development in normal (earth) and hypogravity (space) environments. *Physiologist* **25**(6), S131–S132.

Slocum, R. D., Gaynor, J. J., and Galston, A. W. (1984). Cytological and ultrastructural studies on root tissues. *Ann. Bot.* **54**(3), 65–76.

Smith, A. H., and Abbott, U. K. (1981). Embryonic development during chronic acceleration. *Physiologist* **24**(6), S73–S74.

Smith, G. A., and Portnoy, D. A. (1993). The role of the proline-rich repeats of ActA for the actin based motility of *Listeria monocytogenes*. *Mol. Biol. Cell* **4,** S149a.

Souza, K. A. (1987). Amphibian development in microgravity. *In* "Biological Sciences in Space 1986" (S. Wantanabe, G. Mitarai, and S. Mori, eds.), pp. 61–68. MYU Research, Tokyo.

Spizizen, J., Isherwood, J. E., and Taylor, G. R. (1975). Effects of solar ultraviolet radiations on *Bacillus subtilis* spores on T7 bacteriophage. *Life Sci. Space Res.* **13,** 143–149.

Spooner, B. S. (1973). Morphogenesis of vertebrate organs. *In* "Concepts of Development" (L. Lash and J. R. Whittaker, eds.), pp. 213–240. Sinauer Associates, Stanford, CT.

Spooner, B. S. (1975). Microfilaments, microtubules, and extracellular materials in morphogenesis. *BioScience* **225,** 440–451.

Spooner, B. S. (1992). Gravitational studies in cellular and developmental biology. *Trans. Kansas Acad. Sci.* **95**(1-2), 4–10.

Spooner, B. S., and Thompson-Pletscher, H. A. (1986). Matrix accumulation and the development of form: Proteoglycans and branching morphogenesis. *In* "Regulation of Matrix Accumulation" (R. P. Mecham ed.), pp. 399–444. Academic Press, New York.

Spooner, B. S., Thompson-Pletscher, H. A., Stokes, B., and Bassett, K. (1986). Extracellular matrix involvement in epithelial branching morphogenesis. *In* "The Cell Surface in Development and Cancer" (M. S. Steinberg, ed.), pp. 225–260. Plenum Publishing, New York.

Spooner, B. S., DeBell, L., Armbrust, L., Guikema, J. A., Metcalf, J., and Paulsen, A. (1994a). Embryogenesis, hatching and larval development of *Artemia* during orbital spaceflight. *Adv. Space Res.* **14**(8), (8)229–(8)238.

Spooner, B. S., Hardman, P., and Paulsen, A. (1994b). Gravity in mammalian organ development: Differentiation of cultured lung and pancreas rudiments during spaceflight. *J. Exp. Zool.* **269,** 212–222.

Spooner, B. S., Metcalf, J., DeBell, L., Paulsen, A., Noren, W., and Guikema, J. A. (1994c). Development of the brine shrimp *Artemia* is accelerated during spaceflight. *J. Exp. Zool.* **269,** 253–262.

Sytnik, K. M., Popova, A. F., Nechitailo, G. S., and Mashinsky, A. L. (1992). Peculiarities of the submicroscopic organization of *Chlorella* cells cultivated on a solid medium in microgravity. *Adv. Space Res.* **12**(1), (1)103–(1)107.

Takishima, K., Groswold-Prenner, I., Ingebritsen, T., and Rosner, M. R. (1991). Epidermal growth factor (EGF) receptor T669 peptide kinase from 3T3-L1 cells is an EGF-stimulated "MAP" kinase. *Proc. Nat. Acad. Sci.* **88,** 2520–2524.

Taylor, G. R. (1977). Survey of cell biology experiments in reduced gravity. *In* "Bioprocessing in Space" (D. R. Morrison, ed.), pp. 77–101. NASA TM X-58191, Houston.

Taylor, G. R., Spizizen, J., Foster, B. G., Volz, P. A., Bucker, H., Simmonds, R. C. Heimpel, A. M., and Benton, E. V. (1974). A descriptive analysis of the Apollo 16 microbial response to space environment. *BioScience* **24**(9), 505–511.

Taylor, J. W. R., and Munson, K. (1972). Starting to look at space. *In* "The History of Aviation" pp. 420–423. New English Library, London.

Theimer, R. R., Kudielka, R. A., and Rösch, I. (1986). Induction of somatic embryogenesis in anise in microgravity. *Naturwissenschaften* **73**(7), 442–443.

Tixador, R., Richoilley, G., Templier, J., Monrozies, E., Moatti, J-P., and Planel, H. (1981). Eiude de la teneur intra et extracellulaire des electrolytes dans les cultures de paramecies realisees pendant un vol spatial. *Biophys. Biochim. Acta* **649,** 175–178.

Tixador, R., Richoilley, G., Gasset, G., Planel, H., Moatti, N., Lapchine, L., and Enjalbert, L. (1985a). Preliminary results of Cytos 2 experiment. *Acta Astronaut.* **12**(2), 131–134.

Tixador, R., Richoilley, G., Gasset, G., Templier, J., Bes, J. C., Moatti, N., Lapchine, L. (1985b). Study of minimal inhibitory concentration of antibiotics on bacteria cultivated *in vitro* in space (Cytos 2 experiment). *Aviation, Space Environ. Med.* **56,** 748–751.

Tixador, R., Gasset, G., Lapchine, L., Moatti, N., Moatti, J. P., and Eche, B. (1992). Antibio experiment IML-1 mission. *In* "The World Space Congress" (IAF and COSPAR, eds.), p. 529a. 29th Plenary Meeting of the Committe on Space Research, Washington, DC.

Todd, P. (1989). Gravity-dependent phenomena at the scale of the single cell. *Am. Soc. Gravitational Space Biol. Bull.* **2,** 95–113.

Todd, P., Kunze, M. E., Williams, K., Morrision, D. R., Lewis, M. L., and Barlow, G. H. (1985). Morphology of human embryonic kidney cells in culture after space flight. *Physiologist* **28**(6), S-183–S-184.

Tonks, N. K., and Charbonneau, H. (1989). Protein tyrosine dephosphorylation and signal transduction. *Trends Biochem. Sci.* **14,** 497–500.

Tratner, I., Ofir, R., Verma, I. M. (1992). Alteration of a cAMP-dependent protein kinase phosphorylation site in c-*fos* protein augments its transforming potential. Mol. Cell Biol. **12,** 998–1006.

Treisman, R. (1992). The serum response element. *Trends Biochem. Sci.* **17,** 423–426.

Trinchieri, G., Kobayashi, M., Murphy, M., and Perussia, B. (1987). Immune interferon and cytotoxins: Regulatory effects on myeloid cells. *In* "Lymphokines" (E. Pick, ed.), vol. 14, pp. 267–305.

Tschopp, A., and Cogoli, A. (1983). Hypergravity promotes cell proliferation. *Experientia* **39**(12), 1323–1438.

Tschopp, A., Cogoli, A., Lewis, M. L., and Morrison, D. R. (1984). Bioprocessing in space: Human cells attach to beads in microgravity. *J. Biotechnol.* **1,** 287–293.

Ubbels, G. A. (1992). Developmental biology on unmanned space craft. *Adv. Space Res.* **12**(1), (1)117–(1)122.

Ubbels, G. A., Meijerink, J., Reijnen, M., and Narraway, J. (1992). The effect of gravity on fertilization and development of *Xenopus laevis*. *In* "The World Space Congress" (IAF and COSPAR, eds.), p. 533a. 29th Plenary Meeting of the Committee on Space Research, Washington, DC.

Ushakov, I. A., and Alpatov, A. M. (1992). Possible mechanism of microgravity impact on *Carausius morosus* ontogenesis. *Adv. Space Res.* **12**(1), (1)153–(1)155.

van Kooyk, Y., de Wiel-van Kememenade, P., Weder, P., Kuijpers, T. W., and Figdor, C. G. (1989). Enhancement of LFA-1-mediated cell adhesion by triggering through CD2 or CD3 on T lymphocytes. *Nature* **342,** 811–813.

Van Seventer, G. A., Shimizu, Y., Horgan, K. J., and Shaw, S. (1990). The LFA-1 ligand ICAM-1 provides an important constimulatory signal for T cell receptor-mediated activation of resting T cells. *J. Immunol.* **144,** 4579–4586.

Vernós, I., Gonzalez-Jurado, J., Calleja, M., Carratala, M., and Marco, R. (1987). The effects of short spaceflights on *Drosophila melanogaster* embryogenesis and lifespan. *In* "Biorack on Spacelab D-1" (N. Longdon and V. David eds.), pp. 121–133, European Space Agency, Paris.

Vernós, I., Gonzalez-Jurado, J., Calleja, M., and Marco, R. (1989). Microgravity effects on the oogenesis and development of embryos of *Drosophila melanogaster* laid in the space shuttle during the Biorack experiment. *Int. J. Dev. Biol.* **33,** 213–226.

Volkmann, D., and Sievers, A. (1990). Gravitational effects on subcellular structures of plant cells. *In* "Proceedings of the Fourth European Symposium on Life Sciences Research in Space," pp. 497–501. Trieste, Italy.

Volkmann, D., Behrens, H. M., and Sievers, A. (1986). Development and gravity sensing of cress roots under microgravity. *Naturwissenschaften* **73**(7), 438–441.

Volkmann, D., Czaja, I., and Sievers, A. (1988). Stability of cell polarity under various gravitational forces. *Physiologist* **31**(1), S40–S43.

Wakasugi, H., Bertoglio, J., Tursz, T., and Fradelizi, D. (1985). IL 2 receptor induction on human T lymphocytes: Role for IL 2 and monocytes. *J. Immunol.* **135**(1), 321–327.

Walker, K. R., Hughes-Fulford, M., Schmidt, G. (1992). KC-135 flights for life sciences activities. *Am. Soc. Gravitational Space Biol. Bull.* **6,** 107–112.

Ward, S. G., and Cantrell, D. A. (1990). Heterogeneity of the regulation of phospholipase C by phorbol esters in T lymphocytes. *J. Immunol.* **144,** 3523–3528.

Weiss, A., Manger, B., and Imboden, J. (1986). Synergy between the T3/antigen receptor complex and Tp44 in the activation of human T cells. *J. Immunol.***137,** 819–825.

Weiss, A., Shields, R., Newton, M., Manger, B., and Imboden, J. (1987). Ligand-receptor interactions required for commitment to the activation of the Interleukin 2 gene. *J. Immunol.* **138,** 2169–2176.

Welsh, J. B., Gill, G. N., Rosenfeld, M. G., and Wells, A. (1991). A negative feedback loop attenuates EGF-induced morphological changes. *J. Cell Biol.* **114**(3), 533–543.

Whiteley, B., and Glaser, L. (1986). EGF promotes phosphorylation at Threonine-654 of the EGF-receptor: Possible role of protein kinase-C in homologous regulation of the EGF-receptor. *J. Cell Biol.* **103,** 1355–1362.
Wilfinger, W., and Hymer, W. (1992). The effects of microgravity. *Chem. Br.* **28,** 626–630.
Williams, R. (1987). *In* "JSC Reduced Gravity Program User's Guide. Flight Crew Operations Directorate," JSC-22803, NASA, Houston, TX.
Wilson, E., Olcott, M. C., Bell, R. M., Merrill, A. H., Jr., Lambeth, J. D. (1986). Inhibition of the oxidative burst in human neutrophils by sphingoid long-chain bases. *J. Biol. Chem.* **261**(27), 12616–12623.
Wolf, D. A., and Schwartz, R. P. (1991). Analysis of gravity-induced particle motion and fluid perfusion flow in the NASA-designed rotating zero-head-space tissue culture vessel. NASA Technical Paper 3143. NASA, Houston, TX.
Wolf, D. A., and Schwarz, R. P. (1992). Experimental measurement of the orbital paths of particles sedimenting within a rotating viscous fluid as influenced by gravity. NASA Technical Paper 3200. NASA, Houston, TX.
Woods, K., and Chapes, S. K. (1994). Abrogation of TNF-mediated cytotoxicity by space-flight involves protein kinase C. *Exp. Cell Res.* **211,** 171–174.
Wotton, D., Ways, D. K., Parker, P. J., and Owen, M. J. (1993). Activity of both *raf* and *ras* is necessary for activation of transcription of the human T cell receptor β gene by protein kinase C, *ras* plays multiple roles. *J. Biol. Chem.* **268**(24), 17975–17982.
Yang, W. Y., Mattingly, M. W., and Mulkey, T. J. (1992). Effects of TPA and staurosporine on callus growth and development. *Am. Soc. Gravitational Space Biol. Bull.* **6**(1), 40a.
Yang, W. Y., and Mulkey, T. J. (1993). Activity of calcium, lithium and the activators and inhibitors of PKC on organogenesis of *N. tabacum*. *Am. Soc. Gravitational Space Biol. Bull.* **7**(1), 57a.
Yokota, H., Neff, A. W., and Malacinski, G. M. (1994). Early development of *Xenopus* embryos is affected by simulated gravity. *Adv. Space Res.* **14**(8), (8)249–(8)255.
Zhukov-Verezhnikov, N. N., Maiskii, I. N., Pekhov, A. P., Antipov, V. V., Rybakov, N. I., and Kozlov, V. A. (1965). An investigation of the biological effects of spaceflight factors by experiments with lysogenic bacteria on "Vostok-5" and "Vostok-6." *Cosmic Res.* **3**(3), 382–384.
Zhukov-Verezhnikov, N. N., Volkov, M. N., Maiskii, I. N., Guberniev, M. A., Rybakov, N. I., Antipov, V. V., Kozlov, V. A., Saksonov, P. P., Parfyonov, G. P., Kolobov, A. V., Rybakova, K. D., and Aniskin, E. D. (1968). Experimental genetic investigations of lysogenic bacteria during flight of the AES "Kosmos-110." *Cosmic Res.* **6**(1), 121–125.

Index

A

B

C

D

E

F

G

H

I

K

L

M

N

Q

R

U

V

W

X

Z